Interstellar Travel

Propulsion, Life Support, Communications, and the Long Journey

Interstellar Travel

Propulsion, Life Support, Communications, and the Long Journey

Edited by

Les Johnson

NASA Science and Tech Office, George C. Marshall Space Flight Center, The Interstellar Research Group, Oak Ridge, TN, United States

Kenneth Roy

Retired Professional Engineer, Founding Member and Former Board Member of the Interstellar Research Group, Oak Ridge, TN, United States

ELSEVIER

Elsevier
Radarweg 29, PO Box 211, 1000 AE Amsterdam, Netherlands
125 London Wall, London EC2Y 5AS, United Kingdom
50 Hampshire Street, 5th Floor, Cambridge, MA 02139, United States

ISBN: 978-0-323-91280-8

For information on all Elsevier publications
visit our website at https://www.elsevier.com/books-and-journals

Publisher: Matthew Deans
Acquisitions Editor: Chiara Giglio
Editorial Project Manager: Naomi Robertson
Production Project Manager: Maria Bernard
Cover Designer: Matthew Limbert

Typeset by STRAIVE, India

Contents

Contributors

Grant Allan Anderson
Paragon Space Development Corporation, Tucson, AZ, United States

Gary L. Bennett
Consultant, Boise, ID, United States

John Bradford
SpaceWorks Enterprises, Inc., Atlanta, GA, United States

Edward Chow
Jet Propulsion Laboratory, California Institute of Technology, Pasadena, CA, United States

Eric W. Davis
The Aerospace Corporation, Huntsville, AL; Center for Astrophysics, Space Physics and Engineering Research and Department of Physics, Baylor University, Waco, TX, United States

James E. Dunstan
Mobius Legal Group, PLLC, Fredericksburg, VA, United States

Jennifer Fogarty
Center for Space Medicine, Baylor College of Medicine, Houston, TX, United States

Andreas M. Hein
Initiative for Interstellar Studies (i4is), London, United Kingdom

Michael R. LaPointe
NASA Marshall Space Flight Center, Huntsville, AL, United States

Philip Lubin
Department of Physics, University of California at Santa Barbara, Santa Barbara, CA, United States

Frédéric Marin
Universite de Strasbourg, CNRS, Observatoire Astronomique de Strasbourg, UMR 7550, Strasbourg, France

David G. Messerschmitt
Department of Electrical Engineering and Computer Sciences, University of California at Berkeley, Berkeley, CA, United States

Ian Morrison
International Centre for Radio Astronomy Research, Curtin University, Bentley; Astro Signal Pty Ltd, Perth, WA, Australia

Thomas Mozdzen
School of Earth and Space Exploration, Arizona State University, Tempe, AZ, United States

Simon C. Patané
AIAA Space Settlement Technical Committee, Reston, VA, United States

Mark Shelhamer
Johns Hopkins University School of Medicine, Baltimore, MD, United States

Cameron M. Smith
Biosphere2 Center for Human Space Exploration, University of Arizona, Tucson, AZ, United States

Michael P. Snyder
AIAA Space Settlement Technical Committee, Reston, VA, United States

Introduction

Today, many science fiction stories and a few popular science books and articles discuss interstellar travel, but an overall, comprehensive, nonfiction treatment of the subject is lacking. This book is the second of a three volume monograph series designed to serve as a reference for those aspiring to understand this complex topic. The subject encompasses not just the scientific and technological challenges of the interstellar missions themselves but also the societal, ethical, and risk-related questions that arise from humanity endeavoring to become an interstellar civilization.

Think of the first volume of this monograph series as "on the home front." What type of civilization decides to invest the time, effort, and resources into interstellar missions? How do we decide on a destination and what can we learn about a prospective destination before we actually send out missions? What are the risks associated with such an undertaking? What is the scale of the challenge? Do we send robots, humans, or both? What might this effort do to (or for) humanity? What ethical questions must we address?

The final volume of the series will address the issues and challenges our future explorers will face when they arrive at a new planetary system, including establishing settlements, exploring an entirely new planetary system, terraforming, communications with Earth, and the overall prospects for establishing another interplanetary civilization around a distant star.

This monograph series is intended to serve as a "go to" reference for those aspiring to understand this complex, multidisciplinary field of study, so that they can take the science and technologies required forward toward the next logical step that may one day enable voyages beyond our solar system.

This volume, the second in the series, looks at the actual interstellar journey between stars, which is shown to be an incredibly challenging task. Currently, the prospect for faster than light travel (FTL) is low, and in this volume, it is assumed that humanity will be limited to interstellar voyages at sublight velocities. If FTL is possible, then many of our current scientific theories will need significant revisions and the Fermi paradox becomes far more curious. For those interested, several scientific theories that perhaps permit FTL are discussed in Appendix.

But current science and technology does seem to allow for the possibility of interstellar journeys by large, crewed starships. Building large, crewed ships capable of extended voyages of hundreds or thousands of years and going near the speed of light is certainly difficult and may be impossible, given our current understanding of the rules of the universe. However, starships limited to 10% or 20% of the speed of light on voyages limited to perhaps less than a hundred years appear to be just within our grasp, given significant advances in our current technology. Even then, going to another star will not be inexpensive, nor will it be easy. Such trips will therefore probably be one-way for the crew and passengers.

Herein are chapters that address scientifically feasible propulsion options, energy sources, life support systems, crew health and safety measures, methods of maintaining contact with Earth, and an examination of the roles potentially played by artificial intelligence. Also discussed are the potential for human hibernation and issues associated with ship culture and governance.

The target audience for this series includes scientists, engineers, visionaries, futurists, and writers working on, or interested in, space development and space technology with a far-term vision of interstellar travel and to where such an effort might lead.

We note a clear distinction between humanity going into space to become an interplanetary civilization and humanity going to the stars to become an interstellar one. The former appears to be entirely possible. Up through the 1990s the average cost to launch material into low Earth orbit (LEO) was fairly constant at around $18,500 per kilogram. However, SpaceX's partially reusable Falcon 9 rocket has reduced this cost to around $2720 per kilogram. SpaceX's Starship vehicle, currently in development, is designed to reduce this cost to ~$20 per kilogram.

Two major factors contribute to reducing launch costs: reusability and increased flight rate. SpaceX's Starship, and others being developed, are intended to be reusable. Robert Heinlein is reported to have said, "If you can get your ship into orbit, you're halfway to *anywhere*." Thanks to companies like SpaceX, it will soon be cheap to get halfway to anywhere: lower cost launch and the fact that space agencies are finally seriously considering the use of nuclear power and propulsion in space have the potential to finally open the energy and material riches of our solar system to exploration and exploitation, to the benefit of all humanity. Indeed, our species seems to have within its grasp the very real possibility of becoming a rich and powerful interplanetary civilization spanning the solar system.

But that in no way automatically leads to interstellar missions. The energy requirements, the technologies, the vast distances, and even the risks involved are orders of magnitude beyond anything required for an interplanetary civilization. We do know some of the questions that need to be asked and thoughtfully considered and, in this volume, we asked today's leading thinkers in the field to address them. Have we missed something? Probably. But it's a start.

We want to thank the contributors who offered their time, effort, knowledge, and wisdom to make this volume what it is today.

We sincerely hope that the reader both enjoys and learns something, or many somethings, from this volume. It is also our sincere hope that the reader will be inspired to become part of the conversation and add his/her insights to the discussion, because while we can't know where the interstellar adventure will end, we can at least try to ask all the right questions before it begins.

We also invite the reader to consider becoming part of the community working through these issues today by contacting the Interstellar Research Group (www.irg.space) and attending one of their biannual interstellar symposia.

Ad Astra!

Les Johnson
Oak Ridge, TN, United States

Kenneth Roy
Oak Ridge, TN, United States

CHAPTER

Propulsion options

1

Michael R. LaPointe

NASA Marshall Space Flight Center, Huntsville, AL, United States

Chapter outline

Interstellar Travel. https://doi.org/10.1016/B978-0-323-91280-8.00013-7

1. Introduction

As you read this, humanity's first interstellar probe has left our solar system and is moving at nearly 17 km/s on its journey through interstellar space. Launched in 1977 and carrying a message from Earth on an inscribed golden disk, Voyager 1 completed its grand tour of the outer planets and has since travelled more than 24 billion km on its outward journey; its twin probe, Voyager 2, has also left the heliopause and entered interstellar space, traveling over 20 billion km from Earth. Yet even at these speeds and distances, our intrepid Voyagers have barely moved beyond the influence of our local star. If the distance between our Sun and the closest neighboring star system, Alpha Centauri, were scaled to the size of a meter stick, Voyager 1 would be located just over the ½ mm mark, having traveled 0.06% of the way to the next star (assuming it was pointed in the right direction, which it isn't). At this rate it will take the probe nearly 75,000 years to cover the equivalent distance to Alpha Centauri.

Why is it so hard to get there? The distance to the Alpha Centauri triple star system is immense, approximately 40 trillion km, or roughly 4.2 light years (1 light year is the distance light travels in 1 year). Our fastest spacecraft to date, the Parker Solar Probe, has broken speed records as the fastest human object ever built; in 2025, Parker will be travelling at an astonishing 200 km/s as it hurtles toward another close pass of the Sun. At this speed it could circumnavigate the Earth in less than 4 min, but it would still take over 6 millennia to reach our closest interstellar neighbor. To cover this immense distance we clearly need to go incredibly fast, and to go incredibly fast requires a lot of energy. To send an equivalent of the 720 kg Voyager probe on a fast fly-by of the Alpha Centauri system within 40 years (an average researcher's professional lifetime), the amount of kinetic energy we would need to impart to the probe is around 3.2×10^{19} J, roughly 30% of the total annual energy consumption of the United States in 2022.

Providing kinetic energy to a spacecraft requires a force to act on the vehicle, which is provided by some form of propulsion; for the Parker Solar Probe and our twin Voyager ambassadors, the kinetic energy was provided by chemical rockets. Additional acceleration was provided using well-timed gravity assist maneuvers, which increase the kinetic energy of a passing spacecraft at negligible expense to the orbital kinetic energy of the planet (or Sun). We'll return to these types of trajectories later in this chapter, but for now let's assume that the change in velocity provided to the probe comes from an onboard chemical propulsion system. How much propellant would be required to send a Voyager-type probe on a fast flight to the next star?

1.1 Rocket equation

Near the dawn of modern rocketry, Konstantin Tsiolkovsky developed an equation that describes how a spacecraft accelerates by converting onboard propellant into a high velocity exhaust stream, resulting in a system in which spacecraft mass decreases while its speed increases through the conservation of momentum. This equation, now commonly called the rocket equation, can be written in the form [1]:

$$\frac{M_f}{M_0} = \exp\left[-\frac{\Delta V}{V_e}\right] \tag{1}$$

where M_0 is the initial spacecraft mass including onboard propellant, M_f is the final spacecraft mass once the propellant has been exhausted from the engine, ΔV is the change in velocity provided to the spacecraft, and V_e is the exhaust velocity of the propellant leaving the rocket.

The exhaust velocity is often written in terms of specific impulse (I_{sp}), which is a measure of the impulse (e.g., N-s) produced by the rocket per unit weight of propellant (e.g., N). As such, $I_{sp} = mV_{ex}/mg$, where g is the acceleration of gravity (9.8 m/s^2) and upon cancelling mass in the numerator and denominator is expressed in units of seconds. Returning to the rocket equation and rewriting in terms of I_{sp}:

$$\frac{M_f}{M_0} = \exp\left[-\frac{\Delta V}{g \cdot I_{sp}}\right] \tag{2}$$

Rockets with a higher I_{sp} are more efficient at using propellant and thus require less propellant mass to impart a given change in spacecraft velocity. From Eq. (2), a higher specific impulse will result in a higher fraction of final delivered mass to total initial mass, allowing more payload to be delivered for a given amount of propellant.

2. Chemical rockets

All launch vehicles heading skyward to date have been powered by the chemical energy released when fuels meet oxidizers and chemical bonds are broken and reformed in exothermic reactions. While several fuel and oxidizer combinations are possible, in practice the most energetic reactions result from the use of liquid oxygen (LOX) and hydrogen (H_2). Stored in cryogenically cooled tanks, the measured release and ignition of hydrogen and oxygen in a rocket combustion chamber creates enormous pressure that rapidly expels the heated chemical byproduct (primarily water, in the form of steam) to generate enough thrust to overcome the force of gravity. Large launch vehicles require several engines working in unison, often augmented with supplementary solid propellant rocket boosters, to achieve liftoff and attain orbit. Once in space, chemical propellants are commonly used to maintain spacecraft orbits and change flight trajectories. These engines have successfully propelled humans from the surface of the Earth to orbiting space stations and to the surface of the Moon, and robotic spacecraft to the farthest reaches of our solar system. Chemical propellants currently remain our only means of lifting large payloads from the surface of the Earth and into space, and they have been the mainstay of rocketry from its earliest days.

So why isn't chemical propulsion used to send humans or our robotic emissaries on fast interstellar journeys? The answer resides in the limited amount of energy available in chemical bonds. As noted, the most energetic chemical process created in current rocket propellants involves the controlled combustion of oxygen and hydrogen. The amount of energy released per unit mass during the combustion process is approximately 1.35×10^7 J/kg. Assuming all this energy is converted to kinetic energy, the maximum exhaust velocity imparted to 1 kg of propellant is approximately 5200 m/s, corresponding to an I_{sp} of approximately 530 s. Due to engine inefficiencies the actual I_{sp} for a liquid oxygen-hydrogen engine is closer to 450 s, and lower for other propellant combinations.

Assuming we want to send a probe on a 40-year trip to the nearest star, the required change in velocity imparted to the spacecraft would be approximately 3×10^7 m/s (approximately 10% the speed of light; 0.1 c); at this velocity relativistic corrections are minor and can be safely ignored. Using Eq. (2), achieving this change in velocity with a state-of-the-art chemical engine, I_{sp} of 450 s drives the final to initial spacecraft mass fraction to zero, requiring a nearly infinite amount of propellant to make the trip. If we instead impose a mass fraction of 10^{-9} (where just one billionth of the initial launch mass reaches the target star system), the achievable change in velocity is approximately 9.3×10^4 m/s for a total trip time of nearly 13,700 years. A 1 kg probe would require 10^9 kg of propellant to make even this long duration voyage, which for a Voyager-like probe would require millions of SLS or SpaceX Starship launches just to amass enough propellant in orbit to make the trip. Clearly the use of chemical propellants is unviable for any reasonable excursion to the stars. Even the use of more exotic atomic, gelled, and metallic propellants [2,3] won't change the disappointing calculus of the impossibly large amounts of chemical propellant needed to achieve less than millennia interstellar trip times. Increasing specific impulse using these advanced propellants can significantly improve the mass fraction for flights to orbit or within the solar system but provides negligible improvement to the incredibly long trip times and near-infinite propellant requirements associated with chemically powered journeys between the stars.

3. Getting an assist from gravity

As a spacecraft approaches a planetary body its speed increases due to gravitational attraction and decreases again as it climbs back out of the planet's gravity well. When a spacecraft approaches the planet in the same orbital direction (relative to the Sun) and departs in the opposite direction, it can increase its velocity at the expense of the planet's orbital kinetic energy; approaching the planet in a retrograde direction decreases the spacecraft velocity and adds kinetic energy to the planet's orbital motion. In each case kinetic energy is conserved, and the change in the planet's orbital energy is negligible; however, significant changes in spacecraft velocities and trajectories can be achieved without the additional use of propellant. Such passive gravity assists, first theorized in the early 1920s, have been used by the Parker Solar Probe, the Voyagers, and multiple spacecraft that came before them. At launch, the Parker probe was traveling at approximately 17.6 km/s before performing gravity assist maneuvers at Venus, which positioned it to fall rapidly toward the Sun. Following the latest gravity assist, the probe is expected to reach nearly 200 km/s as it slingshots around the Sun in 2025. However, Parker will also lose a significant amount of this energy as it climbs back out of the deep solar gravity well, reducing its velocity by roughly an order of magnitude as it heads out toward yet another encounter with Venus. The New Horizons Probe that flew past Pluto in 2015 and is now exploring the Kuiper Belt used a similar Jupiter gravity assist to increase its speed by 4 km/s, clocking over 23 km/s on its path toward Pluto. Such passive (nonpropulsive) assists can save significant propellant mass over the life of a mission and, in the case of the Voyagers, provide a well-choreographed tour of the giant planets before a final gravity assist leading out of the solar system.

While passive gravity assists have been used to great effect, as early as 1929 rocket pioneer Hermann Oberth also envisioned the use of propulsive gravity assists to increase spacecraft speeds passing near planets or the Sun [4]. Rather than simply swinging around the Sun or planet, a powered gravity assist fires the spacecraft engine at periapsis to add additional kinetic energy to the spacecraft. Because the spacecraft is already traveling at high speed, the additional increase in speed generates more kinetic energy than could be attained by the same maneuver far from the Sun. For example, a 100 kg spacecraft traveling at 1 m/s has an initial kinetic energy of 50 J; firing its engine to add 1 m/s will increase its kinetic energy by 150 J. The same spacecraft traveling at 100 m/s and firing its engine to add 1 m/s will increase its kinetic energy by 1000 J. As Oberth showed, the energy gained by this propulsive maneuver close to the Sun, when the spacecraft speed is at a maximum, can be significantly higher than attained by passive gravity assists, at a small expenditure of propellant.

The caveat is the spacecraft must travel in close proximity to the Sun, typically within a few solar radii for maximum effect, which introduces significant thermal issues. In 2018 the Johns Hopkins University Applied Physics Laboratory (JHU/APL) initiated a NASA-funded study to evaluate the feasibility of performing an interstellar mission that could be launched by 2030 using existing or soon to be matured technologies [5]. Through a series of workshops and reports, this large, international team of experts investigated the mission designs, technologies, and science opportunities delivered by an interstellar probe that could operate for 50 years and travel to 1000 AU (1000 times the distance from the Earth to the Sun, or around 1.5×10^{11} km). While not an attempt to reach the next star, this distance is significantly further than the current Voyager locations, and the probe would be specifically designed to measure the heliosphere, the circumsolar dust cloud, the local interstellar medium, and targets of opportunity within the Kuiper Belt.

The study evaluated multiple mission scenarios using a notional probe based on the New Horizons spacecraft; several launch vehicle and upper stage combinations were studied, which either alone or together with gravity assists would satisfy the mission design guidelines. In addition to passive Jupiter gravity assists, the team studied Oberth maneuvers at Jupiter and at the Sun to increase the probe velocity. Using launch vehicle and upper stage configurations available in the 2030-time frame and trading distances of closest approach at periapsis, a passive Jupiter gravity assist could reasonably accelerate the probe to a speed of 7 AU/year (33 km/s) while a powered Jupiter gravity assist could increase this speed to 9 AU/year (43 km/s). Both values are lower than the approximately 20 AU/year required to get the probe to 1000 AU in 50 years, leading the study team to perform an extensive analysis of solar Oberth maneuvers. A retrograde Jupiter gravity assist is used to reduce the spacecraft velocity, allowing it to fall toward the Sun; at perihelion a kick stage engine is fired to perform a powered gravity assist and increase the spacecraft velocity as it leaves the vicinity of the Sun. The resulting escape velocity (V_e) is given by:

$$V_e = 7.413\,(\Delta V\,\mathrm{km/s})^{1/2}\left(\frac{R_S}{r_p}\right)^{1/4}\ \mathrm{AU/year} \tag{3}$$

where ΔV is the change in velocity (km/s) imparted by the propulsive kick stage, R_S is the radius of the Sun, and r_p is the perihelion distance.

An estimate of the minimum ΔV required to provide an escape velocity of 20 km/s can be obtained by setting perihelion equal to one solar radius, yielding a minimum propulsive ΔV requirement of 7.3 km/s. Per the rocket equation, using the highest I_{sp} chemical engines currently available would still require propellant loads several times the probe mass. The thermal challenges of skimming through the solar photosphere are of course immense, and the spacecraft will also need to carry a thermal shield whose mass will increase as the perihelion distance decreases. The combination of propellant mass and thermal shielding significantly lowers the achievable escape speed below the theoretical maximum attained in Eq. (3) for perihelion distances more than a few solar radii, providing comparable performance to a powered gravity assist at Jupiter.

Optimizing the perihelion distance to maximize escape velocity, while minimizing thermal shield and propellant mass, is a trade with myriad variables and presents a daunting challenge for mission designers. However, recasting the harsh solar environment as a potential benefit for a close Oberth maneuver, a 2022 study led by Benkoski [6] investigated the design of a solar thermal propulsion system in which the thermal shield used to protect the spacecraft is also used to heat the propellant to a high exhaust velocity, while simultaneously cooling the shield below its melting temperature. The study evaluated a solar thermal propulsion system operated with hydrogen, lithium, lithium hydride, and ammonia propellants at perihelion distances ranging from 3 to 10 solar radii and total spacecraft masses of 1000–5000 kg. Maximum surface temperatures were limited to 2700 K, and both passive emissivity coatings and active propellant cooling through channels in the shield/heat exchanger were employed. The shield design was modeled after the Parker Solar Probe heat shield, with carbon composite panels layered over a carbon foam insulator. Detailed analysis of this multivariate problem identified curves of escape velocity as a function of perihelion distance for the various propellants, with the highest escape velocity of 12.3 AU/year achieved using lithium hydride propellant at a perihelion of 4.5 R_s due to the less complex propellant storage and handling systems and relatively low propellant molar mass. While a significant improvement over the prior estimates of 9 AU/year achievable with a solar Oberth maneuver, this velocity is still too low to reach significantly beyond our stellar neighborhood.

As previously noted, a 40-year mission to Alpha Centauri would require traveling at 10% light speed. Although such velocities are unattainable using passive or powered gravity assists within our solar system, concepts have been proposed in which more exotic celestial objects could anchor the relativistic Oberth maneuvers needed for fast interstellar travel. A detailed examination of one such scenario is included in the JHU/APL Interstellar Study report, which discussed the idealized case of a probe falling from rest toward an uncharged, nonspinning black hole [7]. Using twice the Schwarzschild radius of the black hole as the distance of closest approach, the authors evaluated how an Oberth maneuver would be modified in a strong gravitational field. As it reaches this distance of closest approach the probe is already travelling at 70.7% the speed of light; without an Oberth maneuver the probe would slingshot around the black hole and recede back toward its initial zero velocity state far from the black hole. Instead, by firing its engine to execute an Oberth maneuver at the point of closest approach, the probe can attain a residual asymptotic velocity approaching 2000 km/s as it hurtles back toward interstellar space. While impressively higher than the current Voyager speeds of 17 km/s, it would however still take the probe over 630 years to cover the distance between the Sun and its nearest neighbor.

To achieve truly relativistic speeds, Dyson [8] analyzed a powered flyby of a binary white dwarf star system, later modified to a rapidly orbiting binary neutron star system. Tapping the rotational energy of the white dwarf pair could yield asymptotic velocities approaching 1% the speed of light, while the tighter, faster orbital motion of the neutron star system could generate speeds up to 27% the speed of light (81,000 km/s). At this speed the distance between our star and its neighbor could be covered in a little over 15 years. Recent modifications to the "Dyson slingshot" concept include the use of a binary black hole system [9] in which a light beam sent by a spacecraft a safe distance away from the system would gain energy by travelling around one of the orbiting binary black holes, blue shifting toward more energetic photons that would then provide momentum to the craft that sent the beam. The spacecraft gains energy with each pass of the beam through the system, eventually reaching a terminal velocity approaching 133% of the black hole orbital velocity; for rapidly orbiting black hole pairs, the final spacecraft velocity can be highly relativistic.

While fascinating for the physics they uncover, the lack of a convenient binary neutron star or black hole pair make such relativistic Oberth maneuvers well beyond our current capabilities. With the comparatively limited performance provided by chemical engines and powered solar gravity assists, it appears that Oberth maneuvers within our solar system won't provide the requisite speeds for trips much beyond our local star. To go faster, we clearly need more energy.

4. Nuclear fission propulsion systems

As noted earlier, the energy released in the chemical reactions that power state of the art LOX-H_2 engines is around 1.35×10^7 J/kg, sufficient to provide the high thrust to weight values required to launch and send spacecraft on trajectories throughout the solar system. However, given the limited I_{sp} of these engines, the mass fraction rapidly drops to zero for an interstellar mission. Increasing the exhaust velocity requires increasing the propellant energy, and nuclear reactions provide orders of magnitude more energy per unit mass compared to chemical propellants. If a 1 kg mass of ^{235}U (a fissile isotope of uranium) underwent complete fission, the energy released would be 8×10^{13} J, nearly 6 million times the energy density contained in LOX-H_2 propellants. In reality, only a fraction of the uranium will undergo fission, but even so, the amount of energy released is significantly greater than the energy available through chemical combustion. Is this energy sufficient to propel a spacecraft on a rapid interstellar journey?

4.1 Fission process

When a neutron is captured by a fissile atom it causes the atom to split (fission) into daughter products that separate at high velocity, producing additional neutrons in the process. Nearby fissile nuclei can in turn absorb the additional neutrons, subsequently splitting into energetic daughter products and additional neutrons, etc., leading to a significant release of energy. These reactions are controlled in a purpose-designed chamber (reactor), providing a robust thermal source to heat propellants to higher kinetic energies than can be achieved with chemical rockets.

The most common fission reactions involve the use of ^{235}U, a fissile isotope that makes up 0.72% of natural uranium. To be useful as a power source, natural uranium is processed (enriched) to increase the fraction of fissile isotopes; uranium blends composed of less than 20% ^{235}U are referred to as low enriched uranium (LEU), while higher ratios of fissile to natural uranium are deemed highly enriched uranium (HEU). Commercial power reactors are operated with low enriched fuel, typically 3%–5%, while weapon-grade blends contain greater than 85% ^{235}U. Prior designs for space-based fission reactors for power or propulsion typically used highly enriched uranium to minimize reactor mass and volume; more recent designs are moving toward the use of high-assay low enriched uranium (HALEU) with up to 20% ^{235}U, which mitigates the proliferation risk of weapons grade uranium but still maintains reasonable reactor sizes for in-space applications. Fission reactors are discussed in more detail in a separate chapter by Bennett; below we focus on the applications of nuclear fission for space propulsion, and its potential for powering interstellar travel.

4.2 Fission energy release

When a ^{235}U nucleus absorbs a neutron, it forms a highly unstable ^{236}U nucleus, which immediately fragments and releases prompt neutrons. The fission fragments are formed in very short-lived excited states, which rapidly decay through the release of gamma and beta radiation, and in some decay processes, the delayed emission of additional neutrons, into stable nuclei. This process is represented schematically in Fig. 1A for one possible decay path; hundreds of such decay paths are possible, leading to the distribution of fission product yields shown in Fig. 1B. The asymmetric distribution

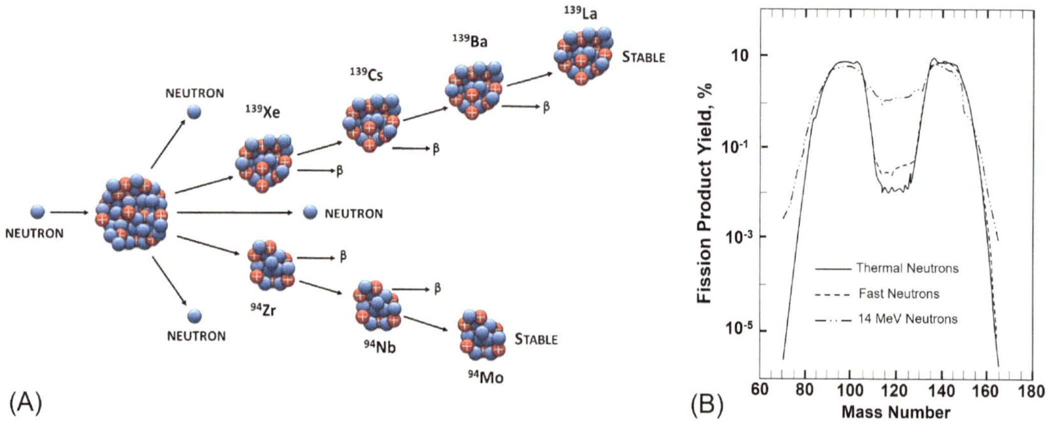

FIG. 1

(A) One of several possible decay paths for ^{235}U fission [10]. (B) Distribution of ^{235}U fission fragments [11].

of product masses is most evident for fission induced by thermal (low energy, sub-eV) neutrons, becoming less so as the incident neutron energy increases.

Roughly 80% of the energy released in a fission reaction is contained in the kinetic energy of the fission fragments, with the remainder contained in the fission radiation decay products, as shown in Table 1.

Except for the energy carried away by the noninteracting neutrinos, the energy contained in the fission fragments and remaining fission products is recoverable, making nuclear fission an efficient energy source for terrestrial and space applications. Various options for using this energy for spacecraft propulsion are briefly discussed below.

Table 1 Average energy release in ^{235}U fission.

Fission product	**Emitted energy (MeV)[a]**
Kinetic energy of fission fragments	168
Kinetic energy of fission neutrons	5
Fission product decay	
β-rays	8
γ-rays	7
Neutrinos	12
Prompt γ-rays	7
Total energy	**207**

[a] 1 MeV = 1.6022 × 10^{-13} J.

Modified from J. Lamarsh, Introduction to Nuclear Engineering, second ed., Addison Wesley Publishing Co., 1983, p. 77.

4.3 Nuclear thermal propulsion

Perhaps the most straightforward application of nuclear fission for propulsion is to absorb the energy of the reaction products to heat a material and then transfer that heat to a propellant to generate thrust. This is the basic principle of nuclear thermal propulsion (NTP), which channels low molecular weight propellants (typically hydrogen) through a hot fission reactor to raise the propellant temperature before expelling it through a nozzle to provide thrust (Fig. 2).

Nuclear thermal engines have a long development history in both the United States and former Soviet Union, an engaging history of which is provided by Dewar [12]. Both the US (Rover/NERVA) and USSR (RD-0410) programs ended with successful ground demonstrations in the 1970s and early 1980s, respectively, before succumbing to budget pressures and changing priorities. However, significant advancements were made under both programs, and numerous design studies have since been conducted to improve the performance demonstrated by these earlier engines. NASA, the US Department of Energy, the Defense Advanced Concepts Research Agency (DARPA), and industry partners are developing the nuclear Demonstration Rocket for Agile Cislunar Operations (DRACO), anticipated to launch into high Earth orbit by 2027 [13], and programs are underway to design and develop NTP engines as part of a robust architecture for crewed Mars exploration [14].

NTP reactors incorporate HEU or HALEU uranium fuel rods, neutron moderators to reduce the energy of fission neutrons and increase their probability of inducing another fission event, reflectors to scatter neutrons back into the reactor core to minimize neutron loss and the amount of fissile fuel needed, and neutron absorbers that capture neutrons to help control the reaction process. Flow channels embedded within the core allow the flow of a low molecular weight propellant (typically hydrogen), which cools the reactor while raising the propellant temperature; the heated propellant is then exhausted through a converging-diverging expansion nozzle. The amount of heat that can be transferred to the propellant is ultimately limited by the material properties of the reactor and flow channels, which will melt above a certain temperature. Refractory metals such as molybdenum and tungsten are used in reactor designs to maximize material temperatures, reaching operating temperatures in the

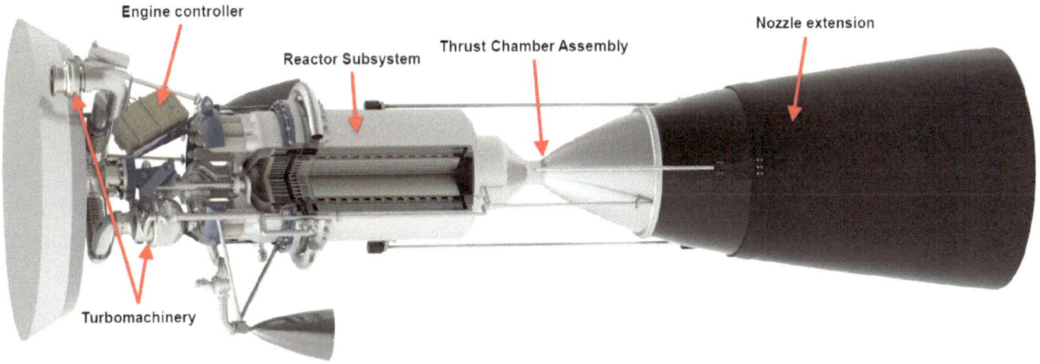

FIG. 2

Nuclear thermal propulsion (NTP) engine assembly (NASA Glenn Research Center).

order of 3000 K. Transferring this heat to the flowing hydrogen propellant increases the propellant temperature to produce a high exhaust velocity, approximated by [15]:

$$V_e = \sqrt{2k\frac{R'}{M}T_0/(k-1)} \propto \sqrt{\frac{T_0}{M}} \tag{4}$$

where V_e is the maximum (ideal) theoretical value of the propellant exhaust velocity, k is the specific heat ratio, R' is the universal gas constant, M is the molecular weight of the propellant, and T_0 is the chamber operating temperature.

The ideal exhaust velocity is proportional to the square root of the temperature and inversely proportional to square root of the propellant molecular weight, emphasizing the need for a high chamber temperature and low molecular weight propellants. For molecular hydrogen propellant and an engine chamber temperature of 3000 K, the ideal exhaust velocity is approximately 9225 m/s, corresponding to a maximum specific impulse of around 940 s. The actual specific impulse will be lower due to real gas flow and nozzle effects; prior NERVA engines demonstrated I_{sp} values up to 850 s, while more modern designs and materials appear capable of achieving specific impulse values approaching 900 s. The thrust generated by NTP engines is also significant. Thrust is related to the mass flow rate of the propellant and the exhaust velocity by:

$$T = \dot{m}V_e = \dot{m}gI_{sp} \tag{5}$$

NTP engines developed and tested under the NERVA program ran at hydrogen mass flow rates exceeding 30 kg/s, achieving thrust values above 250 kN. The combination of high thrust and high specific impulse can provide significant benefits for deep space human and robotic missions over state-of-the-art chemical engines in terms of propellant savings and reduced trip times. However, even a factor of two increase in specific impulse won't noticeably improve the vehicle mass fraction for a trip to the stars, as it still requires a near-infinite amount of propellant to make the trip in decades vs millennia.

4.4 Advanced fission propulsion concepts

The specific impulse achievable with nuclear thermal propulsion is constrained by material thermal properties; above a certain temperature material components will soften and melt, limiting the amount of heat that can be transferred to the propellant. The performance of nuclear fission systems could potentially be improved by removing these material constraints, which has led to several concepts for advanced fission propulsion concepts. While not an exhaustive list, several representative concepts are briefly discussed below.

4.4.1 Liquid core nuclear engine

The thermal constraints limiting solid core NTP engine performance can be relaxed if the uranium fuel is instead kept in a high temperature liquid state. Extensively studied in the early days of nuclear propulsion [16], these liquid core reactor concepts are again being evaluated as alternatives to solid core systems. In most liquid core designs the high temperature liquid fuel is spun to compress it in a thin layer against the inner surface of a porous moderator cylinder. In bubble-through reactors, hydrogen gas is diffused through the cylinder and bubbles through the fuel layer, raising the gas temperature to the temperature of the liquid core before being expelled through the engine nozzle. An alternative design flows hydrogen propellant down the center of rotating liquid fuel tubes, transferring thermal energy to the propellant through radiative heating and surface convection, and other variations use the continuous cycling of liquid fuel droplets within the core to heat the hydrogen. Modern centrifugal reactor designs [17] flow the hydrogen propellant through the moderators and other engine structures to maintain material temperatures below 800 K, then diffuse the gas radially inward through the porous cylinder walls and into the liquid uranium fuel. The liquid uranium near the inner cylinder wall is maintained near 1500 K by the propellant inflow, while the temperature near the center of the rotating cylinder can reach as high as 5500 K. The high temperature liquid uranium layer only contacts the propellant, keeping the material structures well below their melting temperatures. From Eq. (2), these temperatures can provide engine specific impulse values approaching 1800 s for dissociated hydrogen, a significant improvement over chemical and solid core nuclear thermal engines, but still insufficient for an interstellar mission.

4.4.2 Gas core nuclear engines

To achieve even higher propellant temperatures, concepts have been proposed that move from solid and liquid uranium cores to high temperature uranium gas (plasma) cores, which can achieve order of magnitude increases in propellant temperature [18,19]. In a gas core reactor, the uranium plasma is maintained at sufficiently high pressures and densities to keep the neutron and fission particle mean free paths within the fuel volume, maintaining a high gas core temperature and fission criticality. In an open cycle gas core reactor, hydrogen propellant seeded with microscopic particles of graphite or tungsten is injected through a porous wall and flows around the core of uranium gas to create a relatively stable, hydrodynamically confined central fuel region. As the seeded propellant flows around the gas core it is heated through thermal radiation, with the seed particles enhancing the radiative absorption and reducing the radiant heat flux to the pressure chamber walls. The achievable temperatures in open cycle gas core reactors can reach 10^4–10^5 K; transferring this heat to the hydrogen propellant can provide specific impulse values approaching 7000 s with thrust values of several 100 kN. A significant challenge with this engine is the continual loss of uranium gas through entrainment with the propellant flow, which requires a continual replenishment of uranium into the system. A closed-cycle version of the gas core reactor, often called the nuclear light bulb, confines the uranium gas in transparent containers, transmitting the radiant energy through the container walls to the seeded hydrogen propellant flow. An injected buffer gas flows in a vortex pattern around the high temperature uranium to keep it away from the transparent material walls, which must still be actively cooled to remove absorbed radiant heat from the uranium plasma. Preliminary designs of closed cycle systems estimated specific impulse values up to 3000 s, not quite matching the performance of the open cycle gas core systems.

While these next-step fission core engines offer significant performance improvements over solid and liquid core designs, even several thousand seconds of specific impulse isn't adequate for a realistic transit to the next star. A limiting factor with each of the nuclear concepts considered so far is that the fission energy is transferred to a propellant, which is then exhausted to provide thrust; what if the energetic fission fragments could instead be directly used for thrust?

4.4.3 Fission fragment engines

In 1988, Chapline et al. [20] proposed the direct use of reactor fission fragments for propulsion, envisioning the high specific impulse (approaching 10^6 s) produced by these engines could enable trip times to Alpha Centauri in the order of a century, as well as high speed journeys across the solar system (Fig. 3). The conceptual design consists of fissile material coated onto thin fibers, which are in turn

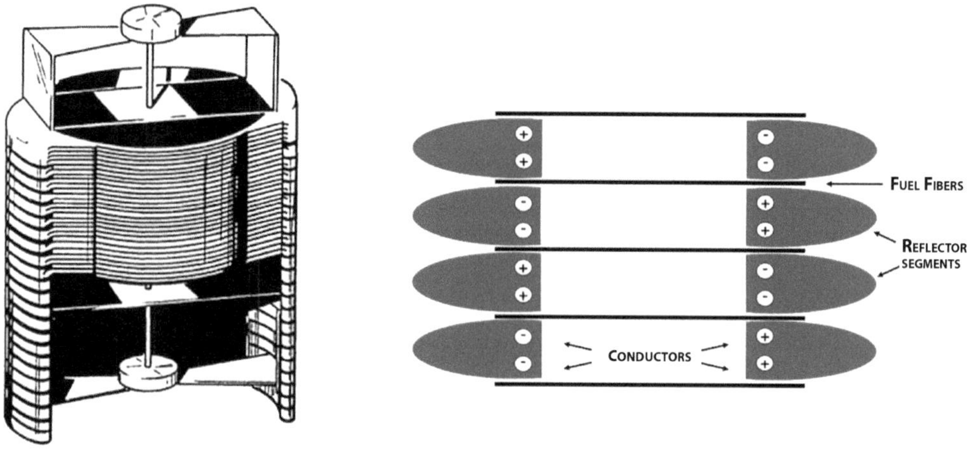

(A) (B)

FIG. 3

(A) Original fission fragment engine design and (B) core design showing conductors for magnetic fields, fuel fibers and reflectors [20].

grouped into thin layers in a reactor core. Current carrying elements within a surrounding moderator material create magnetic fields between the layers, which insulates the moderator and guides the fission fragments out of the core. The separation distance between the layers is determined by the escape probability of the fragments and the average density of material within the core, typically resulting in a few to several centimeter layer separations. Neutron reflectors are used to minimize the fuel requirements and reduce the core size to a diameter less than a few meters. The magnetic field strengths required to guide the fission fragments are found to be modest, on the order of a Tesla or less. Alternate fission fuels were evaluated to maximize engine performance, including highly fissile metastable ^{242}Am and ^{245}Cm, which both require less critical mass for sustained reactions than ^{235}U.

Heat rejection in the fission fragment engine design is enhanced due to the small diameter of the fuel wires, which corresponds to a large fuel surface area to help radiate away heat. Assuming 3 μm diameter carbon fibers coated with 0.4 μm thick americium fuel, Chapline calculates that 10^3 kg of fuel wires will have a surface area of 2×10^5 m^2, sufficient to radiate 20 GW of power at 1100 k. However, the fuel wires within the active core region will see very high temperatures, requiring them to be quickly circulated through the core to prevent them from melting. To do this, the fuel wires are strung out on a large diameter wheel that rotates the fuel layers through the core at a velocity of around 1 km/s. This in turn limits the wheel diameter to ensure the wire tensile strength is not exceeded, resulting in wheel diameters on the order of 200 m. In addition, some fraction of the reactor energy will be absorbed in the moderator material, which will require active cooling.

To demonstrate the potential performance of the engine, Chapline calculates that, for an americium fueled rocket operating at 10-GW, and an escape probability of 50% for the fission fragments, 6 tonnes of mass can be delivered to Alpha Centauri on a fly-by trajectory in about 100 years; 15 tonnes could be delivered in 121 years, and 30 tonnes in 148 years. Increasing the fission fragment escape fraction to 70% would send 10 tonnes in 87 years, and up to 30 tonnes in 113 years.

A variation on the fission fragment concept developed by Chapline was proposed by Clark and Sheldon [21] in which the fuel consists of nanoparticle dust (<100 nm diameter) composed of fissile material. The fuel particles and the fission fragments in the reactor core form a dusty plasma cloud. The dusty plasma fuel elements are electrostatically or magnetically confined within the core, while the higher mass, high velocity reaction products are collimated and channeled out of the core using a magnetic field. Using various fissile isotopes, the dusty fission fragment engine could similarly achieve specific impulse values on the order of 10^6 s. Various mission scenarios were developed to evaluate the potential performance of the engine; for a 10-year mission to the solar gravity lens (550 AU), the required velocity change is about 2% the speed of light. For a rocket I_{sp} of 1.5×10^6 s, the amount of fuel required is about 3% of the total rocket mass. The authors calculate that a 10-year mission to the solar gravitational lens would require approximately 180 kg of nuclear fuel, corresponding to a 350 MW reactor power; a 30-year trip to the Oort cloud, 0.5 light years away, would require a 5.6 GW reactor, and a 50-year trip to Alpha Centauri would require a 208 GW reactor consuming 240 tonnes of fission fuel.

Clearly a significant amount of analysis remains to be performed to better evaluate the feasibility of fission fragment rockets for interstellar missions, and the initial performance estimates outlined above are no doubt highly optimistic. In addition to the reactor and engine design challenges, considerable infrastructure would be required to generate and handle the very large quantities of highly radioactive materials required for these systems. The need to electromagnetically channel the fission fragments out of the engine may limit the space charge density that can be achieved in these systems, which in turn may limit the achievable thrust [22]. Nevertheless, the direct exhaust of highly energetic fission fragments may offer a viable path for high velocity missions, if not to the next star, then perhaps within the solar system and our local interstellar environs.

4.4.4 Nuclear pulse engines

We round out this section with one of the more dramatic uses of nuclear fission for propulsion, in the form of ejected nuclear explosives that detonate behind the spacecraft to push the vehicle forward. The most detailed version of this concept was Project Orion [23,24], based on concepts originally devised in the 1940s and refined into the 1960s. In a pulsed nuclear engine, the nuclear device is jettisoned aft and the resulting explosion produces prompt X-rays followed by the blast of an expanding plasma fireball that consists of high temperature ionized gas and debris, which impinge on a pusher plate attached to the vehicle. The X-rays ablate a layer of the plate's surface, forming a shock wave that propagates into the plate and rebounds back to the plate surface, adding high velocity spall material to the ablated material leaving the plate to provide thrust. The blast wave impacting the plate provides additional momentum to the vehicle, with the transient pulse smoothed out by a spring-like shock absorber. Following the blast, the plate returns to its initial position in time to absorb the next impulse. The explosives are ejected and detonated at a frequency causing harmonic oscillation of the pusher plate to minimize dissipative losses and maximize the coupling between the plate and vehicle. The interaction between the plate material and plasma blast occurs in less than 10^{-4} s, minimizing thermal transport and heating of the plate. Magnetic shielding may also be used to help reflect charged plasma particles and improve plate lifetimes at very high explosive yields. Variations on the pusher plate design include internal pulse designs, in which the detonation occurs within a pressure chamber to harness more of the explosive energy to heat and expel a hydrogen propellant; however, the internal pulse approach is more massive than the external plate approach and suffers from additional radiation heating of the chamber walls.

Originally considered as a launch vehicle to low earth orbit, the Orion vehicle would have an initial mass of 10,000 tonnes and initially operate with 0.1 kton yield pulse units ejected once per second, increasing to 20 kton yield explosives every 10 s as the vehicle accelerated. Recognizing the potential issues with repetitive nuclear detonations to power a launch vehicle, the designs moved toward in-space applications for rapid planetary missions. Various vehicle concepts were developed to provide a range of thrust and I_{sp} values, with typical values of $10–20 \times 10^6$ N and 3000–4000 s, respectively. For interstellar missions the explosive yield was raised from sub-kton to megaton values, and the vehicle size increased dramatically to a total mass of 4×10^5 tonnes with a dry mass of 1×10^5 tonnes (mass fraction of 25%), carrying 300,000 1 MT nuclear devices. The devices were detonated at a rate of 1 every 2.9 s, providing an acceleration of 1.2 g to reach a velocity of 10^5 km/s (3.3% light speed) in just 10 days. At this speed, a trip to Alpha Centauri would take approximately 130 years, carrying a significant scientific payload [25]. While this version of Orion was meant primarily as an existence proof that such trips were possible, the engineering requirements are clearly staggering, as are the political ramifications regarding the repetitive detonation of megaton nuclear devices in space. Development of the Orion concept effectively ended in the 1960s with the signing of the nuclear test ban treaty that prohibited nuclear explosions above ground and in-space, but not before the concept was demonstrated on a small scale using chemical explosives to propel a small vehicle over 100 m into the air off the coast of California [26]. Although restricted by current treaties, modern versions of Orion that employ smaller nuclear fusion devices continue to receive serious attention for interstellar travel, and are discussed later in this chapter.

4.5 Nuclear electric propulsion

In addition to using reactors to heat a propellant or to use the fission byproducts directly for thrust, fission reactors can also be used to generate power for electric propulsion systems. A variety of solar powered thrusters have been used for decades for commercial satellite applications and inner solar system robotic science missions; nuclear powered electric propulsion systems are also being designed for missions farther out into the solar system [27]. Electric (plasma) thrusters generate higher specific impulse but significantly lower thrust than chemical and nuclear thermal engines. Ion thrusters ionize a heavy ion propellant such as xenon and accelerate the ions through high potential electrostatic grids to generate specific impulse values between 2000 and 10,000 s. Hall-effect thrusters use magnetically confined electrons near the thruster exit plane in place of electrostatic grids to set up a potential within the thruster that accelerates the ions; these devices achieve slightly lower values of specific impulse but mitigate wear issues associated with material accelerator grids. Thrust levels for both types of devices are well below 1 N for the kW-class devices flown to date, but 100-kW class laboratory models have achieved up to a few N of thrust [28].

Magnetoplasmadynamic (MPD) thrusters use electromagnetic fields to accelerate a plasma using Lorentz forces; a strong radial current driven between coaxial electrodes ionizes a low molecular weight gas, and the return current through the center electrode creates an azimuthal magnetic field that, in combination with the radial current, axially accelerates the plasma out of the thruster. Additional applied magnetic fields can help stabilize the plasma and allow operation at higher currents, with laboratory devices operating at steady-state power levels of above 250 kW [29]. At MW-class power levels MPD thrusters appear capable of generating several N of thrust with I_{sp} values approaching 10,000 s using hydrogen and lithium propellants. Other high power propulsion engines rely on radio-frequency heating and acceleration; the VASIMR engine [30] employs helicon waves to ionize a propellant followed by ion cyclotron heating of the plasma to high temperature before exhausting through a magnetic nozzle. Thrust values of several N at specific impulses up to 10,000 s may be achievable, depending on propellant type and mission application.

The challenge with using nuclear electric propulsion for interstellar travel again lies in the enormous amount of propellant that must be carried to reach interstellar mission velocities. While I_{sp} values of 10,000 s provide significant advantages for deep space robotic exploration and for efficiently pushing large amounts of mass around the solar system, the achievable specific impulse is still well below the requirement for reasonable interstellar travel times, where mission velocities are fractions of the speed of light.

4.6 Summary: Fission propulsion for an interstellar mission

Fission propulsion can offer significantly improved performance over chemical engines for planetary and deep space missions, although even these energetic engines are limited in capability for rapid interstellar flight. Transferring the fission energy to a propellant generates significant thrust but limits the specific impulse; directly using the fission byproducts provides high specific impulse, but limited thrust and acceleration. Of the fission concepts discussed in this section, those most suitable for interstellar travel appear to be those that use the fission products directly, such as the fission fragment engine and the pulsed fission engines embodied in the Orion concept. As noted, both types of engines face enormous engineering and, in the case of Orion, societal challenges; they do, however, serve as existence proofs that such extraordinary missions may be possible, even if currently far from practical.

5. Nuclear fusion propulsion

Nuclear fusion is the energy source that powers the Sun and the stars we wish to reach, releasing over 25 million times the amount of energy per unit mass available in chemical combustion and several times the energy released in nuclear fission. While the quest continues to control fusion for terrestrial power applications, several studies have investigated designs that, once fusion is achieved, could be used for spacecraft propulsion and power. A detailed discussion doing justice to all the proposed concepts is well beyond the scope of this chapter, and the reader is referred to the literature to more fully investigate the variety of innovative concepts proposed to harness these highly energetic reactions for propulsion. Instead, we provide a brief discussion of the potential benefits of fusion energy for propulsion, outline various types of fusion systems that have been proposed for in-space propulsion applications, and summarize their potential applicability for an interstellar mission.

5.1 Nuclear fusion reactions

With the exception of hydrogen, the mass of any nucleus is less than the total mass of its proton and neutron constituents. When two nuclei are fused together, the resultant total mass is less than the mass of the individual nuclei, and as a consequence of mass-energy equivalence, this missing mass appears

Table 2 Fusion reactions and corresponding energy release [31].

$D + T$	$\rightarrow$	$n + \alpha$	$(Q = 17.6\,MeV)$
$D +^3 He$	$\rightarrow$	$p + \alpha$	$(Q = 18.3\,MeV)$
$D + D$	$\rightarrow$	$p + T$	$(Q = 4.03\,MeV, 50\%)$
	$\rightarrow$	$n +^3 He$	$(Q = 3.27\,MeV, 50\%)$
$T + T$	$\rightarrow$	$\alpha + 2n$	$(Q = 11.3\,MeV)$
$^3He + T$	$\rightarrow$	$\alpha + p + n$	$(Q = 12.1\,MeV, 51\%)$
	$\rightarrow$	$\alpha + D$	$(Q = 9.5\,MeV, 43\%)$
	$\rightarrow$	$^5He + p$	$(Q = 9.5\,MeV, 6\%)$
$p +^6 Li$	$\rightarrow$	$\alpha +^3 He$	$(Q = 4.0\,MeV)$
$p +^7 Li$	$\rightarrow$	2α	$(Q = 17.3\,MeV, 20\%)$
	$\rightarrow$	$^7He + n$	$(Q = -1.6\,MeV, 80\%)$
$D +^6 Li$	$\rightarrow$	2α	$(Q = 22.4\,MeV)$
$p +^{11} B$	$\rightarrow$	3α	$(Q = 8.7\,MeV)$

primarily as kinetic energy of the reaction products. Some fusion reactions of interest and the corresponding energy (Q) produced are shown in Table 2.

Because nuclei are positively charged they repel one another via the Coulomb force. The strength of repulsion increases quadratically as the nuclei are brought closer together until they are sufficiently close for the strong attractive nuclear force to overcome the Coulomb barrier, at which point the nuclei fuse and release energy. The conditions required to create fusion breakeven, when as much energy is created as used to cause the nuclei to fuse, is captured in the Lawson criterion [32]:

$$n_1 n_2 \langle \sigma v \rangle \tau E_f - \tau (P_B + P_S) = \frac{3}{2}(n_1 k T_1 + n_2 k T_2) \tag{6}$$

where n_1 and n_2 are the number densities of the two interacting ion species, $<\sigma v>$ is the velocity averaged fusion cross section (reactivity), τ is the energy confinement time, E_f is the energy created in the fusion process, P_B and P_s are losses due to bremsstrahlung and synchrotron radiation, respectively, k is Boltzmann's constant, and T_1 and T_2 are the temperatures of the interacting ion species.

The Lawson criterion is independent of the confinement method but does depend on the fuel through the dependence on cross section, which in turn is dependent on the temperature. Detailed derivations and discussion of the Lawson criterion for several types of fusion reactors is provided by Wurzel and Hsu [33], and here we simply note that magnetically confined plasmas, with typical number densities in the order of $10^{20}\,m^{-3}$, require confinement times in the order of seconds, while targets that undergo rapid compression can reach orders of magnitude higher fuel densities with concomitant confinement times of tens of nanoseconds to microseconds.

The temperature-dependent reactivity for various fusion fuels is displayed in Fig. 4, showing peak fusion cross sections occurring around 70 keV for D-T reactions and higher for other fusion fuels. Combined with the Lawson criteria, these high temperature requirements create significant design challenges for terrestrial fusion power reactors and in-space propulsion systems.

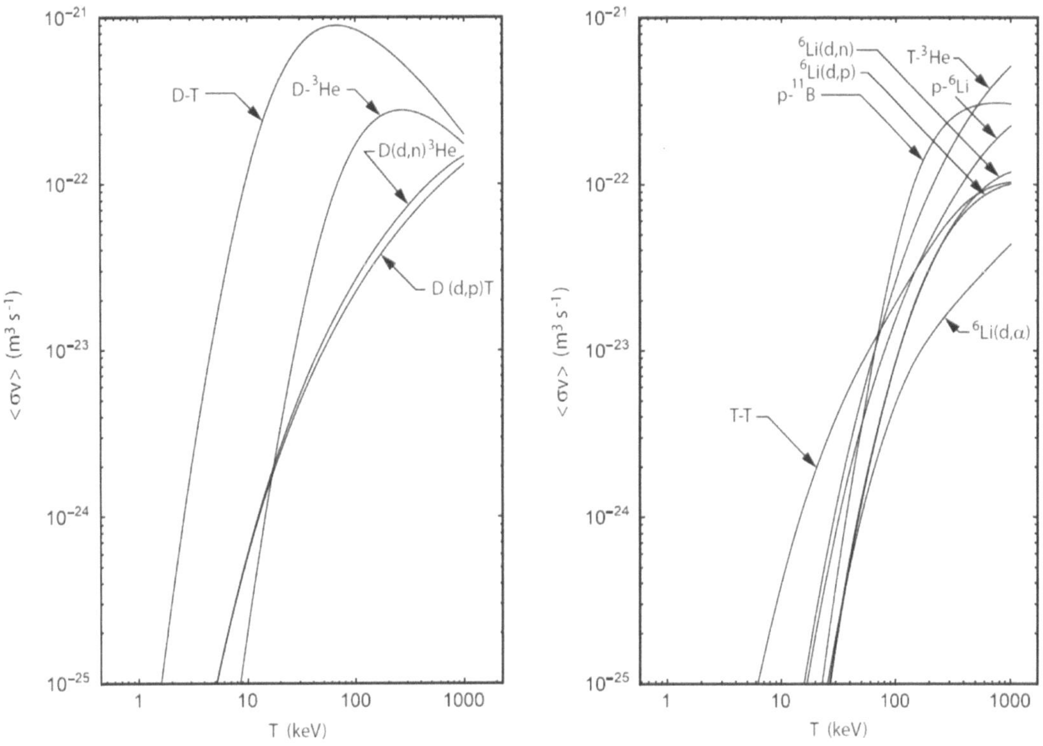

FIG. 4

Maxwellian averaged reactivities for light element fusion reactions [34].

5.2 Approaches to fusion propulsion

A brief review of the approaches used in the worldwide quest to harness fusion power is provided below, together with a short description of several types of fusion propulsion concepts based on their terrestrial counterparts. A succinct history of fusion propulsion is provided by Cassibry et al. [35] and following Cassibry, the various engine concepts are categorized below in terms of the approaches used to confine and heat the plasma.

5.2.1 Magnetic confinement fusion (MCF)

One of the earliest confinement methods studied, MCF uses strong magnetic fields to contain the high temperature ionized plasma; design variations include tokomaks [36], spheromaks [37], stellerators [38], gas dynamic mirrors [39], reverse field plasma devices [40], and other devices that confine the plasma in a sufficiently large volume for the reaction products to transfer their energy via collisions to maintain the requisite fusion reaction conditions. Auxiliary heating may be provided by external (typically radiofrequency) power sources that inject additional energy into the plasma, and plasma losses are replenished through either continual or intermittent injection of the constituent fusion fuels. Several MCF reactors performing basic fusion research are in operation around the world, and an international consortium is building the International Thermonuclear Experimental Reactor (ITER) [41] to study the reactor systems and materials required to generate net energy gain in a terrestrial tokomak fusion reactor.

Adapting a spherical tokamak design, Williams et al. [42] developed a detailed engineering analysis of a compact spheromak propulsion system designed for fast, piloted outer solar system mission. D-^{3}He was used as the fusion fuel, and hydrogen was mixed with the fusion reactor exhaust to increase thrust and reduce specific impulse. The reactor power was designed for 7.9 GW, with approximately 6 GW of power in the charged exhaust products. Trip times to Jupiter were in the order of 4 months with thrust and I_{sp} values of 27 kN and 35,400 s, respectively; a similar crewed mission to Saturn required 7 months, with an engine thrust of 21 kN and an I_{sp} of 47,000 s. In each case, the mass fraction was kept at approximately 10%, with a total initial mass of 1690 tonnes. While not an interstellar mission, the study serves as a primer for the considerations that must go into any fusion powered mission design and demonstrates the significant capabilities of fusion power compared to chemical and fission engines.

A magnetic confinement fusion concept that has been explored for planetary and interstellar travel is the direct fusion drive (DFD) rocket [43–45]. The DFD uses a field-reverse configuration (FRC) to generate a stable, magnetically confined plasma of D-^{3}He, which is radiofrequency heated to fusion temperatures (Fig. 5). Deuterium propellant is tangentially injected to flow around the central reactor region, where it is heated by the reactor products and expelled at high temperature through a magnetic nozzle. The DFD operates at MW-class power levels, and is projected to provide thrust up to 10 N per MW at specific impulse values exceeding 20,000 s. Approximately 35% of the fusion power goes to thrust, 30% is used for electric power, station keeping, and communication, roughly 25% is lost to heat, and the remaining 10% is used for radiofrequency heating of the plasma. Analysis of a 1 MW-class DFD estimated the engine could deliver a 1 tonne payload to Pluto orbit in under 4 years, a considerable improvement over the 10 years it took for New Horizons to reach Pluto on a flyby mission. A mission to the solar gravitational lens at 550-AU estimated that a 1.7 MW DFD engine operating at 4 N and an I_{sp} of 52,000 s could carry a 300 kg payload this distance in approximately 13 years, at a reasonable mass

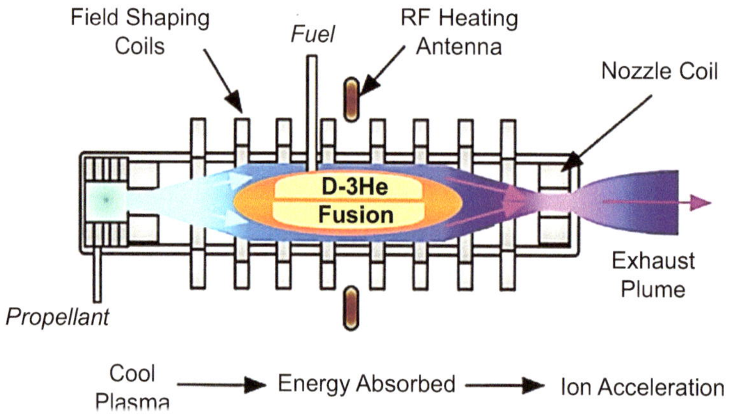

FIG. 5

Reverse field configuration (RFC) used in the direct fusion drive engine [43].

fraction of 6%. As an interesting option, the same spacecraft could be placed into a 550-AU solar orbit in about 18 years, a unique capability of this propulsion system. For an interstellar mission to Alpha Centauri a higher engine power of 160 MW operating at a specific power of 100 kW/kg could reach the star system in approximately 350 years; operating 100 DFD units at 1 MW would reduce the specific power to a more reasonable 5 kW/kg per engine for the same trip time, with the added benefit of reducing mass during transit by jettisoning used DFD engines and spent fuel tanks. For the interstellar mission the DFD engines operate at constant thrust of 4 N and an I_{sp} of 2.4×10^6 s, and both fast flyby and rendezvous missions can be achieved with this concept. While not quite the 40-year mission envisioned at the start of this chapter, the DFD does offer transit times of centuries, vs the multiple millennia required by other propulsion systems. It is a step in the right direction.

5.2.2 Interstellar ramjet

Perhaps the best-known magnetic confinement fusion propulsion system is the Bussard ramjet [46], which replenishes its fuel by scooping up hydrogen from the interstellar medium, and ingesting and compressing it to maintain a continual fusion reaction. The benefit of the engine is that it cheats the rocket equation by not carrying its fuel, which, as previously noted, can rapidly drive the mass fraction to zero for an interstellar mission. A significant challenge, one of many with the concept, is the need to collect a sufficient number of hydrogen ions from the near vacuum of interstellar space to maintain the reactor. Bussard proposed the use of an extended electric or magnetic field that could channel any ions encountered along the path into the reaction chamber, and Fishback [47] later expanded on this concept with a detailed design for the axial component of the magnetic scoop. Schattschneider and Jackson [48] developed an adiabatic analysis of the charged particle dynamics for Fishback's magnetic scoop design, and demonstrated that ionized interstellar hydrogen can indeed be scooped and guided to the fusion reactor using magnetic field strengths of 3–50 T. However, the engineering requirements are formidable. The magnetic field scoop radius is on the order of 2000 km for a reactor inlet radius of 10 m and is assumed to be generated by a superconducting solenoid coil. The faster the ship travels, the higher the magnetic field strength required to channel the ionized hydrogen (protons) into the reactor. The higher fields induce large mechanical stresses due the Lorentz force on the current carrying coils, which in turn increases the mechanical supports and adds mass. For a given thrust, higher mass leads to a lower spacecraft acceleration; as such, the ramjet can only accelerate up to a certain speed, after which the acceleration must be reduced, or the magnetic supports will be stressed beyond their breaking point. Nevertheless, the Bussard ramjet is a truly relativistic engine, capable of providing MN of thrust with accelerations approaching 1 g. Such capabilities make travel to the next star system within a few years' time imaginable, if impractical due to the immense engineering challenges that are currently well beyond our capabilities.

5.2.3 Inertial confinement fusion (ICF)

ICF uses high-powered lasers or heavy-ion beams to rapidly compress and heat a small spherical fuel target. The target structure is typically composed of a hollow shell consisting of various hydrocarbon layers surrounding an inner core of solid D-T ice. The beams may directly impinge on the target, ablating the outer shell material to drive the remaining shell material inward at high velocity to compress the fuel, or use an indirect method, in which the fuel target is placed within a small metal cylinder (hohlraum) and the impinging ion or laser energy excites the hohlraum material to generate a uniform distribution of X-rays. As with a direct drive method, the X-rays ablate the shell material and, somewhat appropriately, this "rocket effect" of the outward moving shell material implodes the fuel target inward at high velocity. As the fuel compresses radially inward it stagnates and generates a "hot spot," which, for proper density and temperature conditions (Lawson criteria), ignites a burn wave that propagates outward into the compressed D-T fuel. Energy deposition from the 3.5-MeV ^4He (α) products created during D-T fusion sustains the fusion burn until the rebounding target radially expands and cools below the conditions required to sustain fusion. The entire process happens literally in the blink of an eye; typical radial implosion velocities are on the order of 350 km/s, compressing the fuel to several hundred GBar and temperatures of several keV within tens of nanoseconds.

One of the first ICF propulsion concepts was VISTA, the Vehicle for Interplanetary Space Transport Application, powered by inertial confinement fusion, developed by the Lawrence Livermore National Laboratory [49,50]. Designed as a single-stage vehicle for interplanetary missions, the propulsion system consisted of a diode-pumped solid-state laser used to drive a spherical array of lasers to uniformly compress D-T fuel pellets at a rate up to 30 pellets/s. The fuel pellets would implode, fuse, and rapidly expand within a magnetic nozzle to provide thrust. A detailed system analysis estimated that a 6000 tonne vehicle could produce 2×10^6 N of thrust at a specific impulse of 27,000 s, capable of performing a round-trip crewed Mars mission in 150 days.

From 1973 to 1977, the British Interplanetary Society sponsored Project Daedalus to design an uncrewed interstellar spacecraft that could reach Barnard's Star (5.9 light years away) within 50 years [51]. The designers conducted a detailed engineering design study of a pulsed ICF engine using D-^3He pellets compressed by energetic electron beams, channeling the resulting fusion products using a magnetic nozzle to provide thrust. The two-stage spacecraft would be constructed in Earth orbit with an initial mass of 54,000 tonnes, including 50,000 tonnes of fuel (7% mass fraction). The first stage would operate for 2 years and reach 7.1% the speed of light, after which it would be jettisoned, and the second stage fired for approximately 1.8 years to attain a final velocity of around 12% light speed. The craft would then cruise for 46 years until it quickly flew by its destination. Detonating 250 fuel targets/s, the first stage engine produced 7.5×10^6 N of thrust at an I_{sp} of 1.08×10^6 s, and the second stage produced 6.63×10^5 N with an I_{sp} of 9.4×10^6 s. Due to the scarcity of terrestrial ^3He, the fuel would be mined over decades from the moon or from the atmosphere of Jupiter and transported to Earth orbit to fuel the spacecraft.

In 2009, an all-volunteer study team revisited the Daedalus concept with Project Icarus, evaluating several alternative pulsed fusion concepts to propel a robotic science mission to Alpha Centauri. Propulsion concepts considered in the Icarus study included a z-pinch variant using D-D fusion (Icarus Firefly); neutron pumped laser ICF with D-D fuels (Icarus Ghost); a laser ICF system using D-^3He fuels (Icarus Endeavor); and additional, less well-developed alternatives. Of these, the Firefly concept received the most engineering analysis, resulting in a spacecraft that was 750 m long with an initial total mass of 23,550 tonnes (compared to the Daedalus 54,000 tonnes) and propellant mass of 21,350 tonnes (mass fraction of 9%). The engine would produce a thrust of 6×10^5 N at an I_{sp} of 1.2×10^6 s, reaching a maximum cruise velocity of just under 5% light speed in 10 years, then coasting for 85 years before decelerating for 5 years to enter the Alpha Centauri system. The spacecraft would eject probes to investigate Alpha Centauri B and enter orbit around Alpha Centauri A for detailed scientific analysis. Significant engineering work remains to be completed on the feasibility of this architecture, but as with the other fusion concepts considered in this chapter, this work shows that interstellar travel in reasonable trip times may be possible, overwhelming engineering challenges notwithstanding.

This section closes with a bright note on the road to ICF that occurred on December 5, 2022, when the US National Ignition Facility (NIF) conducted the first controlled fusion experiment in history to reach scientific energy breakeven [52], where more energy was produced from fusion (3.15 MJ) than laser energy delivered to the target (2.05 MJ). Although NIF is primarily intended to underpin research for the national nuclear stockpile, the successful demonstration of scientific breakeven is a key milestone in understanding the physics of ICF for the design of future commercial fusion power plants and, by extrapolation, potentially more capable in-space propulsion systems for planetary and deep space exploration.

5.2.4 Magneto-inertial fusion (MIF)

The ICF approaches discussed above rely on the rapid spherical implosion of a fusion target by surface ablation to compress the fuel to the pressures and temperatures required for fusion. Magneto-inertial fusion (MIF), also known as magnetized target fusion (MTF), uses magnetic pressure to drive the target implosion [53]. Typically, a conductive liner material surrounds the cylindrical fuel target; an axial magnetic field is applied between the target and liner, and the fuel is preheated prior to sending a high current pulse through the liner. The axial current surging through the outside of the conductive liner generates a strong azimuthal magnetic field, which, combined with the axial current, creates an inward Lorentz force on the liner material. As the liner implodes inward it compresses the internal magnetic field to several thousand Tesla and the fuel to the densities and temperatures required to ignite a fusion burn; the strong internal field decreases radial thermal conduction losses and constrains the motion of the charged fusion products (high energy α particles) to improve the fuel heating. Reducing the energy losses and increasing the energetic particle heating allows a reduction in the required implosion velocity to achieve the necessary fusion conditions, in general making this a more efficient approach than ICF alone [54]. Various versions of MIF have been developed and tested, including z-pinch devices [55] and solid, liquid, and plasma liner implosions of magnetized target plasmas [56]. Significant work continues in this area as a promising alternative to MCF and ICF fusion concepts for terrestrial power generation, including experimental programs at several national laboratories [57].

Adapting terrestrial approaches, several concepts have been proposed that use various forms of MIF for propulsion. The pulsed z-pinch concept employs a high current pulse sent along a cylindrical target to create a strong azimuthal current in the outer layer, driving the outer layers inward at high velocity to compress and heat the fuel to fusion temperatures [58]. Shear flows may be used to stabilize the plasma and improve confinement and burn times [59]. Plasma jets may also be used to generate a compressive liner to implode a spherical fusion target [60], while other devices use rapidly accelerating solid liners or pulsed axial magnetic fields to compress and heat FRC plasmoids to fusion energies [61,62]. Variations on MTF designs include firing fusion pellets into a high gradient magnetic field to induce strong liner currents and a compressive radial Lorentz force [63], as well as devices that boost target fusion by using the fusion neutrons to initiate additional fission reactions in a fissile liner material, further heating the plasma before exhausting the byproducts through a magnetic nozzle for thrust [64,65]. Typical I_{sp} and thrust values appear comparable to the ICF designs, making these concepts highly competitive for rapid planetary and deep space missions; additional analysis remains to be performed regarding their suitability for interstellar missions.

5.2.5 Inertial electrostatic confinement (IEC)

Not to be confused with ICF, IEC devices use recirculating ion beams in a potential well to electrostatically confine electrons and create a confined, quasisteady plasma [66]. Gridded IEC devices use a spherical material mesh grid; in operation, fusion fuel is fed into the grid, and the negative voltage on the grid is raised to 10's of kV to create a plasma discharge between the grid and a grounded outer wall. The large negative voltage extracts the plasma ions from the plasm and accelerates them toward the center of the negatively biased grid at high energy. As they traverse the inner grid region they have a chance of fusing, otherwise they proceed to the opposite side of the grid where they slow, stop, and turn back to accelerate once again toward the center of the grid. The average ion energy set by the grid voltage can approach 100 keV, and ions focused at the center of the grid can in principle achieve densities of $10^{22}\,m^{-3}$, improving the chances for a fusion vs scattering collision to occur. In addition to scattering losses, ions can be lost by collisions with the grid and outer wall material, creating material degradation issues.

An alternative to the spherical grid design is the polywell design by Bussard, in which a multipole cusp magnetic field is used to create a virtual cathode, removing the grid to mitigate electrode collisions and material melting concerns [67]. IEC devices have demonstrated D-T fusion (though so far well below breakeven) and, due to their compact size and inherent simplicity, are being developed as robust neutron sources in addition to continued development toward terrestrial power applications using advanced fusion fuels such as p-^{11}B. An interesting aspect of IEC devices for propulsion is that, by introducing an asymmetry into the grid, a jet mode develops in which a well-collimated beam of fast ions exits the system. The high (multi-keV) potentials used in IEC devices can provide significantly higher ion velocities than other electric propulsion systems, and as such, variations of these devices are also being explored for use in satellite and deep space mission applications [68]. Given the high kinetic energy of the extracted ion beams, specific impulse values well in excess of 10,000 s appear feasible with such systems. However, as with the electric propulsion systems discussed earlier, the extracted ion current and corresponding thrust may be too low for reasonable interstellar trip times.

5.3 Summary: Fusion propulsion for an interstellar mission

Several types of fusion drives have been proposed that take advantage of the significant development work underway in terrestrial fusion power sources. As with the fission concepts discussed earlier, the direct exhaust of the energetic fusion byproducts appears to offer better potential performance for rapid interstellar transport than concepts that transfer heat to a propellant. The pulsed Icarus design, and the constant thrust, high specific impulse operation offered by the direct fusion drive engine, offer possible initial steps on the path toward interstellar flight. The Bussard ramjet, while attractive in its use of interstellar in situ propellants, faces daunting challenges that are beyond known engineering capabilities. IEC devices offer high specific impulse in jet mode, but the thrust appears too low for effective interstellar vehicle accelerations. Fusion concepts employing ICF and MTF require significant additional analysis to determine whether they can realistically be engineered to reach to the next star system.

6. Antimatter propulsion

Moving beyond fusion, the most energetic process currently known is the annihilation of matter with antimatter. Every particle has a corresponding antiparticle (photons, π_0 and μ_0 mesons are their own antiparticles). Compared to its matter counterpart, antimatter has the same mass and lifetime, opposite electrical charge, and opposite magnetic moment. Predicted by Dirac in 1928 as a symmetric solution to his relativistic electron wave equation [69], the antielectron (positron) was discovered in 1932 [70], and the antiproton in 1955 [71]. Just as elements are built up from particles, antielements can be built up from antiparticles. Antihydrogen $(\overline{H})$ consists of an antielectron (e^+) orbiting an antiproton $\overline{p}$; the first $\overline{H}$ atoms were fleetingly created at the CERN accelerator in 1995 [72], and are now routinely trapped and stored for several thousand seconds [73]. In 2011, antihelium nuclei, each containing two antiprotons and two antineutrons, were created at the Relativistic Heavy Ion Collider (RHIC) [74]. The process was exceedingly rare however, generating 18 antihelium nuclei out of nearly a billion high energy collisions. Producing the next stable antielement, antilithium, appears to be beyond the capabilities of today's colliders, but there are no physical principles that prohibit such heavier antielements from existing.

The amount of energy produced when matter meets antimatter is enormous. For one kilogram of combined matter and antimatter the energy released is approximately 9×10^{10} MJ/kg, which is over 250 times the specific energy released in nuclear fusion and over 8 orders of magnitude greater than the specific energy released in chemical combustion. Efficiently using this energy could potentially propel us to the distant destinations envisioned by countless science fiction authors, and antimatter propulsion has received serious attention since the first electron-positron engine was proposed in the 1950s. How this energy might be used for propulsion, and the extraordinary challenges that must be overcome to move these propulsion concepts closer to reality, are briefly discussed below.

6.1 Electron-positron annihilation

The rest mass energy ($m_o c^2$) of an electron or a positron is 0.511 MeV (8.2×10^{-14} J); when an electron and positron meet, they annihilate in a burst of gamma-radiation:

$$e^- + e^+ \rightarrow 2\gamma \left(E_\gamma = 0.511\,\text{MeV}\right)$$

where each γ carries half the energy of the combined e^-e^+ rest mass plus any kinetic energy carried by the pair prior to annihilation.

6.1.1 Electron-Positron propulsion

The earliest antimatter rocket concept was proposed in 1953 by Sanger [75], who suggested redirecting the energetic γ-rays from the e^-e^+ reaction to produce thrust (antiprotons hadn't been discovered yet). Unfortunately, there is no presently feasible method to reflect high energy γ-rays, resulting in a very low engine efficiency. In a later 2006 study, Smith [76] evaluated three positron-based propulsion systems. The first was a solid-core concept that absorbs γ-rays from positron annihilation in a refractory core, which serves as a heat exchanger to transfer energy to a flowing hydrogen propellant. It was estimated that the engine could produce an I_{sp} of 920 s with chamber temperatures at 3000 K, at a thrust and power that matched nuclear thermal fission propulsion systems. For a Mars mission, a spacecraft with three 72-kN solid-core positron engines would require approximately 6–9 mg of positrons, assuming 100% energy conversion efficiency.

The second concept was a gas-core design, where the γ-rays directly heat a propellant through a one- or two-fluid process. It was found that the propellant in a one-fluid system did not efficiently absorb the γ-rays, causing the engine performance to drop below 10%. However, in the two-fluid system, a thrust of 130 kN with an I_{sp} of 3200 s appeared achievable with an engine efficiency of about 85%. A burn time of 30 min could produce a mission velocity of 3.7 km/s using just 25 mg of positrons with a total initial spacecraft mass of about 56,000 kg and final mass of 50,000 kg at burnout.

A third concept used a solid propellant ablating from a surface bombarded by pulses of high-energy photons down converted from the γ-rays. To improve coupling with the material, the γ-rays are shifted to lower wavelengths through an intermediate high-Z material. Initial performance estimates for a Mars mission with a total initial spacecraft mass of 60,000 kg included a propellant mass of 9300 kg and a positron mass of 34 mg to generate 85 kN of thrust, reaching a mission velocity of 3.7 km/s. Positron storage is typically accomplished using Penning-Malmberg traps, which confine positrons in electrostatic potential wells augmented with magnetic fields to help prevent the loss of positrons from the ends of the trap. Trapping and storage of 10^{10} positrons has been demonstrated [77], and trapping up to 10^{15} positrons may be possible in larger traps with higher magnetic fields [78]. Even at this relatively high storage density the mass of positrons is only the order of 9×10^{-10} mg; to store milligram quantities of positrons would require an excessively large number of traps, adding significant mass and complexity to the system.

Rather than store large numbers of positrons, Weed et al. [79] investigated a concept for a radioisotope positron-catalyzed fusion engine that injects pulses of positrons into a substrate containing deuterium. As the positrons annihilate with substrate electrons the annihilation energy couples to lattice-trapped deuterons, giving them sufficient kinetic energy to fuse with other deuterons in the substrate. The D-D fusion reactions are exhausted to provide thrust, while the fusion neutrons are captured in a nozzle blanket to produce a positron-emitting radioisotope (^{79}Kr), which is enriched and deposited onto a source layer to produce more positrons. This "breeder" cycle avoids the need to store large amounts of positrons, greatly simplifying the overall system. Following a detailed cycle analysis, preliminary estimates indicted that a relatively compact engine design might achieve specific impulse and thrust values of 3×10^5 s and 0.2 N, respectively, with a total mission velocity of 60 km/s.

While the positron-based engine concepts discussed above are not suitable for rapid interstellar transit, they do demonstrate the immense power available from the annihilation of small amounts of matter and antimatter. The challenge of trying to efficiently use the high energy γ-rays for propulsion has led to several concepts that instead use the charged energetic products from proton-antiproton annihilation, described in the following section.

6.2 Proton-antiproton annihilation

The higher rest mass energies of the proton-antiproton ($p\overline{p}$) pair yield 1877 MeV per annihilation event, compared with the 1.02 MeV released by electron-positron annihilation. Protons and antiprotons may undergo direct annihilation, or at low energies may form bound states of protonium (Pn) leading to subsequent annihilation. At low energies, $p\overline{p}$ annihilation proceeds via rearrangement collisions in protonium, while direct annihilation dominates at energies above roughly 10 eV.

Unlike electron-positron annihilation, the proton-antiproton annihilation process proceeds through several stages, as shown in Fig. 6. The initial byproduct of $p\overline{p}$ annihilation is the formation (on average) of three charged pions (π^+, π^-), each with a rest mass energy of 139.6 MeV, and two neutral pions (π^0) each with a rest mass energy of 135.0 MeV; hence 689 MeV of the initial annihilation energy is contained in the rest mass of the newly formed pions. The average kinetic energy of each charged pion is approximately 250 MeV/pion; at this energy the charged pions are traveling at 93% the speed of light and their lifetimes are extended from 26 ns (at rest) to approximately 70 ns. Unless acted upon, each

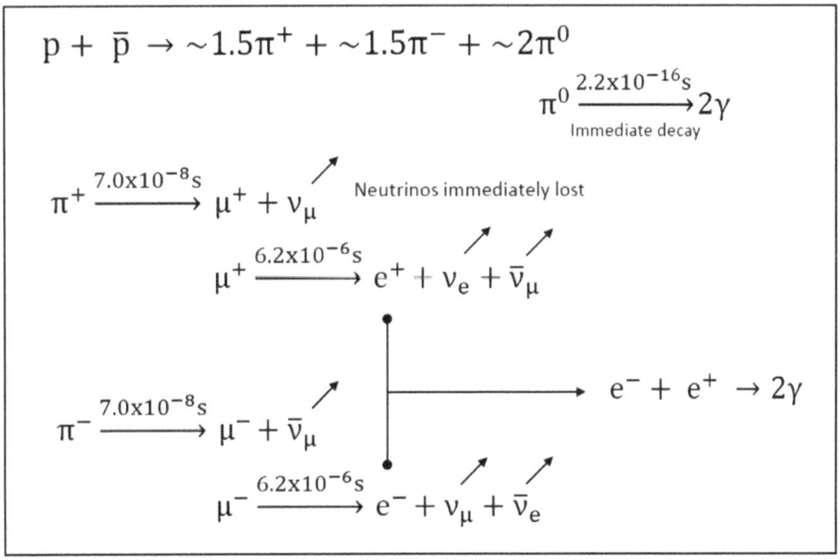

FIG. 6

Proton-antiproton annihilation process.

charged pion will travel almost 21 m before it decays. The remainder of the annihilation energy resides in the kinetic energy of the neutral pions, roughly 220 MeV/pion. At this energy the neutral pion lifetime is extended to a still brief 2.2×10^{-16} s, and upon their decay each neutral pion produces two energetic γ-rays, each with energies between 130 and 300 MeV/γ.

The subsequent decay of each charged pion releases an amount of energy equal to their combined rest mass energy and kinetic energy. The π^+ decays into a charged muon (μ^+) and a mu-neutrino (ν_μ), and the π^- decays into a μ^- and an antimuon neutrino ($\overline{\nu}_\mu$). At rest, the decay energy of a charged pion is distributed into the rest mass of the charged muon (105.7 MeV), a neutrino (or antineutrino) energy of around 30 MeV, and a muon kinetic energy of around 3 MeV (recoiling to conserving momentum). The charged pions have a total rest mass plus kinetic energy of roughly 390 MeV/pion, and each charged pion decay will produce a muon with a kinetic energy of 192.3 MeV and a ~90 MeV neutrino. A charged muon with a kinetic energy of 192.3 MeV is traveling at 94% light speed, and its lifetime is extended from 2.2 to 6.2 μs. Unless acted upon, the muon will travel nearly 1800 m before it decays. The neutrinos, traveling at the speed of light, carry their energy out of the system unimpeded.

The μ^+ decays into a positron, an electron-neutrino (ν_e) and an antimuon-neutrino ($\overline{\nu}_\mu$), while the μ^- decays into an electron, a mu-neutrino (ν_μ) and an antielectron-neutrino ($\overline{\nu}_e$). The total energy available from the decay of a charged muon traveling at 94% light speed is about 300 MeV/muon. Of this energy, 0.511 MeV goes into creating an electron or positron, about 100 MeV into the electron or positron kinetic energy, and the remainder to the noninteracting neutrinos. The decay of the three charged muons provides roughly 300 MeV in kinetic energy of the three electrons/positrons; these are stable against decay but can annihilate with one another to produce 0.511 MeV γ-rays, the ultimate end of the proton-antiproton annihilation process.

6.2.1 Proton-antiproton propulsion

The variety of proposed proton-antiproton ($p\bar{p}$) engines typically fall into one of five categories: solid core, gas core, plasma core, beam core, and antiproton-catalyzed fission-fusion designs. Each of these is briefly described in the following sections.

6.2.1.1 Solid core engines

Solid core $p\bar{p}$ engine designs absorb the antiproton annihilation products in a bulk refractory metal core that acts as a heat exchanger, raising the temperature of a flowing propellant, which is then ejected through a nozzle to provide thrust. Although the neutrinos will carry their energy out of the engine, stopping the charged pions and subsequent muons, electrons, positrons, and γ-rays with the core can capture a high fraction (typically above 70%) of the annihilation energy, providing thrust and specific impulse values comparable to nuclear thermal propulsion systems. In the solid core engine design analyzed by Howe and Metzger [80], the annihilation of 100-mg of antiprotons provided sufficient energy to complete a round-trip, opposition class Mars mission with three times less launch mass than chemical engines and slightly less mass than a nuclear thermal rocket.

6.2.1.2 Gas core engines

Similar to the concept of gas core nuclear thermal engines, gas core antimatter concepts raise the specific impulse by removing the temperature constraints imposed by material walls. This type of engine was also analyzed by Howe and Metzger to perform the crewed opposition-class Mars mission. The engine design assumed a 100 atm hydrogen gas in a central pressure chamber surrounded by high current solenoids producing axial magnetic fields up to 50 T to radially confine the charged annihilation products away from the chamber walls. For antiproton flow rates comparable to the solid core annihilation engine, the gas core engine produced higher specific impulse values of 1000–1500 s with thrust values of 440 kN and 300 kN, respectively. The engine only operated at an efficiency of 8% due to the less efficient transfer of annihilation energy to the propellant gas vs the sold core, wasting a significant amount of the annihilation energy. Cassenti [81] considered the use of a gas-core annihilation engine for a less demanding orbital transfer vehicle (OTV) application, moving a 10 tonne payload from low Earth orbit to geosynchronous orbit and back again. The engine design consisted of a central combustion chamber 2 m in diameter, operating at 100 atm with a hydrogen propellant. A 5 T magnetic field in the chamber radially confined the charged annihilation byproducts, while 50 T fields at the ends of the chamber reflect the products for improved axial confinement. A hydrogen propellant flow rate of 28.7 kg/s and antiproton flow rate of 5.77 μg/s produced an engine thrust of 9.8 kN at a modest specific impulse of 340 s, sufficient for the proposed OTV application but far short of the I_{sp} values of interest for interstellar propulsion. A total of 40 tonnes of hydrogen and 8 mg of antiprotons were used for the mission, with an engine conversion efficiency of around 35%.

6.2.1.3 Plasma core engine

Similar to the gas core design, the plasma core engine utilizes magnetic mirror fields to contain the charged annihilation products, but unlike the gas core, uses an ionized (versus high pressure neutral) gas propellant. Morgan [82] sketched a preliminary design for a pulsed heavy-ion plasma engine, in which antiprotons annihilate with nucleons in a heavy atom nucleus to produce energetic charged nuclides and particles. The charged byproducts are confined by external magnetic field coils and collisionally transfer their energy to the remaining heavy atoms, forming a magnetically confined plasma.

The magnetic mirror field is then relaxed at one end of the system, and the plasma escapes to provide thrust. The antiproton and propellant injection rates can be tailored to provide a variety of thrust and specific impulse values with a single engine. However, the efficiency of the pulsed heavy-ion plasma engine is limited by the amount of energy residing in the kinetic energy of the heavy charged annihilation byproducts, estimated to be less than 20% of the initial annihilation energy; only a fraction of this available energy will be transferred by collisions to the plasma, further reducing the total engine efficiency.

If the charged byproducts produced by antiproton annihilation in hydrogen were used instead, a variable thrust, variable specific impulse engine might be designed without the disadvantages of heavy atom annihilation. LaPointe [83] performed a parametric analysis of an antiproton annihilation plasma engine that uses hydrogen propellant ionized and confined in a strong magnetic field. Injected antiprotons annihilate with the initially neutral hydrogen gas to produce relativistic, charged pions. The pions traverse the hydrogen and transfer some amount of energy before decaying into charged, relativistic muons. The energetic muons interact with the hydrogen over microsecond time scales before decaying into relativistic electrons and positrons. The electrons and positrons transfer energy to the hydrogen over a time scale corresponding to the plasma confinement time, typically a few milliseconds. The electrons and positrons are stable, although positrons may be lost due to annihilation with plasma electrons. Strong (10–20 T) magnetic fields are used for confinement and stability, but even so, particles will be continually lost from the system.

The results of the parametric modeling indicated that the hydrogen plasma core engine suffers from extremely low energy transfer efficiencies at low hydrogen number densities ($10^{20}\,m^{-3}$) and from excessive plasma bremsstrahlung at high hydrogen number densities ($10^{24}\,m^{-3}$). Optimum performance was obtained with a hydrogen number density of $10^{22}\,m^{-3}$ and antiproton number densities between 10^{16} and $10^{18}\,m^{-3}$. Thrust values of several hundred kN could be generated over a wide range of specific impulse values (4600–60,000 s) with 5 ms plasma confinement times and realistic magnetic mirror strengths. However, the annihilation energy transfer efficiencies were generally below 1%, with relativistic electrons and positrons providing most of the plasma heating. Higher thrust could be produced using higher hydrogen number densities, but the plasma was shown to radiatively cool and recombine if not expelled within a few hundred microseconds. Preliminary analysis also indicated that substantial mass will be required to shield the superconducting coils from the high energy gamma radiation emitted in the $p\bar{p}$ annihilation process. Shielding problems are compounded by plasma bremsstrahlung radiation, which will contribute to shield surface ablation and degradation. The total mass of the radiation shield could dominate the system mass, limiting the achievable spacecraft acceleration. Combined with the poor annihilation energy conversion efficiencies, the plasma core engine does not appear to be a viable means for interstellar propulsion.

6.2.1.4 Beam core engines

The solid, gas, and plasma core engine concepts outlined above use the energetic $p\bar{p}$ annihilation products to heat a propellant, with varying degrees of efficiency. Transferring the annihilation energy to a propellant increases the thrust but lowers the specific impulse that might achieved if the relativistic annihilation byproducts were used directly. This is beneficial for planetary and deep space missions, but detrimental for missions requiring relativistic exhaust velocities. To achieve these high exhaust velocities, the beam core engine uses strong magnetic fields to guide the relativistic charged pions created in $p\bar{p}$ annihilation directly out of the engine. Morgan [82] proposed the concept of a relativistic

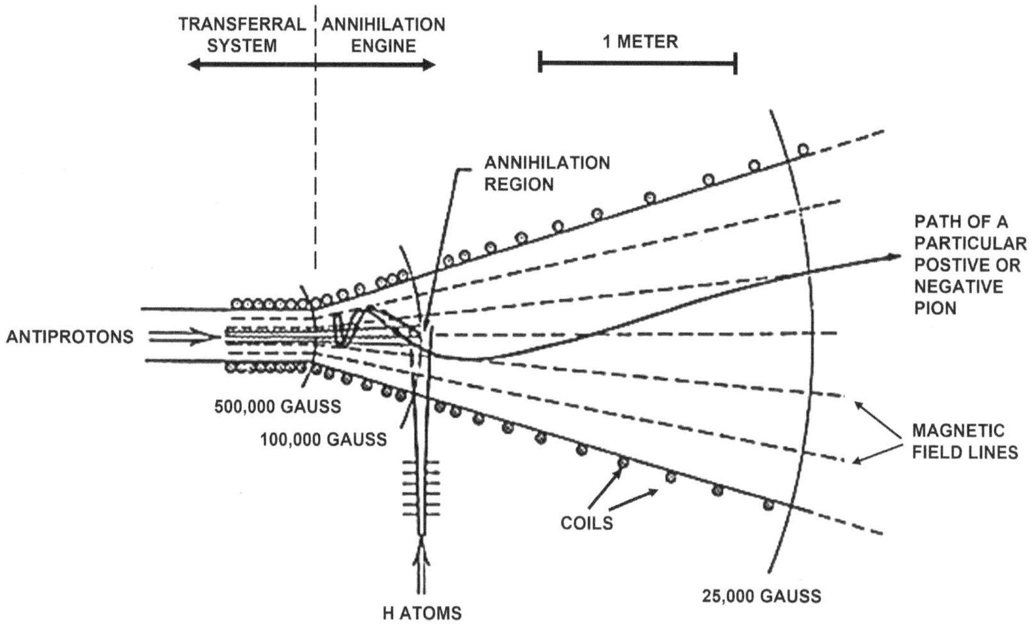

FIG. 7

Antiproton beam core engine design [82].

beam core engine in which antiprotons are injected axially into a strong conical magnetic nozzle, and an equal amount of neutral hydrogen atoms are injected perpendicular to the magnetic field, colliding with the antiproton beam to create the high energy pions, as shown in Fig. 7.

With proper engine design nearly 100% of the antiprotons can be annihilated. The neutral pions will decay almost immediately into high energy γ-rays, but most of the charged pions will be redirected by the magnetic field to exit the engine; about 10% will escape radially across the field lines, and 5% will exit the upstream part of the engine. Depending on the length of the pion path, some will decay into charge muons and neutrinos before exiting the engine, reducing the available energy for thrust. Morgan estimates that 40%–50% of the annihilation energy appears in the kinetic energy of the relativistic charged products leaving the engine; thus an antiproton flow rate of 2×10^{20} $\bar{p}$/s combined with an equal flow of hydrogen atoms will produce about 70 N of thrust at a blazingly high specific impulse of 2.9×10^7 s. The spacecraft mass was estimated to be 7000 kg, including 500 kg of hydrogen and an equal mass of antiprotons; at 70 N the engine would produce an acceleration of around 0.01 m/s^2, and could reach Alpha Centauri in a few hundred years.

In a later analysis, Frisbee [84] developed scaling equations and performed detailed parametric sensitivity analyses on several beam core engine technologies, including the magnetic nozzle, radiation shielding, thermal control systems and radiators, propellant storage and feed systems, electric power systems, payloads, dust shields (as protection for relativistic travel), and miscellaneous spacecraft subsystems. The "stretch" mission analyzed by Frisbee was to send a 100 tonne science payload to a distance of 40 light years at a cruise velocity of 50% the speed of light, which required several million tonnes of antiprotons. An additional analysis by Frisbee [85] for a 70-year rendezvous mission to Alpha

Centauri reduced the amount of antiprotons required to around 1600 tonnes, and an Alpha Centauri flyby mission to 354 tonnes. As discussed below, this amount of antimatter far exceeds our current production capabilities (measured in nanograms per year), but the studies are a tour de force exposition of the technology and supporting infrastructure requirements associated with using antimatter for truly rapid interstellar travel.

6.3 Antimatter-catalyzed fission and fusion

A review of various concepts proposed to use small amounts of antiprotons to initiate fusion or fission-fusion events in the core of a small target is provided by Cassenti et al. [86] These include the use of antiprotons to create fission in heavy nuclei, with the highly energetic byproducts initiating fusion within the target; antiproton heating to directly ignite a fusion reaction in a target; and using the muons created during antiproton annihilation for muon catalyzed fusion.

In the first method, when antiprotons annihilate with fissile nuclei they fission the nucleus and generate high energy nuclides, neutrons, and other particles, with over 20% of the reaction energy contained in the kinetic energy of the nuclides (the basis of Morgan's plasma core engine discussed earlier). In this approach, a fuel target consisting of D-T and fissile material is initially compressed with ion beams or lasers (as in ICF), and a pulse of antiprotons is injected. The antiprotons annihilate with the fissile nuclei, and the energy of resulting nuclides is deposited in the fuel to ignite a fusion reaction. The fission neutrons generate additional fission reactions in the surrounding fissile material, boosting the fusion fuel temperature. The result is an expanding high temperature plasma that can be directed by a magnetic nozzle to produce thrust. An initial analysis of an engine based on this process (ICAN-II) was developed by Gaidos et al. [87] They estimated that a 345 tonne vehicle dry mass with 421 tonnes of propellant could produce an I_{sp} of 13,500 s and a thrust of 100 kN, sufficient to accelerate the vehicle to a velocity of 25 km/s in 3 days using just 30 ng of antiprotons.

The other two approaches appear to have fundamental issues that may preclude their use. In the second method, antiprotons are used to directly ignite a fusion reaction in a D-D or similar fuel target. In this case the annihilation generates mainly neutral and charged pions, with the latter immediately decaying into high energy γ-rays. The highly relativistic charged pions must be absorbed within the fuel for significant energy transfer before they decay into muons and neutrinos (Fig. 6). The muons can also add energy to the fuel before they decay, and the subsequent electrons and positrons will add energy as well. Strong magnetic fields can be used to confine the charged byproducts and enhance the energy transfer. However, as modeled by Morgan et al. [88] each antiproton might only catalyze one fusion reaction; because the energy required to make an antiproton is significantly greater than the energy released in fusion, this may not be a viable means for power or propulsion. The final approach uses muons created in the antiproton annihilation chain to sustain muon-catalyzed fusion of D-T fuel. The antiprotons provide a more compact source to generate the muons compared to an accelerator and would be more suitable for in-space propulsion. However, as pointed out by Cassenti et al., muon-catalyzed fusion is only effective in a narrow temperature range centered around 1200 K, to support a molecular resonance between the tritium and deuterium atoms. These low temperatures would be difficult to maintain, making this a challenging approach.

Building on these results, Cassenti et al. [86] developed a model for a pulsed ICF engine with a fission-fusion burn triggered by antiprotons. The ICF pellet consists of an inner core of D-T fuel. The fuel is coated on one hemisphere with fissionable material, and the core is surrounded by an outer shell consisting of a heavy metal. The inside of the shell consists of neutron reflecting material. A pulse of antiprotons and positrons is injected through a hole in the shell, causing annihilation to occur at the surface of the fissile hemisphere. The resulting annihilation products ionize the fuel, creating plasma within the shell. Electrons created during the annihilation develop a current flow, generating a strong, transient magnetic field within the shell; the magnetic field traps the charged particles (fission fragments, fuel ions, pions, muons, etc.) which heat the core to fusion temperatures. The neutron reflecting material reflects neutrons back into the fuel to boost the fission reactions, and the heavy metal shell provides inertia to contain the plasma long enough for fusion to occur. It was estimated that a pulsed engine based on this design could produce an I_{sp} of 6×10^5 s for complete fusion burn, or about 2×10^5 s for 10% burn, assuming a 3.5 g fuel pellet injected with 2×10^{13} antiprotons. The pellet design parameters were used to estimate planetary mission trip times, with trips to Mars occurring in days and trips to the asteroid belt in weeks using microgram quantities of antimatter.

Each of the antiproton annihilation engine concepts outlined in this section rely on the availability of antiprotons, from the tens of nanogram quantities used in ICAN-II for small robotic planetary exploration to the tonnes required for Frisbee's stretch mission of sending a 100 tonne payload to Alpha Centauri in under a century. How much antimatter can we currently produce and store in our quest to make these trips a reality?

6.4 Antimatter production and storage

Antielectrons (positrons) are produced when a radionuclide decays into a neutron, positron, and electron-neutrino. Cyclotrons are routinely used to create short-lived isotopes such as ^{11}C, ^{13}N, ^{18}F, and others that decay by positron emission, which are commonly used in PET (positron emission tomography) scans for medical diagnostics. Natural positron emitters are rarer; for example, only 0.001% of ^{40}K decays into a ^{40}Ar nucleus via positron emission. An interesting side note, bananas, a rich source of potassium, produce a positron approximately once every 75 min, a rate insufficient to fuel a starship, but a potentially fun fact for the supermarket. Positron-electron pairs can also be created through energetic processes that focus the equivalent rest mass energy of the pair ($2 m_0 c^2$) in a very small volume; this can be accomplished in particle accelerators that collide energetic particle beams with targets; pair creation from the vacuum by energetic photons (laser; γ-rays); and cosmic rays that collide with nuclei in the atmosphere, forming charged muons that decay into electrons and positrons. Positrons can readily be stored in Penning-Malmberg traps, which as noted earlier confine the positrons in electrostatic potential wells with magnetic fields to help prevent the loss of positrons from the ends of the trap.

Antiproton production is not as easy. Given their larger mass, the minimum amount of energy required to create a proton-antiproton pair is 1.876×10^3 MeV, compared to 1.02 MeV for electron-positron pair creation. At present, the only method for us to create proton-antiproton pairs is with high energy particle accelerators, such as those found at the Conseil Européen pour la Recherche Nucléaire (CERN) on the French-Swiss border in Geneva, the Deutsches Elektronen-Synchrotron (DESY) in Hamburg, Germany, and the US Fermi National Accelerator Laboratory (FNAL) in Chicago, Illinois.

To produce antiprotons, a high energy proton beam is collided with a target to produce copious amounts of particles, some fraction of which are antiprotons. FNAL [89] used an incident beam of 120 GeV protons to impact a heavy metal target; typically, less than 5% of the resulting particles spraying out of the target were antiprotons. A magnetic lens downstream of the target was used to focus 10%–30% of the antiprotons, and of these about 1% had the correct energy to be collected and cooled. Once collected, the antiprotons were sent to a decelerator to further reduce their energy, and from there to low energy antiproton accumulator ring, where they could be circulated and stored for several days.

About 1 antiproton was collected for every 10^5 protons incident on the target, and approximately 10^{12} antiprotons could be stored in the accumulator. Between 1985 and 2011, FNAL produced about 17×10^{-9} g (17 ng) of antiprotons, accounting for over 90% of the total world production of antimatter. In comparison, CERN has produced a total of about 2 ng, and DESY about 1 ng. If annihilated, the total amount of antiprotons produced to date would generate 3.2×10^6 J of energy, enough to boil about 1 L of water.

Including the cost of facility operation, producing antiprotons is not cheap; Schmidt et al. [90] calculated the FNAL production cost at approximately \$6.4 $\times 10^{12}$/mg of antiprotons. In fairness, the world's high energy accelerators were not optimized for antiproton production, and they are not dedicated solely to the production of antimatter. The dedicated operation of FNAL for antiproton production would be able to produce nanograms of antiprotons/year. Improvements to the facilities, such as using thinner targets to better focus the antiproton particles, improving the antiproton collection and storage methods, and optimizing the proton acceleration energy and duty cycle, could increase production rates even further, to micrograms/year. Alternatively, a dedicated facility specifically designed for ground-based antiproton production could be constructed. A 1998 Rand Corporation study [91] evaluated the cost of a dedicated facility to produce several mg/year of antiprotons. They estimated the facility cost (in 1988 dollars) at between \$3 $\times 10^9$ and 10×10^9 to build, and including operating expenses the total antiproton production cost could be brought down to around \$6.4 $\times 10^7$/mg, a more reasonable cost for several of the antiproton engine concepts discussed above.

In addition to production, the storage of large quantities of antiprotons presents a unique challenge for propulsion systems. The density of antiprotons that can be stored is limited by space charge effects. As noted, the accumulator ring at FNAL could hold on the order of 10^{12} antiprotons (1.7×10^{-12} g) for long periods of time. Between 2000 and 2003, a portable Penning-Malmberg trap was designed and built at the NASA Marshall Space Flight Center to hold 10^{12} antiprotons for several days to perform ground-based propulsion experiments [92]; though constructed, the project was cancelled before it was used for its intended purpose. The challenge with using electrostatic traps to provide nanograms of antiprotons is that thousands of such traps would be required, adding significant mass and complexity to a vehicle. To circumvent the limits of these space charge storage approaches, concepts have been proposed that use antihydrogen atoms stored as ice in place of charged antiprotons, leading to significantly higher storage densities. As noted earlier, antihydrogen atoms are now routinely made and trapped for long periods of time; Forward [93] suggested that it may be possible to nucleate the antihydrogen to form solid antihydrogen ice or microcrystals that could be more readily stored. Unlike the nucleation of regular hydrogen, antihydrogen can't be allowed to touch the chamber walls and would need to be directly nucleated from cold, trapped antihydrogen atoms. It may be possible to laser cool the antihydrogen atoms to form and trap antihydrogen molecules, which could then condense to form microcrystals of antihydrogen ice. Alternatively, it may be possible to cluster antihydrogen atoms around a single charged antiproton, adding antihydrogen atoms to grow the cluster into microcrystals. While these approaches remain speculative, developing such techniques for high density antihydrogen storage is critical if antiproton engines are to someday become a reality.

Fortunately, once solid antihydrogen is formed there are feasible options to store and transport it [93]. Because antihydrogen is diamagnetic, when it is placed in a magnetic field a weak dipole moment is induced within the antihydrogen opposite to the direction of the applied field. As such, solid antihydrogen could be passively trapped in a magnetic "bottle" formed by current-carrying rings. Alternatively, the antihydrogen could be levitated between two charged electrodes; a weak ultraviolet light source could be used to liberate positrons from the surface of the ice, leaving behind a negative charge that responds to the confining electric fields. To transport antiprotons from storage to the engine, an ultraviolet laser could be used to drive positrons from a spot on the surface, and an intense electric field could then be used to pull off the charged antiprotons; the antiprotons could then be electromagnetically channeled to the annihilation region of the engine. If suspended in the form of a frozen cloud of microcrystals, an ultraviolet light could be used to drive positrons from the surface of a selected microcrystal, leaving a negative charge which would allow an electric field to pull the microcrystal from the cloud. The charged antihydrogen microcrystal could then be electromagnetically channeled to the engine.

6.5 Summary: Antimatter propulsion for an interstellar mission

Clearly there are several challenges associated with use of antimatter for propulsion; in particular, the production rates of antiprotons are woefully inadequate to generate the amount of antimatter fuel needed for an interstellar voyage. Production costs in current facilities are also prohibitive, requiring the construction of a dedicated antiproton factory to reduce production costs and increase production rates. Once generated, the antiprotons would need to be converted to antihydrogen and condensed to form clouds of microcrystals or solid balls of frozen antihydrogen, through methods that have yet to be demonstrated. Once formed there are potentially viable methods for its storage and transport, though scaling these to produce the antiproton beam currents required for interstellar missions remains to be explored. While submicrogram amounts of antiprotons might be used for rapid travel within the solar system, the amount of antiprotons required for interstellar trip times of less than a century borders on hundreds of kilograms to several tonnes, and as noted by Frisbee [84], generating this much antimatter would require a civilization-defining development program.

7. Propulsion without propellant

The propulsion approaches considered so far in this chapter have carried their source of propulsive energy with them: chemicals, fission and fusion fuels, antimatter. Most have also required a significant amount of propellant to be carried as well, the mass of which dramatically reduces their applicability for interstellar missions. Several concepts have been proposed that move the propulsion system off the vehicle, relying on an externally delivered force to accelerate the craft and thereby circumvent the constraints of the rocket equation. Employing energetic beams of photons, particles, and projectiles, these concepts are discussed below.

7.1 Solar sails and sun divers

The Sun is a prodigious source of energy, continually releasing the equivalent of 3.8×10^{26} W. At an average distance of 1 AU (150 million km), the Earth receives about 1.36 kW/m^2, which is tapped to provide power for both terrestrial and space applications. In addition to carrying energy, solar photons also carry momentum, which creates a small but perceptible pressure on objects in space. This radiation pressure must be accounted for to maintain accurate orbital trajectories, and likewise can be used as an off-vehicle source of propulsion. The momentum carried by a photon is equal to its energy divided by the speed of light; the amount of pressure P_s exerted by sunlight normal to a surface is then given by [94]:

$$P_s = \frac{(1+k)I}{c} \tag{7}$$

where I is the intensity of light (W/m^2) falling on the object's surface, c is the speed of light, and the factor k varies between 0 for a completely absorbing blackbody surface and 1 for a fully reflective surface.

Recasting the intensity in terms of the solar intensity at the surface of the Sun and noting the intensity drops off as the square of the distance from the Sun, Eq. (7) can be recast as:

$$P_s = \frac{(1+k)I_o}{r^2c} \tag{8}$$

where I_o is 3.05×10^{25} W/m^2, and r is the distance from the Sun's center to the object's surface.

The resulting force, F, on an object with surface area A normal to the direction of sunlight is then:

$$F = \frac{(1+k)I_oA}{r^2c} \tag{9}$$

and the corresponding acceleration provided to the object with total mass M is given by:

$$a = \frac{(1+k)I_oA}{Mr^2c} \tag{10}$$

To increase the amount of acceleration, the surface area should be as large as possible, the total mass as small as possible, and the material as reflective as possible. This drives the design toward large, lightweight, highly reflective planar sheets, reminiscent of the ship sails that capture the wind for propulsion, giving rise to the equally enchanting vision of solar sailing.

First envisioned by Tsiolkovsky in the 1920s and given a firm theoretical footing in the 1950s [94], the advancement of resilient thin film materials and support structures combined with highly capable, low mass science payloads have ushered in a new era of interest in solar sails for heliophysics research and solar system exploration. Similar to ship sails tacking in the wind, orienting the surface of a solar sail at an angle to the incident sunlight can provide translational motion in addition to the radial motion away from the sun; canting the sail's surface in one direction can increase its tangential velocity and raise its heliocentric orbit, while an opposite orientation can reduce the tangential velocity and move the sail closer to the Sun. In the extreme case of a sail oriented parallel to the Sun's rays, the solar pressure drops to zero and the sail will be pulled inward by the Sun's gravity. Although providing typically low acceleration, the capability to perform such maneuvers without the use of onboard propellant makes solar sailing an attractive option for several types of missions. As discussed by Johnson et al. [95] solar sails can perform heliophysics measurements from unique vantage points out of the ecliptic plane that can't be readily reached with conventional propulsion systems; monitor the Sun to improve space weather predictions and provide timely hazard warnings for crewed space missions; provide sustained measurements of the Earth's magnetotail as well as polar sitting and polar viewing observations; and advanced missions including rapid travel to near-interstellar space.

Solar sail demonstration missions have been flown, starting in 2010 with the IKAROS (Interplanetary Kite-Craft Accelerated by Radiation of the Sun) sail flown by the Japanese Aerospace Exploration Agency [96]. Configured as a square with an area of 196 m^2 and a polyimide resin membrane 7.5 μm-thick, IKAROS was deployed on an interplanetary trajectory, demonstrating successful acceleration using solar pressure and attitude control using liquid crystal panels located at the edges of the sail that could change reflectivity through application of a small electric current. Also in 2010, NASA flew the NanoSail-D2 mission [97], deploying a 10 m^2, 7.5 μm-thick polymer sail from a 3U CubeSat into a 650 km low Earth orbit (the first NanoSail-D was lost in an unsuccessful Falcon 9 launch in 2008.) The mission successfully demonstrated sail deployment and the effects of solar pressure on deceleration, reentering the atmosphere after 240 days. In 2016 the Planetary Society flew LightSail 1, a 32 m^2, 4.5 μm-thick sail, followed in 2019 with the launch and deployment of the twin LightSail 2 [98]. Both sails successfully demonstrated CubeSat deployments and acceleration by solar radiation pressure, countering atmospheric drag for an extended period before eventually deorbiting; LightSat 2 also demonstrated sail control, changing its orientation up to 90 degrees with respect to the Sun.

In 2016 Canada flew the CanX-7 (Canadian Advanced Nanospace Experiment-7) CubeSat [99] that deployed four small drag sails (each about 1 m^2 in area) to demonstrate accelerated orbital decay through atmospheric drag; though not a solar sail demonstration mission per se, the satellite successfully deployed the sail array and demonstrated accelerated orbital decay, reentering the atmosphere in April 2022. The NASA Near Earth Asteroid Scout (NEAScout) spacecraft, designed to use an 86 m^2 solar sail to propel a 6U CubeSat on an asteroid flyby mission, was launched in 2022 but unfortunately communication with the spacecraft was lost following launch and the mission was terminated [100]. NASA's Solar Cruiser mission [101] was designed to fly a 1653 m^2, 2.5 μm-thick polyimide sail containing embedded photovoltaic cells for power generation and reflectivity controls for sail maneuvering to demonstrate station-keeping at the Sun-Earth L1 Lagrange point as well as changing the heliocentric inclination of the spacecraft; although successfully passing several reviews, the mission was not selected for further development and will be closed out in 2024 after testing to demonstrate the technology is sufficiently mature to be proposed for future space science missions. The NASA Advanced Composite Solar Sail System (ACS3) [102] is scheduled to launch to low Earth orbit in 2024, with the key objectives of demonstrating the first deployment of a lightweight composite boom to support the 74 m^2 sail and to evaluate sail performance using solar radiation to change its orbit.

In addition to reflective sails, researchers are investigating diffractive sail concepts that offer potential benefits in reduced sail size and maneuvering efficiency. To perform orbit changes, reflective sails are canted with respect to the incident sunlight to produce a tangential force; however, the amount of sunlight falling on the sail is reduced by the cosine of the angle between the normal to the sail face and the Sun (0 degrees when the sail is perpendicular to the Sun, 90 degrees when oriented parallel to the Sun's rays). As such, the radiation pressure is equally diminished during these maneuvers, requiring large sail areas to maintain a sensible amount of acceleration. In contrast, a diffractive sail can preferentially change the diffraction angle of sunlight incident on the sail, maintaining the sail face perpendicular to the Sun while developing a net force with radial and tangential components. This configuration allows maximum use of the incident sunlight, potentially reducing the required sail area for a given acceleration and allowing for more efficient maneuvering through the use of addressable diffractive sail elements. Underpinned by detailed theoretical models, experimental work is currently underway to develop ultrathin diffractive films with surface structures composed of refractive index materials and birefringent polymeric liquid crystals to evaluate the performance of these new sail materials [103].

At the distance of the Earth from the Sun, the pressure of sunlight pushing on a purely reflective sail is (Eq. 8) approximately 9.1×10^{-6} N/m^2; using the 86 m^2 NEAScout sail as an example, the corresponding force would be around 0.8 mN, and for the total NEAScout vehicle mass of 14 kg, the resulting acceleration would be 5.6×10^{-5} m/s^2. Because the solar intensity drops as the inverse square of the distance from the Sun, the solar pressure and corresponding sail acceleration could be significantly increased if the sail were deployed closer to the Sun. Similar to the powered Oberth maneuvers discussed previously, the sail can be maneuvered such that the solar gravitational force overcomes the radiation pressure, and the sail falls inward to execute a close flyby around the Sun. At perihelion, the sail is maneuvered for maximum tangential acceleration, boosting its speed along its trajectory. This solar-powered gravitational assist can significantly boost the vehicle's velocity at it travels away from the Sun. Known as "sun divers," such maneuvers have been evaluated for rapid transits to the outer planets and near-interstellar space [104,105]. To complete such maneuvers requires significant advancements in sail materials to withstand the intense thermal environment encountered within a few solar radii; such metamaterials, including metallized high temperature substrates and metasurface dielectric films, are the subject of active research [106]. With the potential to reach speeds of 300 km/s, these advanced sun diver sails can significantly reduce travel times for a number of exciting heliophysics and deep space missions, including flights to the 550 AU solar gravitational lens in less than a decade. However, even with such extreme speeds a trip to the nearest star would still take over 4 millennia. The challenge of providing an even higher acceleration to shorten the trip time leads to the consideration of beamed energy propulsion, where high power lasers or microwaves are used to rapidly accelerate a vehicle to fractions of the speed of light. These concepts are briefly discussed in the section below.

7.2 Power beaming

Just as solar sails use the radiance of the Sun for propulsion, light sails can be accelerated by focused laser and microwave beams. The potential advantage of these concepts is that the beams are scalable to significantly higher intensities, and the sails don't require a daredevil dive into the harsh solar environment to achieve high accelerations. The momentum p exerted on a purely reflective sail by a laser or microwave beam with energy E and power P is given by [107]:

$$p = \frac{2E}{c} \tag{11}$$

resulting in a sail acceleration given by:

$$a = \frac{2P}{Mc} \tag{12}$$

for a total vehicle mass M.

Assuming the beam power remains constant, the beam energy is simply the beam power integrated over the time (t) that the beam illuminates the sail. The beam will spread with distance according to the Rayleigh criterion, and as the sail moves away from the beam source the amount of radiant energy incident on the sail will decrease as a function of distance. The maximum distance at which the beam will completely fill the sail is given by [108]:

$$r = \frac{D_S D_A}{2.44\,\lambda} \tag{13}$$

Beyond this distance, the beam spreads beyond the sail area with a concomitant reduction in photon pressure and sail acceleration. Eq. (13), therefore, argues for a large sail area, a large beam diameter, and a short photon wavelength to maximize the acceleration distance.

As a simple example, assume the NEAScout vehicle is accelerated by a laser with a wavelength of 1064 nm at a power level of 100 MW. Assuming the laser is sent through a beam expander with a diameter of $1\,m^2$, the range over which the laser completely fills the sail area of $86\,m^2$ is roughly 3.3×10^7 m. The mass of the NEAScout is 14kg, and from Eq. (12) the acceleration produced by the laser incident on the sail is $0.048\,m/s^2$, over 8500 times the acceleration provided by incident sunlight at 1 AU. The time it takes to traverse this distance at a constant acceleration of $0.048\,m/s^2$ is roughly 3.7×10^4 s, providing a sail velocity of approximately 1.8 km/s in a period of just over 10 h. To increase the sail velocity the power of the laser could be increased, the sail diameter increased, or a combination of both; this scaling has led to the conceptual development of extremely high-power beam arrays on the ground and in space, with large gossamer space structures reflecting the incident beam momentum to provide the high accelerations required to traverse interstellar distances.

One of the first to provide a detailed analysis of laser and microwave power beaming for interstellar travel was Forward, who considered a microwave-powered flyby science mission to Alpha Centauri as well as a laser-powered roundtrip crewed mission to the nearest stars [109,110]. The microwave powered Starwisp consists of a 1 km diameter aluminum wire mesh sail with a mass of just 16 g carrying a 4 g communications payload of microcircuits and optical detectors. A 10-GW microwave beam incident on the wire mesh accelerates the craft in a period of 10 h to 20% light speed, reaching the star in 21.5 years, at which point the wire mesh serves as an antenna powered by the faint microwave energy of the distant transmitter, sending data back to Earth. To accomplish the mission the solar powered microwave transmitter is built as a Fresnel lens with a diameter of 50,000 km, roughly 4 times the diameter of the Earth. Increasing the transmitter power to 1 TW allows higher accelerations, and for a sail diameter of 30 km and a 9000 km diameter transmitter, the acceleration period increases to approximately 18 h and provides a relativistic speed of 0.5 c for the larger, 14 kg sail.

The crewed round-trip laser light sail mission is even more daunting, requiring 43,000 TW of laser power beamed through a 1000 km diameter segmented ring lens, capable of sending a collimated beam of light roughly 44 lightyears before the beam begins to noticeably spread. The sail itself would be built in three sections: an inner payload sail 100 km in diameter, surrounded by a second ring shaped sail 320 km in diameter with a 100-km central hole, and surrounding that, a third ring 1000 km in diameter with a 320 km inner hole. The whole structure is accelerated at 30% Earth gravity for a period of 1.6 years, reaching 50% the speed of light. The intended destination is the Epsilon Eridani star system at a distance of 10.5 light years, which would be reached in about 20 years Earth time (17 years crew time due to relativistic effects). At a distance of 0.4 lightyears from the star, the outer ring sail would separate, and the inner rings would turn to face the separated outer ring. Laser light (still collimated at this distance) would reflect from the outer ring onto the inner ring sails, slowing them down to place them into orbit within the Epsilon Eridani system. To return to Earth, the second ring sail separates from the inner sail and reflects the laser beam onto the inner payload ring, accelerating the payload back toward Earth on a 30-year journey (due to the smaller sail size). Once in the vicinity of the solar system the laser system brings the sail to a halt, completing the roughly 51-year mission (46 years in elapsed crew time).

Forward also considered requirements for a flyby mission of Alpha Centauri by a smaller science probe with a total mass of 1000 kg (1/3 of which is payload) and sail diameter of 3.6 km, powered by a 65 GW laser. As before, the laser is beamed through a large, 1000 km diameter lens to maintains full power on the sail over the acceleration distance, which is calculated to be approximately 0.17 lightyears. Maintaining an acceleration of the probe for a period of 3 years over this distance provides a final probe velocity of around 0.11 c, allowing the probe to complete the trip to Alpha Centauri. These are still enormous structures and power levels, but as winkingly noted by Forward [111], the basic physical principles are known and all that is required is "a lot of engineering and a lot of money." The latter remaining a perpetual challenge, recent research has moved toward potentially more realistic if still daunting beamed propulsion concepts. Landis [112] reevaluated Forward's interstellar probe concept and determined that the use of thin dielectric films in place of the metallized aluminum sails proposed by Forward would allow higher sail temperatures and correspondingly higher power densities, providing higher accelerations and smaller sail sizes. Using dielectric materials the acceleration can be increased to approximately 43 times the acceleration of gravity, nearly 1200 times higher than the acceleration achieved by the metallized sail; at this acceleration, the sail would reach 0.1 c in a little over a week. Keeping the same lens diameter would allow a significant decrease in the sail size by a factor of 1200, or conversely, keeping the sail size the same would reduce the size of the focusing lens by this amount. For the smaller sail size (assuming the payload is again 1/3 of the total spacecraft mass), the laser power would be reduced to around 448 MW, a significant improvement over the 65 GW required in the original concept.

Building on current laser capabilities, Lubin [113] provides a detailed study of both relativistic and nonrelativistic directed energy missions assuming a square diffraction-limited phased laser array illuminating a square dielectric light sail and develops scaling relations for several key performance parameters. The comprehensive analysis includes considerations of payload imaging systems, communications, attitude control, dust impacts, and other aspects associated with an interstellar mission. Options for ground-based vs. space-based deployment of the laser system are discussed, as well as the use of microwave and millimeter beams vs lasers. For an assumed laser array 10 km on a side beaming a power of 70 GW, Lubin evaluates several mission parameters for various payload sizes. Assuming the sail constitutes half the total mass, he finds that for a 1 g wafer-scale spacecraft with a 1 g, 0.85 m dielectric reflector, the wafer-craft will reach 0.14c over a distance of 4×10^6 km in just over 3 min, with a whopping acceleration exceeding 10^4 times the acceleration of gravity. At this speed, a flyby mission to Alpha Centauri would take approximately 30 years. Under continuous illumination, allowing the beam to spread beyond the sail area, the spacecraft would achieve a final velocity of 20% light speed, reaching Alpha Centauri in approximately 20 years.

For increasing values of payload and sail area, the corresponding velocities are reduced and travel times increased; a 100 g payload carried by a 100 g, 8.5 m sail would achieve 0.046c in roughly 1.6 h (0.064 c under continuous illumination), while a 10 kg probe propelled by a 10 kg, 85 m reflector would reach 0.014 c in roughly 6 months (0.02 c under continuous illumination). For crewed missions, assuming the same laser system, a 10^4 kg vehicle attached to a 10^4 kg, 2.7 km reflecting sail would have a small but steady acceleration over roughly 6 years, achieving a velocity of 0.0026 c for a trip time to Alpha Centauri of around 1650 years (or 1200 years under continuous illumination). Clearly the latter is too long for a crewed mission, but such journeys could become feasible with substantial increases in laser array sizes. Lubin concludes his study with a roadmap for technology development for wafer-class interstellar probe missions; this roadmap has been used to help inform the Breakthrough StarShot Initiative [114,115], a privately funded endeavor to send several small gram-size probes ("StarChips") to Alpha Centauri within a generation using a 100 GW laser power station to sequentially illuminate the 4 m sail on each StarChip for short period of time, accelerating them to 0.2 c for a 20-year trip to the next star. During a flythrough of the Alpha Centauri system, the small but capable craft will survey the planet Proxima b orbiting Proxima Centauri, returning enough data to determine the planet's atmospheric makeup and perhaps image large scale features in sufficient detail to determine oceans and large-scale continental features.

Heller et al. [116] evaluate options for slowing down at the target star using a combination of solar radiation pressure and gravitational assists, which increase the total mission time but offers the opportunity for a more leisurely evaluation of the star system; they estimate that probe velocities as high as 5.7% the speed of light can be slowed and captured by their proposed method, requiring a 75-year trip time but allowing more detailed analysis to be performed upon arrival. Finally, Lubin and Cohen [117] have recently investigated the economics of interstellar flight, deriving an analytical cost model driven by the physics and scalability of the directed energy system. They note that the combination of modular and scalable designs, combined with the exponential growth in the capability of photonics, allows for the economic scalability of near-future systems; a promising result as we look to send our robotic emissaries out the stars.

7.3 Particle and pellet beams

We close this chapter with a consideration of directed particle and pellet beams to provide momentum to a target spacecraft. As previously noted, photons carry momentum that can be imparted to a sail to provide thrust; however, generating sufficiently high velocities requires large sails, significant amounts of directed energy, or both. Replacing the solar or laser photons with beams of energetic particles or pellets can also impart a substantial momentum, and several such concepts have been proposed and studied for rapid deep space and interstellar travel. These are briefly discussed below.

7.3.1 Magnetic sails

A concept reminiscent of the solar sails discussed above is the magnetic sail (MagSail) proposed by Andrews and Zubrin [118], that uses the magnetic field created by a large, current-carrying superconducting coil to deflect charged solar wind particles (protons and electrons) to produce thrust. The average solar wind proton number density and speed at 1 AU are approximately $7 \times 10^6/\mathrm{m}^3$ and 450 km/s, respectively; although the momentum per unit area is less than that produced by solar photons, the complexities of fabricating and deploying large, lightweight sails are replaced by the conceptually simpler current loop, whose magnetic field can extend well beyond the coil area. The amount and direction of thrust produced depends on the angle of the MagSail relative to the solar wind, the area of the superconducting loop, and the strength of the magnetic field produced by the current in the coil. Although the energy of the solar wind particles remains fairly constant as they move away from the Sun, the particle density falls off as the inverse square of the distance and, as with solar sails, the force acting on the MagSail will be commensurately reduced as it moves away from the Sun. In their original paper, Andrews and Zubrin calculate that for a 500 tonne, 100 km radius loop operating at 1 AU, with a peak central magnetic field of 10^{-5} T, the interaction with the solar wind will generate a force of 1980 N on the coil, producing an acceleration of around 0.004 $\mathrm{m/s}^2$. Over the period of a year, the coil can achieve a velocity approaching 126 km/s, which is useful for interplanetary missions but too low for reasonable interstellar trip times.

Similar to solar sails, the performance of the magnetic sail could be improved by energizing the sail during a close perihelion maneuver, although the thermal issues flying a superconducting coil close to the Sun would need to be addressed. Even so, as with solar sails, the resulting velocity would be too low for rapid interstellar transport. As noted by Andrews and Zubrin, an option is to instead use the MagSail for braking on approach to the star, using the stellar wind to slow the craft for a rendezvous vs. flyby mission. Additional, detailed mathematical models describing the interactions between the magnetic field and solar wind for single loop MagSails have since been developed, to which the interested reader is referred [119].

As discussed above, the acceleration and final velocity of solar sails can be dramatically increased using high power microwave and laser systems. Similarly, the performance of the MagSail could be improved using a high energy particle beam directed at the sail. As noted by Landis [120], particle beam sails have several potential advantages over solar sails for interstellar travel. In addition to the potentially lower mass of the loop versus large diameter solar sail, the magnetic field itself creates a large target that makes aiming the beam and issues with beam divergence easier to solve. Physical sails have material temperature limitations that in turn limit the incident power of the incoming laser light; there are no such restrictions for the magnetic field, allowing significantly higher beam power densities than can be achieved with physical sails. This in turn increases the acceleration, with the sail reaching the required velocity over a shorter distance (again benefiting beam divergence and stabilization concerns). A particle beam carries significantly more momentum per particle than does a photon beam, transferring more force to the sail at higher efficiency. As noted by Landis, the relationship between momentum (p) and kinetic energy (E) for both particles and photons is given by:

$$p^2 = \frac{E^2}{c^2} + m_o E \tag{14}$$

where m_o denotes the rest mass.

Because photons have zero rest mass, the momentum provided by a particle beam can be significantly higher than a photon beam of similar energy. In addition, while laser efficiencies continue to improve, the efficiency of particle beams is significantly higher (in principle exceeding 90%), reducing the required input power for the beam. Finally, unlike laser beams, particle beams are not diffraction limited, allowing potentially smaller sail sizes. Particle beams would, however, need to be neutralized to avoid electrostatic expansion, and thermal spread would need to be accounted for over large distances. As previously noted, the magnetic sail could also be used to slow down and brake at the destination star system to provide rendezvous options as well as fast flyby trajectories. Beam particle dynamics are further explored by Nordley and Crowl [121] and Benford and Mole [122], the latter providing several examples of particle beam MagSail-driven missions. As a point of reference, a 1 kg MagSail with a diameter of 260 m could be accelerated by an 18 TW beam of mercury ions to an acceleration of $10^6\,\text{m/s}^2$ within 18 s, rapidly reaching a cruise velocity of 0.06 c and completing the mission to Alpha Centauri in just over 70 years. Scaling sail size, beam power, beam particle composition, and acceleration times provides various options for rapidly sending interstellar payloads to the next star within reasonable time frames, using currently known (if challenging to implement) technologies.

To complete the discussion of the magnetic sail concept, it is noted that the charged, highly energetic particles created during a nuclear explosion could also be used to push against the MagSail, similar to the nuclear Orion concept discussed earlier. This concept, termed MagOrion [123], uses small kT-class nuclear devices detonates approximately 2 km behind the spacecraft to create a rapidly expanding plasma, about half of which can be turned into useful thrust. While high specific impulse values of 15,000–45,000 s could be achieved, the challenge of launching and transporting multiple nuclear explosives, combined with the engineering challenges of material survivability over a long duration mission, poses similar issues for the MagOrion as for the original Orion concept. A variation on this concept, dubbed Mini-MagOrion, instead beams particles consisting of fissionable material with a deuterium/tritium core to the spacecraft, where they are rapidly compressed by a z-pinch to ignite and generate a highly energetic charged plasma to push against the magnetic field [124]. Though faced with daunting engineering challenges in terms of beam propagation, capture, and compression, this approach circumvents the issues associated with launching nuclear explosives into space and offers a path toward achieving the high velocities necessary for interstellar travel.

7.3.2 Pellet streams

Similar to particle beams, streams of more massive pellets can also be used to rapidly accelerate a payload. Singer [125] proposed the use of an electromagnetic launcher, roughly 10^5 km long, located in space; superconducting pellets with masses of a few grams up to 100 g would be launched at high acceleration (10^6 times the acceleration of gravity) to impact a target vehicle. While high, these accelerations have been achieved in terrestrial rail guns over much shorter distances. Singer's proposed mission was to send a 450 tonne payload to Barnard's Star, a distance of 5.6 light years, in 50 years (comparable to the Daedalus fusion powered mission discussed earlier in this chapter). Magnetic fields on the target vehicle would be used to reflect the superconducting pellets, providing momentum to the spacecraft. To maintain a collimated beam, the accelerator would be built in multiple sections with trim coils to fine tune the pellet launch direction. Several collimating stations would be placed along the beam path (potentially deployed by the target vehicle along its journey) to monitor and correct the pellet trajectories using electric or magnetic fields. The accelerator power required for the mission was 15 TW (averaged over a firing interval), launching a 2.8 g pellet every 0.5 s at an exit velocity of 0.25 c, for a period of 3 years. Less stringent mission requirements showed that a 10 tonne mass could be sent to Alpha Centauri in 130 years using less than a GW of continuous beam power.

7.3.3 SailBeam

Bridging the gap between laser LightSails, MagSails, and pellet propulsion, the SailBeam concept uses multi-GW laser beams to push small dielectric LightSails (typically 10 cm in diameter with a mass of a few milligrams) to speeds approaching 0.1 c. Millions of these small sails impact the target vehicle, transferring momentum to the vehicle until it asymptotically reaches the sail velocity. The target vehicle uses an onboard laser to vaporize the incoming sails, creating a high velocity plasma that pushes against the vehicle's MagSail to provide thrust; the MagSail is also used to a brake against the stellar wind when the vehicle reaches its destination star. A detailed analysis of the concept developed by Kare [126] considers multiple factors in the design of the system, including the constitution of the dielectric sails, tensile and material limits, sail stability and guidance, MagSail designs, and laser power and optics requirements. A parametric analysis was performed for a nominal sail mission, assuming a 2000 kg payload mass, 2000 kg MagSail mass, and a final vehicle velocity of 0.1 c. The analysis assumed 8.77×10^7 sails, each with a mass of 23 mg, would be accelerated to a velocity of 0.12 c using a high-power laser, with power levels ranging from 35 to 180 GW depending on sail material, areal density, etc. The laser beam would be collimated by focusing optics, providing an acceleration length of 64,800 km and an illumination time per sail of around 3.6 s to provide a sail acceleration of 1×10^7 m/s^2. For these parameters a 0.38 m diameter, two-layered diamond film sail with 98.9% reflectivity offered the best sail performance and required the lowest laser power (35 GW) to meet the mission parameters. With this performance the vehicle would reach Alpha Centauri in just over 40 years, where it would use its MagSail to slow and capture into the system. Options considered by Kare include the use of multiple lower power lasers with multiple focusing optics to reduce the development demands on the beam system. Kare concludes his study with a technology roadmap, defining the key system elements to be developed as well as near-term experiments that could be performed to advance the readiness of the constituent technologies.

8. Concluding remarks

Reaching the next star system in decades or even centuries vs millennia precludes the use of conventional and near-term chemical and nuclear propulsion systems; the mass penalty associated with carrying onboard propellant is too severe to make these systems a feasible option. Fusion ramjets that take advantage of the sparse hydrogen density diffused within the interstellar medium appear to be well beyond our engineering capabilities, and the amount of antimatter required to make an interstellar journey would require a dedicated large-scale production effort, which is prohibitively expensive using current approaches. Concepts that move the propulsion system off the vehicle, such as laser, microwave, and particle/pellet beams, can rapidly accelerate small payloads to the velocities required for reasonable interstellar trip times, at the expense of significant infrastructure on the ground or in space. As noted at the start of this chapter, getting to the next star is incredibly hard given the enormous distance and the amount of energy that must be provided to traverse that distance in relevant time scales. Our Voyager probes will cover an equivalent distance in approximately 75,000 years, and dedicated researchers will continue to design methods to get us there faster by pushing toward evermore energetic propulsion concepts. The enduring quest to reach the stars embodies the ultimate human spirit of exploration; it is ours to make this bold venture a reality.

Acknowledgments

The author thanks the editors for indulging a flagrant violation of recommended page count and for leeway in the choice of propulsion topics. While several key concepts are discussed, the chapter is nevertheless limited in scope and content, and it is recognized that several unique and interesting propulsion concepts have been given insufficient attention or precluded altogether. The interested reader is encouraged to continue their research through the technical journals and papers of professional societies such as the British Interplanetary Society, International Astronautical Federation, and American Institute of Aeronautics and Astronautics; texts such as *The Starflight Handbook* and *Frontiers of Propulsion Science* referenced throughout this chapter; the workshop proceedings of the Interstellar Research Group and its predecessor, the Tennessee Valley Interstellar Workshop; and the pioneering work of Marc Millis and the NASA Breakthrough Propulsion Physics Program [127], which has inspired countless researchers to reach for the (temporarily) unreachable stars. *Ad Astra per Aspara.*

References

[1] G. Sutton, O. Biblarz, Rocket Propulsion Elements, seventh ed., J. Wiley and Sons, 2001, pp. 102–106.

[2] B. Palaszewski, Atomic hydrogen propellants: historical perspectives and future possibilities, in: AIAA-93-0244, 31st Aerospace Sciences Meeting and Exhibit, January 11-14, Reno, NV, 1993.

[3] B. Palaszewski, J. Jurns, K. Breisacher, K. Kearns, Metallized gelled propellant combustion experiments in a pulsed detonation engine, in: AIAA 2004-4191, 40th AIAA/ASME/SAE/ASEE Joint Propulsion Conference and Exhibit, July 11-4, Fort Lauderdale, Florida, 2004.

[4] P. Blanco, C. Mungan, Rocket propulsion, classical relativity, and the Oberth effect, Phys. Teach. 57 (2019) 439–441, https://doi.org/10.1119/1.5126818.

[5] R. McNutt, R. Wimmer-Schweingruber, M. Gruntman, S. Krimigis, E. Roelof, P. Brandt, S. Vernon, M. Paul, R. Stough, J. Kinnison, Interstellar probe—Destination: Universe! Acta Astronaut. 196 (2022) 13–28.
[6] J. Benkoski, Combined heat shield and solar thermal propulsion system for an Oberth maneuver, Phase I Final Report, NASA Institute for Advanced Concepts, February 22, 2023.
[7] R. McNutt, et al., Interstellar probe study 2019 report, appendices: background and beginnings, Oberth maneuver, and escaping the solar system, NASA Contract NNN06AA01C, Johns Hopkins University Applied Physics laboratory, 2019.
[8] F.J. Dyson, Gravitational machines, in: A.G.W. Cameron (Ed.), Interstellar Communication, W.A. Benjamin, New York, 1963, pp. 115–120. Available at: https://arxiv.org/pdf/2305.10470.pdf.
[9] D. Kipping, The Halo Drive: Fuel-Free Relativistic Propulsion of Large Masses via Recycled Boomerang Photons, arXiv:1903.03423 [gr-qc], 2019.
[10] W. Emrich, Principles of Nuclear Rocket Propulsion, Elsevier, 2016, p. 61.
[11] W. Emrich, Principles of Nuclear Rocket Propulsion, Elsevier, 2016, p. 60.
[12] J. Dewar, To the End of the Solar System, The University Press of Kentucky, 2004.
[13] "Nonreimbursable Interagency Agreement Between the National Aeronautic sand Space Administration and Defense Advanced Research Projects Agency for Demonstration Rocket for Agile Cislunar Operations Program" darpa-nasa_draco_nonreimbursable_interagency_agreement_final_2023-01-11_.pdf, 2023.
[14] Space Nuclear Propulsion—NASA.
[15] G. Sutton, O. Biblarz, Rocket Propulsion Elements, seventh ed., J. Wiley and Sons, 2001, p. 54.
[16] W.L. Barrett Jr., Liquid core nuclear rocket, in: AIAA 64-541, AIAA Aerospace Propulsion Meeting, May 4-6, Cleveland, OH, 1963.
[17] D. Thomas, M. Houts, W. Walters, K. Hollingsworth, R. Frederick, J. Cassibry, Establishing the Feasibility of the Centrifugal Nuclear Thermal Rocket, 2021, https://doi.org/10.2514/6.2021-3604. AIAA Propulsion and Energy Forum, August 9-11 (Virtual Event).
[18] T. Kammash, D. Galbraith, Fuel confinement and stability in the gas core nuclear propulsion concept, in: AIAA 92-3010, AIAA/SAE/ASME/ASEE 28th Joint Propulsion Conference and Exhibit, July 6-8, Nashville, TN, 1992.
[19] W. Emrich, Principles of Nuclear Rocket Propulsion, Elsevier, 2016, pp. 273–304.
[20] G. Chapline, P. Dickson, B. Schnitzler, Fission fragment rockets—a potential breakthrough, in: OSTI 6868318, International Reactor Physics Conference, September 18, Jackson Hole, WY, 1988.
[21] R. Clark, R. Sheldon, Dusty plasma based fission fragment nuclear reactor, in: 41st AIAA/ASME/SAE/ASEE Joint Propulsion Conference & Exhibit, July 10-13, Tucson, AZ, 2005.
[22] R. Frisbee, Limits of interstellar flight technology, in: M. Millis, E. Davis (Eds.), Frontiers of Propulsion Physics, Progress in Aeronautics and Astronautics, vol. 227, American Institute of Aeronautics and Astronautics, Inc., 2009, pp. 67–69.
[23] W. Emrich, Principles of Nuclear Rocket Propulsion, Elsevier, 2016, pp. 259–272.
[24] G. Schmidt, J. Bonometti, C. Irvine, Project Orion and future prospects for nuclear pulse propulsion, J. Propuls. Power 18 (3) (2002) 497–504.
[25] R. Frisbee, Limits of interstellar flight technology, in: M. Millis, E. Davis (Eds.), Frontiers of Propulsion Physics, Progress in Aeronautics and Astronautics, vol. 227, American Institute of Aeronautics and Astronautics, Inc., 2009, pp. 66–67.
[26] Propulsion Test Vehicle, Project Orion | National Air and Space Museum.
[27] J. Cassidy, et al., Recent advances in nuclear powered electric propulsion for space exploration, Energy Convers. Manag. 49 (3) (2008) 412–435.
[28] L. Mason, S. Oleson, D. Jacobson, P. Schmitz, L. Qualls, M. Smith, B. Ade, J. Navarro, Nuclear power concepts and development strategies for high-power electric propulsion missions to Mars, Nucl. Technol. 208 (Suppl. 1) (2022) S52–S66, https://doi.org/10.1080/00295450.2022.2045180.
[29] M. LaPointe, E. Strzempkowski, E. Pencil, High Power MPD Thruster Performance Measurements, NASA/TM—2004-213226, September, 2004.

[30] M. Giambusso, F. Chang Díaz, A. Corrigan, L. Dean, M. Warrayat, M. Stoehr, Steady-state testing in the VASIMR VX-200SS project, in: AIAA 2021-4150, ASCEND 2021, November 15-17, Las Vegas, Nevada, 2021.
[31] E. Morse, Nuclear Fusion, Springer Graduate Texts in Physics, 2019, p. 29.
[32] T. Kammash, Principles of fusion energy utilization in space propulsion, in: T. Kammash (Ed.), Fusion Energy in Space Propulsion, Progress in Astronautics and Aeronautics, vol. 167, American Institute of Aeronautics and Astronautics, 1995, pp. 6–7 (Chapter 1).
[33] S. Wurzel, S. Hsu, Progress toward fusion energy breakeven and gain as measured against the Lawson criterion, Phys. Plasmas 29 (2022). pp. 062103-1 to 062103-33.
[34] E. Morse, Nuclear Fusion, Springer Graduate Texts in Physics, 2019, p. 34.
[35] J. Cassibry, R. Cortez, M. Stanic, W. Seidler, R. Adams, G. Statham, L. Fabisinski, The case and development path for fusion propulsion, J. Spacecr. Rocket. 52 (2) (2015), https://doi.org/10.2514/1.A32782.
[36] W. Stacey Jr., Fusion: An Introduction to the Physics and Technology of Magnetic Confinement Fusion, J. Wiley and Sons, 1984.
[37] E. Hooper, et al., Sustained spheromak physics experiment (SSPX): design and physics results, Plasma Phys. Control. Fusion 54 (2012) 113001, https://doi.org/10.1088/0741-3335/54/11/113001.
[38] R. Gross, Fusion Energy, J. Wiley and Sons, 1984, pp. 267–274.
[39] T. Kammash, R. Tang, The gasdynamic mirror fusion propulsion system—revisited, AIP Conf. Proc. 746 (2005) 529–535.
[40] L. Marrelli, et al., The reversed field pinch, Nucl. Fusion 61 (2) (2021) 023001.
[41] ITER Organization, ITER research plan within the staged approach (level III—provisional version), Report No. ITER-18-003, September 17, 2018.
[42] C. Williams, Nuclear Fusion Space Propulsion Research, Experimentation, Theory Development, and Systems Analysis Efforts Led by the NASA Glenn Research Center (1994-2004), 10.2514/6.2020-4084, ASCEND 2020, November 16-18, (virtual event), 2020.
[43] Y. Jain, P. Kakade, Analyzing the engineering feasibility of the direct fusion drive, Acta Astronaut. 206 (2023) 57–71.
[44] S. Cohen, C. Swanson, McGreivy, N. Raja, E. Evans, P. Jandovitz, M. Khodakm, G. Pajer, T. Rognlien, S. Thomas, M. Paluszek, Direct fusion drive for interstellar exploration, J. Br. Interplanet. Soc. 72 (2019) 37–50.
[45] S. Thomas, M. Paluszek, S. Cohen, A. Glasser, Nuclear and future flight propulsion—modeling the thrust of the direct fusion drive, 10.2514/6.2018-4769, in: AIAA Propulsion and Energy, 2018 Joint Propulsion Conference Forum, July 9-11, Cincinnati, OH, 2018.
[46] R.W. Bussard, Galactic matter and interstellar flight, Astronautica Acta 6.4 (1960) 179–194.
[47] J.F. Fishback, Relativistic interstellar spaceflight, Astronautica Acta 15 (1969) 25–35.
[48] P. Schattschneider, A. Jackson, The fishback ramjet revisited, Acta Astronaut. 191 (2022) 227–234.
[49] C. Orth, G. Klein, J. Sercel, N. Hoffman, K. Murray, F. Chang-Diaz, The VISTA spacecraft–advantages of ICF for interplanetary fusion propulsion applications, Lawrence Livermore National Laboratory, Report UCRL-96676, October 2, 1987.
[50] C. Orth, VISTA—a vehicle for interplanetary space transport application powered by inertial confinement fusion, Lawrence Livermore National Laboratory, Report UCRL-TR-110500, March 31, 2005.
[51] K. Long, P. Galea (Eds.), Project Daedalus: demonstrating the engineering feasibility of interstellar travel, British Interplanetary Society, London, England, 2015. ISBN/ISSN/EAN: 5800056982849.
[52] A. Zylstra, et al., Burning plasma achieved in inertial fusion, Nature 601 (2022) 542–548.
[53] R. Siemon, I. Lindemuth, K. Schoenberg, Why magnetized target fusion offers a low-cost development path for fusion energy, Plasma Phys. Control. Fusion 18 (6) (1999) 363–386.
[54] C. Walsh, et al., Magnetized ICF implosions: scaling of temperature and yield enhancement, Phys. Plasmas 29 (2022) 042701, https://doi.org/10.1063/5.0081915.

[55] M. Liberman, J. De Groot, A. Toor, R. Spielman, Physics of High-Density Z-Pinch Plasmas, Springer-Verlag, New York, 1999.
[56] Y. Thio, Status of the US program in magneto-inertial fusion, J. Phys. Conf. Ser. 112 (2008) 042084.
[57] G. Wurden, et al., Magneto-inertial fusion, J. Fusion Energ. 35 (2016) 69–77.
[58] J. Miernik, et al., Z-pinch fusion-based nuclear propulsion, Acta Astronaut. 82 (2) (2013) 173–182.
[59] U. Shumlak, Z-pinch fusion, J. Appl. Phys. 127 (2020) 200901, https://doi.org/10.1063/5.0004228.
[60] Y. Thio, C. Knapp, R. Kirkpatrick, R. Siemon, P. Turchi, A physics exploratory experiment on plasma liner formation, NASA Technical Report Server Document ID 20020066592, January 1, 2002.
[61] J. Slough, G. Votroubek, C. Phil, Creation of a high-temperature plasma through merging and compression of supersonic field reversed configuration plasmoids, Nucl. Fusion 51 (2011) 053008.
[62] A. Hoffman, et al., The TCS rotating magnetic field FRC current-drive experiment, Fusion Sci. Technol. 41 (2) (2002) 92–106, https://doi.org/10.13182/FST02-A205.
[63] M. Lapointe, R. Adams, J. Cassibry, M. Zweiner, J. Gilland, Gradient Field Imploding Liner Fusion Propulsion System, NASA/TP-2018-219996, September, 2018.
[64] R. Adams, J. Cassibry, K. Schillo, Developing the pulsed fission-fusion (PuFF) engine, in: AIAA 2014-3520, 50th AIAA/ASME/SAE/ASEE Joint Propulsion Conference, July 28-30, Cleveland, OH, 2014.
[65] R. Adams, J. Cassibry, K. Schillo, B. Taylor, The pulsed fission-fusion (PuFF) engine—Nacelle concept and development roadmap, in: ASCEND 2020, November 16-18, Virtual Event, 2020, https://doi.org/10.2514/6.2020-4082.
[66] G. Miley, S. Krupakar Murali, Inertial Electrostatic Confinement (IEC) Fusion: Fundamentals and Applications, Springer, 2014.
[67] R. Bowden-Reid, J. Khacha, Electric potential in a magnetically confined virtual cathode fusion device, Phys. Plasmas 28 (2021) 042702, https://doi.org/10.1063/5.0040792.
[68] G. Miley, B. Bromley, Y. Gu, A novel IEC propulsion unit for satellite applications, AIP Conf. Proc. 361 (1996) 1435–1440.
[69] J. Baggott, The Quantum Cookbook, Oxford University Press, 2020, pp. 179–202.
[70] D. Lindley, Landmarks: the first antiparticle, Phys. Rev. Focus 17 (2006) 5.
[71] K. Orrman-Rossiter, Observation and annihilation: the discovery of the antiproton, Phys. Perspect. 23 (2021) 3–24.
[72] G. Baur, et al., Production of antihydrogen, Phys. Lett. B 368 (3) (1996) 251–258.
[73] The ALPHA Collaboration, Confinement of antihydrogen for 1,000 seconds, Nat. Phys. 7 (2011) 558–564.
[74] S. Battersby, Physicists create heaviest form of antimatter ever seen, New Scientist (2011). March 22.
[75] E. Sanger, "Zur Theorie der Photoneraketen" ("The Theory of Photon Rockets"), Ingenieur-Archiv. 21 (1953) 213–226.
[76] G. Smith, Positron Propelled and Powered Space Transport Vehicle for Planetary Missions, NIAC Phase I Final Report, Research Subaward No. 07605-003-048, April 19, 2006.
[77] P. Dupré, A new scheme to accumulate positrons in a Penning-Malmberg trap with a linac-based positron pulsed source, AIP Conf. Proc. 1521 (2013) 113–122.
[78] K. Meyer, et al., Long-term confinement of dense positron plasma, AIP Conf. Proc. 692 (2003) 184–192.
[79] R. Weed, J. Machacek, B. Ramamurthy, Radioisotope positron propulsion: NIAC phase I report, Positron Dynamics, NASA award 80NSSC18K0910, 2019.
[80] S. Howe, J. Metzger, Antiproton-based propulsion concepts and the potential impact on a manned Mars mission, in: AIAA 87-2191, AIAA/SAE/ASME/ASEE 23rd Joint Propulsion Conference, San Diego, CA, June 29-July 2, 1987.
[81] B. Cassenti, Antimatter propulsion for OTV applications, in: AIAA 84-1485, AIAA/ASME/SAE 20th Joint Propulsion Conference, Cincinnati, OH, June 11-13, 1984.
[82] D. Morgan Jr., Concepts for the design of an antimatter annihilation rocket Lawrence Livermore National Laboratory, NASA CR-168660, J. Br. Interplanet. Soc. 35 (9) (1982) 405–412 (Lawrence Livermore National Laboratory, NASA CR-168660, 1982).

[83] M. LaPointe, Antiproton powered propulsion with magnetically confined plasma engines, J. Propulsion 7 (5) (1991) 749–759.
[84] R. Frisbee, How to build and antimatter rocket for interstellar missions—systems level considerations in designing advanced technology vehicles, in: AIAA 2003-4676, 39th AIAA/ASME/SAE/ASEE Joint Propulsion Conference and Exhibit, Huntsville, Al, July 20-23, 2003.
[85] Frisbee, Limits of interstellar flight technology, in: M. Millis, E. Davis (Eds.), Frontiers of Propulsion Physics, Progress in Aeronautics and Astronautics, vol. 227, American Institute of Aeronautics and Astronautics, Inc., 2009, pp. 84–92.
[86] B. Cassenti, T. Kammash, D. Galbraith, Antiproton catalyzed fusion propulsion for interplanetary missions, J. Propuls. Power 13 (3) (1997) 428–434.
[87] G. Gaidos, et al., Antiproton-Catalyzed Microfission/Fusion Propulsion Systems for Exploration of the Outer Solar System and Beyond, CP420, Space Technology and Applications International Forum-1998, edited by Mohamed S. El-Genk, American Institute of Physics, 1998.
[88] D. Morgan Jr., L. Perkins, S. Haney, Antiproton Catalyzed Fusion, UCRL-JC-120149, Lawrence Berkely National Laboratory, May 15, 1995.
[89] S. Nagaitsev, Fermilab antiproton source, recycler ring, and main injector, Report number: FERMILAB-FN-0957-AD; arXiv:1408.0759 [physics.acc-ph], 2014.
[90] G. Schmidt, et al., Antimatter requirements and energy costs for near term propulsion applications, J. Propuls. Power 16 (5) (2000) 923–928.
[91] Proceedings of the Rand Workshop on Antiproton Science and Technology, World Scientific, Singapore, June 1988.
[92] J. Martin, R. Lewis, S. Chakrabarti, W. Sims, J. Pearson, W. Fant, Ion Dynamic Capture Experiments with the High Performance Antiproton Trap (HiPAT), NASA Technology Report Server (NTRS), February 2002, https://doi.org/10.1063/1.1541340.
[93] R. Forward, J. Davis, Mirror Matter: Pioneering Antimatter Physics, Wiley Science, 1998.
[94] E. Mallove, G. Matloff, The Starflight Handbook, J. Wiley & Sons, 1989, pp. 92–95.
[95] L. Johnson, et al., Solar Sail Propulsion by 2050: An Enabling Capability for Heliophysics Missions, 2023, https://doi.org/10.48550/arXiv.2301.01297. arXiv:2301.01297 [astro-ph.IM].
[96] Y. Tsuda, et al., Achievement of IKAROS—Japanese deep space solar sail demonstration mission, Acta Astronaut. 82 (2) (2013) 183–188.
[97] A. Heaton, B. Faller, C. Katan, NanoSail—D Orbital and Attitude Dynamics. https://ntrs.nasa.gov/api/citations/20140000644/downloads/20140000644.pdf.
[98] The Planetary Society, "LightSail, a Planetary Society Solar Sail Spacecraft;" https://www.planetary.org/sci-tech/lightsail.
[99] CanX-7 (Canadian Advanced Nanospace eXperiment-7): https://www.eoportal.org/satellite-missions/canx-7#mission-status.
[100] National Aeronautics and Space Administration, "NEAScout," https://science.nasa.gov/mission/nea-scout/.
[101] National Aeronautics and Space Administration, "Solar Cruiser: Enabling New Vistas for Heliophysics Science," https://science.nasa.gov/heliophysics/programs/technology/solar-cruiser/.
[102] National Aeronautics and Space Administration, "NASA Facts: Advanced Composite Solar Sail System," FS-2019-07-01-ARC, https://ntrs.nasa.gov/api/citations/20190028916/downloads/20190028916.pdf.
[103] G. Swartzlander Jr., C. Johnson, S. Kim, N. Tabiryan, Diffractive Lightsails, NASA Innovative Advanced Concepts (NIAC) Phase II Final Report, NASA Grant No. 80NSSC19K0975, June 2022.
[104] C. Bailer-Jones, The sun diver: combining solar sails with the Oberth effect, Am. J. Phys. 89 (2021) 235–243.
[105] E. Ancona, R. Kezerashvilib, G. Matloff, Exploring the Kuiper Belt with sun-diving solar sails, Acta Astronaut. 160 (2019) 601–605.
[106] A. Davoyan, Extreme solar sailing for breakthrough space exploration, NASA Innovative Advanced Concepts (NIAC) Phase II Annual Report, NASA Grant Number 80NSSC21K0954, July 2023.

[107] E. Mallove, G. Matloff, The Starflight Handbook, J. Wiley & Sons, 1989, p. 75.
[108] E. Mallove, G. Matloff, The Starflight Handbook, J. Wiley & Sons, 1989, p. 74.
[109] R. Forward, Roundtrip interstellar travel using laser-pushed lightsails, J. Spacecraft 21 (2) (1984) 187–195.
[110] R. Forward, Starwisp: an ultra-light interstellar probe, J. Spacecraft 22 (3) (1985) 345–350.
[111] R. Forward, Beamed power propulsion to the stars, in: Advanced Space Propulsion Study, 1987. AFRL TR-87-070, Grant F04611-86-C-0039, Air Force Astronautics Laboratory, Edwards Air Force Base, CA.
[112] G. Landis, Advanced Solar- and Laser-pushed Lightsail Concepts, NASA Institute for Advanced Concepts, Phase I Final Report, May 1999.
[113] P. Lubin, A Roadmap to Interstellar Flight, NASA Innovative Advanced Concepts (NIAC) Program Phase I Final Report, NASA Grant NNX16AL32G S007, January, 2016. See also NASA Technical Report Server, Document 20190001185, 2019.
[114] Breakthrough StarShot website: https://breakthroughinitiatives.org/initiative/3.
[115] G. Popkin, What it would take to reach the stars, Nature 542 (2017) 20–22, https://doi.org/10.1038/542020a.
[116] R. Heller, M. Hippke, P. Kervella, Optimized trajectories to the nearest stars using lightweight high-velocity photon sails, Astron. J. 154 (2017) 115–126.
[117] P. Lubin, A. Cohen, The economics of interstellar flight, arXiv:2112.13911 [econ.GN] (2021), doi:10.48550/arXiv.2112.13911.
[118] D. Andrews, R. Zubrin, Magnetic sails and interstellar travel, J. Br. Interplanet. Soc. 43 (1990) 265–272.
[119] M. Bassettoa, N. Perakis, A. Quarta, G. Mengali, Refined MagSail thrust model for preliminary mission design and trajectory optimization, Aerosp. Sci. Technol. 133 (2023) 108113, https://doi.org/10.1016/j.ast.2023.108113.
[120] G. Landis, Interstellar flight by particle beam, Acta Astronaut. 55 (11) (2004) 931–934.
[121] G. Nordley, A. Crowl, Mass beam propulsion: an overview, J. Br. Interplanet. Soc. 68 (2015) 153–166.
[122] J. Benford, A. Mole, Ultra-high acceleration neutral particle beam-driven sails, J. Br. Interplanet. Soc. 72 (2019) 190–197.
[123] D. Andrews, R. Zubrin, Nuclear device pushed magnetic sails (MagOrion), in: 33rd Joint Propulsion Conference and Exhibit, Seattle WA, July 6-9, 1997, https://doi.org/10.2514/6.1997-3072.
[124] R. Lenard, D. Andrews, Use of Mini-Mag Orion and superconducting coils for near-term interstellar transportation, Acta Astronaut. 61 (2007) 450–458.
[125] C. Singer, Interstellar Propulsion Using a Pellet Stream for Momentum Transfer, Princeton Plasma Physics Laboratory Report PPPL-1583, October 1979. also, J British Interplanetary Society, vol. 33, 1980, pp. 107–116.
[126] J. Kare, High-acceleration micro-scale laser sails for interstellar propulsion, NASA Innovative Advanced Concepts Phase I Final Report, NIAC Grant #07600-070, December 2001 (Revised February 2002).
[127] M. Millis, Prioritizing pioneering research, in: M. Millis, E. Davis (Eds.), Frontiers of Propulsion Physics, Progress in Aeronautics and Astronautics, vol. 227, American Institute of Aeronautics and Astronautics, Inc., 2009, pp. 663–717.

CHAPTER

Electrical power options for interstellar spacecraft

2

Gary L. Bennett*
Consultant, Boise, ID, United States

Chapter outline

Abbreviations

BOM	beginning of mission
BUK	from Russian Бук ("beech tree")
D	deuterium
DIPS	dynamic isotope power system
DOE	United States Department of Energy
GPHS-RTG	General-Purpose Heat Source Radioisotope Thermoelectric Generator

*NASA Retired (formerly Manager of Advanced Space Power Systems and Manager of Advanced Space Propulsion Systems at NASA HQ).

Interstellar Travel. https://doi.org/10.1016/B978-0-323-91280-8.00006-X

H	hydrogen
HEU	highly enriched uranium
LES	Lincoln Experimental Satellite
LILT	low-intensity, low-temperature
MHW-RTG	multi-hundred watt radioisotope thermoelectric generator
MMRTG	multi-mission radioisotope thermoelectric generator
MMW	multi-megawatt
NaK	sodium (Na) potassium (K) alloy
NASA	National Aeronautics and Space Administration
NEP	nuclear electric propulsion
NERVA	Nuclear Engine for Rocket Vehicle Application
PMAD	power management and distribution
PWAR	Pratt & Whitney aircraft reactor
RPS	radioisotope power source (system)
RTG	radioisotope thermoelectric generator
SNAP	Systems for Nuclear Auxiliary Power
SPAR	space power advanced reactor
SPUR	Space Power Unit, Reactor
T	tritium
TES	thermal energy storage
USAEC	United States Atomic Energy Commission
We	watts of electrical power
Wth	watts of thermal power

1 Introduction

To meet its requirements (sensors, lifetime, mass, volume, etc.), a spacecraft requires various subsystems, one of which is the electrical power subsystem that provides the power that allows the spacecraft to operate. To put it bluntly, without electrical power, the spacecraft cannot function for the duration of an interstellar mission.

In designing a power system, the designer needs to know the requirements. For the purposes of this chapter (unless otherwise stated) it is assumed that there are two mission concepts being considered:

- Small robotic craft going all the way to Alpha or Proxima Centauri—or beyond.
- Human-scale colony ship going all the way to Alpha or Proxima Centauri—or beyond.

Such interstellar missions will lay stringent requirements on the power system. Table 1 lists some of the key power system requirements [1]. Generally, power systems and propulsion systems are closely coupled. For example, it may be possible to use the heat from a propulsion system to produce electrical power. And, as shown in this chapter, there will be a lot of waste heat as long as the propulsion system is active. Once the propulsion system is shut down, either during the cruise phase of the journey or after arrival, alternative power source(s) will be required.

Table 1 Key power system requirements for an interstellar mission.

- Must be able to supply adequate power far from the Sun or any other natural energy source.
- Supply electrical power to spacecraft for electronics and provide heating, if necessary, to keep required spacecraft functions warm (if onboard heat is not supplied by another subsystem such as propulsion).
- Must be extremely reliable and meet the mission duty cycle and lifetime requirements, as well as survive the hostile space environment that will likely be encountered on an interstellar mission.
- High specific power (power/mass) is required to keep spacecraft mass manageable, and provide acceptable mission trip times (desire high-energy density or off-board energy-supply-based power systems).

The basic elements of a spacecraft electrical power subsystem (EPS) are shown schematically in Fig. 1. Basically, the power source (which could be solar for inner Solar System operations or nuclear or power beam for more distant flights) produces the electrical power. The power management and distribution (PMAD) subsystem conditions the power so that it is in the proper format (e.g., voltage, current, etc.) for use by the various elements or "loads" of the spacecraft. Energy storage (e.g., batteries, fuel cells, flywheels) is, as its name implies, a subsystem that saves or stores unused electrical power for those times when the spacecraft needs more power than the power source can provide.

Obviously, the power to be produced depends on the mission. As noted earlier, two types of interstellar missions come to mind: robotic explorers and crewed missions. For really small space probes, such as the Starchip envisioned for the Breakthrough Starshot Initiative, powers on the order of milliwatts are envisioned, which would be combined with perhaps 1000 other Starchips. [2]

As a starting point in estimating power consumption for a crewed interstellar mission one can consider terrestrial and International Space Station (ISS) power levels. The average US per capita terrestrial electrical power usage is about 1.2 kilowatts electric (kWe). [3] If one assumes an average crew size of seven for the approximately 90-kWe average power produced by the solar arrays on the International Space Station, then a crewed interstellar vehicle would need at least 13 kWe per crewmember. For the 240-kWe peak, the average would be about 34 kWe per crewmember [4,5]. Given all the

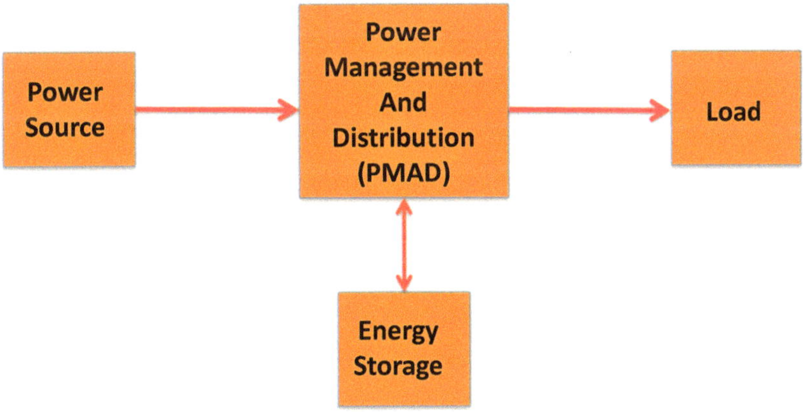

FIG. 1

Basic elements of a spacecraft power subsystem.

Source: NASA.

uncertainties of an interstellar mission one should assume megawatts of power might be needed for a large crewed interstellar mission.

Fig. 2 illustrates qualitatively the regimes in which the different power sources can best be employed.

For the purposes of this chapter, power sources can be divided into current, near-term, far-term, and speculative, as shown in Table 2.

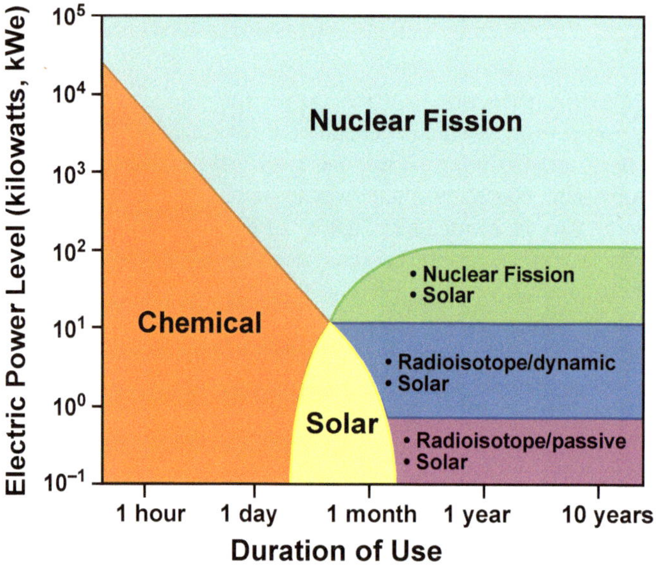

FIG. 2

Qualitative regimes of power source application.

Source: DOE.

Table 2 Qualitative availability of spacecraft power sources.

Current
- Solar photovoltaic
- Radioisotope power sources

Near-term (within about 10 years)
- Solar dynamic
- Nuclear reactors

Far-term (beyond 10 years)
- Power beaming
- Fusion
- Antimatter
- Intergalactic magnetic fields and/or particles

Speculative
- Vacuum energy/dark energy/zero-point energy, multiverse, wormholes, etc.

2 Power sources—Current and near-term

2.1 Solar power sources

Solar power is generally thought of as being produced in one of two ways: (1) using the photovoltaic effect in which sunlight directly produces an electric current in a solar cell or (2) using concentrated sunlight to power a turbine-alternator system (Brayton, Rankine) or linear oscillator system (Stirling) (hence the term "solar dynamic" power). Great strides have been made in the production of photovoltaic electrical power, with conversion efficiencies in the range of 40% having been achieved.

Fig. 3 shows the solar arrays on the International Space Station (ISS), which can collectively produce about 240kWe (peak) in direct sunlight (90-kWe average). Each wing in the figure is 35 meters (m) in length and 12 meters (m) wide. The ISS solar power system shows what can be accomplished with solar power in space [4,5].

Solar-powered spacecraft have already flown into the near outer Solar System with solar-powered missions such as the Jupiter-orbiting Juno spacecraft, which operates at about 793 million kilometers from the Sun. Since sunlight falls off at the reciprocal of the square of the distance from the Sun (see Fig. 4), it quickly becomes apparent that effective solar power is limited to the inner Solar System (although solar power could be used to power electric propulsion for an early departure from the inner Solar System and, hypothetically, could be operated at the star system destination). Solar cells are degraded by radiation (solar, galactic) and by the effects of operating at low intensity and low

FIG. 3

Solar array wings on the International Space Station. When fully extended, each wing is 35 m long by 12 m wide. In direct sunlight, the total power is about 240 kWe.

Source: Boeing, NASA, Wikipedia.

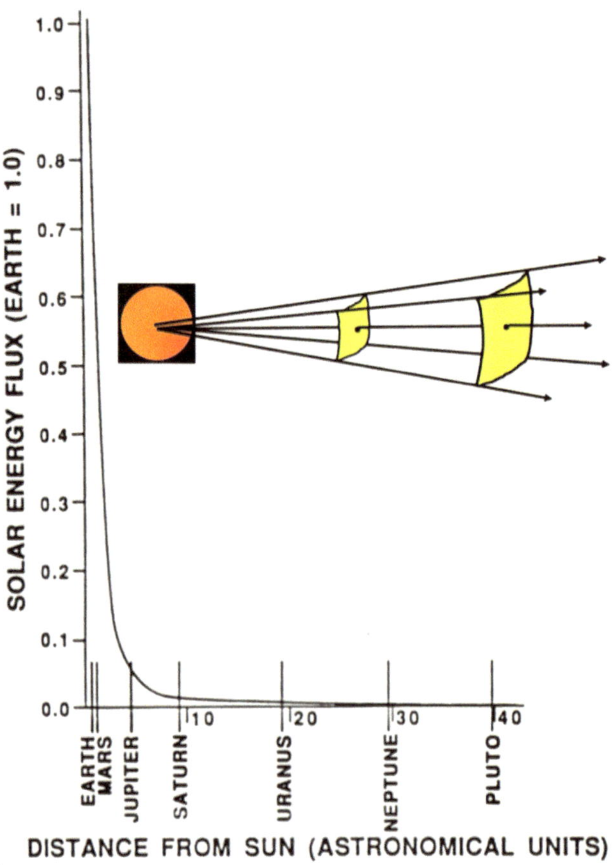

FIG. 4

Diagram illustrating the rapid decrease in solar energy flux (watts per unit area) as the distance from the Sun increases.

Source: DOE and the author.

temperature—the so-called "LILT effect" (for spacecraft traveling far from the Sun, the low incident solar intensity decreases the voltage and, hence, lowers the conversion efficiency of solar cells, while the low operating temperature improves the voltage and hence increases the efficiency; the LILT effect is the anomalous degradation produced by the combination of these defects).

Another solar concept that has been studied for Earth orbiting applications is solar dynamic, in which the incident solar energy flux is focused by a concentrator to heat a working fluid that in turn drives a dynamic conversion system (e.g., Brayton, Rankine, or Stirling) [6]. A thermal energy storage (TES) system (such as lithium fluoride) can be used to store heat for those times when the Sun is eclipsed. Studies have shown that solar dynamic power systems can be very competitive with solar cell arrays in terms of reduced mass and volume [6].

2.2 Nuclear power sources

For applications where there is little sunlight and/or a hostile environment (e.g., radiation) nuclear power sources (radioisotope or reactor), with their self-contained independence, become a natural choice. The current nuclear power source application by the United States is radioisotope thermoelectric generators (RTGs), although space nuclear reactor power systems have been and are being studied. The United States has flown one space nuclear reactor (SNAP-10A in 1965) and continues to do research on reactor power sources. Table 3 provides a summary of space nuclear system types that have been considered in the past as a function of electric power level [7]. Fig. 5 illustrates diagrammatically the "basic" nuclear power source (NPS) options.

Table 3 General classification of nuclear power source options as a function of electric power range.

Electric power range	Power system type	Power conversion option(s)
Up to 500 We	Radioisotope (static)	Thermoelectric
0.5–10 kWe	Radioisotope	Static (to about 1 kWe): Thermoelectric Dynamic: Brayton, Rankine, Stirling
10–1000 kWe	Fission reactor systems	Static: Thermoelectric Dynamic: Brayton, Rankine, Stirling
10–1000 kWe	Fission reactor systems	Static: Thermoelectric Dynamic: Brayton, Rankine, Stirling
10—1000 MWe	Fission reactor systems	Dynamic: Brayton, Rankine, or Stirling Magnetohydrodynamic (MHD)

kWe, *kilowatts electric;* MWe, *megawatts electric;* We, *watts electric.*
Derived from J.A. Angelo, D. Buden, Space Nuclear Power, Orbit Book Company, Malabar, FL, 1985.

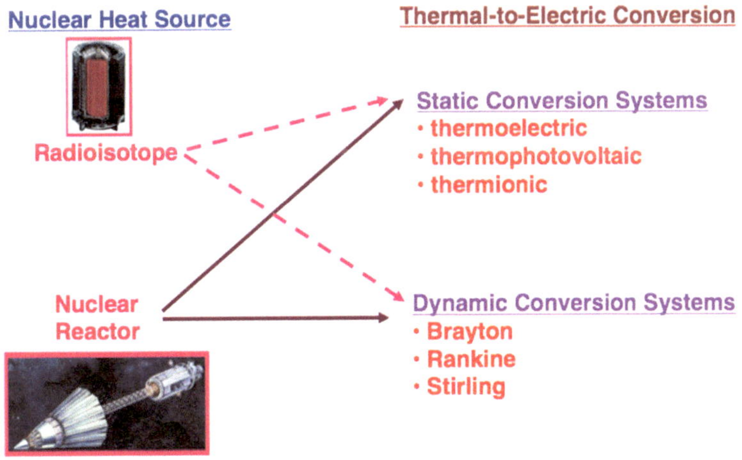

FIG. 5

Basic nuclear power source options.

A rough rule-of-thumb for space nuclear power sources is that RTGs are probably the lowest mass option up to a kilowatt or so, with radioisotope dynamic power sources being the lowest mass option from around 1 kWe to perhaps 10 kWe or so. Beyond that power, the currently available lowest mass nuclear power source will be a nuclear reactor [8,9].

2.3 Radioisotope power sources

2.3.1 Radioisotope thermoelectric generators (RTGs)

A radioisotope thermoelectric generator (RTG) essentially consists of two major parts: the radioisotope heat source (which, as its name implies, provides the heat) and the thermoelectric converter. Where solar arrays produce power from the sunlight striking the solar arrays, an RTG produces power by the heat being conducted to or radiated to the thermoelectric elements. Fig. 6 illustrates the basic parts of an RTG.

The first use of nuclear power in space was with radioisotope power sources (see Table 4). At time of writing, the United States has flown 42 RTGs and one space nuclear reactor (SNAP-10A in 1965) [10,11]. The former Soviet Union flew about 31 (and perhaps 33) BUK thermoelectric reactors and two experimental thermionic reactors [12]. The former Soviet Union also flew RTGs. Thus, the basic efficacy of space nuclear reactor power systems has been demonstrated.

Radioisotope thermoelectric generators have already achieved interstellar flight by moving beyond the heliosphere (the magnetic bubble inflated by the Sun), beginning with the Voyager 1 spacecraft on August 25, 2012 (see illustration in Fig. 7).

The three most recent RTGs flown by the United States are the Multi-Hundred Watt RTG (MHW-RTG), the General-Purpose Heat Source RTG (GPHS-RTG), and the Multi-Mission RTG

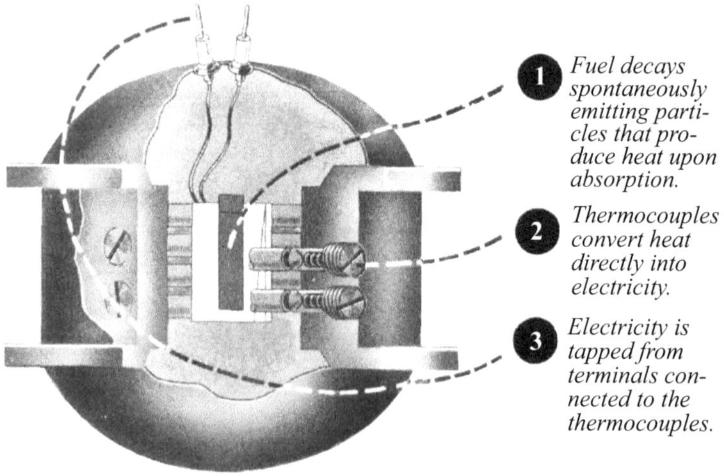

FIG. 6

How a radioisotope thermoelectric generator (RTG) works.

Source: US Atomic Energy Commission.

Table 4 Uses of radioisotope thermoelectric generators (RTGs) by the United States.

(Spacecraft/Year launched/Type of RTG/Beginning-of-mission power)

Transit navy navigational satellites

- Transits 4A and 4B (1961) SNAP-3B (2.7 We)
- Transits 5BN-1 and 5BN-2 (1963) SNAP-9A (≥25 We)
- Transit TRIAD (1972) Transit-RTG (35 We)

Nimbus-3 meteorological satellite

- SNAP-19B RTGs (1969) (2 @ 28 We each)

Apollo lunar surface experiments packages

- Apollos 12 (1969), 14 (1971), 15 (1971), 16 (1972), 17 (1972)
- SNAP-27 (>70 We each)

Lincoln experimental satellites (communications)

- LES-8 and LES-9 (1976) MHW-RTG (2/spacecraft @ ~154 We each)

Interplanetary missions

- Pioneer 10 (1972) and Pioneer 11 (1973) SNAP-19 (4/spacecraft @ ~40 We each)
- Viking Mars Landers 1 and 2 (1975) SNAP-19 (2/Lander @ ~42 We each)
- Voyager 1 and Voyager 2 (1977) MHW-RTG (3/spacecraft @ ~158 We each)
- Galileo (1989) GPHS-RTG (2 @ ~288 We each) (>3-year delay in launch)
- Ulysses (1990) GPHS-RTG (289 We) (>4-year delay in launch)
- Cassini (1997) GPHS-RTG (3 @ ~295 We each)
- New Horizons (2006) GPHS-RTG (245.7 We) (most of fuel >21 years old)
- Curiosity Mars Science Laboratory (2011) MMRTG (~115 We) (>2-year delay)
- Perseverance Mars Rover (2020) MMRTG (~110 We)

GPHS-RTG, *General-Purpose Heat Source Radioisotope Thermoelectric Generator;* MHW-RTG, *Multi-Hundred Watt Radioisotope Thermoelectric Generator;* MMRTG, *Multi-Mission Radioisotope Thermoelectric Generator;* SNAP, *Systems for Nuclear Auxiliary Power.*

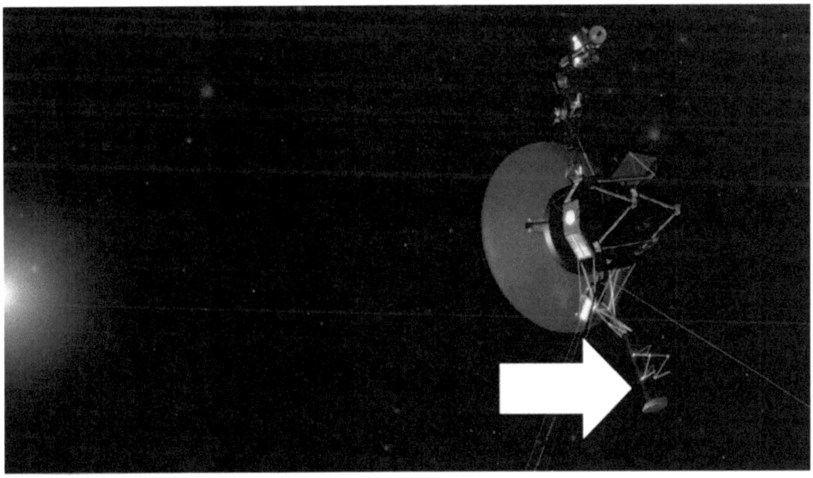

FIG. 7

Artist's illustration of Voyager spacecraft moving beyond the Solar System. The arrow points to the three MHW-RTGs. Note: The three Multi-Hundred Watt Radioisotope Thermoelectric Generators (MHW-RTGs) that power each Voyager spacecraft are shown on a boom below the spacecraft bus. The average beginning of mission (BOM) power produced by each MHW-RTG was about 158 watts electric (We).

Source: NASA/JPL-Caltech.

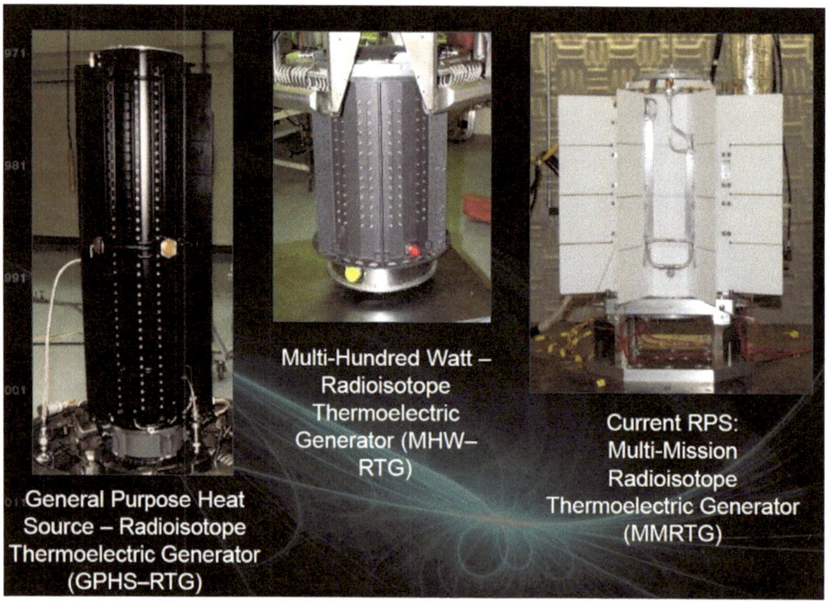

FIG. 8

The three most recent RTGs flown by the United States (not to scale).

Source: DOE.

(MMRTG). These are shown in Fig. 8 (not to scale); each was designed to meet certain specific mission requirements.

2.3.2 Multi-Hundred Watt Radioisotope Thermoelectric Generator (MHW-RTG)

Of the three most recent RTGs flown by the United States, the MHW-RTG was the first to be flown, in 1976 (two each), on two US Air Force communications satellites (Lincoln Experimental Satellites 8 and 9; LES-8/9). The next year, MHW-RTGs were flown (three apiece) on the two NASA Voyager spacecraft, which went to Jupiter, Saturn, Uranus, Neptune, and beyond.

The MHW-RTG was designed to provide a major increase in power for space RTGs. Previous US space RTGs provided at most about 40–70 We, depending on the mission. The MHW-RTG was designed to provide at least 150 We. The MHW-RTG accomplished this by heating 312 thermoelectric elements made from a silicon-germanium alloy. The use of this thermoelectric alloy allowed the MHW-RTGs to operate at a higher temperature than previous RTGs. The higher temperature led to a smaller radiator; hence, a lower mass than may have been achieved with previous thermoelectric elements [13].

The overall diameter of the MHW-RTG was 39.73 cm and its length was 58.31 cm. The average MHW-RTG flight masses were 39.69 kg for LES-8/9 spacecraft and 37.69 kg for the Voyager 1 and Voyager 2 spacecraft [10,13].

After visiting Jupiter, Saturn, Uranus, and Neptune, the two Voyager spacecraft represent the world's first interstellar missions. The two spacecraft, launched in 1977, carried three Multi-Hundred

Watt Radioisotope Thermoelectric Generators (MHW-RTGs) apiece, which provided an average of about 158 We per MHW-RTG at beginning of mission (BOM) for a total BOM power of over 470 We. After about 45 years in space, the MHW-RTGs are producing in the order of 75 We each (about 225 We total per spacecraft). Since the original requirement for the Voyager MHW-RTGs was to produce 128 We per RTG after 4 years in space, the 45-year performance indicates that the prognosis for a 50-year interstellar mission is very good [10,14–16].

2.3.3 General-Purpose Heat Source Radioisotope Thermoelectric Generator (GPHS-RTG)

The Galileo mission to orbit Jupiter and the Ulysses mission to do a polar reconnaissance of the Sun required more power. To meet those requirements, the GPHS-RTG was developed, employing thermoelectric technology from the MHW-RTG program, only this time using 572 of the silicon-germanium alloy thermoelectric elements instead of the 312 elements in the MHW-RTG. The GPHS-RTG was designed such that it could produce 300 We at fueling with a mass of 55.9 kg, making the GPHS-RTG the most powerful RTG with the highest specific power ever flown. The GPHS-RTG has an overall diameter of 0.422 m and a length of 1.14 m. In space, both the MHW-RTG and the GPHS-RTG operated "vented", that is, the space inside the converter was open to the vacuum of space. The silicon-germanium thermoelectric elements had no need for a cover gas [17].

Following the successful Galileo and Ulysses missions, three GPHS-RTGs powered the Cassini mission to orbit Saturn for nearly two decades. The continuing success of the GPHS-RTG made it a natural choice to power the New Horizons mission to Pluto and beyond [17].

For the proposed Interstellar Probe being studied for NASA by Johns Hopkins University Applied Physics Laboratory the lifetime requirement is 50 years for a spacecraft that is to travel about 1000 astronomical units (AUs) from Earth. Using two General-Purpose Heat Source Radioisotope Thermoelectric Generators (GPHS-RTGs) of the type powering the Cassini Saturn orbiter, it would be possible initially to get close to the desired 600 We of power [15,17].

2.3.4 Multi-Mission Radioisotope Thermoelectric Generator (MMRTG)

The two Mars Rovers, Curiosity and Perseverance, are powered by a sealed RTG known as the Multi-Mission RTG (MMRTG). Whereas the MHW-RTG and the GPHS-RTG are open to space, as is fitting for missions operating in the vacuum of space, the MMRTG is sealed with a cover gas so that it can operate in the atmosphere of a planetary surface such as that of Mars.

The MMRTG can provide about 110 We at launch within an envelope of 0.64 m in diameter by 0.66 m long with a mass of about 45 kg [18].

2.3.5 Radioisotopes

Perhaps the major constraint on radioisotope power sources (RPSs) is that the very same radioactive decay of the radioisotope that provides the heat that is converted into electrical power is the same decay that means eventually the RPS will be at very low power. For example, plutonium-238, which has been used in all US RPSs, has a half-life of 87.7 years, which means that after 87.7 years half the thermal power is gone (see Fig. 9). For a 50-year mission, about 33% of the Pu-238 would have decayed. The effect would be a reduced thermal power and a reduced temperature (the temperature difference between the hot and cold sides drives the thermoelectric elements) leading to reduced electrical power.

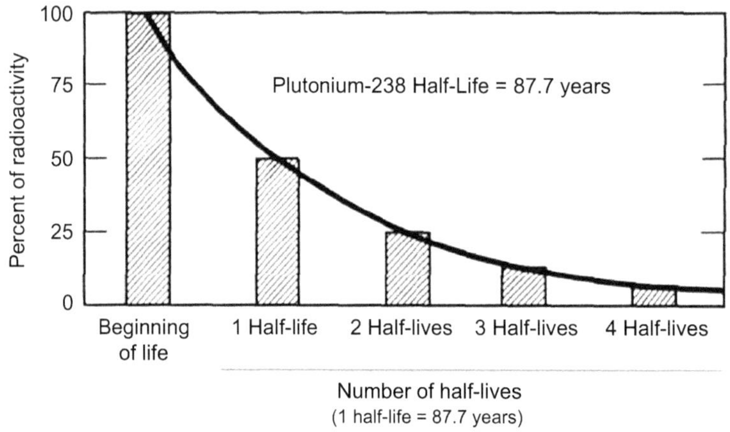

FIG. 9

Radioactive decay pattern of plutonium-238.

Source: USAEC.

One way around the lifetime issue would be the use of a radioisotope with a longer half-life. For example, European researchers have studied americium-241 (Am-241), which has a half-life of about 432 years [19]. Where in 50 years Pu-238 would decay to about 67% of its initial activity, Am-241 would decay to only about 92% of its initial activity—clearly an advantage for interstellar exploration! Obviously one doesn't want to use a radioisotope with too long of a half-life because then the thermal power per unit mass would become too low. As an example, the specific thermal power of Pu-238 is 0.56 watts per gram while that of Am-241 is about 0.11 watts per gram, meaning that a larger quantity of Am-241 would be required to achieve the same initial thermal power as Pu-238.

2.3.6 Radioisotope dynamic power sources

To improve the thermal-to-electric conversion efficiency, the United States has studied using dynamic conversion systems such as linear alternators (Stirling) or turbine-alternators (e.g., Brayton and Rankine). In principle such dynamic conversion systems offer the potential of 30% (or more) efficiency, over four times that of state-of-the-art thermoelectric conversion. Such high efficiencies translate into more power for the same amount of radioisotope, or less radioisotope to produce the same power as an RTG [20]. Turbines on Earth have demonstrated 90+year lifetimes [21].

Fig. 10 illustrates an early concept for a radioisotope dynamic power source using the Brayton (gas) conversion system [22]. As shown diagrammatically in Fig. 11, in the Brayton radioisotope power source, the radioisotope heat source heats the gas (often a mixture of helium and xenon) sending it to the turbine, causing it to rotate. This in turn causes the alternator to rotate, which produces an electric current and hence the power. The Rankine cycle is similar with the addition of a phase change in the working fluid, that is, a liquid is heated until it is vaporized. A Stirling conversion system uses a linear oscillator (imagine a piston moving back and forth) that creates the changing magnetic field that produces the electrical current and hence the electrical power.

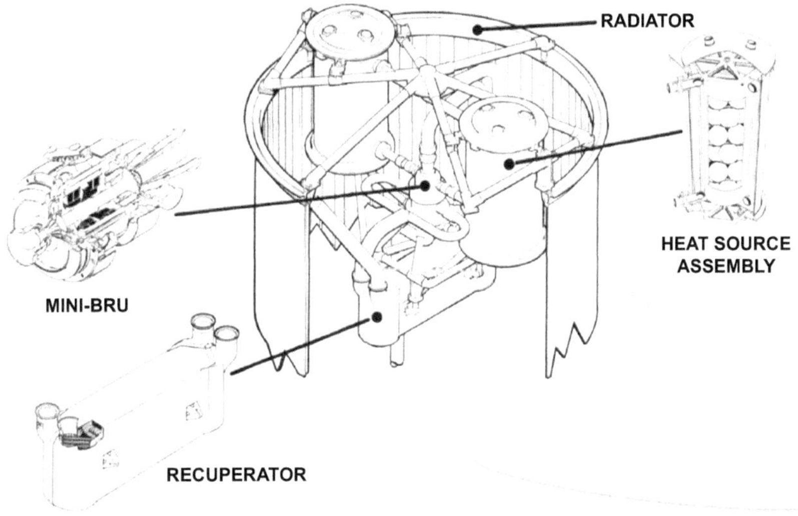

FIG. 10

Brayton radioisotope power source design. *BRU*, Brayton Rotating Unit. The power output was 1.3 kWe with a mass of 204 kg. The length was 0.244 m and the diameter 0.152 m.

Source: Garrett.

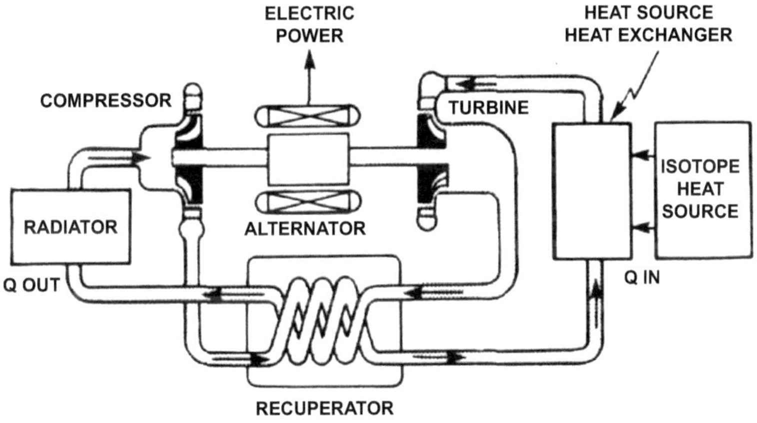

FIG. 11

Flow diagram for a Brayton radioisotope power system.

Source: Garrett.

2.4 Fission reactor power source

An alternative nuclear power source in which the thermal power can be controlled is the nuclear fission reactor. Unlike RPSs, which rely on the natural (and unadjustable) decay of a radioisotope, a nuclear fission reactor power source operates by using the heat produced in the fissioning ("splitting") of a

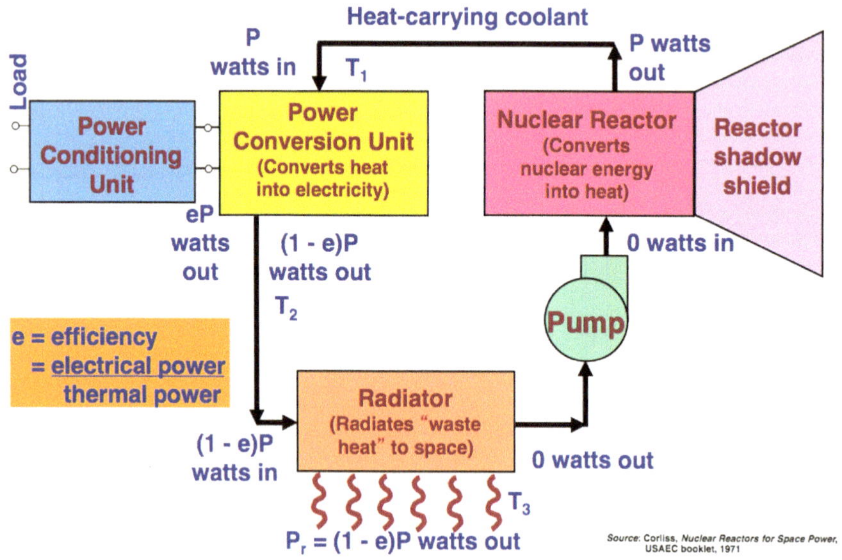

FIG. 12

Schematic diagram of a generalized nuclear reactor space power plant (colorized by author). The nuclear reactor produces heat, which is transmitted to a power conversion unit that converts a fraction (*e*) of the heat into electrical power. The unused heat is eventually radiated into space.

Source: USAEC.

material like uranium-235 (U-235). In addition to the energy ("heat") released in fission, extra neutrons are also released. These "extra" neutrons can cause additional fissioning of nearby U-235 nuclei to produce what is called a "chain reaction". This is the steady-state operation that allows a nuclear reactor to produce power for long periods of time.

Fig. 12 illustrates diagrammatically the general fission reactor power source concept. To put reactors and radioisotopes in context, note that plutonium-238 releases about 5.6 million electron volts (MeV) of energy through radioactive decay. The total energy released in the fissioning of a uranium-235 nucleus is about 200 MeV. Clearly, from a purely energy standpoint, fission reactors are the near-term way to go for missions requiring high power (especially above about 1–10 kWe) and long life. To illustrate the potential of nuclear reactors, on Earth, power reactors producing up to 1000 MWe (1 gigawatt, GWe) have been built.

Table 5 summarizes the principal US space power reactor systems that have been studied.

2.4.1 SNAP-10A: First reactor in space

On April 3, 1965, the world's first nuclear reactor space power system, SNAP-10A, was launched from Vandenberg Air Force Base and placed in a circular polar orbit where it was successfully flight tested [26]. Fig. 13 is an artist's conception of the SNAP-10 space nuclear reactor power system. SNAP-10A was fueled with 4.8 kg of highly enriched uranium-235 (HEU). which was used to heat a sodium-potassium liquid metal alloy coolant that in turn heated 2880 thermoelectric elements. Since SNAP-10A was the first space reactor. The focus was more on remote power performance than on

Table 5 US space power reactor summary table [23–25].

Reactor	Power	Energy conversion	Status
SNAP-2	3 kWe	Rankine	Two reactors tested
SNAP-8	35 kWe	Rankine	Components developed
SNAP-10	300 We	Thermoelectric	Early design
SNAP-10A	500 We	Thermoelectric	First reactor flight (1965)
SNAP-50	100 kWe to 1 MWe	Rankine	Design studies
Advanced Hydride	10–100 kWe	Thermoelectric and Brayton	SNAP-8 improvements
Advanced Liquid-Metal Cooled Reactor	100–500+kWe	Brayton and Rankine	Basic technology program
In-Core Thermionic Reactor	100 kWe to 1 MWe	Thermionic	Technology program
SPAR/SP-100	10–100 kWe	Thermoelectric	Components tested
MMW (Multi-Megawatt)	10s–100s MWe	Brayton, Rankine Thermionic	Some fuel testing
Prometheus	200 kWe	Brayton	Concept studies
Fission Surface Power	40 kWe	Stirling Brayton	Some testing

FIG. 13

Artist's conception of the SNAP-10A space nuclear reactor power system in orbit around the Earth. The reactor is the assembly on the right-hand end of the space vehicle. The reactor produced over 500 watts of electrical power. The dimensions were 3.48 m long by 1.27 m mounting base with a mass of 435 kg (including the shield).

Source: USAEC.

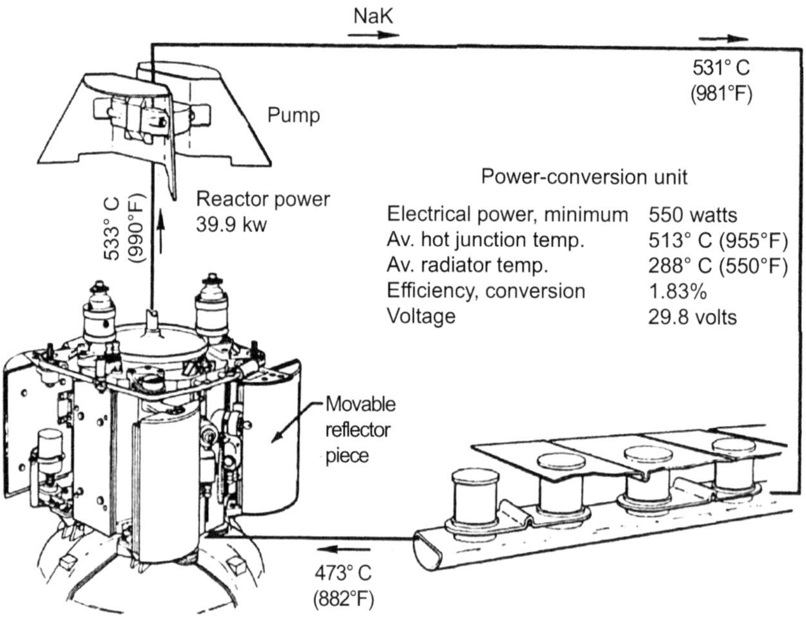

FIG. 14

Schematic diagram of the SNAP-10A and power conversion unit. The sodium-potassium (NaK) liquid metal coolant is heated in the reactor and then sent via the pump to the power conversion unit with its 2880 thermoelectric elements to produce over 500watts of electrical power (We).

Source: USAEC.

increasing power or reducing mass and size. SNAP-10A produced over 500We, operating continuously for 43 consecutive days until a failure of electronic devices on the Agena spacecraft caused the reactor to shut down. A ground test "twin" demonstrated the 1-year operational requirement by operating autonomously for over a year [26].

Fig. 14 illustrates diagrammatically the operation of the SNAP-10A nuclear reactor power system. As stated, the reactor heated a sodium-potassium liquid metal alloy that was pumped to the power conversion unit with its 2880 silicon-germanium thermoelectric elements to produce the electrical power. Excess heat was radiated into space.

2.4.2 Some observations on the future of space nuclear reactor power systems for interstellar missions

Like radioisotopes, with fission reactors (and like just about everything else in this universe), there is no free lunch. As they operate, fission reactors experience various decay modes, such as the buildup of what are termed "poisons" (nuclei that absorb neutrons without producing usable power). However, this problem can be alleviated by designing a fission reactor with highly enriched uranium (HEU) to ensure the maximum number of U-235 nuclei. (By legal definition HEU means that the fuel has been enriched to 20% or more in the isotope uranium-235.) US Navy reactors have operated with HEU and special reactor core designs for 30 years without refueling and the Navy is developing a 40-year reactor. [27] With clever engineering, a 50-year nuclear fission reactor should be possible. (Even longer lifetimes should be possible in the future.) Thus one could conceive of a century-long interstellar mission being accomplished with at least two nuclear reactors, one operating and one in backup ready to operate when the first reactor is shutdown. (A third reactor might be provided for additional backup.) In the case of liquid metal coolant, radioisotope power sources could be used to keep the coolant liquid warm during periods when the reactor(s) aren't operating. An illustration of how multiple space reactors might be assembled is the "Hydra" concept, shown in Fig. 15, for a mission to Mars, assembled from units of 5 MWe reactors that could be assembled for other missions as needed [28].

To illustrate the concept, Fig. 16 shows an artist's visualization of a multireactor NEP transfer vehicle for a piloted Mars mission: "The vehicle utilizes a modular 'Hydra' configuration of three 5-MWe power modules operating in parallel for a total system output of 15 MWe. Each module consists of a roughly 25-MWth reactor, shadow shield, potassium Rankine power conversion elements, and primary and secondary two-sided planar heat pipe radiators. A 100-m separation distance between reactor and crew is enabled by a 4-m square self-deploying boom" [28]. (Note: As needed, reactors with higher powers could be designed and built.)

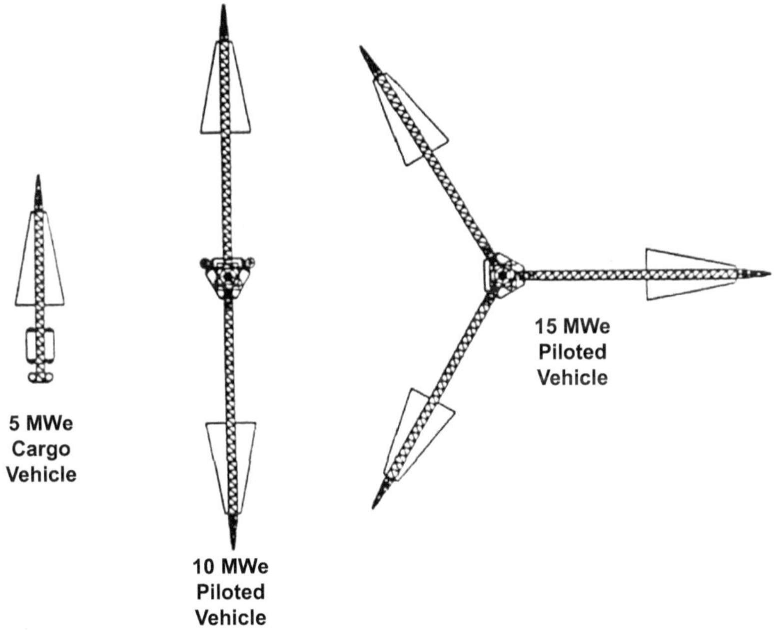

FIG. 15

Family of modular multireactor nuclear electric propulsion (NEP) vehicles.

Source: NASA—J.A. George, Multi-Reactor Power System Configurations for Multimegawatt Nuclear Electric Propulsion, NASA Technical Memorandum 105212, September 1991.

FIG. 16

Conceptual design for a 15 MWe "Hydra" modular multireactor piloted Mars nuclear electric propulsion (NEP) vehicle.

Source: J.A. George, Multi-Reactor Power System Configurations for Multimegawatt Nuclear Electric Propulsion, NASA Technical Memorandum 105212, September 1991; Les Bossinas, NASA/GRC.

As an example of interstellar studies involving nuclear reactors, in the 1980s, researchers such as Kerry Nock at NASA's Jet Propulsion Laboratory (JPL) proposed the Thousand Astronomical Unit (TAU) mission, in which a 1 MWe nuclear reactor would power both ion thrusters and science instruments in the 5000 kg payload module. Fig. 17 is an artist's concept of what the TAU system might look like. (The reactor is shown in red below the spacecraft body.) Operating the ion thrusters for about 10 years would propel the spacecraft on its way to a thousand AUs from Earth from which vantage point exciting astrometric science could be done. RTGs would provide power when the reactor shuts down. [29] A follow-on Ten Thousand Astronomical Unit (TTAU) mission has been proposed [30].

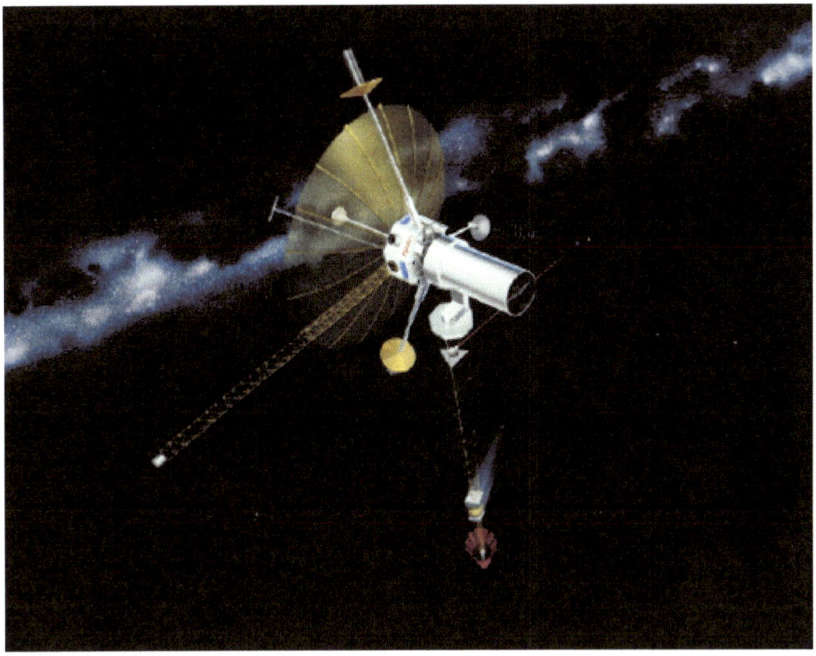

FIG. 17

Artist's illustration of the Thousand Astronomical Unit (TAU) spacecraft. Power would be provided by a 1 MWe nuclear fission reactor (shown in *red*) and, as needed, by RTGs.

Source: NASA/JPL.

The 1 MWe requirement for TAU and the need for megawatts (or more) of power for a human interstellar mission suggest a look at what was at the time the most powerful SNAP space nuclear reactor considered by the US government: the SNAP-50/SPUR reactor [31]. (SNAP-50 was the US Atomic Energy Commission designation; SPUR (or Space Power Unit, Reactor), was the military designation.)

The SNAP-50/SPUR program was initiated "to develop a nuclear-electric powerplant for space use in the 300-kWe to 1000-kWe power level range, having a lifetime of 10,000 hours and a minimum weight." [31] To meet the power requirements, the SNAP-50/SPUR program studied different thermal-to-electric conversion systems, specifically, magnetohydrodynamics, Brayton cycle, Rankine cycle, boiling potassium reactor system, and in-core and out-of-core thermionic systems. Considering such factors as operating temperatures, materials data, masses, and technology readiness, the study pointed toward what has become known as the SNAP-50/SPUR power plant (Rankine cycle) [31,32].

The SNAP-50/SPUR power plant concept consisted of a Rankine cycle power conversion system with a nuclear reactor heat source to convert the working fluid into vapor that can expand through a turbine, which in turn would produce electricity via the connected alternator [31]. Fig. 18 illustrates the

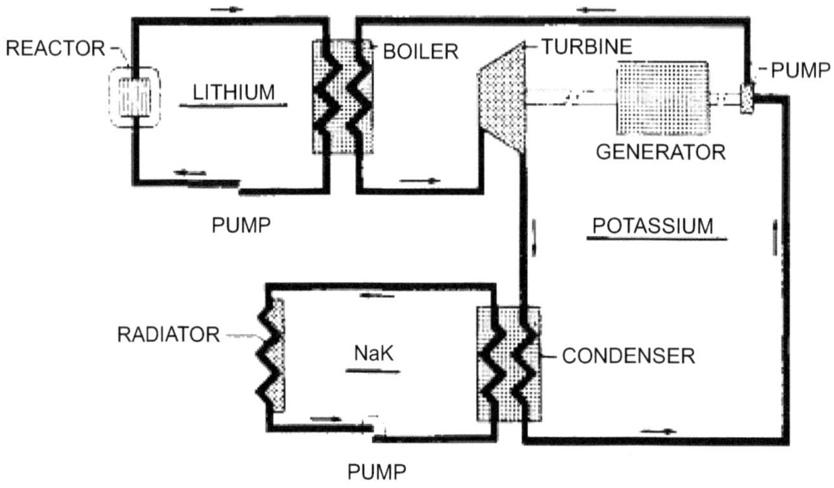

FIG. 18

Simplified flow diagram for the SNAP-50/SPUR power plant.

Source: Pratt & Whitney (AEC-funded).

basic flow diagram, which consists of three basic fluid loops: a primary lithium circuit; a potassium working fluid loop; and a sodium-potassium (NaK) heat rejection loop [33].

Basically, lithium is circulated through the reactor to the boiler, where heat is transferred to potassium, vaporizing it. The potassium vapor then passes through the turbine, converting the energy of the potassium vapor to mechanical power to drive the electrical generator. From the turbine, the potassium flows through the condenser, where it is subcooled. The heat of condensation is transferred to the main radiator loop coolant (a sodium-potassium alloy labeled "NaK"). The radiator then radiates the heat to space [31,33].

The SNAP-50/SPUR reactor was designed for 1370 K liquid metal temperature operation. In the design, the potassium in the boiler enters the turbine at 1280 K and exits at 960 K. The cycle efficiency was 15% (but higher conversion efficiencies are possible with a Rankine system) [31–33].

Fig. 19 is an illustration of the SNAP-50/SPUR reactor (PWAR) and Fig. 20 is an artist's illustration of a SNAP-50/SPUR power plant powering a nuclear electric propulsion (NEP) stage [34]. The PWAR design in Fig. 19 was to provide 300 kWe [31].

Having extra reactors as in the "Hydra" concept would allow the interstellar astronauts to establish a standalone power station on the surface of their planetary destination without relying on the light from an uncertain star. An example of such an arrangement using the SNAP-50/SPUR reactor is shown in the artist's concept in Fig. 21 for a lunar power station [34].

The foregoing discussion provides some ideas for developing a high-powered nuclear reactor power source for interstellar missions. Clearly there is much technical work to be completed (for example, managing two-phase fluid flow in zero-g, protecting against materials corrosion, and vacuum cementing/welding, etc.) and other concepts to be considered [31,32].

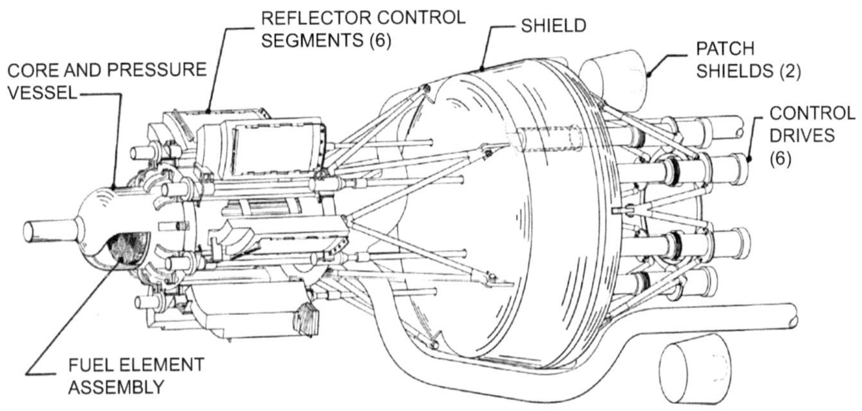

FIG. 19

The 300 kWe PWAR-20 concept was to be a lithium-cooled reactor that would operate in the fast neutron spectrum to produce 2 megawatts (MWt) of thermal output [31]. The length was about 1.5 m (not counting the shield) and the diameter was about 1 m. The mass was projected to be about 820 kg.

Source: Pratt & Whitney (AEC-funded).

FIG. 20

Artist's concept of a SNAP-50/SPUR-powered nuclear electric propulsion (NEP) probe [34]. The nuclear reactor is on the lower right; the payload and thrusters are on the upper left.

Source: Pratt & Whitney (AEC-funded).

FIG. 21

Artist's concept of a 50 kWe lunar nuclear reactor power station [34]. (The reactor is shown on the left mounted on the landing system.) Higher powers could be achieved by employing additional radiator-condenser circuits (SNAP-50/SPUR was to eventually produce 1000 kWe or more).

Source: Pratt & Whitney; AEC.

3 Power sources—Far-term

For decades there have been studies on more advanced power systems that would enable such missions as interstellar exploration. As always, the objective is to have a power source that is long-lived yet compact. One of the ways to power an interstellar spacecraft is by having the power generated on or near Earth and transmitted to the spacecraft. Another way would be to develop an advanced propulsion system that also provides power. Or one could find a way to "live off the land" (e.g., use the interstellar medium).

3.1 Power beaming

Power beaming is the point-to-point transfer of electrical power across space by means of a directed beam such as laser light or microwaves. The advantage of power beaming is clear: the spacecraft mass is greatly reduced because it doesn't have to carry a power source (which can take up a fourth or more of the mass of a spacecraft) [35].

In the case of laser power beaming, the laser could be focused on a receiver, e.g., a solar sail that might also be used for propulsion. With solar cells on the sail or spacecraft, the laser light could be converted into usable electrical power. Power beaming allows delivering power to a spacecraft from

FIG. 22

Schematic showing the subsystems involved in beaming power to a spacecraft [35].

Source: P. Jaffe.

FIG. 23

Beaming to a spacecraft shown flying over the Jovian satellite Io.

Source: NASA/Pat Rawlings, SAIC.

a place where power is readily available to a place (interstellar spacecraft) where power is difficult to get. Fig. 22 illustrates the basic idea behind power beaming [35] and Fig. 23 is an artist's illustration of how beaming might work (the figure is for propulsion but the idea of beaming power is similar).

To illustrate the characteristics of power beaming, consider a 2018 study of an interstellar precursor mission conducted by NASA's Jet Propulsion Laboratory (JPL) [36]. The proposed architecture would use a kilometer-scale, multihundred-megawatt phased-array laser to beam power to a vehicle that converts it to electrical power. The study envisioned developing photovoltaic arrays with an areal density of 200 g/m^2 and efficiencies greater than 50%. In particular, the study noted it would be necessary "to point the laser array with the required accuracy and stability necessary to perform the reference mission…" [36].

In order to reduce the size of the beam array, one study proposed the coupling of a neutral particle beam with a laser beam [37]. Fig. 24 is an artist's conception of how beaming can enable interstellar

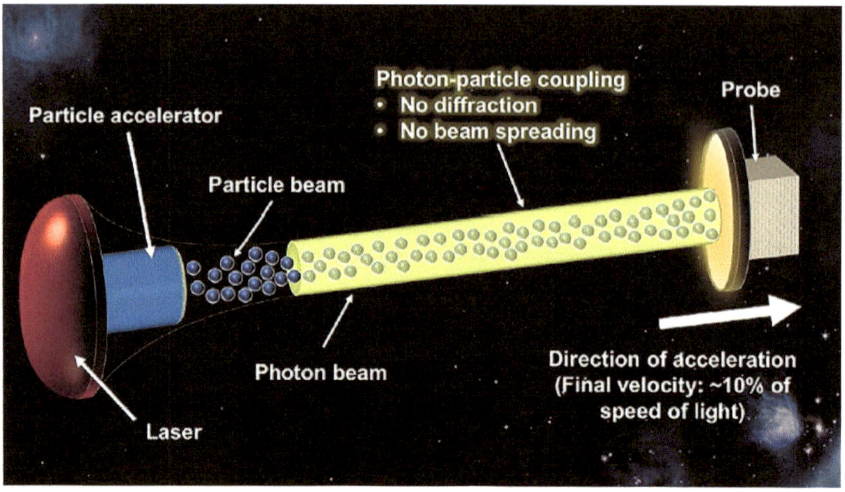

FIG. 24

Graphic depiction of a diffractionless beamed propulsion for breakthrough interstellar missions.

Source: C. Limbach/NASA.

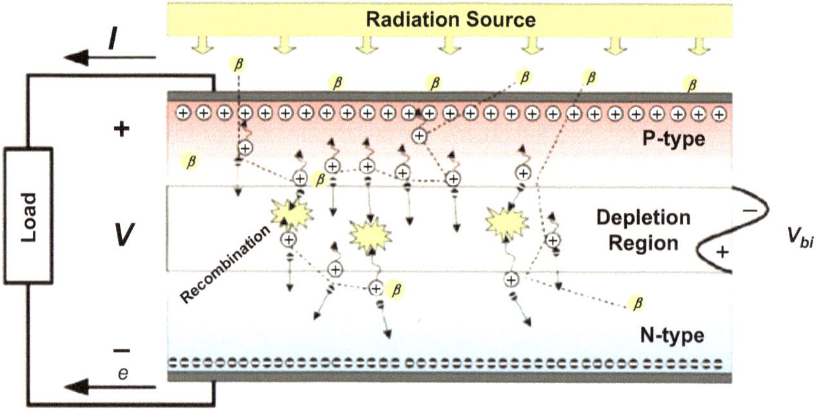

FIG. 25

Diagram of a betavoltaic radioisotope power source.

Source: Geoffrey A. Landis, NASA, 2019.

missions to Proxima Centauri b with a beam diameter at the transmitter of less than 10 m instead of 10s of kilometers. While the concept was to produce propulsion, the same concept can be used to produce power for the spacecraft. [37].

To provide electrical power to a low-mass sail-based interstellar probe, G.A. Landis has suggested using radioisotope direct generation, e.g., collecting the electrons emitted by a radioisotope, as illustrated in Fig. 25 [38]. In this approach, high-energy electrons emitted by the radioactive decay of

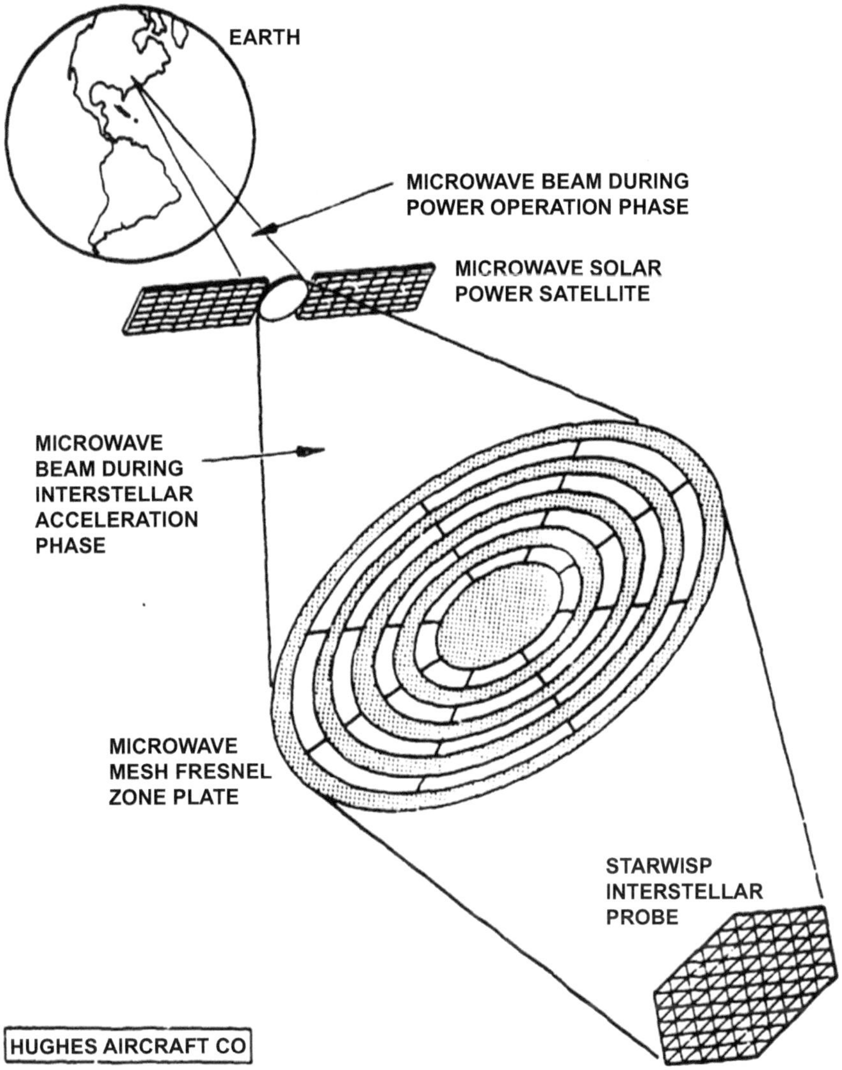

FIG. 26

Diagram of the Starwisp concept for a lightweight, high-speed interstellar probe capable of returning useful information from the nearest stars.

Source: Hughes Aircraft; US Air Force Rocket Propulsion Laboratory.

tritium (half-life 12.3 years) are converted to electrical power by a semiconductor device. For a 20-year cruise to Proxima Centauri, about 32% of the initial power would be available at arrival [38].

Another source of power beaming is using microwave energy. An example of this is shown in Fig. 26 for the Starwisp concept, in which beamed microwaves both push the probe and provide a source of power [39,40]. A number of issues need to be dealt with, including materials, size, and the pointing accuracy of the beam.

3.2 Fusion

Fusion power is theoretically an attractive alternative to fission power. Unlike fission power, which releases energy by the "splitting" of the nucleus of a heavy element like uranium-235, fusion power releases energy by combining or "fusing" two lighter nuclei with masses less than that of iron (e.g., protons or isotopes of helium) to form heavier nuclei. The energy comes from the mass deficit (the nuclei formed from the merger of the two original nuclei have less mass than the fused nucleus; that mass deficit is converted to energy through Einstein's equation ($E = mc^2$)). It has been estimated that a pickup truck filled with fusion fuel would have the equivalent energy of 2 million metric tons of coal or 10 million barrels of oil [41].

The International Thermonuclear Experimental Reactor (ITER), which is based on magnetic confinement (MCF) fusion, is aimed at providing 500 MW of fusion power for 50 MW of input heating power, a power amplification ratio of 10. Such a feat will place ITER as the first fusion device to create net energy. (Note: In Latin, "*iter*" means "journey" or "path" or "way".) To do this experiment, the ITER project is being built on a 180 ha site in southern France. Thirty-nine buildings and technical areas house the ITER. The machine mass is 23,000 MT [42].

Below is a list of some of the light nuclei that could be used in fusion power sources. (In the equations, D represents a deuteron, that is, a hydrogen nucleus with one neutron, and T represents a hydrogen nucleus with two neutrons; He is helium, Li is lithium, and B is boron.) [43]

$$D + T \rightarrow {}^4He + n + 17.6\,MeV$$

$$D + D \rightarrow {}^3He + n + 3.27\,MeV$$

$$D + D \rightarrow T + {}^1H + 4.03\,MeV$$

$$D + {}^3He \rightarrow {}^4He + {}^1H = 18.3\,MeV$$

$${}^1H + {}^6Li \rightarrow {}^3He + {}^4He + 4.0\,MeV$$

$${}^1H + {}^{11}B \rightarrow 3He + 8.67\,MeV$$

In selecting a fusion reaction for a fusion power plant there are a number of considerations. If everything else is equal, it is desirable to maximize the power output. An examination of the above equations shows that the $D + {}^3He$ reaction offers the highest energy per unit of fuel mass and, for that reason, it was selected as the fuel for the Project Daedalus study. [44,45] Since ${}^3He$ is rare on Earth, studies have been conducted to mine the Moon or the outer planets for He [46].

For the ITER research, the hydrogen isotopes deuterium (D) and tritium (T) will be used, as illustrated in Fig. 27, to fuel the fusion reaction because, among other things, tritium is more readily available than ${}^3He$ [42]. And once "breakeven" has been achieved, there is still the issue of coupling the fusion reactor to a conversion system that can turn all of that fusion energy into usable electrical energy.

Fig. 28 illustrates diagrammatically the proposed DEMO (DEMOnstration) nuclear fusion power plant. The key is producing enough energy that can be converted into heat that will in turn drive a turbine-alternator system to produce electrical power. One critical activity is obtaining enough tritium (which has a half-life of 12.33 years) to operate such a power plant for the decades that commercial power plants operate. For an interstellar mission, that could be a challenge!

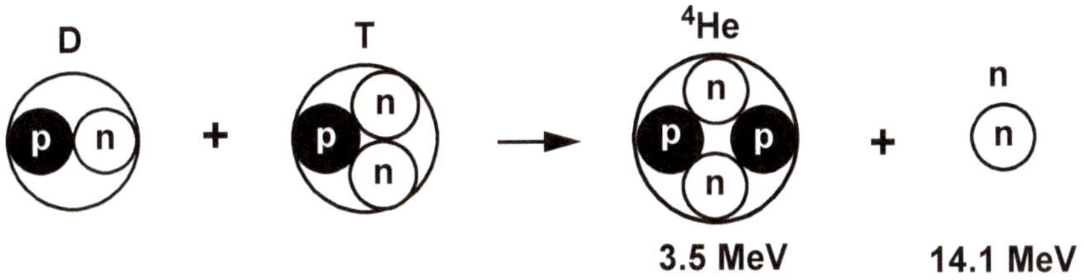

FIG. 27

Diagram of the fusion reaction between deuterium (D) and tritium (T) nuclei. (This is the reaction planned for the International Thermonuclear Experimental Reactor (ITER).) The numbers indicate how the fusion energy is divided between the products in each case. The neutrons (n) are not electrically charged, but the other products carry electrical charges.

Source: USAEC.

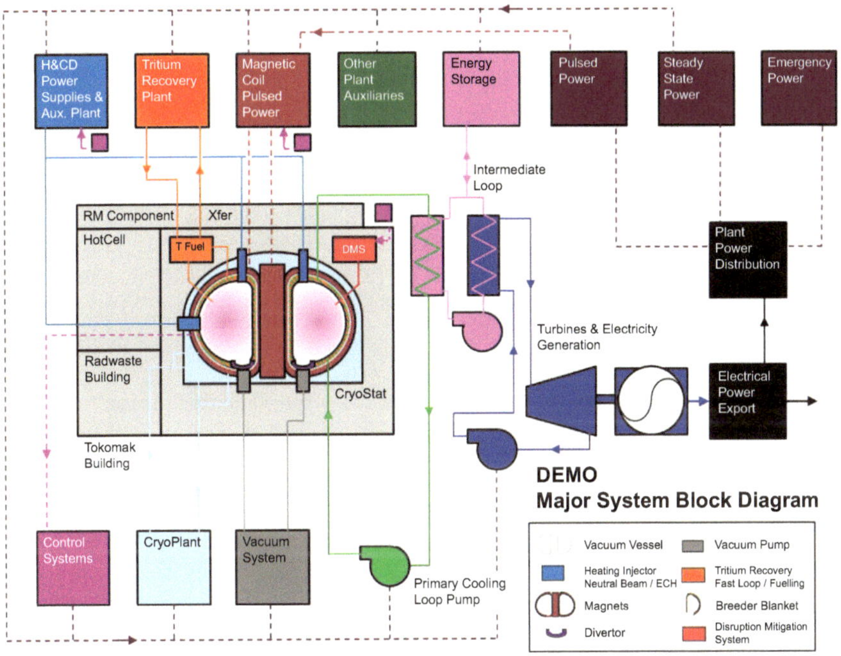

FIG. 28

Schematic of a DEMO(nstration) nuclear fusion power plant.

Source: Wikimedia commons.

The other main method for initiating nuclear fusion reactions is to use inertial confinement fusion (ICF), in which pin-sized targets of thermonuclear fuel (typically D-T) are compressed and heated to achieve fusion. The largest operational ICF experiment is the National Ignition Facility (NIF) at the Lawrence Livermore National Laboratory in the United States. In studies in the 1970s in the United Kingdom to propel the proposed Daedalus starship, it was assumed that ICF would be used for propulsion [44].

3.3 Antimatter

Antimatter has the same mass as its "normal" matter counterpart but it has the opposite electric charge and opposite magnetic moment. The positron was the first antiparticle to be discovered—it is essentially an electron with a positive charge. For the purposes of interstellar travel, the important fact about antimatter is that when it comes into contact with "normal" matter there is a total annihilation of matter, with a release of all the energy, as explained by Einstein's mass-energy equation ($E=mc^2$) [47]. In fact, matter-antimatter annihilation is the most efficient and largest known way of producing energy. (An intriguing sidelight is that the equations describing the behavior of matter and antimatter have both positive energy and negative energy solutions; in fact, the equations would imply an infinite amount of negative energy—something clearly not observed but which will be discussed in the next section.)

In the context of interstellar studies, antimatter is often invoked as a means of producing the energy necessary to propel an interstellar spacecraft; however, if the energy released from the annihilation of matter-antimatter could be hitched to a conversion system, it could be a power source. G.D. Jackson has suggested two ways: capturing charged particles emitted when antimatter annihilates matter and using the annihilation to produce heat. According to Jackson, high thermal-to-electric conversion efficiencies might be obtained [48,49].

4 Speculative concepts

From our spatiotemporal viewpoint, interstellar travel appears daunting. By terrestrial standards, the numbers are indeed intimidating. Proxima Centauri, the closest star to us is still 40,208,000,000,000 km away [50]. In November 2021, humanity's fastest spacecraft, the Parker Solar Probe, screamed past the Sun at about 586,000 km/h [51]. At that speed it would take over 7800 years to reach Proxima Centauri. Such a challenge calls for some serious creative thinking, especially if the spacecraft is to travel to Proxima Centauri in less than a human lifetime. If the human race is to survive, somehow, someway, we have got to find a way around this limit—and that means "new" physics. Speculative concepts such as vacuum energy and wormholes will be discussed in the following sections.

4.1 Negative energy/vacuum energy/dark energy/zero-point energy

In P.A.M. Dirac's theory of the positron, there is an infinite sea of negative energy, something that should intrigue those looking for a way to have unlimited energy for star travel. With reference to Section 3.3, in formulating a quantum theory of the electron, P.A.M. Dirac discovered that the theory led mathematically to another kind of particle, one with the same mass as an electron but with a positive charge. Referring to Fig. 29, this particle, which experimentalists would come to call the "positron," was described by Dirac as a "hole" in an infinite sea of negative energy [52]. When a particle "jumps" out of the sea of negative energy (by absorbing radiation) it leaves a hole, which acts like a positively charged particle. This is the hole-theory interpretation of pair production (the production of an electron and a positron).

If there is an infinite "sea" of negative energy, is there some way to tap into it to provide power for an interstellar spacecraft? If so, it would give new meaning to the concept of "living off the land" that allowed our distant ancestors to colonize the world.

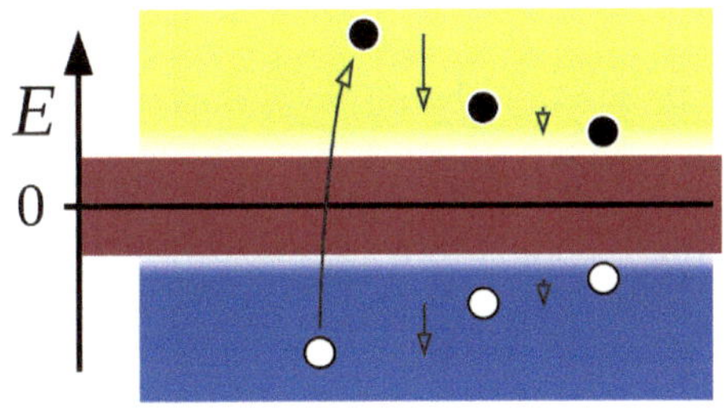

FIG. 29

Illustration of the "Dirac sea". In Dirac's theoretical model the vacuum is an infinite sea of particles with negative energy.

Source: Wikipedia.

Another possible source of energy is the "vacuum energy" that permeates space. Some think that this "vacuum energy" may be the "dark energy" that is thought to be causing the expansion of the Universe to accelerate. ("Zero-point energy," which is the lowest possible energy for a quantum mechanical system, is a special case of "vacuum energy".)

R.L. Forward suggested one way in which energy could be extracted from the vacuum. He began by noting that "Any pair of conducting plates at close distances (<1 μm) experience an attractive Casimir force that is due to the electromagnetic zero-point fluctuations of the vacuum" [53]. One can visualize this force as arising from the greater outside quantum force on the plates [54]. Charging both plates to the same polarity would cause them to repel each other but if the electrostatic force is kept lower than the Casimir force the plates can be drive toward each other such that the Casimir force adds energy to the electric field [53].

Another type of energy is the "dark energy" that permeates space and seems to be causing the expansion of the Universe to accelerate. If there is a field that modifies the expansion of the Universe, could it be harnessed for star travel?

4.2 Wormholes and black holes

In 1988, physicists M.S. Morris and K.S. Thorne mathematically showed how to make a wormhole, that is, an opening in space that would allow rapid transport between distant places [55]. Since then there have been numerous studies of artificial wormholes. One such study indicated that wormholes might possess magnetic fields and be capable of generating synchrotron radiation. If so, some future civilization might be able to extract power from a wormhole [56]. And perhaps that future civilization can find a way to transmit that power to a spaceship (perhaps one that has gone through the wormhole?).

Black holes, on the other hand, offer more promise of being capable of supporting the extraction of rotational energy. According to the theory, when magnetic field lines disconnect and reconnect in just the right way, they can accelerate plasma particles such that large quantities of black hole energy can be extracted [57]. The plasma surrounding a spinning black hole could also be a source of energy. Transmitting that to a spaceship will be an exercise for the future "Black Hole Power Company."

4.3 Biology

While not as exotic as black holes and white holes, biology may offer a means of storing small amounts of energy on a spacecraft, in essence, becoming a microbial fuel cell. NASA has been studying electricity-producing bacteria known as exoelectrogens to power future space missions. (Exoelectrogens are microorganisms, mostly bacteria, which generate electrical energy by the oxidation of organic matter and transferring the electrons to an electron acceptor outside their cells [58,59].) A bacterium called *Shewanella oneidensis* MR-1 being studied by NASA produces energy by using metals such as iron and manganese. In principle, it might be possible to connect many of these bacteria with biological nanowires [59]. Microorganisms like *Shewanella* can "consume organic waste in used water while harvesting the electricity they produce to help power the treatment system itself" [59].

5 Concluding remarks

This chapter has attempted to provide a quick overview of some of the promising space power technologies. The United States continues to fund studies of both radioisotope power sources (RPSs) and fission power sources, which could eventually power interstellar missions. Other nations (e.g., China, the European Union, Russia) are also doing research or have used nuclear power sources (NPSs) in space or on other planetary bodies.

Based on current technology and for a near-term mission, the choice to power an early interstellar spacecraft would be some type of nuclear power source (radioisotope or fission reactor). This is not to ignore or give up on the potential contributions that some of the other power source concepts discussed might contribute to an interstellar mission. Research on Earth continues on various fusion power concepts, and analytical studies of antimatter are also continuing.

Missions to the stars will be a centuries-long undertaking and, if the past is any guide, future technologies as yet undreamed of may enable star flight. One such technology is faster-than-light (FTL) travel. There may be ways around Einstein's "speed limit" (speed of light), perhaps by invoking hypothetical particles such as the postulated tachyons or using wormholes to "jump" from one region of the Universe to another or by using some sort of warp drive to modify spacetime [60]. Maybe some combination will do the trick [61].

As the world comes out of the Covid pandemic, one is reminded of Mary Shelley's apocalyptic novel *The Last Man* (1826), about a worldwide plague. Writing in the early 1800s, Shelley placed her novel in the 21st century. The technology she invoked for the 21st century included horse-drawn carriages and balloons for air travel. Just 60 years later, Carl Benz would apply for a patent for a vehicle powered by a gas engine. And less than 80 years from publication, airplanes would begin flying. To paraphrase the late astronomer J. Allen Hynek [62], "There is a tendency in the 21st century to forget that there will be a 22nd century science, and indeed a 30th century science, from which vantage points our knowledge of the universe may appear quite different. We suffer, perhaps, from temporal provincialism, a form of arrogance that has always irritated posterity."

References

[1] G. Chew, M. Doyle, M. Stancati, Interstellar Spaceflight Primer, Science Applications International Corporation, Schaumburg, IL, February 2001.

[2] Breakthrough Starshot Initiative, https://breakthroughinitiatives.org/initiative/3. (Accessed 28 August 2022).
[3] U.S. Energy Information Administration. https://www.eia.gov/energyexplained/electricity/use-of-electricity.php. (Accessed 28 August 2022).
[4] M. Garcia, About the Space Station Solar Arrays, 03 August 2017. https://www.nasa.gov/mission_pages/station/structure/elements/solar_arrays-about.html. (Accessed 28 August 2022).
[5] M. Garcia, New Solar Arrays to Power NASA's International Space Station Research, 11 January 2021. https://www.nasa.gov/feature/new-solar-arrays-to-power-nasa-s-international-space-station-research. (Accessed 28 August 2022).
[6] W.E. Wallin, J.M. Friefeld, Solar Dynamic Power System Definition Study, March 1988, Final Report, NASA Contractor Report 180877, Rocketdyne Report RI/RD87-250 https://core.ac.uk/download/pdf/42832946.pdf. (Accessed August 2022).
[7] J.A. Angelo, D. Buden, Space Nuclear Power, Orbit Book Company, Malabar, FL, 1985.
[8] V.L. Teofilo, Advanced space power systems, Paper No. 1762, in: Society of Allied Weight Engineers, Inc., 46th Annual Conference, Seattle, Washington, 18-20 May 1987.
[9] G.L. Bennett, A suggested technical approach to the evolutionary development of advanced space nuclear power systems, Space Technol. 17 (2) (1997) 113–122.
[10] G.L. Bennett, J.J. Lombardo, B.J. Rock, US radioisotope thermoelectric generators in space, Nucl. Eng. 25 (2) (1984) 49–58. Reprinted from U.S. Radioisotope Thermoelectric Generator Space Operating Experience (June 1961 – December 1982), Proceedings of the 18th Intersociety Energy Conversion Engineering Conference, American Institute of Chemical Engineers, held in Orlando Florida, 21-26 August 1983.
[11] G.L. Bennett, Radioisotope power: historical review, in: E. Greenspan (Ed.), Encyclopedia of Nuclear Energy, first ed., vol. 3, Elsevier Inc., 2021, ISBN: 9780128197257.
[12] G.L. Bennett, A look at the Soviet space nuclear power program, in: Proceedings of the 24th Intersociety Energy Conversion Engineering Conference, Institute of Electrical and Electronics Engineers (IEEE), Held in Crystal City, Washington DC, 6-11 August 1989.
[13] C.E. Kelly, C.W. Whitmore, Performance of the multi-hundred watt radioisotope thermoelectric generator (MHW-RTG) power source for the LES 8/9 satellites, in: 7th Communications Satellite Systems Conference, San Diego, California, 24-27 April 1978, https://doi.org/10.2514/6.1978-534.
[14] F. de Winter, G. Stapfer, E. Medina, The Design of a Nuclear Power Supply With a 50 Year Life Expectancy: The JPL Voyager's SiGe MHW-RTG, SAE Technical Paper 1999-01-2586, SAE International, 1999, https://doi.org/10.4271/1999-01-2586.
[15] R. McNutt Jr., et al., Interstellar probe—destination: universe! Acta Astronaut. 196 (2022) 13–28, https://doi.org/10.1016/j.actaastro.2022.04.001.
[16] D.F. Woerner, JPL, private communication, 01 March 2023.
[17] G.L. Bennett, et al., Mission of daring: the general-purpose heat source radioisotope thermoelectric generator, AIAA 2006-4096, in: 4th International Energy Conversion Engineering Conference and Exhibit (IECEC), American Institute of Aeronautics and Astronautics (AIAA), San Diego, California, 26-29 June 2006, https://doi.org/10.2514/6.2006-4096.
[18] NASA, Electrical Power. https://mars.nasa.gov/mars2020/spacecraft/rover/electrical-power/. (Accessed 21 January 2023).
[19] R.M. Ambrosi, et al., European radioisotope thermoelectric generators (RTGs) and radioisotope heater units (RHUs) for space science and exploration, Space Sci. Rev. (2019), https://doi.org/10.1007/s11214-019-0623-9.
[20] Dynamic Isotope Power Systems (DIPS) for Space Exploration—Technical Information, BC92-68, Rockwell international, Rocketdyne Division, August 1992.
[21] 90-year-old turbines still spinning at falls, The Associated Press, published in The Idaho Statesman, 24 March 1996.
[22] Garrett Airesearch Manufacturing Company, Brayton Isotope Power System (BIPS) Phase I First Annual Technical Report, 76-311943, SAN-1123-76/1, 27 October 1976.

[23] W.R. Corliss, Nuclear Reactors for Space Power, U.S. Atomic Energy Commission Division of Technical Information, Oak Ridge, TN, 1971.
[24] G.L. Bennett, D. Buden, Use of nuclear reactors in space, Nucl. Eng. 24 (4) (1983).
[25] D. Buden, Space Nuclear Fission Electric Power Systems, Space Nuclear Propulsion and Power Book 3, Polaris Books, Lakewood, CO, 2011.
[26] G.E. Berg, Final SNAPSHOT Performance Report, NAA-SR-11934, Atomics International, 15 August 1966.
[27] S. Magnuson, Nuclear power plants on new submarines may last 40-plus years, National Defense Magazine (01 February 2015).
[28] J.A. George, Multi-Reactor Power System Configurations for Multimegawatt Nuclear Electric Propulsion, NASA Technical Memorandum 105212, September 1991.
[29] K.T. Nock, TAU—a mission to a thousand astronomical units, AIAA-87-1049, in: 19th AIAA/DGLR/JSASS International Electric Propulsion Conference, Colorado Springs, Colorado, 11-13 May 1987.
[30] R.H. Frisbee, Overview of the 10,000 AU (TTAU) mission propulsion requirements, in: Presented at the Ten-Thousand AU Interstellar Precursor Mission, Propulsion Technology Interchange Meeting, Jet Propulsion Laboratory, Pasadena, California, 08 December 1997.
[31] SNAP-50/SPUR Program Summary, CNLM-5889, Pratt & Whitney Aircraft, Middletown, CT, 24 September 1964.
[32] J.E. Allen, Comparison of SNAP-50/SPUR System With MHD, Brayton, Boiling Reactor and Thermionic Systems, CNLM-6249, Pratt & Whitney, 27 April 1965.
[33] E.M. Douthett, The SNAP-50/SPUR Program, Progress in Astronautics and Rocketry, vol. 16, Elsevier, 1966, pp. 541–560. https://doi.org/10.1016/B978-1-4832-3056-6.50026-8.
[34] Presentation of Adapting the SNAP-50 Space Powerplant to a Lunar Base, CNLM-5275, Pratt & Whitney Aircraft, 04 October 1963.
[35] P. Jaffe, Power Beaming and Space Solar, U.S. Naval Research Laboratory (undated charts).
[36] J. Brophy, A Breakthrough Propulsion Architecture for Interstellar Precursor Missions, 06 April 2017. https://www.nasa.gov/directorates/spacetech/niac/2017_Phase_I_Phase_II/Propulsion_Architecture_for_Interstellar_Precursor_Missions/. (Accessed 12 October 2022).
[37] C. Limbach, PROCSIMA: Diffractionless Beamed Propulsion for Breakthrough Interstellar Missions, 30 March 2018. https://www.nasa.gov/directorates/spacetech/niac/2018_Phase_I_Phase_II/PROCSIMA/. (Accessed 16 October 2022).
[38] G.A. Landis, Power system for miniature interstellar flyby probe, in: Presentation at the 6th International Symposium, Wichita, Kansas, 10-15 November 2019.
[39] R.L. Forward, Starwisp: an ultra-light interstellar probe, J. Spacecr. Rocket. 22 (3) (1985) 345–350.
[40] G.A. Landis, Microwave pushed interstellar sail: Starwisp revisited, paper AIAA-2000-3337, in: Presented at the 36th AIAA/ASME/SAE/ASEE Joint Propulsion Conference and Exhibit, Huntsville, Alabama, 16-19 July 2000.
[41] M. Lanctot, DOE Explains....Fusion Energy Science, Office of Science, U.S. Department of Energy, (Accessed 13 October 2022). https://www.energy.gov/science/doe-explainsfusion-energy-science.
[42] Amazing Machine, https://www.iter.org. (Accessed 17 October 2022).
[43] Nuclear Fusion, http://hyperphysics.phy-astr.gsu.edu/hbase/NucEne/fusion.html. (Accessed 13 October 2022).
[44] Project Daedalus Study Group, Project Daedalus: The Final Report on the BIS Starship Study, Space Educational Aids, London, 1978.
[45] A. Crowl, et al., The Enzmann starship: history & engineering appraisal, J. Br. Interplanet. Soc. 65 (2012) 185–199.
[46] B. Palaszewski, Atmospheric Mining in the Outer Solar System: Resource Capturing, Storage, and Utilization, AIAA-2012-3742, NASA/TM-2014-217892, prepared for the 48th Joint Propulsion Conference and Exhibit, Atlanta, Georgia, 30 July-01 August 2012.

[47] What is antimatter? Sci. Am. (24 January 2002). https://www.scientificamerican.com/article/what-is-antimatter-2002-01-24/. (Accessed 18 October 2022).
[48] G.P. Jackson, Antimatter as an energy source, in: AIP Conference Proceedings, International Technical Forum on Space Propulsion and Energy Sources, Huntsville, Alabama, 24-26 February 2009, https://doi.org/10.1063/1.3115551.
[49] G.P. Jackson, Antimatter-Based Propulsion for Exoplanet Exploration, Presentation at NETS 2021, Track 4, hosted by Oak Ridge National Laboratory, 26-30 April 2021. https://nets2021.ornl.gov/wp-content/uploads/gravity_forms/12-b63a96649a525ab5aa39d607840d9d9f/2021/04/jackson_exoplanet_202104261.pdf. (Accessed 18 October 2022).
[50] The Nearest Neighbor Star, NASA Goddard Space Flight Center, https://imagine.gsfc.nasa.gov/features/cosmic/nearest_star_info.html. (Accessed 18 October 2022).
[51] E. Howell, NASA's Superfast Parker Solar Probe Just Broke Its Own Speed Record at the Sun, Space.com, 23 November 2021. https://www.space.com/parker-solar-probe-sun-speed-record-november-2021. (Accessed 18 October 2022).
[52] P.A.M. Dirac, Theory of Electrons and Positrons, Nobel Lecture, 12 December 1933. https://www.nobelprize.org/uploads/2018/06/dirac-lecture.pdf. (Accessed 19 October 2022).
[53] R.L. Forward, Extracting electrical energy from the vacuum by cohesion of charged foliated conductors, Phys. Rev. B 30 (4) (1984). http://www.calphysics.org/articles/Forward1984.pdf. (Accessed 19 October 2022).
[54] H. Casimir, On the attraction between two perfectly conducting plates, reprint from, Proc. R. Neth. Acad. Arts Sci. 51 (1948) 793–795. https://www.mit.edu/~kardar/research/seminars/Casimir/Casimir1948.pdf. (Accessed 19 October 2022).
[55] M.S. Morris, K.S. Thorne, Wormholes in spacetime and their use for interstellar travel: a tool for teaching general relativity, Am. J. Phys. 56 (5) (1988).
[56] A.A. Kirillov, E.P. Savelova, Wormhole as a possible accelerator of high-energy cosmic-ray particles, Eur. Phys. J. C 80 (2020) 45. https://link.springer.com/content/pdf/10.1140/epjc/s10052-020-7609-3.pdf. (Accessed 19 October 2022).
[57] L. Comisso, F.A. Asenjo, Magnetic reconnection as a mechanism for energy extraction from rotating black holes, Phys. Rev. D 103 (2021) 023014. https://arxiv.org/abs/2012.00879. (Accessed 19 October 2022).
[58] C.L. Cheung, A. Tabor, Could Electricity-Producing Bacteria Help Power Future Space Missions, nasa.gov, https://www.nasa.gov/feature/ames/could-electricity-producing-bacteria-help-power-future-space-missions. (Accessed 19 October 2022).
[59] S. Bushwick, Electricity-carrying bacteria lead to new applications—and new questions, Sci. Am. (01 December 2020).
[60] G.L. Bennett, Destination universe: Some thoughts on faster-than-light (FTL) travel, AIAA 2016-4918, in: 52nd AIAA/SAE/ASEE Joint Propulsion Conference, Propulsion and Energy Forum, Salt Lake City, Utah, 25-27 July 2016.
[61] G.L. Bennett, The Star Sailors, Authors Choice Press, New York and Bloomington, 2009 (originally published by St. Martin's Press, 1980).
[62] Time Mag. (04 August 1967) 32–33.

CHAPTER

Crew health—Psychological, biological, and medical issues and the need for a systems approach

3

Mark Shelhamer[a] and Jennifer Fogarty[b]

[a]*Johns Hopkins University School of Medicine, Baltimore, MD, United States,* [b]*Center for Space Medicine, Baylor College of Medicine, Houston, TX, United States*

Chapter outline

Interstellar Travel. https://doi.org/10.1016/B978-0-323-91280-8.00001-0

1 Introduction

Keeping people alive and healthy in space is a difficult task. NASA and other space agencies have been working on this problem for decades and have made significant progress—at least for missions with durations up to approximately 1 year. But what of longer missions—with durations of decades or centuries? In this chapter we present the major risks to human health and performance from prolonged exposure to the space environment, as understood at the time of writing. This chapter also discusses potential mitigation approaches—both those which are feasible in the near future and those which will require longer-term strategies.

In Section 1 we outline the five major hazards inherent to long-duration spaceflight that guide much of the current thinking in this area. In Section 2 we describe the main health and performance concerns that are consequences of these spaceflight hazards. We draw largely on the current NASA taxonomy, which, though based on decades of experience and regular monitoring, is limited to regular flights of approximately 6 months and occasional flights of approximately 1 year. We examine trends based on what we currently know, while recognizing that there may be inflection points beyond our experience. We therefore indulge in some informed extrapolations, and consider other issues that might arise only with longer flights. In Section 3 we present a number of strategies to mitigate these concerns, some of which are well-accepted and currently implemented or under study, while others require further development and evaluation. Finally, in Section 4, we propose a broader systems-oriented approach, which considers the myriad of interactions between the major health and performance concerns and places them in the context of the larger mission.

2 The main health and performance hazards of spaceflight

Before discussing the main issues regarding human health and performance in space, it will prove helpful to establish some context by delineating the major hazards of human spaceflight. We present a common framework of five hazards that has been formally adopted by NASA [1] to aid in planning human exploration missions.

2.1 Altered gravity level

One of the ubiquitous features of spaceflight is altered gravity level. During orbital flight and beyond, the crew experiences weightlessness. This is sometimes termed zero-gravity or $0g$, which are misnomers because, among other things, while in orbit around a planet there is a strong gravitational force which maintains the orbit. Nevertheless, we use weightlessness and $0g$ interchangeably. We completely avoid use of the term microgravity, which implies the deliberate establishment of a level of $\sim 10^{-6}g$, rather than referring to residual net g level from a variety of sources combined. Since the human body evolved and develops in a $1g$ field, it has substantial physiological processes to enable survival and performance at this g level. When the gravity level is reduced, some of those processes no longer need to perform at the same level, since they do not have to work as hard to support bodily functions. This lack of impetus from gravity leads to adaptive processes that are reasonable and appropriate for life in $0g$, but can be problematic upon returning to a gravity field, and might also be detrimental to health.

The most serious effects that result from this hazard are musculoskeletal and cardiovascular deconditioning, changes in neurovestibular processing, alterations of the central nervous system (CNS), and spaceflight-associated neuro-ocular syndrome (SANS). Many of these effects are a result of fluid shifting from the lower to the upper body in the absence or reduction of a hydrostatic gradient (i.e., no or reduced gravity). This altered-gravity hazard can also have secondary effects on immune and microbiome functions (partly through changes in the cellular medium, such as lack of convection and sedimentation) and in the production of kidney stones (through altered calcium balance due to unloading of bones).

2.2 Isolation and confinement

To date, spaceflight has involved small crews in confined spaces with limited volume, separated from other people and many of the comforts of home, with no easy and immediate way to "step outside" for a break. While longer-duration missions might utilize larger spacecraft, engineering practicalities will call for vehicles and habitats with the minimum volume that the crew can tolerate—a psychological need that is likely change over the course of the mission.

The most direct impact of this hazard is, for obvious reasons, related to psychological and behavioral issues. These can include changes in mood and motivation, depression, and challenges with interpersonal relationships. Immune and microbiome function and cardiovascular health might also be affected by increased psychological stress.

2.3 Distance from Earth

Spaceflights to date have been relatively close to the home planet Earth. On average, the distance to Mars is more than 500 times that of the distance to Earth's moon, a distance which pales in comparison to that which would be required for interplanetary journeys and expeditions into interstellar space. These expeditionary distances mean that the crew on board will be almost completely autonomous, as they will be unable to rely on help from Earth. For a mission to Mars, for example, one-way communication delay can be in the order of 20 minutes, emergency mission abort and return will be out of the question, and consumables resupply will be sporadic at best. For more ambitious missions, crews will have to divorce themselves from any sense of Earth-based assistance, especially in urgent emergency situations.

This autonomy also raises issues of crew cohesion and leadership, not to mention possible psychological issues (the "Earth out of view" phenomenon). Practical issues arise, such as how to supply and maintain sufficient stores of food and medicine, which impact the ability to maintain health and fitness across all body systems.

2.4 Hostile and closed environment

This refers not only to the environment of space (e.g., cold, vacuum) but also the spacecraft and habitat itself. Hazards of the spacecraft environment can include elevated carbon dioxide (CO_2), atmospheric contaminants from material off-gassing, and microbial contaminants on surfaces. Further concerns arise on a planetary surface, related to the composition of the atmosphere and soil, which might contain toxic or corrosive substances. This hazard has direct impact on functioning of the immune system and the microbiome.

2.5 Radiation

Radiation in deep space (interplanetary and interstellar) is harsher than that encountered on earth or even in low-earth orbit (LEO). While there is great uncertainty as to the ultimate effect of deep-space radiation on humans, evidence from research to date suggests that this radiation can adversely affect immune and microbiome function, cause cardiovascular degeneration, and impact CNS and cognitive function. Depending on the type and magnitude of the radiation exposure, these adverse health and performance impacts could occur during the mission itself or might only become manifest later in life.

We present the effects of these hazards in an orderly and discipline-specific manner below. Nevertheless, it is important to note (as the comments above suggest) that mapping of these hazards to the main health issues discussed in this chapter is not one-to-one. All of the hazards can contribute to medical and performance issues. There are also interactions between hazards. As an example, the same outcome (e.g., nervous-system damage) might be the result of altered gravity, or radiation, or a combination. These types of additive or multiplicative effects are not yet fully explored and could play a significant role in extended missions. Any and all of these hazards can contribute to various forms of psychological stress, and some stressors such as high workload do not easily fit under one hazard or another.

3 The main health and performance concerns

Human responses to the spaceflight hazards outlined in the previous section range from adaptive to maladaptive and have been documented through evaluation of data from medical operations and dedicated research performed during spaceflight and terrestrial analog missions. Extrapolating about the potential effects expected in multiyear missions from the current database carries substantial risk: some processes may follow initial trends for very long periods, while others may be self-limiting and reach a plateau. The following sections describe the known human responses to spaceflight hazards. The most challenging aspect is that the space-traveling human experiences spaceflight in its totality and begins adapting immediately, such that the hazard exposures are overlapping and influence each other, leading to additive or synergistic effects that are not yet fully elucidated. Acute responses manifest over minutes and hours and, if a person is subjected to a hazard over a long period of time, a longer chronic response process may be activated that includes compensatory mechanisms which may delay or mask overt symptoms and performance impacts [2]. The chronic responses to spaceflight hazards have more significant implications for late in-mission and long-term postmission health and performance. As a result of the holistic nature of the experience of spaceflight, it can be very challenging to identify the first response to spaceflight because of intrinsic dependencies and continuous interaction between molecular and physiologic systems. Given this challenge, it seems logical to describe the human responses based roughly on the order in which they appear at a phenotypic level (i.e., the order in which they manifest as symptoms or behaviors).

3.1 Physiology

3.1.1 Vestibular and sensorimotor disruption and adaptation

The vestibular apparatus is the nonauditory part of the inner ear. It consists of sensors that transduce angular and linear motions (semicircular canals and otolith organs, respectively); the neurovestibular system consists of these sensors and the central neural processes that use the information provided by these sensors to drive posture, eye movements, and sense of orientation. Calibration and adaptation of this system is maintained by neural mechanisms located mainly in the brainstem and cerebellum. Since the otolith organs are direct sensors of linear acceleration, including gravity, the processes that depend on accurate otolith input can be quickly and dramatically altered by spaceflight. These alterations can affect a number of perceptions and reflexive behaviors related to action and orientation with respect to gravity (the net gravitoinertial force vector) [3].

Adverse sensorimotor effects related to the vestibular system can appear immediately on reaching orbit. During the first 2–3 days of spaceflight, 60%–80% [4] of crewmembers experience symptoms of space motion sickness (SMS), which can range from mild nausea to severe vomiting; the latter of these sometimes requires medication to reduce the risk of dehydration. Other symptoms include lethargy, sensitivity to head motion, and orientation illusions. SMS poses a significant risk within the first few days of a mission, when nausea may be accompanied by spatial disorientation, difficulty acquiring and tracking visual targets, and alterations of proprioception [5].

SMS (and related issues) is precipitated most directly by transitions between gravity levels—most notably between 0*g* and other *g* levels. If spaceflight expeditions are long and transitions rare, then the most acute issues that arise as a result of transitioning between gravity levels will not be of major concern, as they resolve quickly with little or no lasting impact. Some issues, however, do take longer to fully resolve; there are, for example, anecdotal reports of "flashbacks" to a 0*g*-adapted state even after weeks or months back on Earth following extended spaceflight [6].

Locomotion and postural recovery are dramatically impacted immediately on return from ISS missions of several months. Extensive testing (the NASA Field Test [7]) immediately on landing has included basic clinical neurological evaluations and structured multisystem research protocols. While there is significant interindividual variability, many crewmembers experience symptoms (dizziness, nausea, vomiting, unstable gate, nystagmus, diminished visual target acquisition, reduced fine motor and gross motor function [8–11]) that can be disabling for the first 24–48 hours, with an asymptotic recovery. Recovery can be facilitated by regular yet limited provocative head movements in the new environment. Therefore, it is important for crewmembers to experience movement in 1*g* around a variety of axes, but it is equally important to allow for rest and recovery to facilitate re-adaptation [9,12].

In flight, operational mitigations (changes in logistics or procedures) are effective, but they can be a challenge in an emergency. Medications can be effective, but many of them have side effects that introduce new problems. The ability to predict SMS susceptibility would provide operational benefits, but is an elusive goal [13,14]: there is little correlation among various forms of motion sickness, but previous SMS in space does indicate the likelihood of SMS on subsequent flights. The purpose of predicting SMS susceptibility would not be to screen out crewmembers, but to temper expectations of a specific individual's capabilities during and immediately after a *g* transition. This could include determining the roles that specific crewmembers should play. When performance in the face of a transition is critical, it may be useful to consider preadaptation procedures [15,16].

Finally, we note that the neurovestibular system has a central role in many aspects of nervous system function [17], and the resulting interactions might be critical, especially in emergencies associated with *g* transitions. While many of these interactions have been delineated in pathological cases, the changes in neurovestibular function that occur in spaceflight might lead to similar effects:

- Patients with vestibular disorders can have alterations in mood [18], which might impact motivation and interpersonal relations.

- Altered vestibular contributions to spatial memory and orientation and to motor reflexes [19,20] might be a source of "space fog" or "space stupids." Extra mental effort might be needed to perform even mundane tasks when usually reliable cues for orientation and reflex control are altered [21].
- Vestibular disorders can change body self-perception [22]. An altered sense of one's self in relation to others (relative personal spaces) might have an effect on behavioral and interpersonal issues [23].
- Lesions of the vestibular system in rats can lead to degradation of weight-bearing bones [24].

The remarkable adaptive capabilities of the neurovestibular system mean that these secondary interactive effects have not, to date, been of great relevance in space. While the neurovestibular system is likely to be equally adaptable over the course of longer missions, an accumulation of these lesser effects, combined with other stressors, might tax this adaptive capacity and lead to a loss of resilience and reduction in the performance margin available to deal with anomalous situations (see Section 4).

3.1.2 Musculoskeletal adaptation

The human body is an amazingly energy-efficient machine. If a physical demand or stressor is removed, the body will adjust as needed so as not to waste metabolic resources. During spaceflight, the "absence of gravity" (weightlessness) calls into play physiological processes that reduce now-unnecessary muscle volume and bone density, as also seen in many terrestrial instances of disuse. These changes in muscle volume and strength have been assessed since the Apollo missions of the 1960s in order to understand the rate and persistence of the changes to skeletal muscle and bone, but it has been challenging to relate any changes to specific functional consequences. Astronauts and cosmonauts have been capable of completing demanding mission objectives, including EVAs (extravehicular activities: spacewalks). Rarely, however, have they been required to self-egress (get out of a spacecraft without assistance) upon landing back on Earth. Anecdotal reports from the few off-nominal cases of self-egress suggest that, in the case of a three-person capsule returning from several months on the ISS, only one of the three has been capable of standing, and often all three must resort to crawling out of the vehicle and resting between activities. Therefore, the realistic concern is for future missions in which the disuse atrophy of skeletal muscle and bone may progress, given the extended duration of weightless exposure and the workload demands on planetary surfaces, and may exceed the capability of the musculoskeletal system, resulting in injuries.

While crewmembers' bodies adapt primarily to the absence of load in 0*g*, new risks are created when the body is rapidly re-exposed to gravitational forces during landing and egress. Changes in muscle strength and volume have been studied during spaceflight for decades, but not until recently have the changes been put into the context of expected mission tasks. The NASA Functional Task Test (FTT) is a multidisciplinary research effort that seeks to systematically assess and quantify changes in performance starting immediately at landing and to monitor the progress of recovery over several weeks [8,25]. Data analysis relates functional changes to the contribution of the various subsystems causing preflight-to-postflight neurovestibular/sensorimotor, cardiovascular, and musculoskeletal deficits. Given the plasticity of the musculoskeletal system, none of the functional changes appear permanent and crewmembers can recover both volume and strength with adequate training within weeks or months of return [26]. The measurable change in muscle volume and strength can be mitigated with countermeasures such as resistance training and artificial gravity (AG) during the transit and by having egress aids and occupant protection to prevent injury at landing.

Changes in bone mineral density (BMD) may put a crewmember at risk for fracture during extended deep-space exploration missions. Changes in BMD are derived by analyzing data collected from preflight/postflight dual-energy X-ray absorptiometry (DXA), which assesses overall bone structure. However, without insight into BMD during the course of the mission, it is difficult to ascertain the true rate of loss. At the moment, rate of bone loss is estimated through a simple calculation of net BMD change divided by mission duration. This oversimplification has led many to believe that the loss rate is linear and would continue indefinitely for the duration of an exploration-class mission of 3 years or more. This, however, is not a reasonable assumption: crewmembers on missions from 8 to 12 months have not demonstrated double the BMD loss as compared to those on missions of 4–6 months. (These data have not been published openly due to the small sample size and the inability to protect the identities of the subjects.) It is more reasonable to hypothesize that the rate of BMD loss is asymptotic, and reaches a plateau at some point based on load exposure, countermeasures, and genetics. It is important to assess the projected loss of BMD and evaluate mission objectives to understand the potential forces that will be encountered and the types of scenarios in which the forces might be applied to the musculoskeletal system.

Medical and research staff measure change in BMD for an individual over time as a clinical surveillance tool. Results for a given bone region are compared to population norms in order to derive a measure of fracture risk that includes traumatic and atraumatic (fragility) factors. (A complicating aspect for interpreting spaceflight data is that population norms for those younger than 65—typical of astronauts to date—are not robust, and the population over 65 often has comorbidities.) To date, skeletal regions with the most significant reductions in BMD are the trochanter, femoral head, and pelvis [27]. A unique effort sponsored by NASA [28] developed an astronaut fracture-risk model and calculated the risk for various missions. The results indicate that fracture risk varies by bone region based on the application of force and the mission duration (manifest in greater BMD reductions).

One of the areas not specifically addressed by this model is the significance of the two different components of bone, which are affected differentially by BMD loss. While both the cortical and the trabecular components lose some BMD, trabecular bone is disproportionately affected and does not appear to fully recover [29]. Total BMD returns to preflight density due to cortical bone exceeding its original volume in order to compensate for loss of trabecular bone, which is problematic because trabecular bone plays a larger role in maintaining overall bone strength and in providing microarchitecture for bone marrow. It has been challenging to assess the differential response of the components, since few technologies are able to distinguish between these components and provide follow-up evaluation over long postmission timeframes. Another complexity associated with the loss of BMD during spaceflight is the resulting increase in circulating calcium, which increases the risk of renal stone formation [30] (see below).

3.1.3 Cardiovascular effects and adaptation

Upon exposure to the weightless condition of space, the body begins adjusting the amount of circulating blood (plasma volume). Without a hydrostatic gradient in 0*g*, less blood is sequestered in the lower body, and blood shifts toward the head and upper body, where sensors in the chest detect this increased torso fluid volume. Blood also more easily returns to the heart from the peripheral circulation, while blood in the head experiences outflow resistance due to venous congestion on the way back to the heart. This leads to puffy faces and sinus congestion, as the additional volume causes an increase in fluid pressure that exceeds the limits of the capillaries; plasma thus leaks into the tissue and becomes trapped within and between cells, creating edema. Eventually, some of the fluid will be removed by lymphatics (a drainage system that parallels the venous system) and excreted through the kidneys, while the rest remains predominantly in the tissue of the upper torso and head. This fluid shift results in an overall plasma volume loss of 10%–20% over the first 4–6 weeks of spaceflight as crewmembers achieve a new normal plasma volume [31,32]. This adaptive process occurs because, even absent a hydrostatic gradient pulling fluids toward the lower body, the cardiovascular system is capable of maintaining normal blood pressure to accommodate the metabolic needs of body tissues at rest. This results in changes in vascular compartments, and in altered performance under stresses such as maximal exercise and re-exposure to gravity.

Components of the vascular system must adjust to the new plasma volume, and to the chronically reduced physical workload associated with living in a weightless environment. Red blood cell (RBC) volume is acutely concentrated due to the rapid loss of plasma volume, and RBCs are then reduced to maintain proper blood viscosity in the new-normal volume condition in spaceflight. The venous system of the lower body begins a remodeling process associated with a reduced blood column in the lower limbs. This remodeling results in a reduction of sympathetic tone and loss of responsiveness to blood volume [31]. The absolute plasma volume loss, relative RBC reduction, and lower limb venous remodeling are not problematic during the weightless portion of a mission unless there is a demand for high aerobic capacity, such as that produced during exercise [33].

Aerobic capacity is a crucial physiologic metric, and is used regularly to assess cardiovascular fitness inasmuch as it pertains to metabolically active skeletal muscle. Changes to maximal aerobic capacity (VO_2 max: uptake of oxygen and production of carbon dioxide at maximal workload) are often tracked to assess training, detraining, and fitness for duty. During long-duration spaceflight, estimated VO_2 max decreases by 15%–20% within the first month, and subsequently recovers by 5%–10%, reaching a plateau after about 2 months [33]. This decrease has not been operationally relevant, since living and working in 0*g* does not engender a high physical workload, but it reflects lost plasma volume and is very relevant to performance capability on landing (especially in the case of unassisted egress or emergency evacuation). Aerobic capacity is significantly impacted immediately on return to Earth, as the body adjusts to gravity, and plasma volume now becomes sequestered in the lower body (veins and tissues), manifesting as a state of dehydration. For long-duration spaceflight, 80% of crew experience orthostatic intolerance (OI) on landing day: inability to stand upright and maintain consciousness. Since dizziness, syncope, unsteady gate, nausea, and vomiting are symptoms of both neurovestibular deconditioning and OI, it is often difficult to identify the dominant cause of postlanding inability to stand or walk. While current mitigation strategies—prelanding fluid loading and lower body compression garment—have limited effectiveness, these symptoms significantly abate by 24 hours after landing.

Another critical cardiovascular component that remodels due to 0*g* exposure is the heart itself. Without gravity exerting a force on the tissues of the body, blood flow to and from the heart is relatively unimpeded, and the heart does not have to produce the same contractile force as it does on Earth to maintain proper cardiac output. In addition, the heart is exposed to slightly lower intrathoracic pressure compared to that on Earth [34]. These factors reduce the cardiovascular burden and can lead to atrophy and weakening of cardiac tissue. Resistive and aerobic exercise can create an acute physiologic stimulus by redistributing blood volume to activated skeletal muscle and creating increased metabolic demand for oxygen supply and carbon dioxide removal. Since this exercise-induced demand may blunt the 0*g*-induced cardiac remodeling, and spaceflight missions routinely include various forms and durations of exercise prescriptions, there is no spaceflight data to determine how extensive this remodeling is.

Cardiac remodeling is also central to another concern: the potential of arrhythmia (abnormal heart rhythm). One of the more critical parameters monitored during spaceflight is the electrocardiogram (ECG), which monitors the heart's electrical activity. Although cardiac function in astronauts is thoroughly tested during medical screening as part of the selection process, and routinely monitored during their careers and training, abnormal heart rhythms can spontaneously develop with age or as a consequence of illness or dehydration. Not all irregular heart rhythms are inherently dangerous, and context is needed in order to understand their genesis as well as their potential to cause sudden cardiac death [35]. A concern has been that spaceflight—a unique combination of physiological stressors—may be conducive to arrhythmia. Although irregular heart rhythms were detected during early NASA missions, there is no evidence that those abnormal rhythms were caused by the complex environment of spaceflight; based on postmission evaluations, they were most likely caused by preexisting undetected disease [36]. Subsequent missions have not demonstrated that crewmembers are more susceptible to arrhythmias during spaceflight compared to before and after flight.

Immediately upon landing, the combination of reduction in plasma volume, changes to vascular compartments, and neuro-vestibular/sensorimotor disruption creates a situation where crewmembers will have measurable deficits in activities such as standing, walking and running, descending and ascending ladders, and operating vehicles. The magnitude and persistence of these deficits somewhat depends on the duration of exposure to 0*g*, with longer flights causing a greater degree of deterioration, which causes a longer duration of deficit. While these effects are measurable upon landing on Earth, there are questions about how these physiological adaptations will affect performance when crewmembers are exposed to partial gravity during planetary landings and EVAs.

3.1.4 Immune system function

Several bacterial species are altered by spaceflight (presumably by extended 0*g* but possibly also by other factors) in ways that can be detrimental to human health. Alterations include increases in virulence and biofilm formation and heightened resistance to antibiotics and stress [37]. At the same time, spaceflight causes changes in some aspects of human immune system function. The prospect of these two trends intersecting is especially troubling when considered in the context of longer and more ambitious spaceflights, since increased virulence and decreased immune function is a bad combination.

Many changes that take place at the cellular level are likely due to altered gravity (weightlessness) in spaceflight. However, studies in Earth-based analog facilities show some similar changes [38,39], which indicates that other stressors can alter immune function at this basic level. This observation coincides with other work that shows that immune function is shaped by environmental exposures and lifestyle choices [40].

It is difficult to know how critical these immune changes will be. On the one hand, the trends are not promising. For example, Space Shuttle and ISS data show that an increase in latent herpesvirus reactivation correlates with increasing mission duration [41]. On the other hand, the clinical relevance of these changes is not clear [42]: even immediately after flight, when astronauts are exposed to their families once again (who have not remained in quarantine during the mission), there does not appear to be a dramatic increase in infectious diseases. While there are many changes in immune-system function in space, so far there have been no disastrous consequences. This may be due to the fact that, although there are many stressors associated with spaceflight, professional astronaut crewmembers follow generally healthy lifestyles that are conducive to maintenance of immune function (exercise, diet). However, such might not be the case when considering spaceflights featuring larger crews on missions to settle and colonize other celestial bodies.

The challenge of understanding the relevance of immune alteration is exacerbated by the diversity of effects, which vary between organisms and between components of the immune system. This makes it difficult to detect general principles by which targeted countermeasures could be designed. "The immune system is remarkably complicated, consisting of a myriad of distinct cell populations, each with unique function" [43]. The complexity of the issue can be seen by the many lengthy review articles that attempt to summarize the findings in a coherent form. We draw especially on recent reviews [37,43] and omit many of the details because, while they provide deep insights into the alterations seen in flight, they provide little in the way of general guiding principles upon which a spaceflight program can act.

Spaceflight induces a broad range of effects on immune function, as demonstrated in human and animal studies. These effects include decreases in lymphocytes, monocytes, and cytokines, decreased function of dendritic cells, reduced T-cell activation, altered distribution of peripheral leukocytes, diminished function of specific leukocyte subpopulations, and skewed cytokine profiles, as well as decrements in white blood cell count. Functionally, a common finding in long-duration flights is reactivation of latent viruses. While not yet clinically relevant (disease-producing) on its own, "it is possible that the reactivation of these latent herpesviruses could be a 'biomarker' of compromised adaptive immune function, particularly cytotoxic T lymphocyte function" [43]. While the changes enumerated above, and others, have been subclinical, health issues such as dental infections, upper respiratory infections, conjunctivitis, influenza, viral gastroenteritis, rhinitis, pharyngitis, and mild dermatologic problems such as rashes have arisen. The presence of these conditions is notable because they occur despite preflight quarantine and isolation from many pathogens. At the cellular level, reduced lymphocyte activation might be a result of changes in interactions between T cells and monocytes as a result of altered gravity. Alterations of the cytoskeletons of actin and microtubules have also been observed, linked to changes in genes related to cytoskeleton structure [44].

Common sense forms the basis of preferred countermeasure approaches so far [45]: premission screening of immune health, additional vaccinations and boosters, stress management in flight, inflight immune surveillance, and administration of immune boosters if indicated. The latter include IG, IL-2, antivirals, antiinflammatories, functional foods, and probiotics. Control of exposure before flight has also proved successful; this includes limiting the number of people with whom the crew comes into contact. Further inflight procedures could include air and water filtering, and use of surfaces that are resistant to contamination.

Given the many complex functions and interactions intrinsic to the immune system, an integrated functional approach is likely to be the most practical and effective. For example, the same overall nutritional measures that are beneficial for overall health and microbiome integrity (which we discuss in the next section) also apply to immune function. Additionally, some cells that underpin natural immunity are derived from hematopoietic stem cells in bone marrow, which thus depend on the proper interplay of bone formation and resorption [46]. Maintenance of bone integrity thus takes on importance for more than reducing the risk of fractures. In some cases, it might be necessary to use specific targeted interventions with pharmaceuticals—catecholamines (EP and NEP), beta blockers, endocannabinoids, and osteoporosis medications—to maintain bone integrity.

3.1.5 Microbiome and microbiota

The population of microorganisms in the gut and on the skin (as well as in the mouth and other locations on the body) make up the microbiome. A healthy microbiome has a critical influence on a number of physiological systems, including immune function as well as cognition and other brain functions [47].

Given this widespread set of influences, microbiome maintenance during spaceflights of any duration should not be neglected. The stressors of spaceflight present special difficulties in maintaining microbiome health, and extended missions even more so.

"Optimally, the microbiota produces essential and beneficial substances, arranges itself in a stable equilibrium, occupies the ecological niches of the gastrointestinal tract, and blocks them from pathogenic intruders" [48]. Given these central physiological roles played by a healthy microbiome, a prime concern in spaceflight is the loss of regular biome turnover and heterogeneity, due to isolation (lack of population diversity) and increased hygiene (and effects of medications like antibiotics), lack of food diversity, and limited exposure to a broad array of natural organisms. Despite several studies in space and in ground-based analogs of isolation and confinement, little is known about the details of what constitutes a healthy and beneficial (presumably diversified) microbiome. There are disparate findings, but a common finding is the large difference between individuals as to the composition of the normal steady-state microbiome. In ISS missions up to 6 months in duration, individual microbiomes became more similar to each other, but each one individually remained diverse, which is desirable [49]. The NASA Twins Study also showed changes in microbiome composition between twin astronauts (one who flew for 340 days and one who remained on the ground), but the individual richness did not change, which again shows good adaptation to spaceflight over these relatively short missions [50].

These results are promising for long-term retention of healthy microbiota, but given the widespread importance of the microbiome and the uncertainties of longer missions, concern remains. Since the relationships between microorganisms and various body functions are complex, and there are associations between changes in the microbiome and many disease and inflammatory processes, it is best to think in terms of functional measures and countermeasures rather than specific targeted interventions [48]. A lack of biome diversity per se does not necessarily reflect an underlying medical or physiological problem; it is the biome composition that matters most (the types and proportions of specific microorganisms). Thus, maintenance of microbial diversity alone is not sufficient as a countermeasure. Similarly, dysbiosis (lack of microbial diversity) can be an equilibrium state, so microbiome instability per se is not necessarily bad either. Reconstitution after a perturbation can lead to a new steady state, which could be different but still healthy. Overall, it's not yet completely clear how to define and identify an undesired microbiome status, and what might be done to mitigate it.

An appealing countermeasure approach to maintaining microbiome integrity is microbiome remodeling [51]: targeted interventions directed at specific biome components via nutrients (probiotics) or bacteriophages. The effectiveness of this general approach is as yet very uncertain, and it runs a significant risk of giving rise to unintended consequences if the wrong components are targeted, or if unexpected interactions occur. In the absence of specific details and a thorough understanding of the biological mechanisms of microbiome action that would make targeted interventions feasible, broader approaches to countermeasures may be called for.

Given the state of knowledge and complexity of biome structure and function, a functional approach is more realistic than tracking details with the goal of maintaining the specific makeup within an individual. Functional approaches should start by monitoring each crewmember before flight in order to establish individual baselines. Functional approaches should also regularly monitor crewmembers and associated metadata (especially food intake) during flight. Given the great intersubject variability in the studies noted above, it is important to understand the normal composition and dynamics of each individual's biome. This individual understanding could help guide medical interventions or quarantines later in flight. Specific countermeasures could be based on the desire to maintain microbiome diversity and promote recovery after a perturbation (stress, change in diet, antibiotic ingestion). It might be sufficient to have a variety of plants and even animals (pets) on the spacecraft so as to provide a rich diversity of microorganisms. This would provide the additional benefit of general psychological stress reduction [52].

One more ambitious approach is autologous fecal microbiota transplantation (aFMT): implantation of properly preserved fecal material into the same individual from which the material was obtained at some earlier time [53]. This procedure could help recover the individual's microbiome after a severe perturbation or from alteration after chronic stressors inherent to spaceflight. Heterologous transfer is also possible, although the risks associated with cross-individual transfer would be significant. The bidirectional nature of gut-brain signaling suggests the intriguing possibility to counteract stress signaling via the action of probiotics on the microbiome [47]. Probiotics, however, are not a panacea, as they might have adverse effects depending, again, on the specific species and person whose microbiome falls under consideration.

Sterilization of the spacecraft might be feasible for short-term flights where cabin integrity can be maintained, but it is probably infeasible and unrealistic for missions involving settlement and colonization. The idea of "contamination" by external factors will become outmoded, since the goal of this sort of mission is essentially exposure to new worlds and hence potentially new organisms. Even so, biofilms generated from crew-indigenous bacteria in a closed environment might be predominant, and should be carefully dealt with so as not to destroy desired microbiome diversity or expose the crew to chemical toxins. Some have suggested using biostatic (inhibiting growth without killing the organism) rather than biocidal disinfection to reduce biological load but maintain diversity [48]. In the same vein, personal-care products (cleansers, soaps, shampoos) might be personalized so that they are effective while minimizing biome damage over the course of long-term use.

3.1.6 CNS effects

Longer spaceflights have drawn more attention to the possibility of changes in brain structure and function resulting from the many physiological and psychological stressors. Chief among these is weightlessness (0*g*), which allows for redistribution of body fluids and changes the forces on the body and brain. While many of the effects that have been directly measured can be attributed to fluid/mechanical effects of this altered gravity, other stressors such as elevated CO_2 and isolation also undoubtedly play a role. Astronauts have long reported cognitive challenges in extended spaceflight, sometimes known as "space fog" or the "space stupids." This refers to a sense of mental slowing wherein normal routine tasks take longer and require more dedicated attention. This is perhaps not surprising given the many mental and cognitive stressors inherent to spaceflight, such as altered light-dark cycles and sleep disruptions, isolation and confinement, interpersonal challenges, weightlessness, radiation, high workload, and a demanding timeline. The reported subjective effects have not been seen in standard cognitive testing [54], but might be manifest in some observed inflight anomalies [55]. Any altered function could be the harbinger of later problems, and neuroanatomical and neurophysiological changes do accompany long-duration flights. There are three main findings that lead to cause for concern for longer flights: the brain shifts upward in the skull, ventricles enlarge, and some rearrangement of neural functions occurs (see review in Ref. [56]).

Cerebrospinal fluid (CSF) surrounds and cushions the brain and spinal cord, along with delivering nutrients and clearing waste from the central nervous system (CNS). CSF is created and stored in the ventricles, which are fluid-filled chambers in the brain. CSF is one of the fluids that shifts toward the head due to lack of a hydrostatic gradient. Indeed, imaging studies have shown an increase in ventricular volume [57] related to mission duration [58,59]. In some cases, these changes persist for months or more after return to earth [60,61]. An increase in CSF volume might lead to increased intracranial pressure (a form of hydrocephalus), which in the long run has the risk of causing neural damage.

Other imaging studies show the possibility of an increase in global white matter in the brain (white matter consists of myelinated neuronal tracts that connect brain areas), but this is not a universal finding across studies [60]. More refined imaging shows white-matter changes in brain areas that perform visual processing, with these changes larger in astronauts who exhibit more problems with postflight balance [62]. Changes in connectivity have also been found in regions related to vestibular processing [63,64] and primary motor cortex [62], which presumably reflect adaptation to the lack of tonic otolith information and to altered use of lower limbs. More detailed methods show structural changes in some white-matter tracts [65] related to sensorimotor function, some lasting for months after flight. On the whole, these results suggest that extended spaceflight increases white matter (neural connecting tracts) in the areas of the brain responsible for motor action and coordination, and that these changes are partly reversed upon returning to Earth [66]. Changes are also seen in the neural gray matter (the neural cell bodies that perform information processing in the brain); specifically, gray matter increases in areas of the brain responsible for sensory and motor function [61,67].

During extended spaceflight, there are changes that are properly adaptive [68] and changes that are maladaptive and can lead to dysfunction. The former reflect the ability of the brain to adjust to new environmental challenges. It is extremely important to maintain the brain's ability to make these appropriate adaptive changes, because if this ability deteriorates then astronauts will not be able to adjust to new planetary settings or overcome the neural changes that lead to deficits in sensorimotor function and in other capabilities. Spaceflight is demanding and crews need all the margin they can get, especially for cognitive function. Countermeasures that impact general health and performance might include not only artificial gravity and lower body negative pressure (AG and LBNP), but also recognition of the potential problem and provision of the capability to reschedule and rearrange cognitively demanding tasks as needed to accommodate maladaptive changes in crewmembers. To this end, it can also be useful to call on concepts of dual-state adaptation [69] and context-specific adaptation [16], as well as adaptive generalization [70]. These would be especially useful if making transitions between *g* levels, as for example between orbit and planetary surface, or between AG and free-floating in the spacecraft. Any mechanism that makes the process of adjustment to a change in the environment more automatic and reflexive will have the additional benefit of helping to maintain cognitive reserve: the more rapid the adaptation to a new setting, the less volitional effort and strategic thought needs to be devoted to maneuvering in that setting.

3.1.7 Kidney stones and a note on medical conditions

Medical issues arise during spaceflights, just as they do on Earth [71]. The point of this review is not to comprehensively detail all possible medical concerns, but to delineate a few that are of specific interest in space. One such concern is renal (kidney) stones. There is meager evidence for the occurrence of renal stones during spaceflight. One such episode occurred on the Soviet Salyut 7 space station in 1982 [72]. In the US program, there have been stones in astronauts after spaceflight at a rate that suggests a causal link [30]: the risk appears to increase with mission duration. While this risk is small, the consequences are potentially serious. There are circumstances consequent to spaceflight that contribute to this risk. Astronauts are typically slightly dehydrated, eat high protein diets, and lose bone calcium due to unloading of gravity. This leads to supersaturation of urine with calcium, which is conducive to the production of some types of stones. Mathematical modeling shows that the factors enumerated above contribute to the risk of developing renal stones [73], suggesting that better hydration and maintenance of bone integrity [74] can be useful countermeasures. Like many of the issues under discussion here, the role of interactions between physiological systems, or the effectiveness of partial *g* (artificial gravity), are largely unknown.

Again, we present this example medical concern, not because it is necessarily the greatest medical risk, but because its (possibly) increased prevalence could be due to the unique and specific stressors of extended spaceflight: reduced gravity, relative dehydration, altered diet, and changes in calcium metabolism resulting from loss of bone. This risk increases with mission duration due to prolonged exposure to the causal stressors and the physiologic responses, which may plateau but not fully abate. Further investigations should be undertaken to determine, based on first principles rather than just flight experience, what other medical conditions might have increased prevalence during extremely long spaceflights. A recent episode of deep vein thrombosis in flight [75] should serve as an impetus for such an effort since it shows that after more than 60 years of spaceflight experience we can still be surprised by novel physiological and medical phenomena. A program of extensive mathematical modeling would be of great benefit in predicting future concerns and designing proactive interventions [76].

3.2 Spacecraft-induced effects

A spacecraft's environmental control and life support system (ECLSS) is responsible for providing clean, breathable air at the appropriate temperature and pressure. The provision of that environment and management of air quality continues to evolve as new technologies are developed and tested. Selection and proper containment of materials used in the fabrication of the vehicle and its on-board supplies are important for controlling toxic exposures, as are monitoring and surveillance to alert the crew of toxic release. The human crew contributes carbon dioxide (CO_2) as a metabolic byproduct of respiration, which must be removed by the ECLSS. This can be a challenge, since scrubbing of CO_2 requires resources, including power, and hence spacecraft have historically had higher than normal levels of CO_2. While the effects of elevated CO_2 have long been appreciated in terrestrial settings, the impact throughout long-duration spaceflight—especially possible synergistic effects when combined with other stressors—is only beginning to be fully appreciated [77,78].

3.3 Cosmic dust, regolith, and environmental toxins

The composition of planetary bodies poses several unknown hazards to potential human explorers. Little can be said about these hazards until destinations are determined and studies made of similar bodies. However, evidence from what we know of the lunar regolith and Martian soil provides some clues. Most notably, in the case of Earth's moon, the regolith has several constituents that are known to be reactive or abrasive, and even toxic in high concentrations [79]. These include silicon dioxide, and iron and other oxides. Silicon dioxide on Earth can lead to silicosis, a life-threatening lung disease. Abrasive effects arise because the lunar surface is not subject to weather erosion as on Earth, or reactions with air, or decomposition through biological interactions. This can lead to health concerns through mechanical action, such as abrasive effects on the lungs, and skin rashes and abrasions [80].

We have limited experience with human exposure to, for example, lunar dust. Apollo astronaut Gene Cernan reported that "dust is probably one of our greatest inhibitors to a nominal operation on the moon. I think we can overcome other physiological or physical or mechanical problems, except dust" [79]. This dust is not only abrasive and possibly toxic, but is so fine that it infiltrates almost everything in sight, and can be suspended in the cabin atmosphere due to static charge and reduced gravity. This makes it easy to breath in, and the finer particles can make their way to deeper parts of the lung, with subsequent longer retention times [81]. While engineering solutions might mitigate exposure and resolve issues with visibility and suspension in the spacecraft cabin, the existing information suggests that this threat be taken seriously, and health monitoring for adverse effects be put into place.

Regolith toxicity per se is uncertain [82]. Many studies have been conducted, including those on simulated regolith. These studies have not produced definitive findings, but results seem to indicate the presence of a general dose-dependent cytotoxicity [83]. For example, a study of neuronal and lung cell cultures with simulated lunar soil showed cell death and DNA damage, which was not correlated with generation of reactive oxygen species in aqueous suspensions nor with the ability to create DNA lesions [84]. Martian soil also has problematic aspects such as an abundance of highly reactive perchlorates [85].

Obviously, these are just two examples, in the local vicinity of Earth, for which we have detailed (although inconclusive) information. There is even greater cause for concern in journeys that take people beyond our solar system. Crews should be prepared with a suite of mitigation strategies. Orbital or flyby phases of flight could provide useful data on atmospheric and surface-soil composition, so that the proper strategy could be put into place as needed. Cell or animal models, or organs-on-a-chip, might be deployed to the surface for "canary in the coalmine" monitoring, after which human health should be monitored for adverse effects during subsequent direct interactions with the planet.

3.4 Multisystem effects

Given that the human body is a system of systems, it is reasonable to treat all body functions and outcomes as the product of multisystem interactions, meaning that each action brings with it both opportunities and risks. While managing a complex endeavor like human spaceflight it is reasonable to articulate the known problems that drive requirements and factors of safety, the known unknowns that represent gaps in knowledge or technology and drive targeted research and development efforts, and the open space of unknown unknowns that drive surveillance efforts and nontargeted research and data analysis. The last category, the unknown unknowns, is of most concern, since within this category lie the "black swan" events that arise from the unique circumstances of spaceflight and will continue to appear with subsequent generations of more ambitious flights. (See Section 4 below for more on this topic.) One of the ways in which these conditions come to light is by what is known in the medical community as a sentinel event, which is often identified only in hindsight.

Spaceflight-associated neuro-ocular syndrome (SANS) is one such nontargeted finding that came to the attention of scientists through a small set of sentinel cases. It is an instructive example of the body's multivariate response to the multiple stressors of spaceflight and its attempts to adapt using several compensatory mechanisms. SANS [86] is most commonly manifest as changes in visual acuity in some long-duration astronauts, and is also associated with a spectrum of findings such as the presence of retinal folds and increased intraocular pressure. Designing countermeasures against SANS is made difficult by the multiplicity of spaceflight stressors, multiple medical findings, and variability of incidence across flight crews.

An underlying mechanism appears to be the head-ward shifting of fluids in space, for which the body has adequate compensatory mechanisms in the short term that are pushed beyond their limits on longer missions (i.e., missions of several months or more). CSF and the brain shift "upward" (head-ward) during flight, crowding the top of the skull and compressing important veins that allow venous blood outflow. CSF propagates down the sheath of the optic nerve, stretching it and impinging on the optic cup where it meets the back of the eye, causing tissue disruption. Fluid pressures exceed the tissue tolerances, causing edema in the delicate choroid and retina, which begin to fold on themselves, creating wrinkles that can affect vision. The ocular globe (eyeball) may flatten due to the CSF pressure, which affects vision to varying degrees. Obstruction in venous outflow exists not just at the top of the brain but is also functionally present in the neck and has been associated with jugular thrombosis (see Section 3.1.7). SANS displays large interindividual variability and a large spectrum of presentations between various crewmembers.

A major concern regarding SANS is that it might be the harbinger of more widespread neural effects (see Section 3.1.6), although this has yet to be definitively determined [87].

3.5 Psychology and behavioral health

Arguably, behavioral concerns (those related to psychology and psychiatry) will be the most critical for long-term missions. These issues, which might include such things as depression, loss of motivation, and interpersonal conflicts, can be directly attributed to the stressors of spaceflight (distance from Earth, duration of mission, remoteness, isolation). They are all the more problematic because it's not clear what, if any, countermeasures will be feasible or effective.

Kanas [88] delineates five major psychological factors inherent in a trip to Mars, all of which would be intensified dramatically on much longer missions. The first of these is the "disappearing Earth phenomenon": the sense of isolation and loneliness that arises as Earth recedes from view. This might be thought of as the converse of the overview effect: the very strong emotional connection to Earth experienced by many astronauts on seeing it in its entirety from space. It is hard to imagine how the inability to see Earth would impact those on such a mission. Virtual reality images of Earth, or a high-power telescope to provide a view for as long as possible, could provide some relief. On the other hand, it is possible that these countermeasures might only serve to intensify the sense of homesickness and the longing to return.

The second factor is time-in-mission, which will of course be significantly longer than that of any human spaceflights undertaken so far. The factor of time is of particular concern because, to date, spaceflight missions do not entail the so-called third-quarter effect: a decrease in mood and motivation that occurs when the halfway point of a long mission has been reached but, because the mission is only half over, the end is not yet in sight. This phenomenon has been observed in a variety of challenging and stressful missions of widely varying durations, and seems to occur as long as the participants know the duration of the mission. Why spaceflight missions to date do not produce this phenomenon is not clear [89], but the absence may be due to salutatory effects of space travel [90], which may override conventional psychological cycles. These positive factors are many: the uniqueness of being in space, the previously mentioned overview effect (at least when in proximity to Earth), the sense of professional accomplishment, the satisfaction of challenging and meaningful work, and even the freedom of weightlessness. It might also be related to the extreme training and levels of dedication of professional astronauts.

A special challenge for longer missions would be the need to accommodate changes in crew interactions over time, and to be aware of withdrawal and formation of subgroups. Of special note is that flight crews often take out their frustrations on mission control, as a sort of shared enemy that can unite the crewmembers. This opportunity will not be available on longer flights far from Earth, when mission control can no longer take the brunt of the crew's displeasure. The loss of this coping mechanism might create increased friction among the crew itself. Some of these concerns might be alleviated through proper crew selection and self-monitoring of behavior.

Also attendant to longer missions (the fourth major factor) is increasing autonomy of the crew: independence from Earth and the need to make and act on decisions without benefit of support from the ground. This would call for a change in leadership roles over time, as a hierarchical structure that is needed in urgent and acute situations gives way to a more democratic community structure that is conducive to crew engagement and long-term performance (also referred to as task versus support leadership). Finally, distance from Earth brings with it increased communication delay, which again calls for increased autonomy. Communication waystations, placed at intermediate points, could provide crews with communication with others with a shorter delay, albeit it would be communication with a fellow crew and not Earth itself.

Some potential countermeasures for the mental health challenges that may result in decreased individual and crew performance have been mentioned and one in particular merits further explanation: meaningful work. It might seem odd to consider that a spaceflight crew would lack for meaningful work. Nevertheless, when one considers the vast distances involved, and hence the vast durations of the missions, it should be apparent that "just being there" would lose its motivational luster after a time. (Even the Apollo astronauts would sometimes comment after the excitement of the lunar phase of their missions about the relative tedium experienced on the trip back to Earth—a trip that took all of 3 days.) Providing the crew with the ability to perform work that is meaningful and intellectually stimulating during slow operational phases of the mission—the otherwise "down times" of a long journey—would be extremely valuable. This might involve scientific work such as observations of planets, asteroids, and other celestial objects as they come (relatively) near. It might also involve continual monitoring (and perhaps occasional experimental interventions) relevant to human health and performance. This would not only provide useful medical information for the mission at hand but could serve to improve the experience for future explorers. Missions should be planned with the requisite facilities to carry out these types of investigations.

Another common countermeasure idea is to provide access to natural stimuli of various kinds (green plants, fresh air) to counteract the sensory paucity experienced by flight crews in the relatively sterile cabin environment and to provide a sense of the comforts of home and of connection to nature [91,92]. This has been shown to reduce stress in laboratory conditions [93], and could be part of a larger set of countermeasures for psychological stress [94].

3.6 Special topic: Reproduction and child development

Consideration of spaceflights of sufficient duration to enable settlement and colonization inevitably leads to questions of reproduction and development, as this would be by definition a multigenerational effort. Very little is known about conception, development, and birth in space, and the issue is fraught with ethical dilemmas [95]. A few experiments with simple animals have been performed, which generally show that it is possible to conceive and give birth through artificial insemination, although this is not without complications. We consider it unlikely that surreptitious experiments with mammals have been performed in any systematic way during spaceflight so far; the requirements for successful reproduction are too complex. The animal studies to date indicate viability of sperm and the ability for fertilization and even birth, at least for simple animals in limited circumstances. Mouse spermatozoa stored on the ISS for 9 months suffered some DNA damage, but the offspring and birthrate associated with these spermatozoa were not impaired [96]. On the other hand, using a clinostat (simulated $0g$ environment) and the same animal model, birthrate was lower and development was slower [97], although fertilization occurred normally. Another clinostat study also demonstrated detrimental effects on the initial stage of development in mouse preantral follicles that had been removed from the organism [98].

Conception, birth, and development would be greatly aided, and some ethical issues avoided, if performed in a gravitational setting (although the gravity level needed for successful conception and development is unknown). This would greatly reduce many of the unknown factors that could inhibit normal growth and development. This might be accomplished with artificial gravity on the spacecraft, or through extended stays on planets and moons expressly for the purpose of procreating the next generation of explorers (who might then venture into space once sufficiently developed). Even with this in mind, radiation presents a likely and unstudied problem for proper development.

3.7 Special topic: Human modification

It is tempting to consider various options to "engineer" the human spaceflight participants themselves in order to make them resistant to one or more of the stressors of extended flight. This prospect could be seen as a natural extension of the rigorous process of screening and selection that is currently in place for professional astronauts—an approach that has been very successful for the major space programs. However, even though current astronauts are thoroughly screened for medical and psychological issues, they still retain conventional human physiology. We have decades of experience in characterizing the response of such people to stressful situations, including spaceflight. Although physiological limits may be pushed and occasionally exceeded, we understand a lot about these limits and the breakdown processes. Artificial human modification would change this dramatically, by placing people with novel and (for the foreseeable future) insufficiently studied modifications into a challenging environment, raising the distinct possibility of unintended consequences as the number of unknowns is increased rather than decreased. This also raises attendant ethical issues, as for instance whether a new species of space-faring human might be created, which would no longer be viable in the previously normal setting of Earth. Further complicating this general approach is the fundamental integrative nature of human biology, where systems are coupled and there is rarely a one-to-one correspondence between a specific physiological subsystem, gene, or metabolic pathway and a spaceflight-relevant maladaptive effect. It is hence unrealistic to assume that targeting just one subsystem will be sufficient, exacerbating the problem of unintended consequences due to unforeseen interactions as well as the unknown effects of the other stressors of spaceflight.

Nevertheless, consideration has been given to some ways in which the human body might be changed to better tolerate extended spaceflight [99–101]. New technologies such as CRISPR (clustered regularly interspaced short palindromic repeats) raise the possibility of making genetic alterations; these could be targeted to gene-expression pathways and metabolic processes for some of the physiological processes that undergo the most dramatic maladaptive changes during prolonged time in space. These would entail permanent changes in an individual's genetics; in the future it might be possible to modify an individual's germline to make genetic changes heritable, enabling the launch of multigenerational expeditions. One of the most appealing targets for genetic manipulation is the set of genes that confers resistance to radiation damage. Several candidates have been proposed [102], including upregulation of endogenous DNA repair, suppression of p53 and apoptosis, and enhancement of p53 tumor suppressor protein. Since current countermeasures for preventing bone loss have been only partially effective, another appealing target is the set of genes involved in remodeling of bone, such as BMP, LRP5, LRP6 [100]; another option is to target proteins relevant to maintaining bone integrity, namely osteoprotegerin, myostatin, activin, and sclerostin.

Perhaps more practical and more feasible in the short term are direct modifications to various sense organs. For example, it might be possible to implant prosthetics that enable vision over a broader range of wavelengths than those that make up the spectrum of visible light for humans [99,103]; this might enable direct perception of radiation sources, or perhaps grant the benefit of "night vision" to individuals on expeditions far from natural sunlight and starlight. Already under development are vestibular prostheses that might provide the "missing" gravity signal during weightlessness, thereby helping to maintain the central neural processes that incorporate gravity and enhance the ability to function once back in a gravity setting (planetary surface).

It is intriguing to consider other possible means of aiding and augmenting human capabilities. Brain-computer interfaces might make the human-machine symbiosis more natural and intuitive, for example. Exoskeletons could aid in heavy lifting and construction work on a planetary surface, while reducing risk of injury to (possibly deficient) muscles and bones. We reserve further comment on these types of interventions, which are more properly regarded as external tools as opposed to actual body modifications (although the line between these two is not as clear as one might think [104]).

3.8 Radiation effects and mitigation

Radiation is a major unknown for extended spaceflights. Deep-space radiation consists of galactic cosmic rays (GCR) and solar particles events (SPE). GCR is composed of high-energy protons and heavy ions (nuclei) moving at relativistic speeds. The broad range of particle types (elements) and energies make it difficult to characterize the biological response. SPEs are associated with solar flares, in which the sun ejects high-energy particles including protons and alpha particles. Exposure to space radiation has the risk of increasing the long-term risk of cancer, CNS effects, and tissue degeneration (especially in the cardiovascular system). There is also the risk of more immediate effects: acute radiation syndrome [105]. It is difficult and often controversial to determine the exact magnitude of the radiation risk, but it is prudent to consider it on a continuum that must be taken seriously and adjusted as evidence is accumulated. This is especially notable in the case of the stream of relativistic heavy ions of GCR, with which we have very limited biological experience, and which are capable of causing extensive cellular damage.

A great deal of research has been conducted on cells and animals, and the results from these studies raise concerns regarding radiation (for a review, see Ref. [106]). There are, however, difficulties in translating these studies into information that is useful in characterizing the risk that radiation poses to humans [107]. Much of the difficulty comes from the fact that there are no good terrestrial analogs of space radiation and its effects on humans. Radiation for medical therapy and imaging has the wrong spectrum of particles and energies, as does radiation from atomic bombs. There are terrestrial laboratories that generate some aspects of space radiation, but until recently these laboratories were limited to exposing the specimen under study to only one species of ion and energy at a time, whereas actual GCR is an amalgamation of a wide spectrum of particles and energies. In addition, the dosing rate from these facilities is typically much different than for real space radiation, so that experimental results can be obtained in a reasonable time frame. Additionally, since animals are not humans, and since the study of cells in vitro may eliminate the action of some repair mechanisms that could mitigate damage from radiation exposure, the results of these experiments might not adequately capture the experience of an astronaut's encounter with space radiation.

Additional problems are involved with attempting to draw conclusions from actual human spaceflight. One is the healthy worker effect, which skews epidemiological studies that compare astronaut morbidity and mortality to that of the general population: astronauts are generally healthier and live healthier lifestyles than the average person. Although astronauts experience increased radiation relative to Earth's surface on the International Space Station, it is not a good analog for radiation studies as the dose is lower than in deep space and the Earth's magnetic field traps and constrains some radiation (Van Allen radiation belt). Finally, interactions of radiation with altered gravity [108] are the reality in space, making it difficult to parse which effects are due to each factor.

Under these circumstances, what can be said about the effect that space radiation might have on interstellar explorers? An increased risk in astronauts for some types of cataracts [109] shows that space radiation does indeed have adverse effects on humans. One study found evidence for an increase in cardiovascular mortality in Apollo lunar (but not solely Earth-orbit) astronauts [110]. However, these results were later called into question for methodological reasons [111], again pointing up the difficulty of getting clear answers in this field. Similarly, a separate study found no excess mortality from radiation in early astronauts (a conclusion complicated by the healthy worker effect) [112]. Overall, there appears to be little if any additional risk of cancer or cardiovascular disease in astronauts, based on missions to date [113].

Given these uncertainties, it is prudent to consider engineering, operational, and biological countermeasures to the effects of radiation. It is difficult to construct shielding that can block GCRs, as some heavy particles can interact with some shielding materials to produce secondary radiation—slower particles on the interior of the spacecraft that can then spend more time traversing the body, causing more biological damage as a result. Effective shielding for SPEs is more feasible, but these events are rare and unpredictable. For interstellar explorers, a good strategy will be to reach a planetary surface and burrow in, using the planet's own material as a shield. Besides shielding, many of the same countermeasures that are relevant to the other stressors and physiological challenges of spaceflight—exercise, diet, and good immune and microbiome function—also provide protection against damage due to radiation. A common path for many radiation effects is oxidative damage or inflammation, for which there are pharmaceuticals or nutritional supplements. Various radioprotective agents are also available, and are commonly employed alongside terrestrial medical radiation [114]. Crews should also be prepared to deal with the possible adverse effects of space radiation with more extreme or unproven countermeasures, such as stem cells, as they may be a last resort in the event of unexpected high radiation or significant medical consequences. Care will be needed to adequately preserve and protect these cells while they make the journey with the crew. In the long run, it might be possible to select for people who have natural (relative) radiation immunity, or to engage in a program of preflight radiation habituation [115].

In the final analysis, missions in the reasonable future may result in astronauts having an increase in lifetime risk of cancer. This might be considered a problem to be addressed by terrestrial medicine, since the cancer and its treatment would likely occur after return to Earth. Referring to this problem as "terrestrial" is not to be cavalier about astronaut health; rather, it acknowledges that increased health risks are already inherent to spaceflight, and radiation risks should be seen as part of the spectrum of risks. This calculus is vastly different for those who will undertake interstellar missions, for which any long-term health consequences will take place away from Earth, and hence for which the terrestrial standard of care might not be available.

4 Mitigation possibilities

4.1 Nutrition

There is a vast amount of knowledge on nutrition for high-performing individuals in extreme environments. As a general rule, the standard requirements based on terrestrial populations also apply to crews in space, and have been reviewed extensively (e.g., Ref. [116]). In this section we focus on three aspects of nutrition that are especially relevant to extended spaceflights, and are extremely relevant to interstellar voyages [117].

1. **Foods that address the specific deconditioning aspects of spaceflight.** It is fairly straightforward (although not trivial) to understand the basic principles of nutrition and their importance for human health. The stressors of spaceflight exacerbate some of the normal health concerns that are addressed by adequate nutrition. For example, the final common pathway for several stressors involves inflammation and/or oxidative damage via reactive oxygen species (ROS). Thus, foods that serve as antioxidants and antiinflammatories—which are already recommended for terrestrial diets—will have increased importance in the diets of astronauts. Many of these foods also serve as radioprotectants, and can benefit cardiovascular function and bone and muscle retention. The foods in question include fresh fruits and vegetables as well as foods high in omega-3 fatty acids and polyphenols. Vitamin D deficiency is a common problem among spacefaring crews, due in part to the lack of natural sunlight (a deficiency that can also occur on Earth during the extensive training period that precedes a mission); although supplementation is typically not as effective as natural food sources, vitamin D supplementation has effectively addressed this deficiency in flight crews. Conversely, it will be wise to minimize foods that produce adverse effects, such as sugar, which when eaten in excess can lead to neuroinflammation and cognitive disorders.
2. **Provision of foods that enhance health and performance.** In the concept of functional foods, diets are designed to have beneficial effects beyond the maintenance of essential nutritional needs. The intent is to maintain high performance and avoid degradation, as well as to aid repair mechanisms. Again, many of the same principles as above are applicable, but some aspects could be emphasized. As an example, a ketogenic diet has been proposed for enhanced physical performance [118], although there are concerns with the diet's high protein and resulting impact on calcium excretion [119].
3. **Logistical challenges of providing basic and enhanced nutrition within the constraints of spaceflight.** Given the limited storage life of many food items, the uncertain impacts of extended radiation on food, and the importance of palatability for psychological reasons, it is crucial to

supply astronauts with a varied, nutritious, and enjoyable diet. While crews in early spaceflights exhibited caloric insufficiency, this difficulty has been remedied by providing more interesting and varied foods in flight. Head-ward fluid shift blunts the sense of taste, but spicier foods help overcome this. Supplementation to supply some essential nutrients is possible and can be effective in some cases, but the bioavailability is not as good as for nutrients in natural form. Providing the proper food will be a challenge, especially in the case of fresh fruits and vegetables; growing plants in space requires valuable resources but has many advantages [120].

4.2 Exercise

Once crewmembers adapt to the weightlessness of spaceflight, they function and navigate well. While films from Apollo surface operations often captured astronauts falling, it is challenging to identify the role played by various factors such as sensorimotor deficit, EVA suit configuration (high center of mass), and EVA tool function in these falls, making it difficult to determine which factors play a dominant role. Experience during weightless operations indicates that crewmembers adapt well to the new environment within the first few weeks, successfully accomplishing tasks that require fine motor control (e.g., performing cell culture and dissection experiments or telerobotic operations) as well as activities that require gross motor activity (e.g., EVA or unloading cargo). Measures to ease the effects of the adaptation period consist of medications to manage symptoms, and physical exercise such as walking and running on a treadmill and resistive training. Physical exercise is imperative to maintain physical, physiological, and psychological balance throughout space flights, including landing and immediately postlanding, given that these can be some of the most physically demanding parts of a mission.

The critical importance of exercise as a countermeasure against the deconditioning effects of spaceflight on various physiological systems has been mentioned in corresponding places throughout this text. Exercise also serves as a psychological countermeasure for many astronauts, and this multivariate aspect should be seriously considered when making accommodation for inflight exercise. As with nutrition, spaceflight exercise generally follows terrestrial guidelines. Special attention must be paid to providing an appropriate counterforce with a harness system when using a treadmill or bicycle device, and proper technique must be used to mitigate the possibility of injury.

While exercise has many benefits [121], it also presents logistical challenges. Time must be taken from other activities and devoted to exercise, the equipment must be cleaned and maintained, improper mounting can lead to vibrations conducted throughout the spacecraft structure, and exercise increases interior CO_2 levels, which then must be mitigated. It may be wise to consider alternative forms of exercise to supplement traditional aerobic and resistive exercise regimens or to replace them in the event of equipment breakdowns. Supplementing traditional exercise might be especially suitable for extended spaceflights. Some of these alternatives include exercise during blood-flow restriction [122], exercise during muscle heating or cooling [123], and isometric exercises [124]. It is unclear to what extent exercise will be needed on a planetary surface, or in the presence of artificial gravity, as both of these provide an ambient force that must be overcome by muscles, bones, and the cardiovascular system.

4.3 Artificial gravity (AG)

Artificial gravity (AG) is often proposed as a means to mitigate a number of gravity-related maladaptive processes, including neurovestibular, musculoskeletal, and cardiovascular issues. AG is a multisystem countermeasure [125,126]. It can be implemented in different manners, typically involving the generation of centripetal/centrifugal force by rotating all or part of the spacecraft. Rotation of the entire vehicle provides AG throughout the spacecraft but can be an engineering and operational challenge. Rotation of a subset of the station would allow for transitions between AG and $0g$, and rotation of just the sleeping compartment or beds might suffice if it is determined that overnight exposure is sufficient. The great unknown in AG is the "dosing" needed to ameliorate processes of physiological deconditioning: what g level and for how long [127,128]. This unknown is related to the question of how much a low-g planetary gravity level would mitigate this deconditioning, and if that mitigation would be sufficient to later permit a return to the $1g$ of Earth.

Another way to present a gravity surrogate is to provide a constant linear acceleration to the spacecraft. This linear acceleration would impart a reaction force on the occupants indistinguishable from that of an actual gravity field. This solution has advantages over the rotating-structure solution. Under rotation, head movements in certain directions lead to cross-coupled stimulation of the angular-rotation sensors of the human vestibular system (the semicircular canals). Cross-coupled stimulation can be debilitating, and may lead to nausea, vomiting, and similar symptoms, including lethargy (sopite). Studies have shown that humans can adapt to this rotating environment, but the adaptation can be specific to the direction of rotation (clockwise or counterclockwise), and the resulting forces on the body depend on the person's distance from the center of rotation (centripetal force is inversely proportional to the radius or distance from center) [129]. These effects can be troubling and would contribute to the adaptive burden of long-duration flight. Prolonged linear acceleration avoids these issues by avoiding rotation. The propulsion requirements to maintain even a low level of constant acceleration are, however, likely to be prohibitive.

There is a potential drawback to prolonged artificial gravity of any sort. On an extended flight—certainly of the duration of anything needed for interstellar travel—crews will face unprecedented levels of stress from many factors discussed in this chapter. It is possible that one of the few things that would bring pleasure to these explorers is the experience of weightlessness, as they might have few other sensory pleasures to help sustain morale. Although AG might provide many benefits, it might be unwise to totally remove weightlessness as a constant reminder of the uniqueness of the venture they are undertaking, in order to combat the inevitable tedium that might otherwise prevail.

4.4 Lower body negative pressure (LBNP)

Lower body negative pressure (LBNP) has some appeal as an alternative to, or augmentation of, artificial gravity. LBNP is generated by a device that forms a partial vacuum around the lower body, pulling bodily fluids from the upper to the lower body, simulating some of the effects of living in a gravity field [130]. This technology is cumbersome; however, recent developments show promise for making a mobile LBNP suit [131] or sleeping bag (where nightly application may be beneficial in reducing some of the effects of fluid shift [132]). LBNP has a long history in terrestrial medicine as well as in spaceflight [133]. It has not been tested systematically in spaceflight except in a few cases. It appears to be in regular use near the end of extended flights, to prepare for return to Earth, in the Russian (and previously Soviet) space program [133]. It has been shown to partially alleviate inflight increases in intraocular pressure (a possible contributor to SANS), but has not mitigated the rate of choroidal thickening (a related change in retinal structure) [134].

Regular exposure to LBNP might be an effective countermeasure for at least some of the adverse effects of reduced gravity. However, since gravity itself is not restored, vestibular alterations would not be countered by LBNP, except for those vestibular alterations that might result from fluid shift. LBNP might be very useful in countering the effects of fluid shift, including vision decrements associated with SANS, some changes in brain structure and function, and cardiovascular deconditioning. LBNP might also provide musculoskeletal and proprioceptive benefit by pulling the subject toward the bottom supporting surface (the effective "floor"), thereby providing ground-reaction forces.

4.5 Touch

Another countermeasure against the negative psychological effects of spaceflight is based on the role of human touch. The calming effects of touch are widely, if anecdotally, recognized. Prolonged lack of touch and tactile stimulation can have adverse consequences for development and psychological health [135]. The stress-reducing effects are well-known [136]. Unfortunately, human spaceflight is not conducive to even normal daily interactions that would involve touching others. On the relatively short missions to date, there are serious concerns about adverse effects on morale, stemming from the possibility of misinterpreting another's touch. Cause for concern increases as crews become more diverse, in terms of both ethnicity and gender. Different cultures can have different views on physical touch and personal space [137] that contribute to the complexity of this issue. Recent research has elucidated the importance of touch by exploring its strong neural underpinnings: mechanosensitive C fibers woven throughout skin carry signals related to touch, and "represent the neurobiological substrate for the affective and rewarding properties of touch" [138].

Given the established importance of touch, it is fortunate that alternative mechanisms are available to promote the positive aspects arising from this sensory modality despite the aforementioned practical difficulties. For example, even gentle stroking from an inanimate object can provide some of the same benefits as regular human touch by triggering the neural activation patterns from which these benefits arise [139]. This form of implementation might be practical in the confines of a spaceflight. On the other hand, when considering longer flights with the intent to colonize and settle, it is likely the crews will be selected with procreation in mind, in which case touch per se becomes a moot issue. Nevertheless, the underlying lesson regarding importance of touch as a means to maintain and further human interaction and to reduce stress retains its force.

4.6 Other approaches

Several more speculative countermeasure approaches are under investigation. These countermeasures have been even less thoroughly tested in extended spaceflight than the ones discussed so far, such that the prospects for unintended consequences and interactions are unknown. Nevertheless, they may prove useful in specific situations where no alternative exists or in emergency situations as a last resort. The following countermeasures fall into this highly speculative category:

- **Slowed metabolism through induced torpor** [140,141]. By lowering core body temperature, metabolic processes can be dramatically slowed, with attendant reductions in the need for consumables such as food, water, and oxygen. Slowing metabolism might halt ongoing processes of degradation, and might mitigate radiation damage [142]. There are also obvious benefits for psychological and interpersonal issues. Successful use would require continuous monitoring of vital functions, management of waste products, and a demonstrated ability to reverse the torpor state with

minimal adverse long-term effects. The entire crew would not have to undergo torpor induction all at once, and torpor could be induced in stages or on a rotating schedule over various lengths of time.
- **Stem cells** show promise for repairing or mitigating damage caused by radiation and other spaceflight stressors [143,144].
- **Induced mild hypoxia** is another way to reduce metabolism, by altering mitochondrial activity and thereby reducing reactive oxygen species, a major source of cell damage [145].

5 The bigger picture

What's missing in this discussion so far is a systems approach. Without a larger overarching view, we run the risk of designing a constrained ecosystem (the spacecraft and habitat) without understanding how the parts interact (dependencies, conflict, adaptation, habituation, microscale evolution). We discuss several aspects of a systems approach here.

The presentation in this chapter, to this point, has been mostly discipline-specific, with some points of interaction noted. This largely follows the current NASA approach. Drawing on the main spaceflight hazards enumerated by NASA's hazards framework, the major risks to human health and performance are discerned, and research is organized to mitigate those risks with countermeasures. This risk-based approach provides rigor, but also has a serious shortcoming: only those risks that have been previously identified are addressed. The list of known concerns is going to get longer with more extensive missions, and we should be alert for the emergence of new risks and issues that might impact health and performance. As part of this plan, a system of continuous monitoring and assessment should be implemented: every flight should be treated as an experiment and as an opportunity to add to our knowledge base in a systematic manner. Monitoring will also facilitate the design and implementation of personalized mitigations and countermeasures.

The approach of only addressing those risks that have already been identified must be augmented with an approach based on fundamental understanding of the physics of spaceflight and the physiology of the human body [76,146]. This would permit reasonable predictions of problems that could occur in longer, farther flights, which have therefore not yet been observed. It could also identify and accommodate cases where biology might limit the number and extent of adverse outcomes. The risk-based approach might identify the key domains of concern, while physiology and biochemistry might further refine them, and personalized medicine approaches may inform the boundary cases.

As noted, the presentation heretofore has been mostly discipline-specific. We can do a better job, in many respects, if disciplinary interactions are explicitly considered. "Body parts will not travel on exploration missions. Instead, the individual space traveler's body must be viewed realistically, with all parts connected and fully interacting" [146]. On Earth, we have the luxury of treating complex medical issues by way of one specialty at a time, addressing interactions and complications as they arise, and depending on referral to a series of specialists and specialized-care facilities. In space, this model is not viable, for many reasons: (1) multiple adaptive (and maladaptive) processes occur simultaneously, and current clinical measures are inadequate to interpret how these processes might influence a medical emergency or concern; (2) there will be neither adequate volume nor sufficient power to accommodate unique specialty testing equipment on a spacecraft; (3) side effects and complications raise the stakes exponentially and will not be accommodated by a resource-limited medical system. Interactions between body systems must be explicitly addressed, for the sake of better medical outcomes and more nearly optimal crew performance, which are critical for mission success.

In fact, it is the needs of the mission that must take priority over any individual crewmember's health. Health and performance must be considered within the context of the larger mission. Health exists on a continuum and can be measured and compared to the performance and functionality that is needed to carry out the mission. This is not to be cavalier with human lives, but to recognize that spaceflight missions for the foreseeable future will call on highly trained and motivated crews, living under environmental extremes, for long periods of time, with particular goals (exploration, colonization, settlement) driving their activities. In the long term, it should be expected that medical care will approach, or even (by taking these considerations into account) exceed the best available on Earth. In the interim, during the early phases of interplanetary/interstellar voyaging, realistic consideration must be given to constraints on resources and crew expertise. If the goal is to maintain health and wellbeing at all conceivable costs, then it is better not to undertake the journey in the first place.

The systems approach can also, if properly implemented, allow for early detection of disease processes through monitoring of key parameters [147–149]. Detection could encompass not only conventional medical measures, but also parameters that are easy to monitor continuously and unobtrusively, such as time to perform (and errors in performing) routine mission tasks. Such data could provide indications of impaired function as a manifestation of an underlying and developing medical problem, which could trigger more focused measurements and more conventional diagnostics as needed. Another aspect of the systems approach is the ability to provide personalized medical care and countermeasure interventions [99]. Close monitoring of each individual crewmember can provide longitudinal information from which a decision can be made about what interventions (medications, nutrients, exercise, rest) are most effective and have the strongest effects on individual physiology and psychology [150]. Even before flight, establishing a comprehensive understanding and record of an individual's baseline health is extremely important to inform risk assessment and risk acceptance processes during flight.

We might also consider when it is best—or at least acceptable—to let an inflight physiological process take its course, even if it creates a physical or medical impairment that would preclude a return to the previous normal life [101]. The overall goal should not necessarily be to prevent all physiological change. While some adverse consequences of adaptation, and some forms of maladaptation, may warrant mitigation, other forms of adaptation to the space environment are beneficial and should be permitted or even encouraged. For example, if it is known with certainty that some portion of the crew will settle (to stay) on a planet with very low gravity, then it may be acceptable to allow bone loss to proceed in those individuals without countermeasures—countermeasures that consume resources that might be better used by the rest of the crew. This decision could prevent that crew segment from ever again living on a planet with higher gravity. This is not a decision to be made lightly, but it is a possibility present when thinking in terms of a systems approach that considers resource constraints in addition to mission goals and individual health and welfare. Decisions like these are part of the larger balance that must be considered between trying to maintain an Earth-like environment versus helping people break the connection to home that might lead to chronic stress and homesickness.

This systems-integration approach should extend to the design of the spacecraft/habitat and the design of the mission. The need for an individualized and personalized approach to medicine and countermeasure interventions has been noted: individual crewmembers bring unique combinations of genetics, environmental exposures, and lifestyle factors that contribute to medical risks and could lead to impaired function that might affect the mission. However, most decisions about vehicle and mission design are made long before specific crewmembers are assigned to a flight, a situation that is likely to persist for a long time. These design considerations encompass medical and countermeasure apparatus (diagnostic and exercise equipment, LBNP, etc.). The likelihood then arises of a mismatch between the needs of an individual and the capabilities provided on board the spacecraft. To date, this has been resolved by using the best available data from populations of previous astronauts on similar missions, but both the demands of more ambitious missions and the desire to develop more refined interventions will require a personalized approach. The vehicle and mission can mitigate risk by accommodating specific medical and countermeasure hardware, custom protocols for existing hardware, and personalized pharmaceuticals in custom assemblies. This is a form of human-system integration (HSI) and requires significant communication, evaluation of data, and integration by many specialists from the medical, engineering, and risk-management communities.

The overarching need for these missions—and the goal of a systems approach—is to maintain resilience, so that health and performance can be maintained at a level that is appropriate to the needs of the mission, even in the face of unexpected perturbations and anomalies [151]. It might be useful to develop a quantitative approach to assess acceptable risk with various parameters, including interactions and mission impacts [152]. The definition of successful outcome has a direct bearing on the risk that is acceptable for a given class of flights, a determination that will change as flights become more ambitious in terms of duration and distance, and will be different for early versus later settlers. It is, after all, these unknown unknowns that are part of the reason for undertaking the interstellar journey in the first place.

References

[1] D. Francisco, E. Romero, NASA's human system risk assessment process, in: NASA Human Research Program Investigators Workshop, Galveston TX, Feb 8-11, 2016.

[2] A.E. Nicogossian, C.L. Huntoon, S.L. Pool, Space Physiology and Medicine, Lea & Febiger, Philadelphia, 1989.

[3] J. Carriot, I. Mackrous, K.E. Cullen, Challenges to the vestibular system in space: how the brain responds and adapts to microgravity, Front. Neural Circuits 15 (2021).

[4] M. Heer, W.H. Paloski, Space motion sickness: incidence, etiology, and countermeasures, Auton. Neurosci. 129 (2006) 77–79.

[5] J.R. Lackner, P. DiZio, Space motion sickness, Exp. Brain Res. 175 (2006) 377–399.

[6] L.R. Harris, M. Jenkin, H. Jenkin, J.E. Zacher, R.T. Dyde, The effect of long-term exposure to microgravity on the perception of upright, npj Microgravity 3 (2017) 1–9.

[7] M. Reschke, I.B. Kozlovskaya, I.S. Kofman, E.S. Tomilovskaya, J.M. Cerisano, M.J. Rosenberg, J.J. Bloomberg, M.B. Stenger, S.M. Lee, S.S. Laurie, I.V. Rukavishnikov, Field test: results from the one year mission, in: NASA Human Research Program Investigators Workshop, Galveston TX, Jan 23–26, 2017.

[8] C.A. Miller, I.S. Kofman, R.R. Brady, T.R. May-Phillips, C.D. Batson, E.L. Lawrence, L.C. Taylor, B.T. Peters, A.P. Mulavara, A.H. Feiveson, Functional task and balance performance in bed rest subjects and astronauts, Aerosp. Med. Hum. Perform. 89 (2018) 805–815.

[9] A.P. Mulavara, B.T. Peters, C.A. Miller, I.S. Kofman, M.F. Reschke, L.C. Taylor, E.L. Lawrence, S.J. Wood, S.S. Laurie, S.M. Lee, Physiological and functional alterations after spaceflight and bed rest, Med. Sci. Sports Exerc. 50 (2018) 1961.

[10] S.T. Moore, V. Dilda, T.R. Morris, D.A. Yungher, H.G. MacDougall, S.J. Wood, Long-duration spaceflight adversely affects post-landing operator proficiency, Sci. Rep. 9 (2019) 1–14.
[11] G. Clément, S.J. Wood, Eye movements and motion perception during off-vertical axis rotation after spaceflight, J. Vestib. Res. 23 (2013) 13–22.
[12] M.F. Reschke, E.F. Good, G.R. Clément, Neurovestibular symptoms in astronauts immediately after space shuttle and international space station missions, OTO Open 1 (2017). 2473974X17738767.
[13] S.G. Diamond, C.H. Markham, Prediction of space motion sickness susceptibility by disconjugate eye torsion in parabolic flight, Aviat. Space Environ. Med. 62 (1991) 201–205.
[14] M.F. Reschke, Statistical prediction of space motion sickness, in: G.H. Crampton (Ed.), Motion and Space Sickness, CRC Press, Boca Raton, FL, 1990, pp. 263–316.
[15] M. Shelhamer, K. Beaton, Pre-flight sensorimotor adaptation protocols for suborbital flight, J. Vestib. Res. 22 (2012) 139–144.
[16] M. Shelhamer, D.S. Zee, Context-specific adaptation and its significance for neurovestibular problems of space flight, J. Vestib. Res. 13 (2003) 345–362.
[17] M. Shelhamer, Trends in sensorimotor research and countermeasures for exploration-class space flights, Front. Syst. Neurosci. 9 (2015) 115.
[18] R.T. Bigelow, Y.R. Semenov, S. Du Lac, H.J. Hoffman, Y. Agrawal, Vestibular vertigo and comorbid cognitive and psychiatric impairment: the 2008 National Health Interview Survey, J. Neurol. Neurosurg. Psychiatry 87 (2016) 367–372.
[19] A. Palla, B. Lenggenhager, Ways to investigate vestibular contributions to cognitive processes, Front. Integr. Neurosci. 8 (2014) 40.
[20] P. Smith, Y. Zheng, From ear to uncertainty: vestibular contributions to cognitive function, Front. Integr. Neurosci. 7 (2013) 84.
[21] S. Besnard, C. Lopez, T. Brandt, P. Denise, P.F. Smith, The vestibular system in cognitive and memory processes in mammalians, Front. Integr. Neurosci. 9 (2015) 55.
[22] C. Lopez, A neuroscientific account of how vestibular disorders impair bodily self-consciousness, Front. Integr. Neurosci. 7 (2013) 91.
[23] D. Deroualle, C. Lopez, Toward a vestibular contribution to social cognition, Front. Integr. Neurosci. 8 (2014) 16.
[24] G. Vignaux, S. Besnard, J. Ndong, B. Philoxène, P. Denise, F. Elefteriou, Bone remodeling is regulated by inner ear vestibular signals, J. Bone Miner. Res. 28 (2013) 2136–2144.
[25] N.M. Arzeno, M.B. Stenger, J.J. Bloomberg, S.H. Platts, Spaceflight-induced cardiovascular changes and recovery during NASA's Functional Task Test, Acta Astronaut. 92 (2013) 10–14.
[26] L. Ploutz-Snyder, J. Ryder, K. English, F. Haddad, K. Baldwin, Evidence Report: Risk of Impaired Performance Due to Reduced Muscle Mass, Strength, and Endurance, NASA Human Research Program, 2015. https://humanresearchroadmap.nasa.gov/evidence/reports/Muscle.pdf?rnd=0.425593686953929. (Accessed 24 March 2022).
[27] A.D. LeBlanc, E.R. Spector, H.J. Evans, J.D. Sibonga, Skeletal responses to space flight and the bed rest analog: a review, J. Musculoskelet. Neuronal Interact. 7 (2007) 33.
[28] E.S. Nelson, B. Lewandowski, A. Licata, J.G. Myers, Development and validation of a predictive bone fracture risk model for astronauts, Ann. Biomed. Eng. 37 (2009) 2337–2359.
[29] R.D. Carpenter, A.D. LeBlanc, H. Evans, J.D. Sibonga, T.F. Lang, Long-term changes in the density and structure of the human hip and spine after long-duration spaceflight, Acta Astronaut. 67 (2010) 71–81.
[30] R.A. Pietrzyk, J.A. Jones, C.F. Sams, P.A. Whitson, Renal stone formation among astronauts, Aviat. Space Environ. Med. 78 (2007) A9–A13.
[31] J.V. Meck, W.W. Waters, M.G. Ziegler, H.F. deBlock, P.J. Mills, D. Robertson, P.L. Huang, Mechanisms of postspaceflight orthostatic hypotension: low α1-adrenergic receptor responses before flight and central autonomic dysregulation postflight, Am. J. Physiol. Heart Circ. Physiol. 286 (2004) H1486–H1495.

[32] W.W. Waters, M.G. Ziegler, J.V. Meck, Postspaceflight orthostatic hypotension occurs mostly in women and is predicted by low vascular resistance, J. Appl. Physiol. 92 (2002) 586–594.
[33] A.D. Moore Jr., M.E. Downs, S.M. Lee, A.H. Feiveson, P. Knudsen, L. Ploutz-Snyder, Peak exercise oxygen uptake during and following long-duration spaceflight, J. Appl. Physiol. 117 (2014) 231–238.
[34] P. Norsk, A. Asmar, M. Damgaard, N.J. Christensen, Fluid shifts, vasodilatation and ambulatory blood pressure reduction during long duration spaceflight, J. Physiol. 593 (2015) 573–584.
[35] H.W. Khine, K. Steding-Ehrenborg, J.L. Hastings, J. Kowal, J.D. Daniels, R.L. Page, J.J. Goldberger, J. Ng, B. Adams-Huet, M.W. Bungo, Effects of prolonged spaceflight on atrial size, atrial electrophysiology, and risk of atrial fibrillation, Circ. Arrhythm. Electrophysiol. 11 (2018) e005959.
[36] R.S. Johnston, L.F. Dietlein, C.A. Berry, Biomedical Results of Apollo, Scientific and Technical Information Office, NASA, 1975.
[37] J.L. Cervantes, B.-Y. Hong, Dysbiosis and immune dysregulation in outer space, Int. Rev. Immunol. 35 (2016) 67–82.
[38] B. Crucian, R.J. Simpson, S. Mehta, R. Stowe, A. Chouker, S.-A. Hwang, J.K. Actor, A.P. Salam, D. Pierson, C. Sams, Terrestrial stress analogs for spaceflight associated immune system dysregulation, Brain Behav. Immun. 39 (2014) 23–32.
[39] S. Ponomarev, S. Kalinin, A. Sadova, M. Rykova, K. Orlova, B. Crucian, Immunological aspects of isolation and confinement, Front. Immunol. 12 (2021) 2449.
[40] P. Brodin, V. Jojic, T. Gao, S. Bhattacharya, C.J.L. Angel, D. Furman, S. Shen-Orr, C.L. Dekker, G.E. Swan, A.J. Butte, Variation in the human immune system is largely driven by non-heritable influences, Cell 160 (2015) 37–47.
[41] B.V. Rooney, B.E. Crucian, D.L. Pierson, M.L. Laudenslager, S.K. Mehta, Herpes virus reactivation in astronauts during spaceflight and its application on earth, Front. Microbiol. (2019) 16.
[42] B. Crucian, C. Sams, Immune system dysregulation during spaceflight: clinical risk for exploration-class missions, Brain Behav. Immun. 19 (2009) 235–242.
[43] B.E. Crucian, A. Choukèr, R.J. Simpson, S. Mehta, G. Marshall, S.M. Smith, S.R. Zwart, M. Heer, S. Ponomarev, A. Whitmire, Immune system dysregulation during spaceflight: potential countermeasures for deep space exploration missions, Front. Immunol. 9 (2018) 1437.
[44] M.L. Lewis, L.A. Cubano, B. Zhao, H.-K. Dinh, J.G. Pabalan, E.H. Piepmeier, P.D. Bowman, cDNA microarray reveals altered cytoskeletal gene expression in space-flown leukemic T lymphocytes (Jurkat), FASEB J. 15 (2001) 1783–1785.
[45] G. Makedonas, S. Mehta, A. Choukèr, R.J. Simpson, G. Marshall, J.S. Orange, S. Aunon-Chancellor, S.M. Smith, S.R. Zwart, R.P. Stowe, Specific immunologic countermeasure protocol for deep-space exploration missions, Front. Immunol. (2019) 2407.
[46] F.E. Mercier, C. Ragu, D.T. Scadden, The bone marrow at the crossroads of blood and immunity, Nat. Rev. Immunol. 12 (2012) 49–60.
[47] M.H. Mohajeri, G. La Fata, R.E. Steinert, P. Weber, Relationship between the gut microbiome and brain function, Nutr. Rev. 76 (2018) 481–496.
[48] T. Kuehnast, C. Abbott, M.R. Pausan, D.A. Pearce, C. Moissl-Eichinger, A. Mahnert, The crewed journey to Mars and its implications for the human microbiome, Microbiome 10 (2022) 1–14.
[49] A.A. Voorhies, C. Mark Ott, S. Mehta, D.L. Pierson, B.E. Crucian, A. Feiveson, C.M. Oubre, M. Torralba, K. Moncera, Y. Zhang, Study of the impact of long-duration space missions at the International Space Station on the astronaut microbiome, Sci. Rep. 9 (2019) 1–17.
[50] F.E. Garrett-Bakelman, M. Darshi, S.J. Green, R.C. Gur, L. Lin, B.R. Macias, M.J. McKenna, C. Meydan, T. Mishra, J. Nasrini, The NASA Twins Study: a multidimensional analysis of a year-long human spaceflight, Science 364 (2019) eaau8650.
[51] H.L. Lee, H. Shen, I.Y. Hwang, H. Ling, W.S. Yew, Y.S. Lee, M.W. Chang, Targeted approaches for in situ gut microbiome manipulation, Genes 9 (2018) 351.

[52] S. Häuplik-Meusburger, R. Peldszus, V. Holzgethan, Greenhouse design integration benefits for extended spaceflight, Acta Astronaut. 68 (2011) 85–90.
[53] J. Suez, N. Zmora, G. Zilberman-Schapira, U. Mor, M. Dori-Bachash, S. Bashiardes, M. Zur, D. Regev-Lehavi, R.B.-Z. Brik, S. Federici, Post-antibiotic gut mucosal microbiome reconstitution is impaired by probiotics and improved by autologous FMT, Cell 174 (2018) 1406–1423.
[54] G.E. Strangman, W. Sipes, G. Beven, Human cognitive performance in spaceflight and analogue environments, Aviat. Space Environ. Med. 85 (2014) 1033–1048.
[55] A.C. Stahn, S. Kühn, Brains in space: the importance of understanding the impact of long-duration spaceflight on spatial cognition and its neural circuitry, Cogn. Process. 22 (2021) 105–114.
[56] M. Roy-O'Reilly, A. Mulavara, T. Williams, A review of alterations to the brain during spaceflight and the potential relevance to crew in long-duration space exploration, npj Microgravity 7 (2021) 1–9.
[57] D.R. Roberts, M.H. Albrecht, H.R. Collins, D. Asemani, A.R. Chatterjee, M.V. Spampinato, X. Zhu, M.I. Chimowitz, M.U. Antonucci, Effects of spaceflight on astronaut brain structure as indicated on MRI, N. Engl. J. Med. 377 (2017) 1746–1753.
[58] D. Roberts, D. Asemani, P. Nietert, M. Eckert, D. Inglesby, J. Bloomberg, M. George, T. Brown, Prolonged microgravity affects human brain structure and function, Am. J. Neuroradiol. 40 (2019) 1878–1885.
[59] A. Van Ombergen, S. Jillings, B. Jeurissen, E. Tomilovskaya, A. Rumshiskaya, L. Litvinova, I. Nosikova, E. Pechenkova, I. Rukavishnikov, O. Manko, Brain ventricular volume changes induced by long-duration spaceflight, Proc. Natl. Acad. Sci. 116 (2019) 10531–10536.
[60] L.A. Kramer, K.M. Hasan, M.B. Stenger, A. Sargsyan, S.S. Laurie, C. Otto, R.J. Ploutz-Snyder, K. - Marshall-Goebel, R.F. Riascos, B.R. Macias, Intracranial effects of microgravity: a prospective longitudinal MRI study, Radiology 295 (2020) 640–648.
[61] K.E. Hupfeld, H.R. McGregor, J.K. Lee, N.E. Beltran, I.S. Kofman, Y.E. De Dios, P.A. Reuter-Lorenz, R.F. Riascos, O. Pasternak, S.J. Wood, The impact of 6 and 12 months in space on human brain structure and intracranial fluid shifts, Cereb. Cortex Commun. 1 (2020) tgaa023.
[62] J.K. Lee, V. Koppelmans, R.F. Riascos, K.M. Hasan, O. Pasternak, A.P. Mulavara, J.J. Bloomberg, R.D. Seidler, Spaceflight-associated brain white matter microstructural changes and intracranial fluid redistribution, JAMA Neurol. 76 (2019) 412–419.
[63] A. Demertzi, A. Van Ombergen, E. Tomilovskaya, B. Jeurissen, E. Pechenkova, C. Di Perri, L. Litvinova, E. Amico, A. Rumshiskaya, I. Rukavishnikov, Cortical reorganization in an astronaut's brain after long-duration spaceflight, Brain Struct. Funct. 221 (2016) 2873–2876.
[64] E. Pechenkova, I. Nosikova, A. Rumshiskaya, L. Litvinova, I. Rukavishnikov, E. Mershina, V. Sinitsyn, A. Van Ombergen, B. Jeurissen, S. Jillings, Alterations of functional brain connectivity after long-duration spaceflight as revealed by fMRI, Front. Physiol. (2019) 761.
[65] A. Doroshin, S. Jillings, B. Jeurissen, E. Tomilovskaya, E. Pechenkova, I. Nosikova, A. Rumshiskaya, L. Litvinova, I. Rukavishnikov, C. De Laet, Brain connectometry changes in space travelers after long-duration spaceflight, Front. Neural Circuits 16 (2022).
[66] S. Jillings, A. Van Ombergen, E. Tomilovskaya, A. Rumshiskaya, L. Litvinova, I. Nosikova, E. Pechenkova, I. Rukavishnikov, I.B. Kozlovskaya, O. Manko, Macro-and microstructural changes in cosmonauts' brains after long-duration spaceflight, Sci. Adv. 6 (2020) eaaz9488.
[67] V. Koppelmans, J.J. Bloomberg, A.P. Mulavara, R.D. Seidler, Brain structural plasticity with spaceflight, npj Microgravity 2 (2016) 1–8.
[68] K. Hupfeld, H. McGregor, P. Reuter-Lorenz, R. Seidler, Microgravity effects on the human brain and behavior: dysfunction and adaptive plasticity, Neurosci. Biobehav. Rev. 122 (2021) 176–189.
[69] R.B. Welch, B. Bridgeman, S. Anand, K.E. Browman, Alternating prism exposure causes dual adaptation and generalization to a novel displacement, Percept. Psychophys. 54 (1993) 195–204.
[70] R.D. Seidler, Multiple motor learning experiences enhance motor adaptability, J. Cogn. Neurosci. 16 (2004) 65–73.

[71] M.E. Walton, E.L. Kerstman, Quantification of medical risk on the international space station using the integrated medical model, Aerosp. Med. Hum. Perform. 91 (2020) 332–342.
[72] V.V. Lebedev, Diary of a Cosmonaut: 211 Days in Space, Phytoresource Research, 1988.
[73] D.A. Goodenow-Messman, S.A. Gokoglu, M. Kassemi, J.G. Myers, Numerical characterization of astronaut CaOx renal stone incidence rates to quantify in-flight and post-flight relative risk, npj Microgravity 8 (2022) 1–17.
[74] A. Okada, T. Matsumoto, H. Ohshima, T. Isomura, T. Koga, T. Yasui, K. Kohri, A. LeBlanc, E. Spector, J. Jones, Bisphosphonate use may reduce the risk of urolithiasis in astronauts on long-term spaceflights, JBMR Plus 6 (2022) e10550.
[75] S.M. Auñón-Chancellor, J.M. Pattarini, S. Moll, A. Sargsyan, Venous thrombosis during spaceflight, N. Engl. J. Med. 382 (2020) 89–90.
[76] J.I. Leonard, A Systems Analysis Approach to Understanding the Physiological Adaptation to Spaceflight, NASA/CR-2020-220487, 2020.
[77] H.R. McGregor, J.K. Lee, E.R. Mulder, Y.E. De Dios, N.E. Beltran, I.S. Kofman, J.J. Bloomberg, A.P. Mulavara, R.D. Seidler, Brain connectivity and behavioral changes in a spaceflight analog environment with elevated CO2, NeuroImage 225 (2021) 117450.
[78] J.K. Lee, Y. De Dios, I. Kofman, A.P. Mulavara, J.J. Bloomberg, R.D. Seidler, Head down tilt bed rest plus elevated CO2 as a spaceflight analog: effects on cognitive and sensorimotor performance, Front. Hum. Neurosci. (2019) 355.
[79] M. Braddock, Hazards of lunar regolith for respiratory, central nervous system, cardiovascular and ocular function, in: The Human Factor in the Settlement of the Moon, Springer, 2021, pp. 141–157.
[80] R.A. Scheuring, J.A. Jones, J.D. Polk, D.B. Gillis, J. Schmid, J.M. Duncan, J.R. Davis, The Apollo Medical Operations Project: Recommendations to Improve Crew Health and Performance for Future Exploration Missions and Lunar Surface Operations, NASA, 2007.
[81] G.K. Prisk, Effects of partial gravity on the function and particle handling of the human lung, Curr. Pathobiol. Rep. 6 (2018) 159–166.
[82] J.T. James, C. Lam, R.R. Scully, Comparative Benchmark Dose Modeling as a Tool to Make the First Estimate of Safe Human Exposure Levels to Lunar Dust, NASA, 2013.
[83] C. Lam, V. Castranova, P.C. Zeidler-Erdely, R. Renne, R. Hunter, R. McCluskey, R.R. Scully, W.T. Wallace, Y. Zhang, V.E. Ryder, Comparative pulmonary toxicities of lunar dusts and terrestrial dusts (TiO2 & SiO2) in rats and an assessment of the impact of particle-generated oxidants on the dusts' toxicities, Inhal. Toxicol. (2022) 1–17.
[84] R. Caston, K. Luc, D. Hendrix, J.A. Hurowitz, B. Demple, Assessing toxicity and nuclear and mitochondrial DNA damage caused by exposure of mammalian cells to lunar regolith simulants, GeoHealth 2 (2018) 139–148.
[85] A.F. Davila, D. Willson, J.D. Coates, C.P. McKay, Perchlorate on Mars: a chemical hazard and a resource for humans, Int. J. Astrobiol. 12 (2013) 321–325.
[86] A.G. Lee, T.H. Mader, C.R. Gibson, W. Tarver, P. Rabiei, R.F. Riascos, L.A. Galdamez, T. Brunstetter, Spaceflight associated neuro-ocular syndrome (SANS) and the neuro-ophthalmologic effects of microgravity: a review and an update, npj Microgravity 6 (2020) 1–10.
[87] K. Marshall-Goebel, B.R. Macias, L.A. Kramer, K.M. Hasan, C. Ferguson, N. Patel, R.J. Ploutz-Snyder, S. M. Lee, D. Ebert, A. Sargsyan, Association of structural changes in the brain and retina after long-duration spaceflight, JAMA Ophthalmol. 139 (2021) 781–784.
[88] N. Kanas, D. Manzey, Space Psychology and Psychiatry, Springer, 2008.
[89] N. Kanas, V. Gushin, A. Yusupova, Whither the third quarter phenomenon? Aerosp. Med. Hum. Perform. 92 (2021) 689–691.
[90] E.C. Ihle, J.B. Ritsher, N. Kanas, Positive psychological outcomes of spaceflight: an empirical study, Aviat. Space Environ. Med. 77 (2006) 93–101.

[91] N. Salamon, J.M. Grimm, J.M. Horack, E.K. Newton, Application of virtual reality for crew mental health in extended-duration space missions, Acta Astronaut. 146 (2018) 117–122.
[92] B.N. Neilson, C.M. Craig, G.C. Altman, A.T. Travis, J.A. Vance, M.I. Klein, Can the biophilia hypothesis be applied to long-duration human space flight? A mini-review, Front. Psychol. 12 (2021).
[93] A.P. Anderson, M.D. Mayer, A.M. Fellows, D.R. Cowan, M.T. Hegel, J.C. Buckey, Relaxation with immersive natural scenes presented using virtual reality, Aerosp. Med. Hum. Perform. 88 (2017) 520–526.
[94] G.M. Sandal, G.R. Leon, Psychological countermeasures, in: Stress Challenges and Immunity in Space, Springer, 2012, pp. 363–372.
[95] A.E. Ronca, Mammalian development in space, Adv. Space Biol. Med. 9 (2003) 217–251.
[96] S. Wakayama, Y. Kamada, K. Yamanaka, T. Kohda, H. Suzuki, T. Shimazu, M.N. Tada, I. Osada, A. Nagamatsu, S. Kamimura, Healthy offspring from freeze-dried mouse spermatozoa held on the international Space Station for 9 months, Proc. Natl. Acad. Sci. 114 (2017) 5988–5993.
[97] S. Wakayama, Y. Kawahara, C. Li, K. Yamagata, L. Yuge, T. Wakayama, Detrimental effects of microgravity on mouse preimplantation development in vitro, PLoS One 4 (2009) e6753.
[98] S. Zhang, D. Zheng, Y. Wu, W. Lin, Z. Chen, L. Meng, J. Liu, Y. Zhou, Simulated microgravity using a rotary culture system compromises the in vitro development of mouse preantral follicles, PLoS One 11 (2016) e0151062.
[99] C.E. Mason, The Next 500 Years: Engineering Life to Reach New Worlds, MIT Press, 2021.
[100] M. Braddock, Limitations for extraterrestrial colonisation and civilisation built and the potential for human enhancements, in: Human Enhancements for Space Missions, Springer, 2020, pp. 71–93.
[101] K. Szocik, M. Shelhamer, M. Braddock, F.A. Cucinotta, C. Impey, P. Worden, T. Peters, M.M. Ćirković, K. C. Smith, K. Tachibana, Future space missions and human enhancement: medical and ethical challenges, Futures 133 (2021) 102819.
[102] F. Cortese, D. Klokov, A. Osipov, J. Stefaniak, A. Moskalev, J. Schastnaya, C. Cantor, A. Aliper, P. Mamoshina, I. Ushakov, Vive la radioresistance!: converging research in radiobiology and biogerontology to enhance human radioresistance for deep space exploration and colonization, Oncotarget 9 (2018) 14692.
[103] M. Shelhamer, Human enhancements: new eyes and ears for Mars, in: Human Enhancements for Space Missions, Springer, 2020, pp. 95–103.
[104] A.M. Paul, The Extended Mind: The Power of Thinking Outside the Brain, Eamon Dolan Books, 2021.
[105] J.C. Chancellor, G.B.I. Scott, J.P. Sutton, Space radiation: the number one risk to astronaut health beyond low Earth orbit, Life 4 (2014) 491–510.
[106] A.R. Kennedy, Biological effects of space radiation and development of effective countermeasures, Life Sci. Space Res. 1 (2014) 10–43.
[107] J.C. Chancellor, R.S. Blue, K.A. Cengel, S.M. Auñón-Chancellor, K.H. Rubins, H.G. Katzgraber, A.R. Kennedy, Limitations in predicting the space radiation health risk for exploration astronauts, npj Microgravity 4 (2018) 1–11.
[108] J.S. Willey, R.A. Britten, E. Blaber, C.G.T. Tahimic, J. Chancellor, M. Mortreux, L.D. Sanford, A.J. Kubik, M.D. Delp, X.W. Mao, The individual and combined effects of spaceflight radiation and microgravity on biologic systems and functional outcomes, J. Environ. Sci. Health Part C Toxicol. Carcinog. 39 (2021) 129–179.
[109] L.T. Chylack, L.E. Peterson, A.H. Feiveson, M.L. Wear, F.K. Manuel, W.H. Tung, D.S. Hardy, L.J. Marak, F.A. Cucinotta, NASA study of cataract in astronauts (NASCA). Report 1: cross-sectional study of the relationship of exposure to space radiation and risk of lens opacity, Radiat. Res. 172 (2009) 10–20.
[110] M.D. Delp, J.M. Charvat, C.L. Limoli, R.K. Globus, P. Ghosh, Apollo lunar astronauts show higher cardiovascular disease mortality: possible deep space radiation effects on the vascular endothelium, Sci. Rep. 6 (2016) 29901.
[111] F.A. Cucinotta, N. Hamada, M.P. Little, No evidence for an increase in circulatory disease mortality in astronauts following space radiation exposures, Life Sci. Space Res. 10 (2016) 53–56.

[112] S.R. Elgart, M.P. Little, L.J. Chappell, C.M. Milder, M.R. Shavers, J.L. Huff, Z.S. Patel, Radiation exposure and mortality from cardiovascular disease and cancer in early NASA astronauts, Sci. Rep. 8 (2018) 1–9.
[113] R.J. Reynolds, I.V. Bukhtiyarov, G.I. Tikhonova, S.M. Day, I.B. Ushakov, T.Y. Gorchakova, Contrapositive logic suggests space radiation not having a strong impact on mortality of US astronauts and Soviet and Russian cosmonauts, Sci. Rep. 9 (2019) 1–7.
[114] M. Lalkovičová, Neuroprotective agents effective against radiation damage of central nervous system, Neural Regen. Res. 17 (2022) 1885.
[115] S.M. Mortazavi, J.R. Cameron, A. Niroomand-Rad, Adaptive response studies may help choose astronauts for long-term space travel, Adv. Space Res. Off. J. Comm. Space Res. COSPAR 31 (2003) 1543–1551.
[116] S.M. Smith, S.R. Zwart, M. Heer, G.L. Douglas, Human Adaptation to Spaceflight: The Role of Food and Nutrition, National Aeronautics and Space Administration, 2021.
[117] S.R. Zwart, A.P. Mulavara, T.J. Williams, K. George, S.M. Smith, The role of nutrition in space exploration: implications for sensorimotor, cognition, behavior and the cerebral changes due to the exposure to radiation, altered gravity, and isolation/confinement hazards of spaceflight, Neurosci. Biobehav. Rev. 127 (2021) 307–331.
[118] N.E. Murphy, C.T. Carrigan, L.M. Margolis, High-fat ketogenic diets and physical performance: a systematic review, Adv. Nutr. 12 (2021) 223–233.
[119] S.R. Zwart, S.M. Smith, The impact of space flight on the human skeletal system and potential nutritional countermeasures, Int. SportMed J. 6 (2005) 199–214.
[120] C.M. Johnson, H.O. Boles, L.E. Spencer, L. Poulet, M. Romeyn, J.M. Bunchek, R. Fritsche, G.D. Massa, A. O'Rourke, R.M. Wheeler, Supplemental food production with plants: a review of NASA research, Front. Astron. Space Sci. 8 (2021).
[121] D. Moosavi, D. Wolovsky, A. Depompeis, D. Uher, D. Lennington, R. Bodden, C.E. Garber, The effects of spaceflight microgravity on the musculoskeletal system of humans and animals, with an emphasis on exercise as a countermeasure: a systematic scoping review, Physiol. Res. 70 (2021).
[122] K.J. Hackney, M. Everett, J.M. Scott, L. Ploutz-Snyder, Blood flow-restricted exercise in space, Extreme Physiol. Med. 1 (2012) 1–13.
[123] R.D. Hyldahl, J.M. Peake, Combining cooling or heating applications with exercise training to enhance performance and muscle adaptations, J. Appl. Physiol. 129 (2020) 353–365.
[124] J.C. Longhurst, C.L. Stebbins, The isometric athlete, Cardiol. Clin. 10 (1992) 281–294.
[125] W.B. Scott, Artificial gravity, Aviat. Week Space Technol. 162 (2005) 62–64.
[126] G.R. Clément, A.P. Bukley, W.H. Paloski, Artificial gravity as a countermeasure for mitigating physiological deconditioning during long-duration space missions, Front. Syst. Neurosci. 9 (2015) 92.
[127] T. Vogt, V. Abeln, H.K. Strüder, S. Schneider, Artificial gravity exposure impairs exercise-related neurophysiological benefits, Physiol. Behav. 123 (2014) 156–161.
[128] L.R. Young, Artificial gravity considerations for a Mars exploration mission, Ann. N. Y. Acad. Sci. 871 (1999) 367–378.
[129] L.R. Young, H. Hecht, L.E. Lyne, K.H. Sienko, C.C. Cheung, J. Kavelaars, Artificial gravity: head movements during short-radius centrifugation, Acta Astronaut. 49 (2001) 215–226.
[130] N. Goswami, A.P. Blaber, H. Hinghofer-Szalkay, V.A. Convertino, Lower body negative pressure: physiological effects, applications, and implementation, Physiol. Rev. 99 (2019) 807–851.
[131] N. Ashari, A.R. Hargens, The mobile lower body negative pressure gravity suit for long-duration spaceflight, Front. Physiol. (2020) 977.
[132] C.M. Hearon, K.A. Dias, G. Babu, J.E. Marshall, J. Leidner, K. Peters, E. Silva, J.P. MacNamara, J. Campain, B.D. Levine, Effect of nightly lower body negative pressure on choroid engorgement in a model of spaceflight-associated neuro-ocular syndrome: a randomized crossover trial, JAMA Ophthalmol. 140 (2022) 59–65.
[133] M.R. Campbell, J.B. Charles, Historical review of lower body negative pressure research in space medicine, Aerosp. Med. Hum. Perform. 86 (2015) 633–640.

[134] S.H. Greenwald, B.R. Macias, S.M. Lee, K. Marshall-Goebel, D.J. Ebert, J.H. Liu, R.J. Ploutz-Snyder, I.V. Alferova, S.A. Dulchavsky, A.R. Hargens, Intraocular pressure and choroidal thickness respond differently to lower body negative pressure during spaceflight, J. Appl. Physiol. 131 (2021) 613–620.

[135] M. Wade, N.A. Fox, C.H. Zeanah, C.A. Nelson, Effect of foster care intervention on trajectories of general and specific psychopathology among children with histories of institutional rearing: a randomized clinical trial, JAMA Psychiatry 75 (2018) 1137–1145.

[136] T. Kinnunen, M. Kolehmainen, Touch and affect: analysing the archive of touch biographies, Body Soc. 25 (2019) 29–56.

[137] M.S. Remland, T.S. Jones, H. Brinkman, Interpersonal distance, body orientation, and touch: effects of culture, gender, and age, J. Soc. Psychol. 135 (1995) 281–297.

[138] F. McGlone, J. Wessberg, H. Olausson, Discriminative and affective touch: sensing and feeling, Neuron 82 (2014) 737–755.

[139] R. Pawling, P.D. Trotter, F.P. McGlone, S.C. Walker, A positive touch: C-tactile afferent targeted skin stimulation carries an appetitive motivational value, Biol. Psychol. 129 (2017) 186–194.

[140] Z. Shi, M. Qin, L. Huang, T. Xu, Y. Chen, Q. Hu, S. Peng, Z. Peng, L. Qu, S. Chen, Human torpor: translating insights from nature into manned deep space expedition, Biol. Rev. 96 (2021) 642–672.

[141] A. Choukér, T.J. Ngo-Anh, R. Biesbroek, G. Heldmaier, M. Heppener, J. Bereiter-Hahn, European space agency's hibernation (torpor) strategy for deep space missions: linking biology to engineering, Neurosci. Biobehav. Rev. 131 (2021) 618–626.

[142] T. Squire, A. Ryan, S. Bernard, Radioprotective effects of induced astronaut torpor and advanced propulsion systems during deep space travel, Life Sci. Space Res. 26 (2020) 105–113.

[143] S. Ohi, A. Roach, S. Ramsahai, B.C. Kim, W. Fitzgerald, D.A. Riley, S.R. Gonda, The hematopoietic stem cell therapy for exploration of deep space, AIP Conf. Proc. 699 (1) (2004) 938–950.

[144] P. Huang, A.L. Russell, R. Lefavor, N.C. Durand, E. James, L. Harvey, C. Zhang, S. Countryman, L. Stodieck, A.C. Zubair, Feasibility, potency, and safety of growing human mesenchymal stem cells in space for clinical application, npj Microgravity 6 (2020) 1–12.

[145] D.C. Fuhrmann, B. Brüne, Mitochondrial composition and function under the control of hypoxia, Redox Biol. 12 (2017) 208–215.

[146] R.J. White, M. Averner, Humans in space, Nature 409 (2001) 1115–1118.

[147] X. Li, J. Dunn, D. Salins, G. Zhou, W. Zhou, S.M. Schüssler-Fiorenza Rose, D. Perelman, E. Colbert, R. Runge, S. Rego, Digital health: tracking physiomes and activity using wearable biosensors reveals useful health-related information, PLoS Biol. 15 (2017) e2001402.

[148] T. Mishra, M. Wang, A.A. Metwally, G.K. Bogu, A.W. Brooks, A. Bahmani, A. Alavi, A. Celli, E. Higgs, O. Dagan-Rosenfeld, Pre-symptomatic detection of COVID-19 from smartwatch data, Nat. Biomed. Eng. 4 (2020) 1208–1220.

[149] M. Scheffer, J. Bascompte, W.A. Brock, V. Brovkin, S.R. Carpenter, V. Dakos, H. Held, E.H. Van Nes, M. Rietkerk, G. Sugihara, Early-warning signals for critical transitions, Nature 461 (2009) 53–59.

[150] R. Chen, G.I. Mias, J. Li-Pook-Than, L. Jiang, H.Y. Lam, R. Chen, E. Miriami, K.J. Karczewski, M. Hariharan, F.E. Dewey, Personal omics profiling reveals dynamic molecular and medical phenotypes, Cell 148 (2012) 1293–1307.

[151] M. Shelhamer, A call for research to assess and promote functional resilience in astronaut crews, J. Appl. Physiol. 120 (2016) 471–472.

[152] R.J. White, J.I. Leonard, J.A. Rummel, C.S. Leach, A systems approach to the physiology of weightlessness, J. Med. Syst. 6 (1982) 343–358.

CHAPTER

Interstellar life support

4

Grant Allan Anderson
Paragon Space Development Corporation, Tucson, AZ, United States

Chapter outline

Interstellar Travel. https://doi.org/10.1016/B978-0-323-91280-8.00011-3

1 Introduction

In the human endeavor to reach the stars, there will be one constant. While the technology to get us there will evolve and, perhaps, be revolutionized in the coming years, centuries, and even millennia, the human body will not adapt so quickly that it will not require a life support system. In many ways, you can see the need for life support to be a perpetual need (and business) as long as humans can be described as, well, human. Sure, at some time in the future, humans may be able to download their characteristics (knowledge, personality, and even quirks) into some future quantum-based supercomputer and be sent on a journey among the stars. But is this really "Interstellar Travel"? There is a unique human need to explore, to see what is over the next hill, and experience it for ourselves. As long as this need is present life support will be needed to allow the journey among the stars. In this chapter, I will endeavor to bring the reader up to speed on what a life support system really is, and what the current and near-future technologies or approaches are that can be extrapolated into the future of multigenerational exploration expeditions to reach further and further from the cradle of civilization that we call Earth. For it is a fact that there is no proven life support system that can even get us to Mars, let alone to the stars. It remains the "long pole in the tent," [1] preventing humans from venturing more than a few months away from supply and emergency return.

2 The big picture: Humans with a life support system is the payload

Life support, to put it simply, is the systems, accommodations, and elements of a spacecraft that allow humans to survive when otherwise surrounded by the very inhospitable environment that the universe—in all its vast, cold, unforgiving vacuum-permeated nature—really is.

The human body has basic needs. In the world of life support, we have a "law of threes." This stands for:

(a) Three seconds: The time the human body can be exposed to zero pressure. Despite the many movie and TV tropes, a human doesn't "hold their breath" when exposed to a vacuum. The human lung tissue cannot hold that type of pressure, and your cells will embolize and you will drown in your own body fluids if exposed to greater than about 3 seconds of vacuum.

(b) Three minutes: This is the time that most people can go without oxygen. In short, how long you can hold your breath? But note that if exposed to an oxygen-low environment (e.g., pure nitrogen), a human will pass out within seconds. This is because those few breaths you take before you hold your breath can purge your lungs of oxygen and lead to almost immediate loss of consciousness.

(c) Three days: This is the nominal amount of time that a human can go without water. After this time the human tongue swells and the body's osmotic potential processes (differentials of concentration of chemicals across the cell membranes) start to fail, resulting in metabolic shutdown.

(d) Three months: This is the nominal amount of time humans can go without food. The body, after only a few tens of hours, starts to switch metabolic processes to start burning stored energy (fat), but after 3 weeks much of the stores are gone and the metabolic system again starts to shut down.

Of course, no good life support system would starve the human participants, and, in fact, it is very important psychologically that each person within a life support system never worries that the above needs are not going to be met each and every day. NASA has developed a table of these needs, expressed in kilograms needed per crew member, a per mission day (kg/CM-d) that lists baseline human consumption, shown in Fig. 1, and outputs from humans that must be processed by the long-duration life support system (where negative values "−" denote inputs and positive values "+" denote outputs [2]). While this information has been updated since [3], the 2004 version has a nice summary table, shown in Fig. 1.

Balance [36]	**Interface**	**Units**	**Nominal Value**
	Basis		
	Overall Body Mass	kg	70.0
	Respiratory Quotient		0.869
	Air		
− m	Carbon Dioxide Load	kg/CM-d	0.998
+ m	Oxygen Consumed	kg/CM-d	0.835
	Food		
+ m	Food Consumed; Mass [37]	kg/CM-d	0.617 [38]
+ E	Food Consumed; Energy Content	MJ/CM-d	11.82
+ m	Potable Water Consumed [39]	kg/CM-d	3.909 [(1)]
	Thermal		
− E	Total Metabolic Heat Load [40]	MJ/CM-d	11.82
	Sensible Metabolic Heat Load	MJ/CM-d	6.31
	Latent Metabolic Heat Load [41]	MJ/CM-d	5.51
	Waste		
− m	Fecal Solid Waste (dry basis)	kg/CM-d	0.032
− m	Perspiration Solid Waste (dry basis)	kg/CM-d	0.018
− m	Urine Solid Waste (dry basis)	kg/CM-d	0.059
	Water [42]		
− m	Fecal Water	kg/CM-d	0.091
− m	Respiration and Perspiration Water [43]	kg/CM-d	2.277
− m	Urine Water	kg/CM-d	1.886 [(1)]

FIG. 1

Human metabolic needs.

Courtesy of NASA (A. Hanford, Advanced Life Support Baseline Values and Assumptions Document. Johnson Space Center, NASA, Houston, 2004, NASA/CR-2004-208941.).

The life support system encompasses all the systems needed to make sure that, as a first priority, none of the above "rules of three" are violated. However, as shown in Fig. 1, there are seemingly secondary (but also fully necessary) accommodations of the human payload that also must be satisfied to truly be able to operate productively as a human. These include maintaining the correct temperature (thermal control), the collection, handling, storage, and processing of waste products from the body, and the psychological needs of safety, space, and human interaction. Entwined with all of this are other basic human needs of clothing, hygiene, and mental stimulation.

3 Life support major elements

Life support for spacecraft is referred to, traditionally, as environmental control and life support systems (ECLSS). For the systems in operation today, for both transportation to orbit and in low Earth orbit (LEO), the typical ECLSS encompasses six major elements, as shown in Fig. 2. In order of the hierarchy of importance as described above, these systems are:

- Air management
- Water management
- Food management
- Thermal control
- Waste management
- Crew systems/human accommodations

4 Major elements—Definitions, descriptions, and examples

The Fig. 2 shows not only a breakdown of the six major elements of an ECLSS system but also some of the major subsystems that make up those elements. Note that, for interstellar travel, some of the major subsystems will be different than the more generic diagram labels shown, and this is the focus of this chapter. But in all cases of the functional makeup of the elements, the figure is correct.

4.1 Air management

Air management, as noted in the introduction, is the most time-critical vital element of the "law of threes" as it ensures that the two most immediate needs of the human metabolic system are met: air pressure and oxygen; thus, the need to store and provide atmospheric gases (oxygen and, in most cases, nitrogen) at sufficient pressure to be able to maintain the desired cabin pressure and to provide oxygen into that same cabin environment. Since humans have evolved at roughly sea-level pressure on Earth—at a level of 14.7 pounds per square inch (psi) (0.101 MegaPascals or MPa)—most crew cabins today provide pressure at that level, and hence the storage pressure of the gases has to be greater than this value (gas will only flow from higher pressure to lower pressure). Nominal gas storage is at 3000–5000 psi [4], though gas systems up to 10,000 psi have been built and tested [5]. (Note that oxygen, at such high pressures, is especially dangerous, as even slight, microscopic contamination at that level in the plumbing may result in fires or damage to the gas delivery system.) As gas storage systems trend larger, pressures need to drop to ensure structural integrity of the containment vessel at an

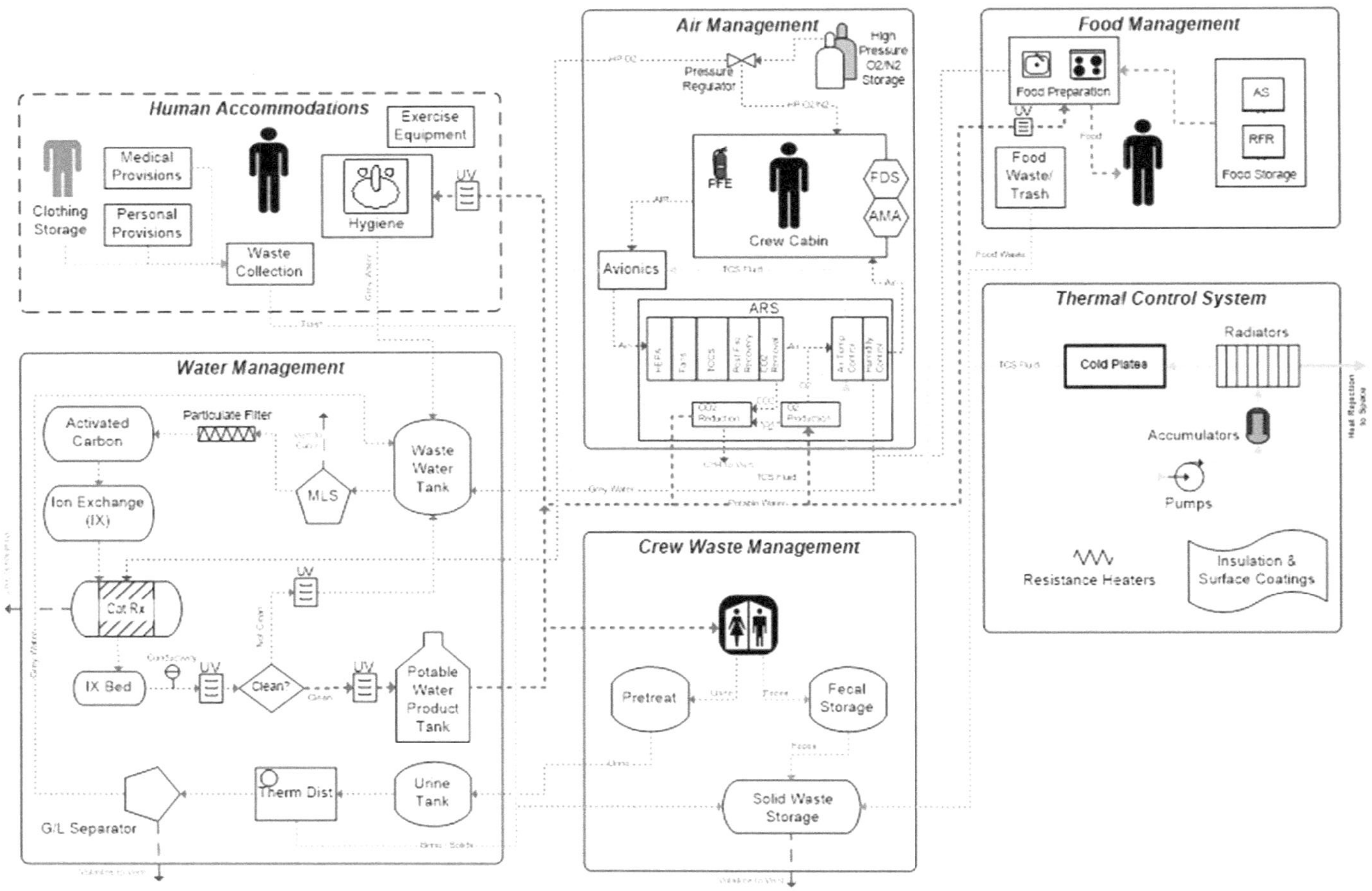

FIG. 2

Major elements of an ECLSS system.

Courtesy of NASA.

acceptable weight. (If pressure is constant, the larger diameter of the pressure vessel requires a thicker wall of similar strength material.) In the case of very large systems, maintaining the gas storage at just slightly above the cabin level is sufficient. Science fiction writers have speculated that city-wide gas storage system for a moon base, for example, might be constructed from subsurface chambers or "bubbles" left over from lunar volcanic activity. Robert Heinlein described such a system in his 1957 short story *The Menace from Earth*, contained in an anthology of short stories of the same name [6].

An alternative to voluminous gas storage is liquefied storage at cryogenic temperatures. Though storage insulation adds more mass and cooling systems would decrease overall maintainability within the spacecraft. External storage, though theoretically solving the cooling problem, also has the issue of exposure to the interstellar environment, such as micrometeoroids, which is wholly unknown at this time.

The ECLSS air management system (AMS) must provide for storage, but also the reduction of the gas pressure from the storage pressure to that required for the cabin. This is usually done through a series of regulators known as the primary and secondary regulators. For those readers that are certified for use of SCUBA diving equipment these are analogous to the primary and secondary regulators on that apparatus.

The primary regulator reduces the pressure from the tank storage pressure to one much closer to the cabin pressure, but still with enough pressure to provide for a force for distribution from the storage tanks to the point of cabin injection. Nominally, the primary regulator reduces pressure to about 5–20 times the cabin pressure, or approximately 75–300 psi. The need for this first reduction is to both reduce the mass of the distributing plumbing—the larger the pressure, the thicker the walls of the (usually metal) distribution tubing and fittings must be—and to allow the primary regulator to be the first line of defense against a runaway pressure loss from the tanks due to a faulty system. In that regard, the primary pressure regulator can also accommodate a restrictor, which limits the flow in case of such a failure.

The secondary regulator controls the pressure of the cabin and, either through a mechanical or electrical system, can be set to the desired cabin pressure and therefore make up gas as it is used (metabolized) by the human crew or through leakage to space. Of course, the latter is always minimized but is not insubstantial in today's spacecraft.

There are two elements of the cabin atmosphere that the pressurization system must control. First is overall absolute pressure; but the second is the ratio of breathable oxygen to the "buffer gas"—in the case of Earth, this gas being nitrogen (The buffer gas is only called that for human missions but performs functions both as a mitigation for runaway oxygen reactions (fire) by keeping the percentage of O_2 down and for plant and human physiology, as pure O_2 atmospheres are not clearly safe for humans beyond a few days—and very little data is available for long-term exposure over more than a few days such as the Apollo missions. Pure oxygen is also toxic to plants.). So the AMS must also sense the amount of oxygen and the total pressure, and then chose to inject the right type of gas depending on which parameter has fallen below a control limit. Nominally, full sea-level pressure is 14.7 psi, with a 21% oxygen mixture, or a partial pressure of 3.1 psi of oxygen. There is an acceptable error on the partial pressure of oxygen that spans from roughly 2.6 to 3.7 psi, but the regulating systems must be able to detect that partial pressure. Oxygen sensors are therefore a critical element of the AMS, and in discussion of an interstellar system below, we will go into why these sensors are a vital technology in need of some development.

Secondary, but no less important to atmospheric control, is the control of trace contaminants. These are comprised of airborne compounds put out both by the human body and due to off gassing of the materials of construction. The latter must be limited and so the materials from which you can build the interior of a spacecraft are restricted such that they don't off gas (emit gases integral to their manufacture) nor react to produce gases (such as candles) or contaminate the environment (perfumes for example). Typically, the amounts of trace contaminants that are allowed in the atmosphere are measured in "SMAC levels," where SMAC stands for spacecraft maximum allowable concentrations. A typical SMAC table is shown in Fig. 3 and illustrates that SMACs are usually defined by exposure time (the shorter the exposure time, the higher the concentration allowed) [7].

4.2 Water management

Being necessary to survival for almost all biological life that we know of, and surely required for life propagation, water, in our hierarchy of needs, has a priority just under the air we breathe. In a life support system, though, the provisions for water can be very simple for short durations. A squeeze bottle of water often gets astronauts up to and back from LEO. And, if needed, the human processing and expulsion of water handled by diapers and overboard dumps.

However, over a mission of just a few days, water management quickly becomes much more complex, both from a standpoint of technologies to store and deliver it and from a standpoint of how to collect, process, and deliver potable (drinkable) water back to the human crew. There is a reason that the "water management" box in Fig. 2 has arguably the most elements. As the human body is mostly water, almost everything that we do consumes, produces, or processes water. Thus, the water management system receives inputs from human accommodations, air management, food management, and crew waste management, and delivers water back to the crew. Like air, there are limits to the contaminants for human potable (drinkable) water, called spacecraft water exposure guidelines (SWEGS) [8], an example of which is shown in Fig. 4.

Chemical	1 hr		24 hr		7 d		30 d		180 d		1000 d	
	ppm	(mg/m^3)	ppm	(mg/m^3)	ppm	(mg/m^3)	ppm	(mg/m^3)	ppm	(mg/m^3)	ppm	(mg/m^3)
Chloroform	2	(10)	2	(10)	2	(10)	1	(5)	1	(5)	Not Set	(Not Set)
	Organ	Effect	Organ	Effect	Organ	Effect	Organ	Effect	Organ	Effect	Organ	Effect
CAS #: 67-66-3	CNS	Depression	CNS	Depression	CNS	Depression	CNS	Depression	CNS	Depression		
REFERENCE: Garcia, Hector D. (2000), Chloroform, Spacecraft Maximum Allowable Concentrations for Selected Airborne Contaminants, Vol 4:264-306, National Academy Press, Washington, DC	Kidney	Nephrotoxicity	Kidney	Nephrotoxicity	Liver	Hepatotoxicity	Liver	Hepatotoxicity	Liver	Hepatotoxicity		
REMARKS:					Kidney	Nephrotoxicity						

FIG. 3

Example of SMAC levels from JSC 20584 April 2020.

Chemical	1 day		10 days		100 days		1000 days		Remarks:
Formaldehyde	20 mg/L		20 mg/L		12 mg/L		12 mg/L		
	Organ	Effect	Organ	Effect	Organ	Effect	Organ	Effect	
CAS #: 50-00-0	G.I.	Gastric upset	G.I.	Gastric upset	G.I.	Gastric upset	G.I.	Gastric upset	
Reference: McCoy, J. Torin (2007), Formaldehyde, Spacecraft Water Exposure Guidelines for Selected Contaminants, Vol. 2: 300-341. The National Academies Press, Washington, DC.									

FIG. 4

Example of SWEGs levels from JSC 63414 July 2017.

For example, water not already processed and stored for launch comes from the crew hygiene (washing, brushing teeth), urine, feces (yes, feces by weight is mostly water), humidity condensed from the air, food waste processing (again, a lot of food waste is water), and even food processing. All these inputs to the water processing stream have different biological contaminants, chemical contaminants (some good, some bad—though good design limits the bad), and even elemental differences. And to make the job more complicated, the concentrations vary over the course of the day, the course of the mission, and even the course of human stages of life. (Your urine in the morning has a different concentration of salts, metabolites, and even some dietary compounds than at other times of the day. This "first void" is often the stressing system design requirement for storage and processing systems for urine.)

Most water processing systems have common elements. First, the neutralization of biological activity through sterilization or biocides (chemicals that kill life). In the spirit of the famous line from the movie Jurassic Park, "life will find a way"—wherever water is present and not extremely toxic, life will find it and utilize it to propagate. Molds, fungus, algae, and bacteria all thrive in water with any amount of nutrients, and urine and even cabin condensate have all sorts of nutrients. Thus, the processing of water must first control the contamination of the system through depressing the propagation of growth of "unwanted" lifeforms. Nominally in US spacecraft this has been elemental iodine, though elemental silver is also a candidate [9] and phosphoric and chromic acid, or sulfuric and chromic acid, are utilized for pure urine.

The second element of water processing is methods to remove unwanted contaminants or convert unwanted contaminants to desirable, or at least not harmful, potable water elements as defined by the SWEGs. This can take the form of gross filters, fine filters, chemically active beds, electronic filters, and finally biologically killing processes such as ultraviolet light or heating.

And finally, the third element is the final conditioning, or "polishing" of the water to remove nonharmful, but undesirable elements that, though no longer dangerous for humans, can decrease the palatability of the water. This includes filters to reduce turbidity (suspended particles—causing cloudiness), and addition of key elements that effect the taste of water. Contrary to popular opinion, pure water is not good. It is toxic to many forms of life, and because of water's proclivity to hold almost any contaminant (Water is often referred to by general scientists as the "universal solvent."), pure water leaches out chemicals from containers, plumbing, and processing equipment. This will become more important as we discuss long-duration systems.

4.3 Food management

In our hierarchy of needs, after air and water, food is the next needed element for human crew sustainment. On short missions—as on short hikes in the wilderness—the answer to food needs is to bring along those wrapped nutrition bars and bags of dried foods. Just unwrap and eat or pour in a bowl and combine with water, and voila, you have an instant meal. But for those that have any backpacking experience, you know what's left—all the food, packaging, and other waste from cleanup. Thus, food management, though seemingly simple, becomes more complex quickly—even ignoring the nutritional requirements [10]. MOST of the weight of items leaving the International Space Station (ISS) today is not experiments, new high-tech materials, or even the crew returning, it's the waste collected on the ISS from humans and their food production, processing, and consumption. Much is burned up in the atmosphere and falls gently to Earth or dissipates in the atmosphere, though some is also returned for science—to study metabolic impact and responses of the humans that have produced the waste.

How much food each human needs is usually measured in calories. But health also includes the variety of food and the inclusions of many compounds (vitamins and minerals). This is why whole menus are prepared by the astronaut kitchens on Earth and the ISS Food Systems group at the NASA Johnson Space Center. There is an element of psychological need of food too, with the consistency (smooth, grainy, crunchy), taste (salty, spicy, sour, etc.) and visual appeal all being important. (In one instance, beef stroganoff was used in a hazing incident on the Space Shuttle and smeared on the privacy curtain of the toilet prior to a "space rookie's" turn to clean the commode. Yes, it can look the same as human feces right out of the package in the right context.)

Food preparation usually includes the introduction of moisture, very often taking the form of the injection of hot or very warm water to reconstitute dried food for a warm meal, or even coffee or tea. With the aforesaid menu providing variety by having been packaged individually to allow for the ability to "mix and match" different meals—a long way from the Apollo-era squeeze tubes of paste.

Today we also have ovens or other heating means to add another element of variety for dryer foods not sufficiently heated by addition of water.

Of course, food "management" also becomes food production on long-duration missions where no resupply is available. That's a subject for a latter section.

4.4 Thermal control

From the human perspective, thermal control is easy to quantify. A person is either comfortable, or not. However, due to the activity of humans, ranging from sleeping to vigorous exercising, maintaining a comfortable environment is one of the most difficult functions of a life support system. Temperature of the air is not enough. Depending on the airflow movement about the body, 70°F (21°C) can feel pleasant, cold (if in a cooling environment such as an air-conditioned cabin), or warm (if in a heating environment.); and that is also dependent on clothing (more on that in Section 4.6). And of course, almost everyone has experienced the relative comfort of people in a room. The comedy skits about one spouse wanting the heat turned up and the other wanting the room cooler are not imaginary, but very real. Humans perceive the surrounding air differently. This may be due to mass—weather muscle or fat, the more mass a person has the greater their insulation—metabolic rate, sweat gland activity, and even sickness. Despite all these variables, NASA has attempted to define what is acceptable. NASA's requirements for human comfort include a "comfort zone" bounded by a minimum and maximum temperature on the one hand, and the relative humidity on the other. In general, the box spans dry bulb temperatures of 64.5–80.5°F and 25%–75% relative humidity [11], as shown in Fig. 5.

But thermal control in human life support systems is not only about the human. The equipment surrounding the humans is also important to keep within operating limits and, due to condensation considerations, interfaces with the human cabin in ways that could harm equipment. Thus, it's not enough to just control air temperature, but also the temperature of equipment. Usually this is done through both convective air cooling—blowing air over the equipment like the fan of a computer—and conduction cooling—setting the electronics on a cold plate or actively cooled surface. For all human spacecraft, a pump looped system is used to carry the heat from its source (the air, electronics, thermal radiation from the sun coming in the window, etc.) to where the heat can be rejected—usually outside of the spacecraft pointed mostly away from the sun. The most common heat rejection device on a human spacecraft is a radiator. But rather than a ground-based radiator that relies on the heat being expelled to the surrounding air, a space radiator can only reject heat through electromagnetic radiation of a certain wavelength.

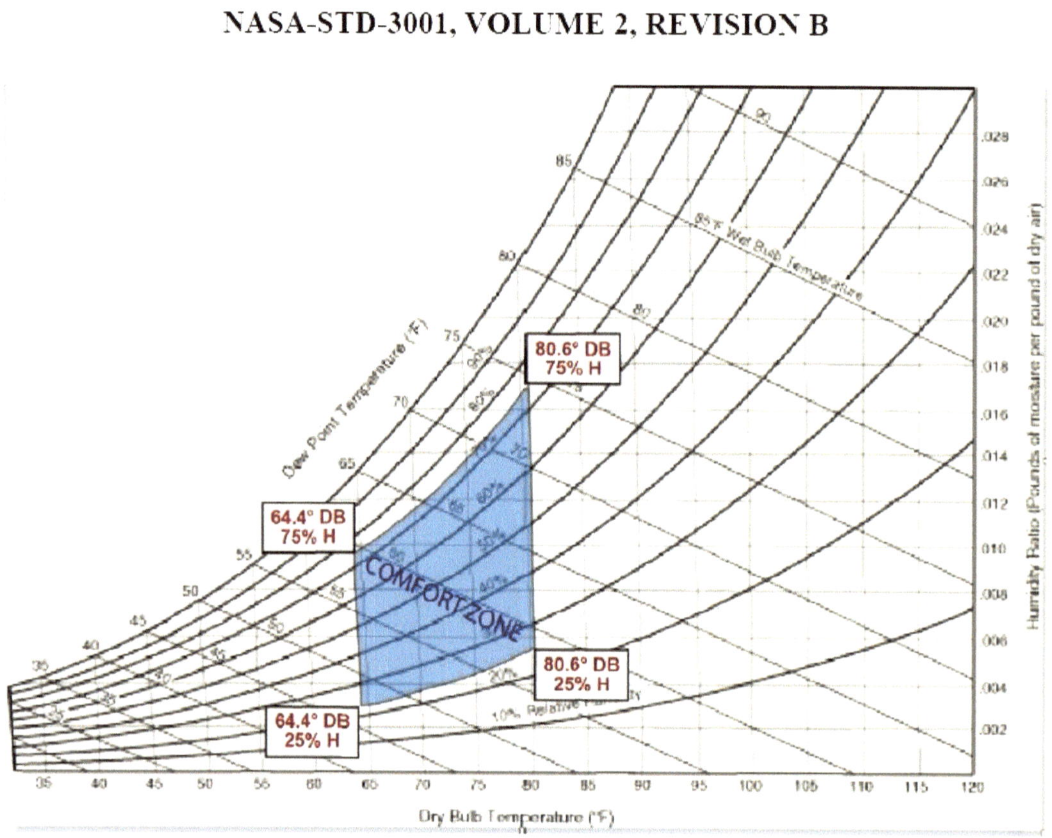

FIG. 5

NASA comfort zone.

Figure 2 of NASA-Std-3001 Volume 2.

This can often cause issues for spacecraft in LEO, because the Earth is a relatively warm heat sink (average temp about 300 K compared to the 4 K background of space). In interstellar travel, this will not be a problem.

One other subtlety should also be mentioned. Spacecraft can be designed, thermally, in two ways; cold biased and hot biased. In cold biased, the spacecraft in general operation will lose heat faster than it's generated. Thus the purpose of the thermal control system is to keep things warm by turning on heaters or pointing the spacecraft toward heat sources (the sun, the planet you are orbiting, etc.). Hot biased spacecraft must actively reject heat as described above (collect, transport, and radiate) or they will eventually become too hot. Human spacecraft are now designed with the latter method. However, this requires that the spacecraft maintain enough heat generation inside to be hot biased. A sudden shutdown of internal heat generation equipment, such as happened on Apollo 13, can result in runaway cooling and a very uncomfortable environment inside the spacecraft. This will also become pertinent to interstellar life support.

4.5 Waste management

Humans are, when it comes to food, "batch processors" (For air, we are more continuous, in that we never stop breathing. Review Fig. 1 and try to identify "batch" vs continuous metabolic needs.). We intake food at regular intervals (mealtimes), and then expel the wastes in another, hopefully regular, interval (Luckily, nature has given us the nominal ability to control those waste intervals. Otherwise a spacecraft (or a house) would become quite unpleasant. Anyone that has trained a puppy knows this to be true.). So naturally the waste collection and management system for a spacecraft also has to accommodate those "batch mode" operations. Luckily, also, humans can bifurcate their water waste and their solid waste due to both our biological plumbing and the plumbing accommodations of the spacecraft. However, during times of sickness or stomach upset, we also have mixed digestive waste (diarrhea) and off-nominal acidic waste (vomit) that must be dealt with. This strains the design of the method of collection, processing, and recovery of the useful parts of the waste.

For fecal systems, a typical advanced commode (toilet) is shown on the left in Fig. 6. These units are designed for a microgravity environment and represent a major advancement from the Apollo-era systems, which were mainly a collection tube and reservoir for expulsion of urine and—universally hated by astronauts—a specially designed plastic bag with a thumb accommodation to "separate" the fecal matter from the astronaut's body—needed with no gravity or air flow to promote separation (The popular 1960s and 1970s songwriter Donovan (of "Mello Yellow" fame) got it mostly right in his song "Intergalactic Laxative." https://youtu.be/mn3Jec3V0As.). For the short term, astronauts wear a high-tech diaper in case they cannot reach the accommodations for waste collection.

As noted, collection of liquid waste and solid wastes separately is ideal, as this reduces the need for an initial process to separate the wastes for processing. The designer must consider the batch mode of collection and allow for storage until the system is ready to process the waste. How large the waste tanks are will depend on the number of crew, capacity of the processing system, and the length of storage time. (Note the need to reduce potential for microbe and bacterial growth during this storage time mentioned in Section 6.3.1.)

The collection of wastes poses special consideration for those with male plumbing and those with female plumbing, as also shown in Fig. 6. Due to geometry of the sexes, urine collection of males is

FIG. 6

Flexible membrane commode (left two pictures) and male and female urine collection (right top to bottom).

Courtesy of NASA.

seen as more straightforward, while female collection must consider not only the geometry, but the proximity of the vagina to the urethra, and the need to prevent contamination of the former during use of the latter. Though it should be said that the male collection system (also used by long-duration pilots such as U2 spy plane pilots) have been known to fail allowing urine to enter the leg and boot of the garment thereby causing great discomfort (and sometimes injury) by the exposure of the skin for a long period (hours) to the acidic and chemically laden urine. Mostly these failures are caused by the overestimation of the size needed for the collection device (Thus, the instructions to male pilots to size the collection device to the actual geometry and not their ego.).

4.6 Crew systems/human accommodations

Crew systems or human accommodations tends to be a catch-all that includes such disparate items as lighting, clothing, hygiene facilities, medical stations and capabilities, exercise, and other things that not only support life but make it pleasant and tolerable in a small, enclosed space. There are many images online of astronauts doing normal things like shaving, brushing teeth, and running on a treadmill. All of these are necessary (a partial list is shown in Fig. 7), as they are during Earth-bound life. In a microgravity environment, some of the accommodations take on life-threatening importance, as the deterioration of the human physical body is very real and has deleterious effects on the functioning of the human body's systems and must be counteracted with vigorous and daily exercise. Thus, any interstellar system MUST provide some sort of gravity vector—the only argument among professionals being the magnitude of the artificial gravity: Lunar, Martian, or Earth normal?

Psychologically, lighting, and its effects on both orientation (lights are usually "up") and mood, is more understood than at the beginning of the space age. The modern traveler today will notice that light intensity and color are an integral part of the experience of travel. The advent of light emitting diodes (LEDs) has revolutionized this aspect of interior design of airplanes and is now making its mark on

	Mass Required (kg/CM-d)		
	Water Rich	Nominal	Water Lean
H_2O – Shower	2.30	0.00	0.00
H_2O – Body Wash	2.00	2.00	0.91
H_2O – Urine Flush	0.50	0.50	0.30
H_2O – Oral Hygiene	0.37	0.37	0.00
Totals:	5.17	2.87	1.21

FIG. 7

Levels of water use for crew accommodations.

Derived from M. Anderson, M. Ewert, et al., Life Support Baseline Values and Assumptions Document EC2 NASA, Houston, 2015, p. 51 NASA/TP-2015-218570.

Item	Mass (kg/CM-d)
Feminine Health	0.008
Toilet Paper	0.028
Gloves	0.007
Dry Wipes	0.013
Detergent	0.058
Disinfectant	0.056
Paper	0.077
Tape	0.033
Clothing	0.486
Total (Female)	0.766
Total (Male)	0.758

FIG. 8

Crew personal provisions.

Derived from M. Anderson, M. Ewert, et al., Life Support Baseline Values and Assumptions Document EC2 NASA, Houston, 2015, p. 51 NASA/TP-2015-218570.

spacecraft. Limitations of fluorescent lighting provided for a harsher, more clinical environment in the Apollo-era spacecraft as well as Shuttle-era spacecraft.

Clothing is also another major issue of crew accommodations. For long durations, the storage, cleaning, and reuse of clothing takes on added importance. The types of materials used for clothing also has a large impact on the filtering of cabin air (lint of natural clothes—such as cotton—is a cause for failure of many air handling systems) as well as the safety of the crew due to flammability and outgassing concerns. Nomex was developed due to these concerns and is a versatile and combustion-resistant material (made especially important in high oxygen environments). Of course, there are other necessary items to support humans. Fig. 8 [3] gives some examples.

Washing of clothing is also a concern, as it utilizes a precious commodity (clean water) and introduces soaps (surfactants) that then must be removed from the waste stream, as is true for Earth-bound waste processing plants. Chemical cleaning (dry cleaning) uses harsh and very persistent chemicals that are also not compatible with a closed system such as on a spacecraft.

Medical provisions are very difficult to design for and install on spacecraft. What type of diseases, nutritional effects, injuries, or hereditary situations must be accommodated? Couple this with galactic radiation (high energy particles permeating the vacuum of space) and the need for medical capability is inescapable, but difficult to implement. With rapid advances in medical technology on Earth, when is it ready for spacecraft use? Any reader today that has submitted to medical diagnostics knows that the medical equipment is bulky, needs high power, and must have special architectural accommodations. Something as simple as an X-ray requires mass allocations and power allocations that are much higher than a nominal spacecraft's capabilities. Moving to MRI requires liquid helium temperatures, superconducting magnets, and intense magnetic fields and the associated shielding.

Drugs are also a problem. If you try to take a big enough supply for a decades' long trip, they will expire before being used. This indicates an ability to fabricate drugs as needed, which could lead to other problems of contamination, not to mention the source of the chemicals used in the drugs.

And finally, in space, exercise in a microgravity environment requires resistance machines that both pull and push on muscles. Even in partial gravity (Moon gravity is 1/6th of Earth, Mars 1/3rd of Earth) does not necessarily provide enough resistance for a human body that has evolved over billions of years in Earth's gravity.

Now that the basics of ECLSS are understood by the reader, the special needs of long-duration, multigenerational ECLSS will be discussed in more detail.

5 Introduction to long-duration life support

Long-duration life support is a special branch of ECLSS that considers the need to provide life support without resupply from Earth, and with systems and processes that can be counted upon to last for generations during long trips outside of the solar systems and to the stars. Just a trip to our nearest star (Proxima) will require tens, if not hundreds, of years with present technology. Before diving into those considerations, a brief background of the history of "long-duration" life support is warranted.

5.1 Some history—Space stations, ground simulations

Almost as long as humans have been flying into space, the desire for more permanent outposts in space has been a driving factor for space technologists, scientists, and science fiction authors. Gerard O'Neil, in his book *The High Frontier* [12], and even earlier, Werner Von Braun's *Collier* series [13] speculated about hundreds, thousands, and even millions of people living and working in space. In fact, Jeff's Bezos' company Blue Origin's goal used to be stated as "Millions of People Working in Space."

During the first decades of the Space Age, after the landings on the Moon, the United States orbited a research laboratory called "Skylab." Derived from the third stage of a Saturn V moon rocket, this outpost was the first to have large, free volume in which astronauts could live and work (and play!). Fraught with some technical difficulties due to a design issue which resulted in a failure of the protective fairing during launch, Skylab also demonstrated the adaptability and tenacity of humans in space to do real, unplanned, and dangerous work. However, Skylab's longest duration mission was just over 84 days [14], and though much science (and living) was completed, it still relied on resupply from Earth for each mission. No meaningful recycling was done, nor attempts to optimize many first-off ECLSS systems that were designed to accomplish the mission of three people living and working in space.

The Soviet Union (now Russia), China, and the United States have embarked on more ambitious LEO habitats, including Salyut (prior to Skylab, but small comparatively), Mir, the International Space Station, and recently, China's Tiangong [15], among others (Fig. 9).

But all of these attempts at longer human duration in orbit have been designed, operated, and even repaired with extensive Earth-based support and supply. As such they are designed—when something goes wrong—for resupply and installation of often large replacement parts (a pump package, for example, rather than the resistor on the electronic board that may have failed in the pump controller). In that regard, though the general public view of "we've done this before" prevails when talking about longer mission, we have not really designed a life support system that can truly survive away from Earth and Earth's support industries and technologies for any appreciable length of time.

On the ground, Earth analogs have been used to simulate long-duration life support and missions in general. However, most have been designed to simulate on-planet surface operations of future

Moon and Mars bases. One example [16] that stands out is the Biosphere 2 project in Southern Arizona. Designed as a true analog of a closed ecological life support system, this experimental apparatus, started in the 1980s, endeavored to demonstrate what the human species knew about the stability and operation of a closed, heavily biological system much smaller than Earth (Biosphere 1—hence the name). As it was, while a tour-de-force of technical innovations and creative thinking, the

FIG. 9

Past and current long-duration orbital space stations.

(Continued)

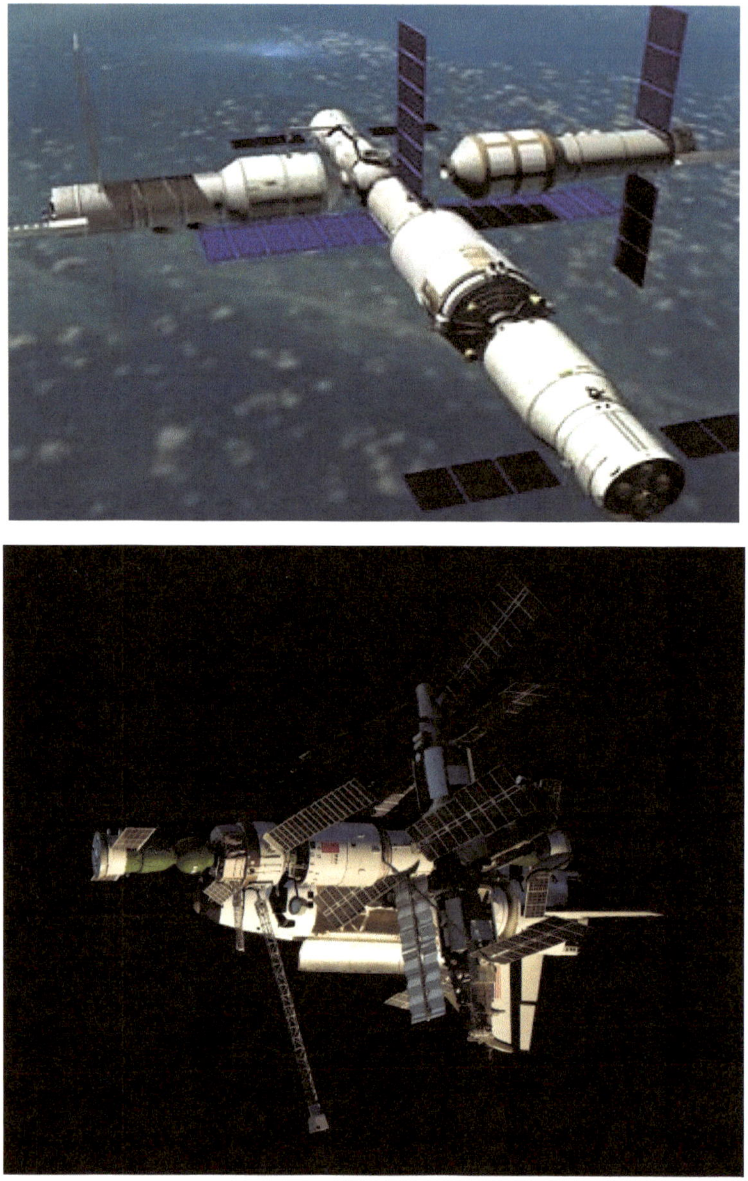

FIG. 9, CONT'D

experiment once populated with Humans did exhibit the limitations of technology and our understanding of closed biological systems to stay stable and productive in its ability to support 8 humans for extended durations. As documented by one of the participants [17], ants ate the insulation on wires, cockroaches overwhelmed the facility, and oxygen was absorbed into the concrete in the form of carbon dioxide, thereby reducing oxygen available to the crew. As such, for the subject of long-duration life support, Biosphere 2 can be seen as a cautionary tale of the hubris of humans believing that they can "engineer nature" to create a smaller-scale, robust, reliable, and efficient life support system.

5.2 Considerations for long-duration vs short-duration life support

From the overview of ECLSS provided, perhaps the insightful reader has already observed that life support is complex, fraught with unknowns and unexpected potential failures and biological incompatibilities, and heavily reliant on the interactions of that life support systems with its valuable cargo—the humans within it. And they would be right. Long-duration life support systems, however, have the added challenge of needing to be robust and resilient over periods of time measured in years and decades—made even harder by the lack of resupply from Earth in the cases contemplated in this monograph on interstellar flight.

Three areas emerge from long-duration life support considerations that dominate this complexity: (1) closure, meaning recycling of vital elements, molecules, and materials; (2) maintenance and repair of the systems to perform these functions; and (3) the presence of, or lack of, sufficient uniform acceleration that influence almost all processes and the biology of humans themselves.

5.3 Artificial acceleration or microgravity?

The third item above should be discussed first as it influences almost all aspects of human survival in space for long durations. As mentioned before, the human body (and all life on Earth) has evolved in a gravity-induced acceleration environment of $9.8\,m/s^2$ ($32.2\,ft/s^2$). In biological terms, humans are evolved to take advantage of (and survive against) a "1*g*" environment. Our bodies have responded to this constant in our lives by evolving a skeletal structure, muscles, connective tissue and tendons, organs, glands, and cellular processes that both rely on and adapt to different stages of life. Babies' bones are soft at the outset, allowing for passage through the birth canal, and harden over time as they adapt from lying to crawling to standing, walking, running, jumping, and so forth. At this time, there is a substantial amount of doubt as to whether humans can procreate or even survive without an acceleration environment. For example, certain processes to produce human immune T-cells rely on a gravity vector, and indeed, astronauts experience a rather large decline in white T-cells over the course of a long microgravity mission [18]. For this reason, and others explained below, a generational vehicle such as an interstellar spacecraft is assumed to have some level of gravity. Speculation continues whether sufficient gravity levels are a fraction of 1*g* or the full 1*g*. But in most life support questions—as the saying goes in the industry—1/6th gravity is a lot closer to one gravity than zero.

While not mathematically correct, from the perspective of designing systems this is a truism. For purposes of the below discussions, we assume that some gravity exists on an interstellar transport. Whether this is provided by constant acceleration and deceleration of the system (reference

Chapter 1), or a physical means to provide acceleration, such as a spinning spacecraft, we will assume that it exists. Whether reduced acceleration is sufficient for the survival and reproduction of humans on board will be left to another chapter (or even, book).

5.4 The concept of closure

As will be explained later, not everything is recoverable back to a state that is usable in an ECLSS system. Whether a system is "closed" or not is a matter of degree. How much a life support system is closed or not is often expressed in the percentage of recovery during each cycle. For example, perhaps you can recover 98% of the water from urine. That means that if a human consumes 100kg of water over a period of time, and then expels the water as urine, what is returned to the human over the next period of time is 98kg of water. After 10cycles, that original 100kg of water would be reduced to just under 82kg (0.98^10=0.817).

This concept extends to almost any element, molecule, or chemical compound: oxygen, carbon, nitrogen, sodium chloride (salt), calcium, iron, etc. How complex this can be is easily understood by reading the label of a vitamin and mineral supplement, which is probably in your kitchen at this time.

Closure also extends to the loss of material to the outside environment. A spacecraft is a very well-sealed pressure vessel. But even highly sealed containers leak; either through flaws in the seal (even welds can leak), or through permeation of gases through some types of metal or plastics. In today's spacecraft, a nominal leak rate can be 0.05%/day [3], as shown in Fig. 10. While this seems like a low number, again, over a period of 10years, almost 75% of your atmosphere would be gone. (In 5years, 60%, and half your atmosphere would be vented to space in just over 3.75years.) This gas, of course, must be replaced. While oxygen is present in food and recovered from the carbon dioxide that is exhaled, nitrogen (if that is your buffer gas) is an issue (Forget precious metals, the rich space miners of tomorrow will be mining nitrogen.) as it must be replaced from stores, of which internal, pressurized nitrogen containers are seen as the most robust at this time (An external tank at interstellar distances, if isolated from the spacecraft, may be very efficient, as deep space at 4K will liquefy almost anything. However, as nitrogen freezes at 63K and oxygen freezes at 54K, the external tanks must be kept warm, also a failure mode—a failure that is perhaps now on the outside of the spacecraft.).

Closure becomes more complex when the food sources are brought into play. But this will be discussed in Section 6.5.

Parameter	Value	Unit
Leak Rate		
Lower	0.00	%/d
Nominal	0.05	%/d
Upper	0.14	%/d

FIG. 10

Leak rates of spacecraft.

Derived from M.S. Anderson, M.K. Ewert, J.F. Keener, S.A. Wagner, NASA/TP-2015–218570. Life Support Baseline Values and Assumptions Document. Responsible National Aeronautics and Space Administration Official: Molly S. Anderson CTSD, Lyndon B. Johnson Space Center National Aeronautics and Space Administration Mail Code EC22101 NASA Road One Houston, Texas 77058; March 2015, Table 4.1.1.

5.5 The importance of maintenance and repair

It is often said that we live in a throwaway society. Even when we buy a "durable good"—defined by the United States government as something that is storable, can be inventoried, and has an average life of more than 3 years—we do expect at some point for that dishwasher to break, that car to die gasping on the side of the road, and that must-have cellphone to be obsolete or overtaken by new technology within a few years and replaced. Not getting into the shortsightedness of assuming endless resources on Earth, in a long-duration spacecraft our expected capability to replace critical hardware on which our life depends would be a deadly delusion. Ignoring Star Trek "making matter" replicators, the capability to bring a whole world-wide industry of capabilities along with us to the stars is not possible in the foreseeable future. Just a few generations ago, in the pioneering days of westward United Sates expansion, the ability to fix, or kluge a solution together, for a critical piece of survival gear often determined survival or death (or at least economic prosperity or ruin). On an interstellar journey, death is the inevitable outcome.

There are whole service industries that determine the critical supplies needed on long sea voyages. But these voyages have a clear range, with assumed redocking at destinations or intermediary ports along the journey to replace critical supplies. And what is needed on a voyage has changed with the technology of the seagoing vessel—oil, pumps, and seals have replaced sails and masts—and the list goes on and on. An interstellar spacecraft is a wholly new type of ship, and before one sets out, what will need to be maintained and how it will be maintained will all have to be determined and the planning worked out in detail. Often, additive manufacturing (3D printing in the vernacular) is held out as a panacea for this conundrum. However, 3D printers are limited in their materials, and the machines themselves need to be maintained (what will break—do you have a 3D printer to fix the 3D printer?) So, the design of the entire spacecraft needs to be limited by the materials that can be printed, and the raw materials that can be carried along (or recycled). This flies in the face of design optimization by using the material best suited for the function at hand.

The overarching question is, "How do you determine what will break, how critical that item is, and how will you replace it?" In the design of an interstellar spacecraft, and especially for a life support system that needs time-critical repair before violating one of the "rules of threes," this is an overarching, critical, and necessary question that needs to be answered.

6 Summary of state of the art of long-duration life support

Life support has not progressed as far as many assume. Over the last 30 years of spaceflight, the really "new" technologies brought to bear on the needs of life support can probably be counted on one hand. Despite the consistent investment of NASA in innovation funds, there have been many derivatives of Earth-based technologies that have been adapted for space, and frankly that has been the driving force for many innovations in the space sector. In fact, in 2015, NASA came out with a slogan that stated, "Technology Drives Exploration" [19] and in an update "Innovation and Diversity Drive Exploration" [20]. But seen in the light that very often innovations are adaptive from industry, the opposite might be true: that exploration drives innovation. Intuitively this is true. NASA wants to explore the solar system, and technologists and scientists propose innovations that will help them reach their goals. In reality, there is a bit of back-and-forth and a very much "chicken and the egg" philosophical discussion. So it is with life support...

6.1 General limitations of current technologies

Innovation and technology advancement in the realm of life support is slow because the criticality of the hardware coupled with thorough testing that, in the case of many life support technologies, cannot be "hurried" by quick cyclic testing. While an aircraft pressurization system and structures can be tested on the whole aircraft at a rate of many "simulated" flights per day, thereby staying ahead of in-service aircraft undergoing these same cycles, the biological nature of life support in its totality dictates the pace of testing. Bacterial growth, mold growth, seal deterioration, permeation of gases into equipment, and even leaching of metals from coolant pipes, are processes dictated by biology and physics and cannot be accelerated. This means that a life support technology cannot be considered to be truly "certified for flight" until it has either undergone testing that is of equal length to the flight, or, at least, determined to be repairable and maintainable (see Section 5.5). In the case of a multigenerational interstellar craft, the latter method is the only one available to the technologist.

Cost and availability of true flight tests is another issue. While the frequency of space-bound flights is on an upward swing, it still costs many thousands of dollars per kilogram to launch an experiment into orbit to experience the true environment of space. NASA calls this the Technology Readiness Level Nine technology [21]. Couple this with the length of test needed, and new life support technology has a long and tortuous path to truly be adopted. Thus, many of the technologies discussed below may be state of the art, but yet decades old, with only refinements in the implementation.

6.2 Air management

Air management, as described in the prior sections, involves the management of overall pressure, constituents of gases, and the removal or remediation of byproducts of, mostly, metabolic activity of the humans within the spacecraft. Concentrations of each gas are important and the effects of which have emerged as the length of time within the current spacecraft have increased combined with overall volume of the cabin.

6.2.1 Pressure control

Most of the recent innovations in overall air management have come in the electronics of control that meter out the replacement gases (oxygen and, if the buffer gas, nitrogen). Prior to the Space Shuttle, most spacecraft were operated on pure—but reduced pressure—oxygen. This made regulation of overall pressure as simple as having the proper setting on the stage one and two regulators. However, with the advent of the space shuttle, and the desire for "shirt sleeve" environmental conditions, a standard atmospheric mix of nitrogen and oxygen was dictated by the higher, sea-level pressure. Commensurately, the control of both nitrogen and oxygen introduction into the cabin had to be coordinated. As oxygen is used up, and the exhaled carbon dioxide is removed, the pressure of the cabin is reduced. But if nitrogen is introduced to make up the pressure, a gradual depletion of oxygen would result. In modern systems, this is performed by monitoring the concentrations of oxygen and overall pressure, and an electronic system deciding which gas to introduce. However, this is prone to some critical issues in an emergency scenario (a micrometeoroid hole in the cabin, for example) and some spacecraft have replaced, or are contemplating replacing, pure nitrogen storage with a mixed gas "nitrox" (70% nitrogen/30% oxygen, for example) primary storage tank so that a failure that only introduces nitrogen and displaces oxygen is precluded [22]. Utilizing the assumption that the buffer gas for an interstellar spacecraft is nitrogen, the ability to combine and store the gas is a necessary technology on the spacecraft.

6.2.2 Oxygen production

Regenerable oxygen product is most efficiently performed by electrolysis, of the splitting of water molecules (two hydrogen atoms and one oxygen) into their constituents. This is now performed by a form of reverse fuel cell technology [23,24]. Whereas a fuel cell produces electricity through controlled combination of hydrogen and oxygen, therefore making water while releasing bond energy that is converted to electricity, the reverse process of using electricity to divide the water into hydrogen and oxygen through membrane fuel cells has proven to be reliable and operable on the ISS. Limitations include what to do with the excess hydrogen, and sufficient control to ensure hydrogen doesn't become concentrated or leak causing an explosion hazard. For long-duration life support, the processing of water to the correct cleanliness level to not foul the electrolysis system is important. On the ISS, this is done by processing the condensation water from the condensing heat exchanger.

On the horizon are higher pressure electrolysis systems that ameliorate the need to take lower pressure oxygen up to the storage tank pressure [24]. Oxygen pressurization pumps are maintenance-intensive and can be a source of fire and explosion hazard.

6.2.3 CO_2 collection and processing

The collection of carbon dioxide (CO_2) from the atmosphere, where it has been introduced by human metabolic output, has probably seen the most advancement over the last years. This has been driven by the need to maintain the concentration of carbon dioxide in future spacecraft at below 2000 parts per million (ppm). Whereas the ISS and Space Shuttle allowed nominal concentrations of 4000 ppm, recent physiological sciences question whether the long-term effects of a 4000 ppm level of CO_2 is detrimental to the cognitive capabilities of those exposed [25]. (As a contrast, the Earth's atmosphere is now at a human-era historically high level of over 400 ppm, or almost an order of magnitude less.)

This has driven innovation because the efficiency and speed of scrubbing the CO_2 out of the air is a function of the concentration already in the atmosphere. Given chemical amine systems or molecular sieve systems that absorb CO_2 and then desorb using a reduced pressure (known as pressure-swing absorption) [26,27] or collection under a lower pressure, the efficiency, or the size of the system, needs to be much higher to remove CO_2 at the same rate as from the higher starting concentration atmosphere. As humans produce CO_2 at a constant rate, the rate of CO_2 removal to support the same number of people must be equivalent and not reduced in the new systems.

In order to achieve this better atmospheric concentration level, advances in Amine formulations are continuing, and other systems, such as liquid contact ionic sorbents, are also being proposed [24].

The processing of CO_2 to return the oxygen to its elemental form and recover the carbon that was derived from food and combined with oxygen in the human body for metabolic energy takes on two forms. Direct electrolysis and Sabatier processes. Direct electrolysis is much like water electrolysis (dividing the water molecule into elemental hydrogen and oxygen) but takes much more energy and higher temperatures to achieve. One method used—and presently used on the MOXIE experiment now on Mars and hailed for producing oxygen from the Martian atmospheric CO_2—is solid oxide electrolysis (SOE), where an electric field is presented over a ceramic membrane [28] at high temperatures, thereby disassociating the oxygen and producing oxygen and carbon monoxide (CO). The CO is then further "cracked" by lowering the temperature to release the final oxygen and leaving a carbon powder.

The second method is to introduce the CO_2 into a Sabatier reactor [29]. This reaction, named after the discoverer, combines CO_2 and hydrogen at elevated temperatures and pressures to produce methane (CH_4) and water. This reaction has the advantage of utilizing the hydrogen produced from the oxygen-generating electrolysis method (see above) but has the disadvantage of producing methane—a combustible and trace contaminant. A Sabatier system has been flown on the ISS with varied success, with thermal management and continuous operation being some of the difficulties [30,31].

A hybrid method of combining the SOE and Sabatier processes was attempted under a NASA contract in 2014 in which hydrogen was introduced directly after the SOE breakdown. The experiments were successful [32], with the added benefit that the lower-temperature Sabatier process (300–400°C) was applied after the SOE process (600–650°C) and therefore removed the production of elemental carbon from the process.

6.2.4 Humidity collection and processing

Humidity collection has not changed appreciably since the Apollo days. The concept is familiar to anyone that's ordered and iced tea in Texas in the summer. A cold surface in a high humidity environment will cause moisture to condense out of the atmosphere when the temperature of the surface is less than the dew point of the atmosphere. In engineering terms, the carrying capacity of moisture in air is a function of the temperature, and at any given relative humidity, if the air is reduced to this dew point, water will condense out (Weather services at airports use this to calculate the bottom of the cloud layer. The air lapse rate in a standard atmosphere is 5°F for every 1000 ft. So, if the ground level temperature is 80°F, and the dewpoint is 65°F, then the cloud layer starts at $(80–65)/5 = 3000$ ft.). The problem with condensing heat exchangers is that the water condenses out while air is blowing over the heat exchanger, thereby risking the generation of airborne water particles and water blowing out of the vents. The Apollo Lunar Module exhibited this problem. To remove the moisture, a "slurper bar" sucks in the moisture from a hydrophobic (repels water) surface, after which the wet, gas-filled water is spin separated into water and air. The complexities of this system, and the danger of contamination of the hydrophobic surface, result in deterioration of the collection method over time, and are significant. Moving parts, motors, and continuously wet surfaces all conspire to introduce failure modes in the system which will be hard to repair/replace in an interstellar mission.

Two technologies have been developed that control humidity. One is a membrane system called ionomer water processor technology, which uses a membrane of ionomer material to draw water from the humid air to a lower humidity environment such as vacuum or dry air [33]. Membrane technology was discouraged in the first 40 years of spaceflight due to the potential leak path around the multiple edges of the membrane. Paragon developed methods for ensuring the membrane edges are sealed. One unique aspect of this technology is that the water is selectively passed through the membrane while other atmospheric gases such as oxygen and nitrogen do not pass through. The downside of this technology is that once the water has passed through the membrane it still has to be condensed, collected, and stored. However, in the partial gravity environment, this is a simple process. (In short term missions, on both Dragon2 and the CTS-100 Starliner, this technology expels the water to space.)

The second technology utilizes centrifugal motion (spinning) to separate the water droplets coming off the condensing heat exchanger while simultaneously "pumping" the collected water through a pitot pump to a holding tank. This "spin" method [34] can uniquely be utilized in both microgravity and full Earth gravity. This second technology is state of the art for long-duration systems and will be coming online for LEO stations in the future.

The aforementioned need to maintain and repair systems will apply to these technologies. IWP technology requires rejuvenation of the membrane sheets, which is well understood but requires specific chemicals (acids) that will need to be manufactured or stored on long-duration spacecraft if the mission time exceeds the ability to carry along sufficient replacements. On the plus side, sheet membranes can be stored very compactly over indefinite times. For the spin technology, maintenance and repair is much more straightforward [34].

6.2.5 Trace contaminants

Trace contaminants put out by both humans and the off gassing of materials used in the construction of the vehicle put a load of chemical contaminants in the atmosphere. As mentioned in the opening sections of this chapter, the levels of allowable concentrations, or SMAC levels, vary by each compound and by the duration of exposure. Events that increase concentrations of trace contaminants (a burned-out circuit board, a small fire, or biological production by spoiling food left out) must be taken into account and, though allowed to increase the contamination, must be brought back within SMAC levels in a timely fashion.

Trace contaminant systems are usually multitier as not all compounds respond to the same methods of removal. For example, ammonia (NH_3) is a common contaminant produced by humans, but it is water soluble and usually removed by the humidity control system. But methane is explosive in concentration (as is ammonia), not very soluble, and very often removed by high temperature catalytic oxidizers (HTCOs), which is just an engineering way to say, "It's burned on a high temperature material surface." (The catalytic converter on your internal combustion car works under the same principal, oxidizing nitrogen oxide (NO) and uncombusted hydrocarbons into less harmful compounds with help of a platinum catalyst.) Of course, those compounds produced by the HTCO must also be removed if harmful. For example, some fluorinated compounds when passing through the HTCO will produce hydrogen fluoride (HF). This, when combined with water, is one of the more powerful acids. However, HF is easily removed by activated carbon [35], which is a high surface area material important to almost all purification systems for both air and water.

The state of the art in trace contaminants control technology is mostly concentrated around the latest and improved inorganic impregnates for activated carbon to take advantage of the latter's high surface area and to target the adsorption characteristics for the specific contaminants—silver and iodine are examples of impregnates. (Note that the advent of nanoparticles offers opportunities for higher surface area to weight ratios, and future improvements may be substantial in trace contaminant control systems due to this relatively new technology.) Cations can also be used as an impregnate for removing other compounds. These pretreatments are highly proprietary and usually obscured in literature if they can be found at all. Another area of emerging expertise is different containment methods to prevent powdered activated carbon from exiting the system and clogging downstream filters.

For long-duration life support, the making and rejuvenating of these activated carbon systems involves heating, exposure to vacuum, or chemical means, which requires a nominally equipped lab. However, activated carbon itself is easily producible from almost any organic plant material (see Section 4.3). On Earth, activated carbon is made from corn husks, wheat stalks, and even coconut shells.

6.2.6 Measurement and verification

By far the most pressing technology is detecting the air (and water) contaminants. As the SMAC and SWEG levels illustrate in the references of Figs. 3 and 4, the number and variety of compounds that must be detected to monitor the effectivity of the trace contaminant control and water processing systems is substantial. On Earth, the most general chemical monitoring system are gas chromatographs (GCs) and differential mobility spectrometers (DMS), which can identify elemental and molecular concentrations of multiple species. However, reducing the MS size and power to those acceptable to a spacecraft has been difficult, with the first test units flying to the ISS in 2013 [36]. There is also the problem of how to calibrate the MS, which is usually done by injecting known compounds in very precise concentrations into the MS. The production of these calibrated gases requires highly sophisticated gas companies that, unto themselves, have highly sophisticated chemical detection technology. Thus, for long-duration life support, the calibration gas production has to be placed onboard the spacecraft. At this time, these capabilities are nascent and in need of full development.

As mentioned before, measuring the true oxygen concentration of the atmosphere is important to ensure a healthy breathing environment. The method of measurement involves three basic types of detection [37]:

(1) A chemical reaction that gives off electrons in the presence of oxygen;
(2) A change in light intensity given off by a fluorescing material when exposed to oxygen; and
(3) A change in the wavelength of sound, light, or electromagnetic radiation as oxygen (or other gases) passes through it.

The issues with technology using methods one or two is that the detectors drift over time (measured in months) and must be recalibrated or rejuvenated. This presents an obvious drawback for long-duration mission. However, method three has promise in not needing calibration beyond ensuring the operation of the basic electronics. Thus, wavelength modulation spectroscopy (one of the technologies utilizing the third method) is gaining attention for long-duration missions [38].

6.3 Water management

Water management technologies are mainly concentrated around the cleaning, storage, and delivery of potable water, with, like air, the measurement and verification of the cleanliness of the water being critical. Luckily, most of these technologies are mature, well understood, and in use. However, certain limitations on the filtering medium, prevention of microbial growth, and scrubbing of dissolved compounds are still evident in long-duration systems.

6.3.1 Water processing

Water processing is a mature technology for spacecraft. The ISS has demonstrated operability and robustness of the filtering technologies including particulate filters, activated carbon filters, and catalytic reactors to remove organic compounds. Sterilization techniques using ultraviolet exposure are also well understood (and commercially available). However, clogging of filters by microbial growth and contaminants continues to be a problem. This is solved by resupply of filters and changeout on the ISS.

However, for long-duration missions, where resupply is impossible, the ability to clean these filters continues to be an area of development. One of NASA's latest technology developments is "backflushing" technologies, which reverse the flow of the various filtration systems periodically to remove gross contaminants. As mentioned before, activated carbon filters can be rejuvenated through production of additional activated carbon and repacking of the filter systems. Backflushing may improve the filter life but quantification of the life-extension of this method has not been tested in a realistic environment. UV-emitting elements are a consumable that are not quite yet producible on a spacecraft, though light emitting diodes (LEDs) have very extended lives. As these are small, carrying of spares may be the easiest solution.

The most development over the last decades in water processing technology for spacecraft revolves around the prevention of microbial growth in the potable water. Iodine was the preferred method of the US space program, but discovery that the iodine in the water then lead to thyroid issues in astronauts (due to radiation absorption) has led NASA to now develop ways to remove the iodine from the water prior to consumption. The Russian space program has used silver microbial growth inhibitors. Chlorine systems are not allowed on spacecraft due to the off gassing of chlorine during the delivery process (The aerator on the typical American faucet serves to allow the chlorine to come out of solution and dissipate in the air in the home. This lower-level contaminant would be a big problem on a spacecraft due to contamination of sensitive equipment and the buildup to SMAC levels very quickly without scrubbing in the TCSS.). Silver has the potential to be a recoverable and safe method of microbial growth inhibitor.

6.3.2 Water storage and delivery

Storage of potable water must be done in tanks that resist any leaching, and thus most metallic tanks must contain a lining (usually a Teflon-type lining) that is inert to water.

Pumping systems that can deliver the water around the spacecraft are a life-limited item that must be planned in the maintenance, sparing, and repair cycle. As stated previously, the spares must be at the smallest level possible (e.g., individual electronic components, O-rings, bearings) to minimize spares storage. Peristaltic pumps [39] have shown operational life measured in decades due to their use in medical systems and resistance to contamination. Their drawback is lower pressure than positive displacement or centrifugal pumps. However, for most delivery needs, except for showers, their capacity and pressure are acceptable. (Whether showers are utilized, given their large water requirements, is a debatable subject for long-duration spacecraft. In 1g environments, a gravity-fed shower would be more than adequate, as they have been in nondeveloped countries for millennia.)

6.3.3 Measurement and verification

Unlike air measurement, water measurement technologies are rather mature and compact. Necessary monitoring, including total dissolved solids (TDS), pH (acid/base) levels, and turbidity are all well understood and require little consumable products. All these sensors are commercially available and are perfectly adequate for long durations with minimal accommodations, even on decades-long travel if properly maintained.

6.4 Waste management

Given the requirement, driven by health considerations, that a long-duration spacecraft have clear acceleration levels at least in parts of the spacecraft, waste management becomes much easier and straightforward. Restriction of the waste collection facility to the portions of the spacecraft with acceleration vectors solves most problems, as noted in the introductory sections of this chapter. Waste management then becomes an exercise in dividing the wastes at the source (e.g., excrement not combined with urine or hygiene water) and processing to prevent microbial growth in storage, and maintaining separation of these systems. Often this separation is described as "black water" systems and "gray water" systems, where urine and liquid derived from excrement are "black water," and hygiene, hand washing, clothes washing, and food preparation water are "gray water". The necessary processing of gray water includes the prevention of contaminants such as certain inorganic soaps (surfactants), perfumes, or dyes from being allowed on the spacecraft.

6.4.1 Waste collection

Waste collection is dependent on two factors, the amount of liquid versus solids involved, and the geometry of the waste elimination systems. The latter, of course, is related to the sex of the individual and must be handled separately.

The male urination geometry lends itself to a "condom like" urine waste collection method as shown in Fig. 6: for microgravity, which is unnecessary for any significant gravity environment. The geometry of females is such that, while not impossible, the separation of the functions of urination and defecating is an act of control. However, during a long-duration mission, the probability for mixing the two functions due to sickness (diarrhea) or the potential for solid productions in the urinary system (kidney stones, infection) is almost 100%.

6.4.2 Female-specific considerations

Biological waste products from both sexes deserve special consideration. As mentioned in the overview at the beginning of this chapter, the ability to recycle almost everything is important. It is highly unlikely that the nominal sicknesses that result in organ failure, etc. will not afflict the population of the spacecraft. Therefore, removed or expelled appendices, cancerous tumors, placentas, teeth, foreskins (assuming such to be done), fingers, toes, and other appendages must be accommodated. However female reproductive-specific biological contaminants (uterine linings, blood, etc.) from menstruation (periods) provide a continuous load on the biological waste stream. While this can be nominally handled by the solid waste collection and recycling, the modern female accommodations of tampons, pads, and other collection devices would overwhelm most waste systems on a spacecraft. Therefore, nonwaste-producing products or methods must be utilized. While there are cups and other methods utilized today, they are not ubiquitous. Experimentation and utilization of those, or new, methods must be encouraged on future exploration spacecraft.

6.4.3 Stabilization/biohazard control

While females are the source of continuous biological waste, generational spacecraft will have to deal with the aforementioned occasional introduction of human biological waste beyond nominal urine and feces. Additionally, if any plant-based food production is used, the nonedible plant must also be broken down and recycled. However, before recycling, almost all wastes will need to be stabilized and sterilized. The potential for pathogenic propagation if sterilization is not performed approaches certainty over any significant (years) mission length. The danger to the crew must be mitigated, and if the food system includes bioregenerative food production, the danger to the plants within the system is also extreme. (While every care may be taken to not introduce any pathogenic species into the spacecraft during construction and stores delivery, the probability of humans being the carrier for "bugs" is not zero and years of galactic radiation exposure may mean the even benign biology will become hostile over decades through genetic mutation.) Radiation and heating are the two major sterilization techniques, with the former requiring a radiation source such as X-ray generators, and the latter requiring sufficient power to bring what is mainly water to sterilization temperatures—thereby driving off much of the water in the process. At this time, radiation sterilization may be the most utilitarian, although see Section 4.3 for further discussion.

6.4.4 Storage

As mentioned in prior sections, humans are by nature batch processors. Combined with sporadic biological waste generation (and if growing plants—harvest wastes), storage of wastes must be accommodated until such time as they can be processed for recycling. For a significant number of people, this is not a trivial volume or mass. (When Elon Musk revealed his starship design and said he could fly 500 people to Mars, the author was asked about this feasibility. My response was that he did not have the volume for the bathrooms needed for that many people.) Assuming sterilization is thorough, almost indefinite storage can be achieved, but the size of the storage system must be calculated for worst case scenarios of the breakdown and repair of the recycling system (throwing excess overboard is not advisable due to mass loss and the potential for high-speed recontact with the spacecraft as it decelerates.)

6.4.5 Recycling

Recycling will be the key technology, after production of metabolic needs, for interstellar travel. In the past 40 years of spaceflight, with the advent of long-term LEO spacecraft, many technology development programs have been executed by NASA and other spacefaring entities to "close the loop." As calculated in previous sections, even small amounts of "nonrecyclable wastes" will compound over time to greatly increase the needs for replacement stores. For example, the best water recycling technology recovers 98% of the water from urine. The leftover water is locked up in the salts and minerals in the urine. Retrieving those lost percentages of water may be impossible, as salt-bound water is very hard to extract. Fortunately, calculating the "extra" water that you have to launch with is easy—assuming the recycling system does not break down for appreciable amounts of time.

Harder to predict and utilize is the material locked up in feces and nonedible waste products. The best current technology for recycling this is a membrane aerated biological reactor (MABR), which has also been paired with other systems in an integrated water recovery assembly (IRA) [40]. This system combines a wet oxidizer with membrane distillation, membrane separation, and photocatalytic systems to recover CO_2, N_2, and H_2O. While feces may not be compatible with the system as now designed, separation of feces water to introduce into the IRA system to complete the water cycle to a great extent is achievable.

Feces and solid wastes not handled by an IRA can be pyrolyzed (burned) for raw production of activated carbon to be utilized in a trace contaminant control system, and also be introduced into the food production systems described below.

6.5 Food production and management

Food, of course, is the stuff of life. Probably the most consequential tradeoff of an interstellar-capable spacecraft is the amount of food brought from Earth vs what is produced in an endless recycling program. Often, bioregenerative life support (BLS) is seen as a necessary component of long-duration life support. This is the idea that growing food—often plants like potatoes, peanuts, and wheat—to provide the majority of caloric intake for a person is desirable. The VEGGIE system on space station is an experiment to test out the idea of growing edible products in space [41] (although be it in microgravity). Hybrid systems have also been studied, utilizing a 50% BLS/50% package food system, which trades well against full BLS systems [42]. However, the contingency of a breakdown in food production will mean a need for stores (how substantial is a complex calculation) and the inefficiency of plant food is very high (the best plants are 80% edible) and taxing on the recycling system. However, some plant supplement to the production of food is necessary—if not for nutrition, then for variability of the diet.

However, the state of the art of food production on Earth has gone well beyond that of the current space technology [43]. But rather than food from plants, food technology is rapidly evolving to the use of basic building blocks of nutrition (amino acids, starches, fats) being grown in vats and processed, combined, and reconstituted into "fake" food. In fact, the printing of tissue is coming out of the laboratory now, and there is no reason that this technology can't be used, combined with the millennia-old technology of vat processing of bacteria, yeast, and algae, to produce any basic building block of food. (It should not be lost on the reader that this is the same technology used for making beer, wine, and other adult beverages—hence the millennia-long experience. And the reader can be assured on a generations-long journey that a FEW vats will be dedicated to this ancient art soon after leaving the solar system.) Advances in this technology will be improved to introduce the necessary vitamins, minerals, and elemental substances to make this a true "food substitute." On an interstellar voyage, it may be possible to order your food a few days or weeks in advance—not unlike making a grocery list—to plan your meals and have them delivered (or ready in the eating area) when you want them.

In fact, this same system may be used to make other products for the interstellar ship such as edible or biodegradable wrapping, containers, and even medical instruments and bioplastic replacement parts.

6.5.1 The importance of packaging

This brings us to a further discussion on food, and other items: packaging. Packaging is the landfilling curse of the current human race. But on an interstellar journey a landfill is both wasteful and volume-consuming—as well as an indication of the breakdown of recycling. The production of recyclable packaging is key to a healthy food system—protecting the food from manufacturing, storage, and delivery damage and loss. How packaging is solved for long-duration spacecraft is a key technology that is nascent in today's world but making rapid advancements. Right now, the state of the art of LEO missions is plastic, but this is impractical for the purposes of interstellar flight. The ability to produce bioregenerative (and perhaps edible) packaging must be a key technology advancement.

6.5.2 Metabolic differences of different diets

Often, the type of diet chosen for a crew can have a high impact on the closure of the life support system. Meals with caloric intake dominated by carbohydrates are metabolized differently by humans than meals primarily with fat caloric sources. For a spacecraft, the most important impact is the metabolic respiratory quotient [44], which means that the ratio of oxygen input for a given output of CO_2. Fat diets reduce the CO_2 metabolic respiratory quotient by close to 10%, thereby reducing the load on CO_2 removal systems and CO_2 reduction systems. But the production of fats, except for peanut oils, is not adapted to vat production at this time. However, genetic engineering advancements have made this production possible and should be researched and developed.

6.6 Thermal control

Thermal control of the spacecraft is probably one of the most well-developed technologies for long-duration spaceflight. Because of the lack of proximity of the sun or the destination star for the vast majority of the flight, the spacecraft will have an almost perfect view of the 4K space environment surrounding it. Depending on the power source for the spacecraft (see Chapter 2), substantial waste heat from the power system will most likely be available.

6.6.1 Active thermal systems

In case of loss of power for some time due to breakdowns, the spacecraft will most likely be "hot biased" to mitigate the interior of the spacecraft plunging into a deep freeze. Hot biased means that an active method for removing heat from the spacecraft and rejecting it to space will be required. This is done routinely today on Mars-bound spacecraft to cool the radioisotope thermal generators on the rovers (which always produce heat within the heat shield, thereby necessitating active rejection). While pumps, plumbing, heat exchangers, and radiators are all potential failure points, most of the reliability and maintenance concerns revolve (pun intended) around the pump. Therefore, for interstellar travel, these pumps must be accessible to the interior of the spacecraft and operate with fluids that are nontoxic and highly unreactive with the materials of construction. The state of the art today is diaphragm pumps. Replacements of the flexible diaphragms of such pumps must be considered, but these are easily stowed and probably best to be brought from Earth. In today's small spacecraft, these are adequate. However, for larger spacecraft, the pressure capability of these pumps limits their utility, unless used in series arrangements. This may be a good idea, as a series diaphragm pump allows for failure of one part, and replacement, without shutting down the entire system, which may be detrimental to the active system if it results in freezing of the radiator.

The active thermal control fluid is also a well understood state of the art. Perfluoroether fluids are used today on Dragon2 and the CTS-100, and are highly stable, nontoxic, low vapor pressure (they don't evaporate easily), and low reactivity fluids that are inert with almost all metallic materials and nonfluorinated compounds used for other components of the active thermal system. NASA spacecraft utilize propylene glycol and water mixtures that have properties that deteriorate over time and must be "doped" with corrosion inhibitors (if using aluminum tubing) that are consumed by contact with the aluminum tubing and hence are not considered for longer-term spacecraft by most designers today.

6.6.2 Passive thermal systems

Passive thermal systems include multilayer insulation, shades, and louvers, with the latter two driven by thermomechanical systems that respond to heat sources. Again, these current technologies are considered mature, long lasting, and adequate to govern the protection of exterior spacecraft components with some parasitic heat source, such as the heat emanating from the interior of the spacecraft.

6.7 Crew systems

Crew systems may be the difference between a successful mission and one that ends in failure. Thirty years ago, when NASA listed the top 50 issues of going to Mars, psychology was not on the list. After the first set of Shuttle/Mir missions, it had risen to number five on their list. The change of focus was due to the long-duration mixing of cultures—not of nationality, but of astronaut cultures. Astronauts were chosen early in the space program for "the right stuff," which was outwardly communicated as test pilot, A-type personalities. However, NASA and the Russian space program discovered that both of their astronaut selection criteria tended to choose not those that are able to reduce tension among the crew, but those that are quick decision makers who lack emphasis on soft skills such as empathy, co-operation, and thoughtfulness. It was only during the near-disastrous US/Russia Mir missions that these features were on display to show NASA that a new type of astronaut was needed for long duration missions [45].

Tension among the crew is not just a product of the workload, but of the implementation of crew systems such as hygiene, exercise, lighting, space, and the accommodation of personal tastes through the allocation of personal provisions.

6.7.1 Personal provisions

Personal provisions include those items that spacefarers can take with them that remind them of home, as well as clothing, grooming products, entertainment, and downtime activities such as talking to their loved ones at home. The latter item will be a problem as once the spacecraft is even a few light seconds from Earth, conversations with those back on Earth will become impossible, and conversations will take on a much more "letter to home" feel; or, for today's generation, a series of podcasts back and forth. The spacecraft crew systems must allow for this capability, and maintenance of those capabilities, which includes microphones, screens, computers, and whatever peripheral technology is invented between today and the embarkation to the stars. But who hasn't experienced crashed computers, inoperable microphones, and faulty screens? Thus, the personal provisions will put a burden of maintenance parts on the total enterprise as most electronic parts will probably not be producible or recyclable on the interstellar journey.

Other personal provisions, as mentioned previously, will need to be restricted to ensure that harmful chemicals or materials are not introduced into the air or water of the spacecraft, thereby harming the recycling and recovery systems.

Clothing can be problematic for many reasons. You need to clean clothes and produce new ones after their useful life is complete. Modern clothing today, especially "natural fiber" clothing, deteriorates with use. No one wears hand-me-downs from their grandparents except for precious items specifically preserved—wedding dresses for example. On a generations-long journey, unless the spacecraft becomes a "clothing optional" space (not outside of the realm of possibility), clothing will need to be produced. How will it be replaced? There are really two options: clothing made from "cloth" that is a recyclable product from within the spacecraft, or clothing made from materials brought along from Earth. However, at the destination, we will need to produce clothing of some type, so production of material for clothing will need to be developed for spacecraft use. This is another reason for vat technology to grow genetically modified bacteria: to produce the raw material for clothing, as well as food products.

6.7.2 Hygiene

Provisions for hygiene have been developed for the ISS and are considered the state of the art in today's space industry. However, they are produced and shipped up to LEO from Earth. The interstellar spacecraft must have the ability to produce hygiene products that are not toxic to the environment nor contaminating to the recycling system for both water and air. Frequency of cleaning will probably have to be reduced, which may be strange to, for example, the average US citizen. The myriad of products utilized by the average US resident—toothpaste, deodorant, tissue, toilet paper, soap, shampoo, conditioner, razor blades, shaving cream, to name a few—need to be produced on the spacecraft. At this time, the substitute materials—if any—do not exist.

6.7.3 Exercise

Exercise on a spacecraft (or at least one not providing one gravity acceleration vectors) is necessary for human health. Exercise machines on the ISS are mechanically complex (to prevent excessive movement that disturbs scientific instruments) or rely on material products not easily produced on a spacecraft (elastic cords). In partial gravity, the amount of mass for exercise equipment must be provided for (squats of 600 kg!). In 1g, isometric exercises may be sufficient, but a nominal gym equipped vehicle will be a necessary allocation to the spacecraft design.

6.7.4 Lighting

As mentioned previously, adequate lighting is both important for operation of a spacecraft and the psychological wellbeing of the passengers. Humans are biologically conditioned to respond to light in both direction (lights come from "above") and for circadian rhythms of night and day. Light also provides for human health through the production of vitamin D. In northern Siberia, children put on goggles and stand in front of an ultraviolet light source as part of their health regime. While in the United States lack of vitamin D resulted in nominally vitamin D supplemented milk and pills, passengers on long voyages will probably need to be exposed (in tanning beds?) to sufficient light of the correct wavelength. Adequate attention will need to be paid to both the type of lighting (can the lighting source be replaced or fixed?) and the geometry, color, and intensity of the lighting to maintain the proper biological response such as eye performance and psychological wellbeing. Optimization, or at least adequate attention, must be paid to the lighting provisions for the spacecraft.

6.7.5 Free volumes

Finally, adequate free volume geometry may be a critical area of both research and design. If an accelerate magnitude is chosen that is less than Earth, will the ceiling height of hallways and rooms still be adequate if the humans in a few generations are all taller? Analogous to the increased stature of Japanese males after the introduction of the westernized diet following World War II, and the resulting reproportioning of their dwellings, the spacecraft will either need to allow for raising of ceilings or the ceilings must be predesigned for anticipated heights.

And finally, the activities anticipated on the generational spacecraft should be provided adequate room. Over the past two decades, NASA-STD-3000, which allocated free volumes for spacecraft, has been discovered to be inadequate for design. With activity-based design [46], this has been corrected, but activities on a generational spacecraft will need to be carefully predicted and designed for.

7 Holes in the picture

When looking far into the future, there are unknowns, and there are unknown-unknowns (Unk-Unks). The former are those areas where we know there are gaps in our knowledge. The latter are items where we don't know the correct questions to ask in the first place, and therefore can only solve the unknown when encountered in the development and implementation of an engineering enterprise. While the

Unk-Unks are not addressable in this chapter, the unknowns can be highlighted for future thought and solution. The following is an attempt to highlight some of those unknowns.

7.1 The inadequacies in closure

As shown in some of the calculations of closure previously in this chapter, even 98% recyclability—or even 99.99%—may not be adequate to achieve a truly interstellar design. Depending on the mass limitations put on the spacecraft—usually determined by the propulsion capability—it may be adequate to just provide enough material at the onset of the mission to handle the shortfall in closure of the myriad of environmental control and life support system's materials, components, and subsystems [42]. Barring that, ECLSS technology will need to progress beyond the state of the art before a trip to the stars can be contemplated.

7.2 The consideration of nonrecyclable consumables

Inevitably, there will be consumables that just cannot be supplied in the spacecraft facilities. Over the course of the COVID-19 pandemic, the general public has become more aware than ever that the industry of the world is connected. Supply chain issues have highlighted this, as even seemingly simple items have run in short supply. To produce many things that we consume today whole industries with sophisticated processing and manufacturing equipment have utilized massive amounts of capital equipment, whole factories and chemical processing plants, and highly sophisticated tools to produce something as simple as a pencil. Any design of an interstellar spacecraft will have to be very carefully audited for those consumables that just cannot be produced on a spacecraft but yet may be critical to the survival of the passengers. Some of this issue is being tackled now, such as Cislunar Industries' ability to "recycle" spent materials into useful secondary products (but not to their original use), but something as simple as replacement of a light diode may be beyond any on-spacecraft manufacturing capability. One possible consideration is sending faster spacecraft behind the stellar vehicle to "catch up" to the passenger vehicle and dock to provide some critical components. But after a generation of travel, any "resupply" would be leaving Earth too late to catch up with the initial vehicle, and something could go wrong in the navigation, propulsion, or docking activities that would jeopardize the mission success. Small things may be manifested in the cargo (resistors, chips, sheets of material, etc.) but ultimately the passengers just may have to make do, or do without, something that has originally been launched with them from Earth. Thus, a robust design takes on criticality like never before in the history of spacecraft design.

7.3 Some hard ethical questions

Finally, the social aspects of a long duration spaceflight must be considered. For example, does someone ever "retire" on a spacecraft? At what age is a passenger considered "no longer functional" or "no longer needed"? And what do you do with them? What are the ethics of medical care being provided to a clinically not dead—but on life-support and immobile—passenger? And when they die, how much of the expired passenger do you recycle? Do you perform a ceremonial burial of a small, hard-to-recycle body part (a tooth perhaps) and eject it out of the airlock analogous to burial at sea? Closure would also mean that most of the remaining corporeal body be recycled. And do you sacrifice internal spacecraft volume to establish a memorial space for each expired passenger?

Also to be considered is what is the minimum number of passengers to prevent the genetic problems of inbreeding, and to mitigate those problems do you have a prescribed "who reproduces with who" regulation? Or do you bring along frozen embryos of nonrelated humans to supplement the genetic pool, thereby restricted some of the potential pairing couples from producing just their own kids from a genetic standpoint?

All these are difficult, but necessary, answers that are needed for the ultimate design of the vehicle.

8 Speculations on the future evolution of life support

Technology is moving very quickly in our world today, so there may be some items "just over the horizon" that would make items of this monograph obsolete or at least no longer pertinent in the form discussed. This section attempts to predict the "near future" of major advances that may change the equations for long duration ECLSS in the future.

8.1 Magnetic fluids

Much of the discussion of long-duration ECLSS centers around the repair and maintenance of components and subsystems. Of these, pumps hold special attention. Everyone has experienced the failure of a pump of some sort, whether a fuel pump on a car, a coolant pump, or a medical pump. Elimination of pumps, in a mechanical sense, can be achieved by development of magnetic fluids (such as nanoparticle doped fluids) that would preclude moving parts of a pump, replacing them with electronic motivators that could maintain circulation of the desired fluid. While this introduces electrical failures, it is highly possible that these would be much more reliable that their mechanical equivalents (which can also fail electronically).

8.2 3D printing advances

Currently, 3D printers (additive manufacturing machines) are rather simplistic. Development of 3D printers capable of producing multiple alloys of material (6061 aluminum has a different elemental makeup than 7075 aluminum) by combining the constituent parts (and heat-treating to specific tempers) may make them more useful for long duration spacecraft. However, the raw material needs to be recycled from the original part to make the breakeven equation work. Otherwise, you still need to bring along the raw material, in the right form, to operate the machine. And don't forget that the machine must also be repairable and maintainable during the journey.

8.3 Genetic engineering of simple lifeforms

Some of this monograph has assumed genetic engineering of bacteria, yeast, algae, or some other cellular lifeform will advance in the near term to provide for production of not only edible product building blocks, but perhaps also synthetic plastic materials that can be used as a basis for manufacture of simple replacement parts. The extent of genetic engineering advances makes this speculation less uncertain. Further advances, where pharmaceuticals can be produced to purity and concentrations with simple enough equipment to formulate almost any medicinal need, would be highly valuable.

8.4 Genetic modifications of humans

Of perhaps more ethical uncertainty, but also technological uncertainty, is the ability to genetically modify humans to make them more compatible with spacecraft. This could take the form of synthesis of alternative energy sources (photosynthesis in the skin or enzymatic breakdown of algae that are heretofore indigestible, for example) to produce metabolic needs without food—or at least more complex food; probably not to replace digestion, but to supplement the capability of the human to reduce the needs for caloric production and consumption by other means. Other genetic modification may be to increase the ability of the human body to survive or thrive in significantly lower gravity. And the advances in the understanding of epigenetic driven changes in lifeforms have indicated that the humans themselves may undergo heretofore unforeseen substantial changes in their generic expression due to the radically changed environment.

9 Conclusion

Given the state of ECLSS systems for space, humans are not ready yet to even travel to Mars—the veritable back yard of Earth—let alone venture out toward even the nearest stars. ECLSS has been known to be the "long pole in the tent" for space travel for some time [1] for destinations even as close as Mars.

Major tradeoffs that need to be determined prior to attempting any design of such as system include: (1) the level of acceleration provided continuously to the passengers via either propulsion or rotation of spacecraft elements, (2) The number of potential passengers, and (3) the actual duration of the mission. A peripheral consideration is whether the vehicle will decelerate into the new solar system or "drop off" landing spacecraft to mitigate deceleration of the whole system (and therefore required solutions to the ECLSS needs for those terminal landing spacecraft that may be dormant for generations).

No matter whether this monograph has anticipated all considerations, there are surely unknowns that only can be discovered by further study, experimentation, and actual operation through human exploration of the current solar system.

References

[1] G. Anderson, ISPCS Conference 2019 Video Recording 2019, October 25, 2019 (Online] [Cited: September 2022, 2022) https://www.youtube.com/watch?v=Tj3FwIfpdmI.

[2] A. Hanford, Advanced Life Support Baseline Values and Assumptions Document, Johnson Space Center, NASA, Houston, 2004. NASA/CR-2004-208941.

[3] M. Anderson, M. Ewert, et al., Life Support Baseline Values and Assumptions Document, EC2, NASA, Houston, 2015, p. 51. NASA/TP-2015–218570.

[4] M. Lamantea, O. Faure, et al., The Orion MPCV-ESM consumable storage system, in: International Conference on Environmental Systems, Charleston, 2017. ICES-2017-338.

[5] The Auto Channel, The Auto Channel, 2002, [Online] [Cited: January 15, 2023] Press release citation https://www.theautochannel.com/news/2002/07/29/144622.html.
[6] R.A. Heinlein, The Menace from Earth, Signet, New York, 1962, pp. 92–115.
[7] NASA, Spacecraft Maximum Allowable Concentrations for Airborne Contaminants, 2020. NASA standard. JSC 20584.
[8] Spacecraft Water Exposure Guidelines (SWEGs), Johnson Space Center, NASA, 2017. JSC 63414.
[9] W. Li, L.M. Calle, A.J. Hanford, I. Stambaugh, M.R. Callahan, Investigation of silver biocide as a disinfection technology for spacecraft—an early literature review, in: 48th International Conference on Environmental Systems, ICES, Albuquerque, 2018. ICES-2018-82.
[10] H. Lane, V. Kloeris, et al., Food and nutrition for the Moon base: what we have learned in 45 years of spaceflight, Nutr. Today 42 (3) (2007) 102–110.
[11] NASA, NASA-STD-3001 Vol2 Version B, Johnson Space Center, NASA, 2019. NASA-STD-3001 Vol2.
[12] G. O'Neill, The High Frontier: Human Colonies in Space, William Morrow and Company, 1976.
[13] W. Von Braun, F. Whipple, et al., Space Stuff, 1952–1954 (Online] [Cited: January 15, 2023) https://www.rmastri.it/spacestuff/wernher-von-braun/colliers-articles-on-the-conquest-of-space-1952-1954/.
[14] Wikipedia, Skylab, Wikipedia, March 6, 2022 (Online] [Cited: March 6, 2022) https://en.wikipedia.org/wiki/Skylab.
[15] Wikipedia, Past Space Stations, Wikipedia, February 26, 2022 (Online] [Cited: February 2022, 2022) https://en.wikipedia.org/wiki/List_of_space_stations#Past_stations.
[16] Biosphere2, University of Arizona. Biosphere2 About, Biosphere2, March 6, 2022 (Online] [Cited: March 6, 2022) https://biosphere2.org/about/about-biosphere-2.
[17] J.E. Poynter, The Human Experiment, 2 Years and 20 Minutes Inside Biosphere2, American First Edition, Basic Books, 2006, p. 384, ISBN: 978-1560257752.
[18] C. Woods, The Effects of Actual Microgravity and Vector-Averaged Gravity on the Development of T Cells, University of Arizona, Tucson, AZ, 2004, p. 156. This is Chris Cory Woods PhD Thesis.
[19] NASA, Technology Drives Exploration, NASA.gov, May 13, 2015 (Online] [Cited: March 6, 2022) https://www.nasa.gov/feature/technology-drives-exploration.
[20] NASA, NASA Space Tech, NASA.gov, December 14, 2017 (Online] [Cited: March 6, 2022) https://www.nasa.gov/offices/oct/feature/innovation-and-diversity-drive-exploration.html.
[21] Technology Readiness Level, NASA.gov, October 28, 2012 (Online] [Cited: January 15, 2023) https://www.nasa.gov/directorates/heo/scan/engineering/technology/technology_readiness_level.
[22] J. Silverman, A. Irby, et al., Development of the Crew Dragon ECLSS, Virtual, in: International Conference on Environmental Systems (ICES), 2020, p. 11.
[23] T. Ball, B. FInger, et al., Inspiration Mars ETDU air management system test results, in: International Conference on Environmental Systems, Tucson, 2014. ICES-2014-273.
[24] P. Henson, S. Yates, et al., An Environmental Control and Life Support System (ECLSS) for deep space and commercial habitats, in: International Conference on Environmental Systems, St. Paul, 2021. ICES-2021-212.
[25] R.R. Scully, M. Basner, J. Nasrini, et al., Effects of acute exposures to carbon dioxide on decision making and cognition in astronaut-like subjects, npj Microgravity 5 (2019) 17.
[26] C. Chullen, C. Campbell, W. Papale, et al., Design and development comparison of rapid cycle amine 1.0, 2.0, and 3.0, in: International Conference on Environmental Systems, Vienna, Austria, 2016. ICES-2016-073.
[27] D. Wickham, K. Gleason, J. Engel, et al., Advanced supported liquid membranes for carbon dioxide control in cabin applications, in: International Conference on Environmental Systems, Vienna, Austria, 2016. ICES-2016-77.
[28] C.S. Iacomini, Enabling air revitalization systems capable of 100% oxygen regeneration, in: Conference on Habitation Research and Technology Development, HABITATION 2006, Orlando, Florida, February 5–8, 2006, 2006.

[29] Wikipedia, Sabatier Reaction, Wikipedia, February 4, 2022 (Online] [Cited: February 4, 2022) https://en.wikipedia.org/wiki/Sabatier_reaction.

[30] P. Hintze, H. Guardado, TFAWS Website, NASA TFAWS Website, 2018 (Online] [Cited: March 6, 2022) https://tfaws.nasa.gov/wp-content/uploads/5_ISRU-Sabatier-Reactor-for-TFAWS-2018.pdf.

[31] P. Yu, J. Woods, M. Corcoran, et al., "Getter" development for International Space Station Sabatier Assembly, in: International Conference on Environmental Systems, St. Paul, 2022. ICES-2022-54.

[32] C. Iacomini, P. Benjamin, D. Milobar, Feasibility tests of oxygen recovery technology as an internally reforming regenerative fuel cell, in: International Conference on Environmental Systems, Bellevue, 2015. ICES-2015-152.

[33] C. Iacomini, J. Harrell, J. Lopez, Nafion bundle technology for space flight humidity control, in: International Conference on Environmental Systems, Bellevue, 2015. ICES-2015-153.

[34] B. Zimmerman, T. Cognata, Condensate Separator for Microgravity Conditions (COSMIC) for two-phase separation, in: International Conference on Environmental Systems, Boston, 2019. ICES-2019-243.

[35] Wikipedia, Activated Carbon, Wikipedia, March 6, 2022 (Online] [Cited: March 6, 2022) https://en.wikipedia.org/wiki/Activated_carbon.

[36] W. Wallace, T. Limero, L. Loh, et al., Monitoring of the atmosphere on the International Space Station with the air quality monitor, in: International Conference on Environmental Systems, Charleston, 2017. ICES-2017-103.

[37] Gaslab, How Do Oxygen Sensors Work, Gaslab.com, August 19, 2021 (Online) https://gaslab.com/blogs/articles/how-does-an-oxygen-sensor-work.

[38] P. Mudgett, M. Skow, T. Limero, et al., Seeking the tricorder: evolution of the NASA anomaly gas analyzer, in: International Conference on Environmental Systems, Boston, 2019. ICES-2019-93.

[39] Wikipedia, Peristaltic Pump, Wikipedia, February 12, 2022 (Online] [Cited: February 12, 2022) https://en.wikipedia.org/wiki/Peristaltic_pump.

[40] A. Jackson, B. Finger, B. Zimmerman, et al., Development of a hybrid integrated water recovery assembly for exploration habitats, in: International Conference on Environmental Systems, ICES, Albuquerque, 2018. ICES-2018-269.

[41] J. Buncheck, A. Burry, M. Romeyn, Sustained VEGGIE: a continuous food production comparison, in: International Conference on Environmental Systems, Lisbon, 2021. ICES-2021-229.

[42] H. Jones, Starship life support, in: International Conference on Environmental Systems, SAE, 2009. 2009-01-2466.

[43] M. Roman, A. Herblet, J. Broyan, et al., NASA centennial challenges deep space food challenge competition to incentive innovation in food systems for long-duration space exploration missions, Virtual, in: International Conference on Environmental Systems, Lisbon, 2021. ICES-2021-392 Centennial Food Challenge.

[44] Medical Dictionary for the Health Professions and Nursing © Farlex 2012, The Free Dictionary—Medical, The Free Dictionary, Sept 25, 2022 (Online] [Cited: Sept 25, 2022) https://medical-dictionary.thefreedictionary.com/metabolic+respiratory+quotient.

[45] B. Burrough, Dragonfly: NASA and the Crisis Aboard Mir, Harper Collins Publishers, 1998. 1841150878 (ISBN13: 9781841150871).

[46] L. Wickman, G. Anderson, Activity-Based Habitable Volume Estimating for Human Spaceflight Vehicles, IEEE, Big Sky, 2009. IEEE 978-1-4244-2622-5/09/. Paper#1098.

CHAPTER

Timing relationships and resulting communications challenges in relativistic travel

5

David G. Messerschmitt[a], Ian Morrison[b,c], Thomas Mozdzen[d], and Philip Lubin[e]

[a]Department of Electrical Engineering and Computer Sciences, University of California at Berkeley, Berkeley, CA, United States, [b]International Centre for Radio Astronomy Research, Curtin University, Bentley, Perth, WA, Australia, [c]Astro Signal Pty Ltd, Perth, WA, Australia, [d]School of Earth and Space Exploration, Arizona State University, Tempe, AZ, United States, [e]Department of Physics, University of California at Santa Barbara, Santa Barbara, CA, United States

Chapter outline

Interstellar Travel. https://doi.org/10.1016/B978-0-323-91280-8.00012-5

Consider a spacecraft cruising between two at-rest locations (an origin and a destination) separated by large interstellar distances. Maintaining communications with the origin, destination, or with other spacecraft brings many benefits, from practical mission engineering and management, to scientific data, and to the psychological benefits of astronauts remaining in touch with the civilization left behind. However, interstellar travel with human astronauts necessitates high spacecraft velocities approaching the speed of light. Relativistic effects thereby come into play, and these can have extreme (and sometimes surprising) impacts on communications systems and capabilities.

In particular, a central result of relativity is that clocks in relative motion run at different relative rates. This means that a relativistic traveler's measurements of time are inconsistent with that of observers at the origin or destination. In isolation everyone is content with their own sense of time, but this affects the traveler's sense of connection to their environment. This includes affecting communication among travelers in motion with relativistic relative velocities, where clock inconsistency is directly apparent. In particular, it strongly affects perceptions of relative timing, which has to be taken into account in both the formulation and interpretation of communicated messages. This concept is explored analytically and quantitatively in this chapter and applied to two simple mission profiles. See Appendix A for a compilation of variables employed in the following.

1 Travelers' clocks in relativistic travel

Participants in interstellar travel, such as spacecraft (including instrumentation and possibly astronauts) and civilizations that launch such spacecraft, perceive time in terms of clocks that they carry along with other payloads and supplies. Suppose for one such participant $\mathcal{A}$ the time measured by a clock that it carries is τ_a. This is called the *proper time* associated with $\mathcal{A}$, meaning "the time observed by a clock at rest relative to participant $\mathcal{A}$." Following Ref. [1], a more descriptive term for τ_a is $\mathcal{A}$'s *traveler's time*.

1.1 Clock images

For many purposes there is no concern about other participants' traveler's times. However any activity that involves two or more such participants, such as navigation or communication of information among them, emphasizes a relationship among their traveler's times.

Consider two spatially separated travelers in a one-dimensional (1D) space with spatial dimension x. These travelers are denoted by $\mathcal{A}$ and $\mathcal{B}$, they have possibly distinctive velocities and accelerations as measured relative to x.

These travelers each carry their own clocks with them, and these clocks measure traveler's times τ_a and τ_b. In a communication context (the primary concern here) this relationship between $\mathcal{A}$ and $\mathcal{B}$ is created by *photon exchange*, in which $\mathcal{A}$ emits a photon at its traveler's time τ_a and that photon is detected by $\mathcal{B}$ at its traveler's time τ_b. Thus $\mathcal{A}$ is in control, since this traveler chooses τ_a, and τ_b is determined by the position of $\mathcal{A}$ at the instant of photon emission as well as the time-dependent position of $\mathcal{B}$ (and the related measurement of its local clock) during the photon's propagation through spacetime.

We can define two related functions $\Psi_{\mathcal{A}\Rightarrow\mathcal{B}}$ and $\Omega_{\mathcal{A}\Rightarrow\mathcal{B}}$ that establish a causal relationship between these times as

$$\tau_b = \Psi_{\mathcal{A}\Rightarrow\mathcal{B}}[\tau_a] \text{ and } \tau_a = \Psi^{-1}_{\mathcal{A}\Rightarrow\mathcal{B}}[\tau_b] = \Omega_{\mathcal{A}\Rightarrow\mathcal{B}}[\tau_b]. \tag{1}$$

The interpretation of $\Psi_{\mathcal{A}\Rightarrow\mathcal{B}}$ is the time τ_b at which $\mathcal{B}$ detects a photon that was emitted by $\mathcal{A}$ at time τ_a, assuming that this photon propagates from $\mathcal{A}$ to $\mathcal{B}$, assumed to be at the vacuum speed of light c. The inverse $\Omega_{\mathcal{A}\Rightarrow\mathcal{B}}$ always exists because on physical grounds $\Psi_{\mathcal{A}\Rightarrow\mathcal{B}}$ is a strictly monotonically increasing function. The domain and codomain (range) of $\Psi_{\mathcal{A}\Rightarrow\mathcal{B}}$ are a subset of the real numbers. The domain of $\Omega_{\mathcal{A}\Rightarrow\mathcal{B}}$ is the range of $\Psi_{\mathcal{A}\Rightarrow\mathcal{B}}$ and vice versa.

The relationship $\Psi_{\mathcal{A}\Rightarrow\mathcal{B}}$ is determined by kinematic variables (relative positions, velocities, and accelerations of $\mathcal{A}$ and $\mathcal{B}$). Similarly, the inverse function $\Omega_{\mathcal{A}\Rightarrow\mathcal{B}} = \Psi^{-1}_{\mathcal{A}\Rightarrow\mathcal{B}}$ infers, under the same conditions, the time of emission τ_a corresponding to an observed detection time τ_b. Function $\Psi_{\mathcal{A}\Rightarrow\mathcal{B}}$ is called the *clock image* [1] because of its analogy to a picture or image of $\mathcal{A}$ taken by $\mathcal{B}$.

Example 1: If $\mathcal{A}$ wants to send $\mathcal{B}$ a "happy birthday" greeting, the relevant birthday is defined in accordance with τ_b. If $\mathcal{A}$ wants that greeting (sent as a message carried by photons) to arrive at $\tau_b = \tau_r$, corresponding to say 8:00 a.m. on $\mathcal{B}$'s next birthday, then $\mathcal{A}$ should originate that message at its own traveler's time $\tau_a = \Omega_{\mathcal{A}\Rightarrow\mathcal{B}}[\tau_r]$.

1.2 Reversing direction of propagation

The direction of the arrow in the designation $\mathcal{A}\Rightarrow\mathcal{B}$ specifies the direction of photon transfer from $\mathcal{A}$ to $\mathcal{B}$ and hence establishes a cause-and-effect relationship between the emission and detection of a photon and the resulting timing relationship. The alternative $\mathcal{B}\Rightarrow\mathcal{A}$ implies a transfer from $\mathcal{B}$ to $\mathcal{A}$ and is markedly different due to a reversal of the direction of propagation for a photon relative to the motion of $\mathcal{A}$ and $\mathcal{B}$. Thus $\Psi_{\mathcal{B}\Rightarrow\mathcal{A}}$ cannot be inferred from $\Psi_{\mathcal{A}\Rightarrow\mathcal{B}}$ in any straightforward way; that is, $\Psi_{\mathcal{A}\Rightarrow\mathcal{B}}$ and $\Psi_{\mathcal{B}\Rightarrow\mathcal{A}}$ must be determined independently.

Example 2: Continuing Example 1, $\mathcal{B}$ receiving a birthday greeting message at $\tau_b = \tau_r$, can infer that the message was sent at $\mathcal{A}$'s traveler's time $\tau_a = \Omega_{\mathcal{A}\Rightarrow\mathcal{B}}[\tau_r]$. Further, if $\mathcal{B}$ immediately replies to that message at $\tau_b = \tau_r$, then the time that message will arrive at $\mathcal{A}$, according to $\mathcal{A}$'s local clock, can be inferred by $\mathcal{B}$ to be $\tau_a = \Psi_{\mathcal{B}\Rightarrow\mathcal{A}}[\tau_r]$ (note the switch from $\Omega_{\mathcal{A}\Rightarrow\mathcal{B}}$ to $\Psi_{\mathcal{B}\Rightarrow\mathcal{A}}$). Traveler $\mathcal{B}$ may want to take this arrival time into account in any return message, such as relating it to $\mathcal{A}$'s own age at the traveler's time τ_a at which $\mathcal{A}$ receives the message.

1.3 Compact notation for clock image compositions

In the following it will often be the case that a given scenario of interest involves two or more photon exchanges. Anticipating that need, we employ a standard notation for the composition of functions.

1.3.1 Communication among three travelers

This is illustrated by a configuration in which there are three travelers $\{\mathcal{A},\mathcal{B},\mathcal{C}\}$ along the spatial axis x in this order. If $\mathcal{A}$ emits a photon at time τ_a in the $+x$ direction, which is detected by $\mathcal{B}$ at time τ_b, suppose that $\mathcal{B}$ in response to that detection immediately emits a photon which is in turn detected by $\mathcal{C}$ at time τ_c. In that case the relationship among $\{\tau_a, \tau_b, \tau_c\}$ is summarized by

$$\tau_b = \Psi_{\mathcal{A}\Rightarrow\mathcal{B}}[\tau_a], \tag{2}$$

$$\tau_c = \Psi_{\mathcal{B}\Rightarrow\mathcal{C}}[\tau_b], \tag{3}$$

$$\Longrightarrow \tau_c = \Psi_{\mathcal{B}\Rightarrow\mathcal{C}}[\Psi_{\mathcal{A}\Rightarrow\mathcal{B}}[\tau_a]]. \tag{4}$$

In the following we denote the composition of two clock images by the equivalent but arguably more transparent representation:

$$\tau_c = (\Psi_{\mathcal{B}\Rightarrow\mathcal{C}} \circ \Psi_{\mathcal{A}\Rightarrow\mathcal{B}})[\tau_a]. \tag{5}$$

With the composition operator "$\circ$," the invocation of the functions is from right to left. Note that there is an assumption of immediacy; namely, $\mathcal{B}$ responds to its own detection of a photon by immediately emitting a new photon. In addition, the domain and range of the clock image functions has to be considered, since it is possible to establish nonsensical conditions that violate these constraints.

Example 3: $\mathcal{A}$ wants to wish $\mathcal{B}$ a happy birthday, with the message arriving at $\mathcal{B}$'s traveler's time $\tau_b = \tau_r$ appropriate for the birthday. As in Example 2, $\mathcal{A}$ should originate that message at time $\tau_a = \Omega_{\mathcal{A}\Rightarrow\mathcal{B}}[\tau_r]$. Now suppose $\mathcal{A}$ wishes to remind a third traveler $\mathcal{C}$ that it is $\mathcal{B}$'s birthday by sending to $\mathcal{C}$ a message "Please immediately send a birthday greeting to $\mathcal{B}$, and if you do so it will arrive at the appropriate time." To satisfy that promise, $\mathcal{C}$ should receive $\mathcal{A}$'s message at $\tau_c = \Omega_{\mathcal{C}\Rightarrow\mathcal{B}}[\tau_r]$. Thus $\mathcal{A}$ should originate that reminder message to $\mathcal{C}$ at its own traveler's time:

$$\tau_a = \Omega_{\mathcal{A}\Rightarrow\mathcal{C}}[\tau_c] = \Omega_{\mathcal{A}\Rightarrow\mathcal{C}}[\Omega_{\mathcal{C}\Rightarrow\mathcal{B}}[\tau_r]] = (\Omega_{\mathcal{A}\Rightarrow\mathcal{C}} \circ \Omega_{\mathcal{C}\Rightarrow\mathcal{B}})[\tau_r]. \tag{6}$$

This requirement becomes nonsensical if either τ_r falls outside the domain of function $\Omega_{\mathcal{C}\Rightarrow\mathcal{B}}$ or $\tau_c = \Omega_{\mathcal{C}\Rightarrow\mathcal{B}}[\tau_r]$ falls outside the domain of function $\Omega_{\mathcal{A}\Rightarrow\mathcal{C}}$. This will occur whenever $\{\mathcal{A},\mathcal{B},\mathcal{C}\}$ are too distant and the travelers' clocks are such that it is impossible to get one or more messages to the appointed destinations by the appointed time.

The "$\circ$" version in Eq. (6) is quicker and easier to interpret, and with a bit of practice also becomes straightforward to specify directly.

1.3.2 Query-response communication

Another common situation is *query-response* communication between two travelers. Suppose that $\mathcal{A}$ wishes to receive a response from $\mathcal{B}$ to its own query. This involves two one-way communications, first $\mathcal{A} \Rightarrow \mathcal{B}$ conveying the query followed by $\mathcal{B} \Rightarrow \mathcal{A}$ conveying the response back to $\mathcal{A}$. Suppose the original query traveler's time is at traveler's time τ_{a1}, $\mathcal{B}$ responds immediately, and the response arrives back at $\mathcal{A}$ at traveler's time τ_{a2}. Then we can infer that $\tau_{a2} = (\Psi_{\mathcal{B} \Rightarrow \mathcal{A}} \circ \Psi_{\mathcal{A} \Rightarrow \mathcal{B}})[\tau_{a1}]$. Note that the direction of photon travel is reversed in the two invocations of clock image Ψ.

Typically, $\mathcal{A}$ will be interested in the elapsed time $(\tau_{a2} - \tau_{a1})$ (as measured by its local clock) between the time it initiates a query and the time that it receives a response. This suggests a new function called the *query-response latency*, which is denoted by $\mathfrak{L}_{\mathcal{A} \Leftrightarrow \mathcal{B}}[\tau]$

$$\begin{aligned} \mathfrak{L}_{\mathcal{A} \Leftrightarrow \mathcal{B}}[\tau_{a1}] &= (\Psi_{\mathcal{B} \Rightarrow \mathcal{A}} \circ \Psi_{\mathcal{A} \Rightarrow \mathcal{B}})[\tau_{a1}] - \tau_{a1} \\ &= (\Psi_{\mathcal{B} \Rightarrow \mathcal{A}} \circ \Psi_{\mathcal{A} \Rightarrow \mathcal{B}} - \mathbb{1})[\tau_{a1}], \end{aligned} \tag{7}$$

where $\mathbb{1}$ is the identify function. This latency depends on the traveler's time τ_{a1} corresponding to the initiating query. The notation $\mathcal{A} \Leftrightarrow \mathcal{B}$ denotes a bi-directional communication between $\mathcal{A}$ and $\mathcal{B}$ that is initiated by $\mathcal{A}$.

Example 4: If $\mathcal{A}$'s query is "how do I repair or circumvent my broken CO_2 scrubber," then circumstances favor a quick (low latency) response!

There is an implicit traveler's time $\tau_b = \Psi_{\mathcal{A} \Rightarrow \mathcal{B}}[\tau_{a1}]$ at which $\mathcal{B}$ receives the query and is expected to immediately generate a response. This value of τ_b will often influence $\mathcal{B}$'s response.

Example 5: If $\mathcal{A}$'s query is "what time do you have," the response would be something like "my current time is $\tau_b = 1.645$ years following my launch." From that response, if $\mathcal{A}$ is knowledgeable of the navigation plan for $\mathcal{B}$, then trajectory coordinates $\{x_b, t_b\}$ of $\mathcal{B}$ at the instant of response could also be inferred (see Section 4.1).

1.4 Factors determining clock images

There are three factors that collectively influence and determine the clock image $\Psi_{\mathcal{A} \Rightarrow \mathcal{B}}$ in the context of communication by photon exchange:

Propagation delay. In photon exchange there is a light-speed propagation delay between emission and detection. If $\mathcal{A}$ is the photon emitter, the delay depends on $\mathcal{A}$'s position at the time of emission, but not on $\mathcal{A}$'s subsequent motion. The delay then depends on $\mathcal{B}$'s motion throughout the photon propagation, since that motion affects $\mathcal{B}$'s distance at the subsequent photon detection.

Relativistic time dilation. In the event of any relative motion, τ_a and τ_b advance at inconsistent rates.

Clock initialization. Should the clocks τ_a and τ_b have arbitrary initializations, this would result in an unknown or unspecified mutual offset. To eliminate this ambiguity we assume that the clocks are *synchronized* in a sense that depends on the mission context (see Section 3.3).

2 Spacecraft trajectories

In order to infer the clock image functions, it is necessary to know the *trajectory* of the two travelers exchanging photons. In the case of a spacecraft traveler, its *mission* describes this trajectory in the context of a concrete objective (such as scientific investigation or exploration). The clock image $\Psi_{\mathcal{A} \Rightarrow \mathcal{B}}$ and query-response latency $\mathfrak{L}_{A\Leftrightarrow B}$ are powerful tools in understanding the limits to communication in the context of an interstellar mission, one that has the objective of reaching the vicinity of stars other than our Sun.

Communication challenges are best illustrated in the context of concrete interstellar mission examples. The goal here is to emphasize communications to and from a spacecraft traveling at relativistic speed (a significant fraction of the speed of light), while avoiding complications wrought by complex mission scenarios. Therefore two simple (and yet recognizably idealized) trajectories and associated missions are considered in the following:

Indefinite acceleration. Following launch from some stationary origin, the spacecraft and its payload experience a constant acceleration that continues indefinitely. In this case "acceleration" is that measured by an accelerometer carried by the spacecraft.
Canonical mission. The spacecraft is launched from a stationary origin and subsequently lands on a stationary destination. To this end it experiences a constant acceleration for half the distance from origin to destination, followed by a constant deceleration (with the same magnitude) for the remaining distance.

The assumptions and details of these two trajectories are now considered further.

2.1 Indefinite acceleration assumptions

The *self-acceleration* (sometimes called *proper acceleration*) of a spacecraft is that measured by an accelerometer carried by the spacecraft. Significantly, this is the acceleration imparted by the propulsion subsystem, and as well is the acceleration experienced by the payload, including possibly a human crew.

Indefinite acceleration makes the assumption of a constant self-acceleration with a numerical value α starting at the instant of launch (from a state of rest) and continuing thereafter, with no defined ending time. Of course, maintaining acceleration indefinitely is not physically realizable (since it would require infinite energy expenditure), but this assumption results in a simple and analytically tractable *hyperbolic* trajectory (see Section 5). Further, a canonical mission trajectory is readily inferred based on a limited-time initial segment of a hyperbolic trajectory combined with the symmetry of the acceleration profile (see Section 6).

2.2 Canonical mission architecture

A typical interstellar mission has the objective of landing on some alien body, and is also physically realizable since its total energy expenditure is finite. In such a mission, a spacecraft is presumed to be launched from one stationary body, the *origin*, and to subsequently land on another stationary body, its *destination*. Typical origins and destinations are planets (e.g., Earth or Moon for an origin and an exoplanet for a destination).

The *canonical* mission is the simplest mission with this objective and is illustrated in Fig. 1. A spacecraft is propelled during a *cruise* phase with a constant self-acceleration for the first half of the journey and a constant (equal-in-magnitude) deceleration for the second half of the journey.

Although the origin and destination will generally be in relative motion, their relative velocity will generally be insignificant compared to the speed of light. Thus, in our canonical mission scenario, any

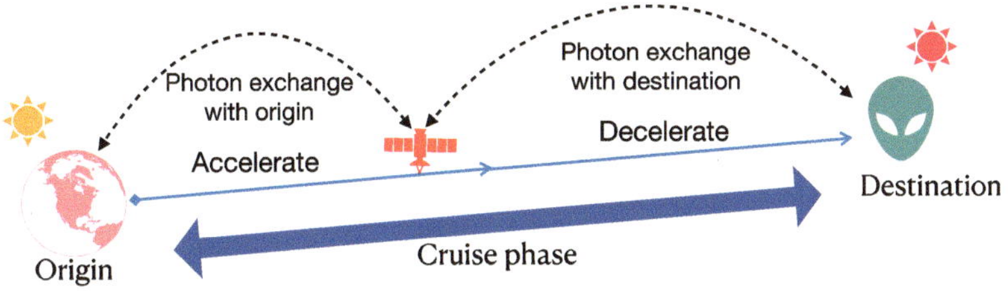

FIG. 1

Illustration of the cruise phase of a canonical mission for a self-propelled spacecraft. At rest at the origin, beginning with launch it experiences constant self-acceleration for the first half of the distance to the destination followed by a constant self-deceleration (with the same magnitude) for the second half, at rest at the instant of landing on the destination. The origin and destination are themselves assumed to be indefinitely at rest relative to a common inertial frame S.

relative motion between origin and destination is neglected and they are assumed to be inertial (not accelerating) relative to a common inertial frame S.

Our primary concern here is communication between the spacecraft and either origin or destination during the cruise phase. Although such a mission is presently beyond human technological capabilities and available energy resources, it is still worthwhile and interesting to consider (as we do here) the implications of timing relationships among these entities during the mission. The physical obstacles to communications uncovered is one among many important considerations in determining whether the large investments required to enable and execute interstellar travel are justified.

In the case of a mission carrying human astronauts, those astronauts (as well as any biological material they carry, such as plants, animals, or food) have to survive the accelerations endemic to the journey. While equipment can be designed to withstand large accelerations or weightlessness, biological material originating from a planetary environment is vulnerable to conditions not endemic to that environment, including large accelerations and weightlessness. To address this, in numerical examples, we assume a so-called 1-g mission, which chooses a constant acceleration magnitude equal to Earth gravity, or $\alpha = \pm g$, thereby creating a spacecraft environment that replicates the Earth's surface in this respect.

3 Relativistic effects on spacecraft trajectories

When traveling near the speed of light c, relativistic effects strongly affect the relationship between the spacecraft and its environment.

3.1 Travelers' clocks

Following the description in Section 1, the canonical mission involves three travelers of interest: the origin $\mathcal{O}$, the spacecraft $\mathcal{C}$, and the destination $\mathcal{D}$. (Although $\mathcal{O}$ and $\mathcal{D}$ are assumed to be inertial, they can still be considered "travelers" in a degenerate sense.) The three travelers carry clocks measuring, respectively, traveler's times τ_o, τ_c, and τ_d. (If this is not physically true of $\mathcal{D}$ because it is technology-sterile, this remains useful for analytical purposes.)

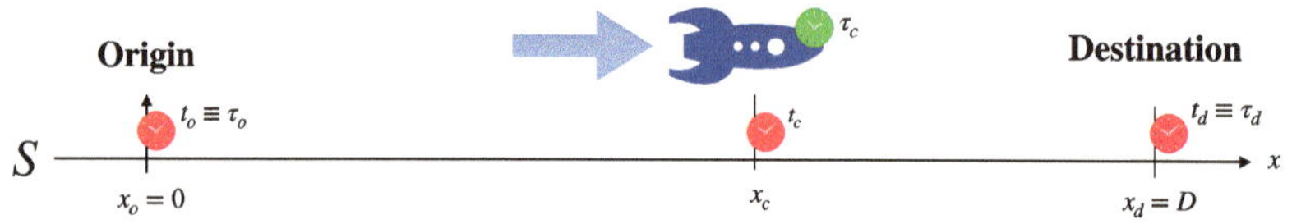

FIG. 2

Illustration of the coordinates of a spacecraft's trajectory as a function of distance x and coordinate time t with respect to inertial frame S. The spatial coordinate x of the origin, spacecraft, and destination are labeled $\{x_o = 0, x_c, x_d = D\}$. Clocks that measure coordinate times t, at those positions measure times labeled $\{t_o, t_c, t_d\}$. While t_o will likely be measured by a physical clock, $\{t_c, t_d\}$ may be notional (rather than physical) clocks. In addition, there are three traveler's (proper) times $\{\tau_o, \tau_c, \tau_d\}$ carried by the respective travelers. The coordinate and travelers' clocks can be considered to be one and the same at the origin and destination, but differ for the spacecraft position x_c due to their relative motion.

3.2 Inertial frame and coordinate times

The location of the clocks of interest and a notation for their respective times are illustrated in Fig. 2. The inertial frame S defined previously can be presumed to have 1D spatial dimension x, as well as coordinate time t defined as the time measured by any clock at rest with respect to S. The speed of light as measured by $\{x, t\}$ is a constant c, and this allows for the determination of propagation delays. In the case of $\mathcal{O}$ and $\mathcal{D}$, their traveler's clocks are also at rest with respect to S, and thus they can be considered to have coordinate times $t_o \equiv \tau_o$ and $t_d \equiv \tau_d$. The coordinate time t_x at an arbitrary intermediate position $0 < x < D$ is defined as the time that would be measured by a clock at rest at that location, and generally $\tau_c \neq t_c$ due to the relative motion of the respective clocks. All clocks of interest can have a fixed offset dependent on their initialization. Taken together, the location of a traveler is quantified by its three coordinates $\{\tau, x, t\}$.

3.3 Clock synchronization

While $\{t_o, t_c, t_d\}$ run at identical rates (since they possess no relative motion), there remains the question of their appropriate initialization, and likewise the initialization of traveler's times $\{\tau_o, \tau_c, \tau_d\}$. The presumption of $\tau_o \equiv t_o$ and $\tau_d \equiv t_d$ defines the synchronization of two of the six clocks of interest. It is natural to assume that $t_o = \tau_o = \tau_c = 0$ at the instant of launch, since the respective clocks are colocated and thus can be simultaneously initialized.

The remaining two times $\{t_c, t_d\}$ are not measured by clocks colocated with one another or with t_o, and thus their initialization is more challenging. We assume that both are *synchronized* with t_o, meaning that a measurement of the light-speed c using $\{t_o, t_c, t_d\}$ yields the correct value c. This synchronization is also suitable for characterizing clock images. That is, a photon emitted at time $t_o = 0$ at the origin is detected at time $t_c = x_c/c$ at x_c and $t_d = D/c$ at the destination.

Example 6: If $\mathcal{O}$ sends a message to $\mathcal{D}$ as it initializes its local clock to $t_o = 0$, that message traveling at the speed of light can upon arrival trigger an initialization of the destination clock to $t_d = D/c$, assuming that $\mathcal{D}$ has knowledge of message propagation distance D.

When the propagation distance is unknown to the travelers, clock synchronization requires bidirectional photon exchange protocols [1].

3.4 Measures of mission-elapsed time

The mission-elapsed time (MET) is the cumulative time elapsed for the spacecraft to reach some position x_c during the spacecraft cruise phase. There are two versions, depending on what clock is used to measure that elapsed time. The mission-elapsed coordinate time (MECT) is measured by coordinate time t_c, whereas the mission-elapsed traveler's time (METT) is measured by the spacecraft's traveler's time τ_c, where we follow the convention that $\tau_c = 0$ at the instant of launch.

The spacecraft speed must be less than light-speed c relative to S, and thus MECT $> x_c/c$. Due to relativistic time dilation, the spacecraft's traveler's clock runs more slowly than its respective coordinate clock, and thus METT $\ll$ MECT (see Section 5.2). Thus it is entirely possible that METT $\ll x_c/c$, so that, as measured by the traveler's clock, the light-speed limit has been exceeded.

An alternative trajectory could include a ballistic (constant speed) phase in its center. In this case the ballistic phase would be bracketed by equal durations of acceleration and deceleration. Our canonical mission includes no such ballistic phase. This alternative would have minimal impact on the MECT at the time of landing, because with or without the ballistic phase the spacecraft speed is very near to c for most of the cruise. However, as will be seen, omitting the ballistic phase beneficially and substantially reduces METT (to the benefit of resources consumed on the spacecraft and biological aging).

4 Communication with a spacecraft

Especially for a spacecraft with human astronauts, persistent communication during the entire mission would be desirable during the cruise phase, as opposed to following landing only. However, except for relatively brief time periods shortly after launch and before landing, the spacecraft will be traveling at nearly the speed of light. This presents special issues as we will see. Postlanding the spacecraft is assumed to be at rest in S, so the situation is far simpler.

During the cruise phase there are four instances of communication represented by $\mathcal{O} \Rightarrow \mathcal{C}$, $\mathcal{C} \Rightarrow \mathcal{O}$, $\mathcal{C} \Rightarrow \mathcal{D}$, and $\mathcal{D} \Rightarrow \mathcal{C}$. For each of these cases the clock image Ψ and its inverse Ω characterize the timing of the communication as affected by both propagation delay and relativistic time dilation (see Section 1.1).

The query-response latency $\mathfrak{L}_{\mathcal{A} \Leftrightarrow \mathcal{B}}$ characterizes the delay experienced by a traveler in two-way communication. Instances of these involving the spacecraft include $\mathcal{O} \Leftrightarrow \mathcal{C}$, $\mathcal{C} \Leftrightarrow \mathcal{O}$, $\mathcal{C} \Leftrightarrow \mathcal{D}$, and $\mathcal{D} \Leftrightarrow \mathcal{C}$. This latency can be determined by the composition of one clock image Ψ for the query and another for the response. Since travelers may not be fully cognizant of one another's trajectories, both query and response messages are likely to include a time-stamp confirming or revealing the local traveler's time.

Here we summarize the results of this analysis absent mathematical justification, followed in Section 5 by an analysis employing the special theory of relativity (SR).

4.1 Spacecraft trajectory

The spacecraft trajectory reveals the evolution of its position x_c in S versus time, specifically $\{\tau_c, x_c, t_c\}$. This can be interpretated as space time $\{x_c, t_c\}$ relative to S coordinates as parameterized by the traveler's time τ_c. S provides a common reference for measuring photon propagation delays, time dilation effects for the spacecraft traveler's clock, and clock images.

In the case of a spacecraft, τ_c governs the operation of the spacecraft instrumentation and is very significant to any human astronauts (governing their heart rate, circadian rhythm, etc.). On the other hand, t_c matters greatly to navigation systems aboard the spacecraft and at the origin (which measure and control spacecraft kinematics relative to an inertial frame S), to controllers and operators at the origin and destination, and determines photon propagation delay.

4.2 Event horizon

The simpler indefinite self-acceleration trajectory is illustrated in Fig. 3. A convenient parameter describing this trajectory is the *event horizon* $t_H = c/\alpha$. The physical significance of t_H is the earliest coordinate time for which a photon leaving the origin never intersects the spacecraft trajectory, and hence will never be detected onboard the spacecraft.

Three photon trajectories (each linear with unity slope) are compared to the spacecraft in Fig. 3. The first is a photon emitted from the origin at the instant of spacecraft launch. The spacecraft has to travel

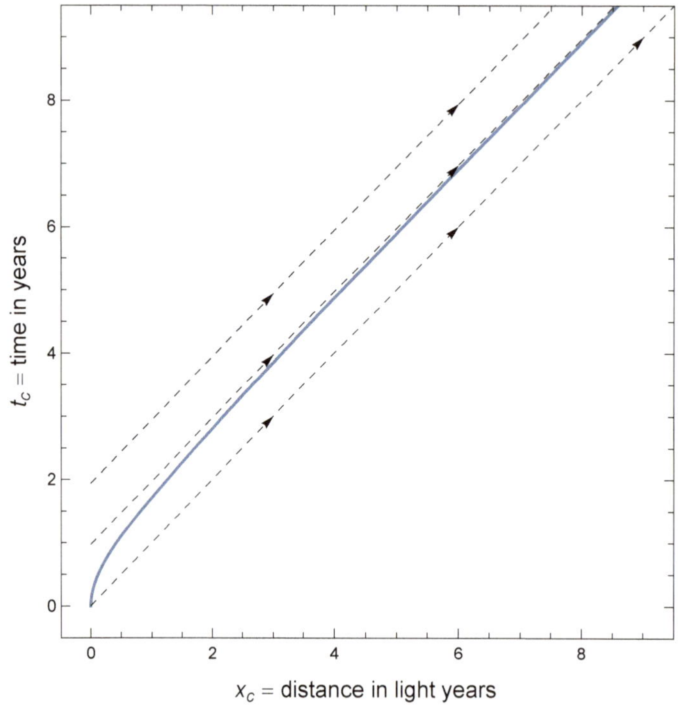

FIG. 3

The trajectory of an indefinite 1-g self-acceleration spacecraft with $\alpha_c = +g$ and event horizon $t_H = 0.97$ years is plotted over an elapsed traveler's time of $0 \le \tau_c \le 3$ years. See Section 5.4 for a derivation of this trajectory. Also shown as *dashed lines* are the trajectories of a photon emission from the *origin* at three coordinate times $t_o \in \{0,\ t_H,\ 2t_H\}$ in the + x-direction. Each is a straight line with slope +1 on a plot with time units of years and distance units of light years.

more slowly than this photon, since it cannot exceed light speed c, and thus its trajectory always falls above the photon's trajectory (its coordinate time to reach any position x_c has to be larger). As the spacecraft's speed approaches light speed c, its trajectory exponentially approaches (but never intersects) that of a photon emitted from the origin at $t_o = t_H$. The trajectory of any photon emitted from the origin for $t_o \geq t_H$, which falls inside the shaded region on Fig. 3, never intersects the spacecraft trajectory, and hence no such photon can ever be detected by the spacecraft. A typical case is the photon emitted at $t_o = 2t_H$ as shown.

In summary, photon emissions originating from the origin beyond the event horizon (shown as the shaded area) are invisible to spacecraft experiencing indefinite acceleration. The result is a communication *blackout* from the origin to the spacecraft. While such a blackout is not present for the canonical mission, the value of t_H has a significant impact on clock images in that case as well.

4.3 Spacecraft communication with origin and destination

The trajectory of a spacecraft on the canonical mission is shown in Fig. 4 together with some typical trajectories for photons exchanged to and from the origin $\mathcal{O}$ and destination $\mathcal{D}$. Due to the time-reversal symmetry of the acceleration/deceleration, the second half of the canonical mission trajectory is simply a mirror image of the first half (see Section 6).

The shaded regions in Fig. 4 represent photons that intercept $\mathcal{C}$'s trajectory, and thus were either emitted at that point of intercept or detected at that point of intercept. These photons are deemed to be *productive*, in the sense that they can be employed as a basis for communication during $\mathcal{C}$'s cruise phase. Photon trajectories outside these shaded regions must be emitted or detected while $\mathcal{C}$ is at rest at $\mathcal{O}$ prelaunch or at $\mathcal{D}$ postlanding. $\mathcal{C}$'s payload (including possibly human astronauts) is governed by a local clock measuring traveler's time τ_c (represented only by parameterization in Fig. 4).

In the direction $\mathcal{O} \Rightarrow \mathcal{C}$, a photon emitted from the origin over the coordinate-time interval $0 < t_o < {\sim}2t_H$ can be detected by the spacecraft during its cruise phase (later-emitted photons reach the spacecraft following landing at the destination). The factor ${\sim}2t_H$ is the composition of ${\sim}t_H$ for the first half and another ${\sim}t_H$ (due to the mirror-image symmetry) for the second half. (The precise range of times is slightly smaller than ${\sim}t_H$, since the positive self-acceleration of the spacecraft is terminated at the halfway point.)

A set of important distinctions and conclusions can be drawn from Fig. 4:

- From the perspective of spacecraft $\mathcal{C}$, detection of photons originating from both $\{\mathcal{O}, \mathcal{D}\}$ and emission of photons toward both $\{\mathcal{O}, \mathcal{D}\}$ are productive during the entirety of the cruise phase. This constitutes $\mathcal{C}$'s entire cruise-phase METT.
- Consider the perspective of $\mathcal{O}$ and $\mathcal{D}$ and photons either emitted or detected that propagate in the same direction as $\mathcal{C}$'s velocity (which is the direction $\mathcal{O}$ to $\mathcal{D}$). That includes directions $\mathcal{O} \Rightarrow \mathcal{C}$ and $\mathcal{C} \Rightarrow \mathcal{D}$. These photons are productive during a relatively short coordinate-time duration ${\sim}2t_H$ in both cases. Since MECT ${\sim}(D/c + 2t_H)$ for the entirety of the cruise phase, there is a long coordinate-time duration ${\sim}D/c$ overlapping the MECT over which no photon emissions and detections can be productive.
- Consider again the perspective of $\mathcal{O}$ and $\mathcal{D}$, but in this case photons either emitted or detected propagate in the opposite direction to the spacecraft $\mathcal{C}$'s velocity. This includes directions $\mathcal{C} \Rightarrow \mathcal{O}$ and $\mathcal{D} \Rightarrow \mathcal{C}$. Those photons are productive during a long coordinate-time duration, which is about *double* the METT.

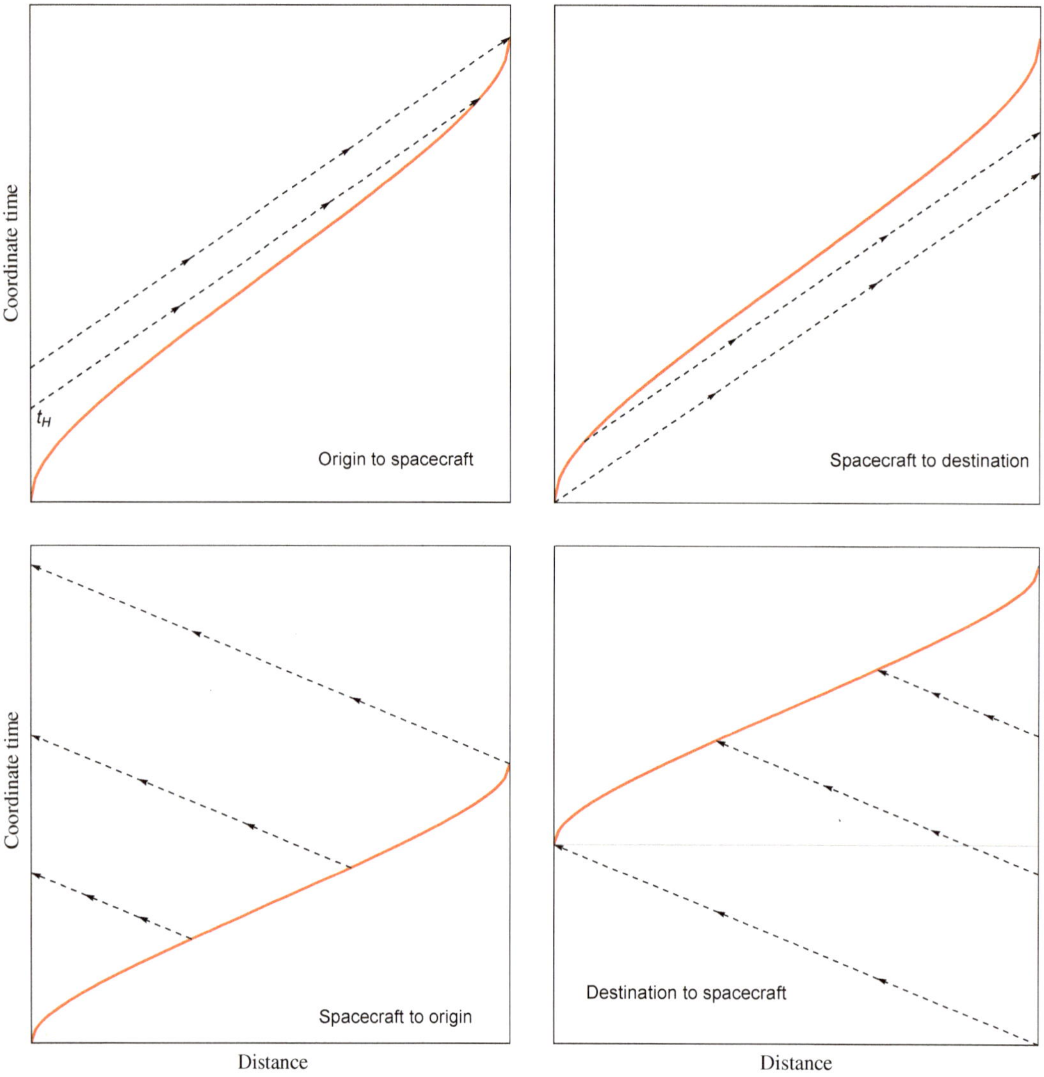

FIG. 4

Illustration of the time constraints on communication during a spacecraft's travel from launch to landing for the canonical mission. The spacecraft trajectory is the *solid line*, and photon trajectories to and from the spacecraft and to and from the origin and destination are shown as *dashed lines*. Each *shaded region* represents the ensemble of all feasible photon trajectories that intersect the spacecraft trajectory during its cruise phase. The most significant influence on this ensemble shape is the direction of photon propagation.

Table 1 Communication blackout conditions.

$\mathcal{O} \Rightarrow \mathcal{C}$	Communications from $\mathcal{O}$ does not reach the spacecraft $\mathcal{C}$ during the cruise phase except for a coordinate-time duration $\sim 2t_H$ immediately following launch. Such communications do arrive following landing, while the spacecraft is at rest at the destination.
$\mathcal{C} \Rightarrow \mathcal{O}$ and $\mathcal{D} \Rightarrow \mathcal{C}$	Communication is possible during $\mathcal{C}$'s entire cruise phase
$\mathcal{C} \Rightarrow \mathcal{D}$	$\mathcal{D}$ receives no communication from $\mathcal{C}$ initiated during its cruise phase except for a coordinate-time duration $\sim 2t_H$ immediately preceding landing.

- There is a substantial distinction between the productive coordinate-time duration of communication in both directions as compared to the spacecraft's METT. This represents a *time-warping* effect, in which communication signals always appear to have distinctive time axes at transmitter and receiver (see Section 4.7). Time warping is distinct from relativistic time dilation (see Section 5.2), and is considerably more extreme in its effect. While time warping is affected by time dilation, it is primarily a manifestation of the observation that photons propagating in the same direction as the spacecraft have difficulty "catching up with" or "getting out ahead of" a spacecraft traveling at near the speed of light c.

The MECT at landing, from the perspective of the origin $\mathcal{O}$ or destination $\mathcal{D}$, defines a coordinate-time period coinciding with the cruise phase of spacecraft $\mathcal{C}$. Ideally, communication to and from the spacecraft would be possible during the entirety of this MECT. Whenever such communication is not possible (resulting from nonproductive photon emissions or detections), this constitutes a *communication blackout*, which is summarized in Table 1. The practical manifestation of these blackouts is that $\mathcal{C}$ (or $\mathcal{D}$) cannot be informed of any events at $\mathcal{O}$ (or $\mathcal{C}$) except for a short period of coordinate-time duration $\sim 2t_H$ (about 2 years for a 1-g canonical mission) immediately following launch (or preceding landing).

4.4 Mission-elapsed time at landing

The METs can be considered and compared from a traveler's and coordinate-time perspective. The traveler's perspective is simpler, since METT $\equiv \tau_c$ for cruising distance $0 < x \leq x_c$. There are three coordinate times $\{t_o, t_c, t_d\}$ that differ in their initialization, but MECT $\equiv t_c$ is the correct choice for the MECT when the spacecraft is at position x_c, since it defines the coordinate time at that position. The initializations of $\{t_o, t_c\}$ are set to correctly measure the speed of a photon, which implies that they also correctly measure a photon's propagation time $t_c - t_o$ from $x = 0$ to $x = x_c$. It follows that those initializations are also the correct measure of MECT for a spacecraft.

The MECT and METT at the time of landing are quite different. The landing-time MECT of the canonical 1-g mission is MECT $\sim(D/c + 2t_H)$. This is an asymptote for large D, with the precise value plotted in Fig. 5 (see Section 4.3). The METT is plotted in Fig. 6 and is much smaller (METT $\ll$ MECT) due to relativistic time dilation.

This distinction between MECT and METT is very significant for spacecraft designers, since METT governs all time-dependent aging and resource utilization, including the time period over which self-acceleration has to be maintained, onboard energy, wear and tear of equipment, life support

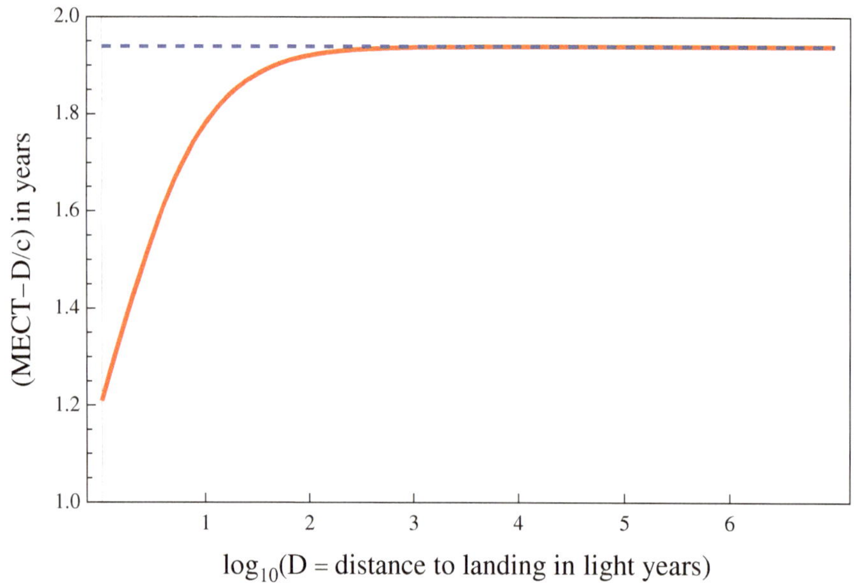

FIG. 5

A plot of the excess of the MECT over the time for a photon to reach the destination for the canonical mission of Fig. 1. This is compared to twice the event horizon ($2t_H$) as the *dashed line*. The horizontal axis is the log of distance $\log D$ for $1 \leq D \leq 10^7$ ly. The longest distance exceeds the distance to the Andromeda galaxy. At greater distances, twice the event horizon $2t_H$ (the *dashed line*) is a good approximation to that excess MECT.

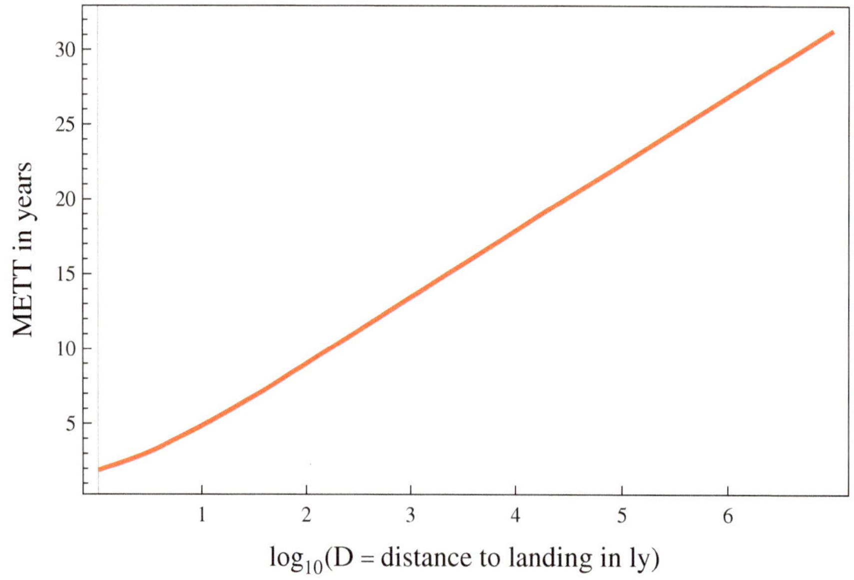

FIG. 6

A repeat of Fig. 5 for the traveler's time (METT). The canonical-mission METT falls well within a typical single human lifetime even at the greatest distance shown (10^7 ly).

supplies for human astronauts including food and oxygen, and the biological aging of the astronauts. Even for a journey to the nearest galaxy Andromeda ($D \approx 2.5 \times 10^6$ ly), the total canonical-mission METT at landing falls within a typical human lifetime.

Another consideration is the timekeeping aspect of communications. This affects the perception of the communication from the receiving end, but the transmitting end should also be aware of any time limitations on their ability to convey information or events.

Example 7: Consider the uplink (direction $\mathcal{O} \Rightarrow \mathcal{C}$). A spacecraft crew that travels $D = 1000$ ly will have received, at the instant of landing at the destination, ~1.9 years of transmissions from the origin (as measured by the origin's coordinate clock) during a ~13.5 years journey (as measured by the spacecraft's onboard traveler's clock). Thus the crew perceives that they have lost visibility into ~11.6 years of current events at the origin during their journey. Thereafter the communication from the origin will stream in real time, since both transmitter and receiver are at rest in the same inertial frame S. Thus the crew will, after landing, be kept up to date with events at the origin as they happen, although ~11.6 years after the fact according to their own traveler's clock. Any denizens at the destination will, in contrast, have the perception that their knowledge of events at the origin is 1000 years after the fact, which is dramatically greater.

The downlink (direction $\mathcal{C} \Rightarrow \mathcal{O}$) is very different due to the relatively rapid advance of the coordinate time. Photons emitted from the spacecraft will arrive at the origin with an x_c/c propagation delay as measured by the origin's coordinate clock, which becomes large for large D.

Example 8: Continuing Example 7, at the origin, news of a successful landing will be delayed for ~2002 years following the spacecraft launch, which amounts to many generations for a human population. During those intervening centuries, ~13.5 years of transmissions from the spacecraft will have been slowly received, hopefully serving to maintain awareness of (and interest in) the mission and its outcome. Thereafter, after landing, the origin will receive real-time transmissions from the spacecraft, albeit with the perception that they are arriving

$$\sim (2002 - 13.5) = 1988.5 \text{ years}$$

after the fact.

Rather than communicate back to the origin, an alternative scenario for the spacecraft would be to immediately turn around and travel back to the origin. This would effectively implement time travel ~2004 years into the future, at the expense of ~27 years total elapsed traveler's time.

A general conclusion of the preceding Examples 7 and 8 is that the impact of effects like propagation delay and time horizon is minimal for the spacecraft payload in isolation, largely due to compensating relativistic time-dilation effects. However, the impact of these conditions on denizens of the origin and destination is great, and as will be seen (see Section 4.6), all parties encounter severe obstacles to two-way communication among themselves.

4.5 One-way communication: Clock images

The method of calculating the clock image function (see Section 1.1) from two trajectories is illustrated in Fig. 7. The trajectories are plotted with respect to inertial frame S, and photon trajectories are plotted as unit-slope dashed lines. Each photon trajectory that intersects the two traveler's trajectories in the direction $\mathcal{A} \Rightarrow \mathcal{B}$ defines the clock image $\tau_b = \Psi_{\mathcal{A}\Rightarrow\mathcal{B}}(\tau_a)$ for the pair of traveler's times $\{\tau_a, \tau_b\}$ at the intersection with that photon trajectory.

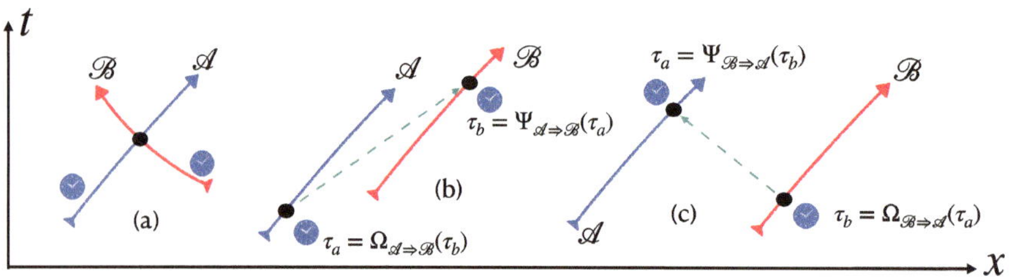

FIG. 7

Illustration of the clock image functions for photon transfer between traveler $\mathcal{A}$ with traveler's clock τ_a and traveler $\mathcal{B}$ with traveler's clock τ_b with specified spacetime trajectories represented relative to a common inertial frame S. For the degenerate case of an origin or destination at rest, the corresponding trajectory will be vertical.
A photon trajectory in S, which intersects both trajectories, creates a mapping $\tau_a \Rightarrow \tau_b$, which can be represented by a clock image function $\Psi_{A\Rightarrow B}(\cdot)$, as well as a mapping $\tau_b \Rightarrow \tau_a$, which can be represented by an inverse function $\Omega_{A\Rightarrow B}(\cdot)$. Note that one or both of the travelers may be at rest (say at the origin or destination), and for such a traveler the trajectory is vertical (with $\tau \equiv t$ within an arbitrary initialization). (a) If the trajectories intersect at some point in spacetime, the instantaneously contiguous traveler's clocks can be directly compared, and if desired, initialized to achieve synchronization. (b) A photon trajectory in the +x-direction. (c) A photon trajectory in the $-x$-direction.

Over the course of a mission there will be a considerable time-warping of the time base of the received signal transmissions in relation to the transmitter time axis (see Section 4.7). This is manifested by the clock image functions shown in Figs. 8 and 9 for a range of launch-landing distances, and for an uplink from origin to spacecraft and downlink from spacecraft to origin.

On the uplink $\mathcal{O} \Rightarrow \mathcal{C}$ shown in Fig. 8, there is a sudden increase in the uplink clock image $\Psi_{\mathcal{O}\Rightarrow\mathcal{C}}$ at the event horizon t_H because photons will no longer intercept the spacecraft trajectory during the acceleration phase and thus delay their arrival at the spacecraft to the deceleration phase. Postlanding $\Psi_{\mathcal{O}\Rightarrow\mathcal{C}}$ becomes linear with unit slope since the two clocks increase in lockstep. On the downlink $\mathcal{C} \Rightarrow \mathcal{O}$ shown in Fig. 9, this transition is more gradual, because the photon trajectory is oblique (rather than nearly tangent) to the spacecraft trajectory, but $\Psi_{\mathcal{O}\Rightarrow\mathcal{C}}$ is exponentially larger (requiring a log plot) since the origin clock does not benefit from time dilation. Thus the photon arrival latency (as observed by the local clock) is exponentially larger on the downlink. As with the uplink, postlanding $\Psi_{\mathcal{C}\Rightarrow\mathcal{O}}$ becomes linear.

4.6 Two-way communication: Query-response latency

Much can be accomplished by one-way telemetry from spacecraft to origin (e.g., monitoring the status of mechanical systems or human crew) and a stream of current events from the origin to the spacecraft. However, if the spacecraft is not to be totally autonomous, there is need for two-way communication. This can take the simple form of a query followed by a response to that query (see Section 1.3.2). In the case of a "conversation," a query followed by a response followed by another query following up on that response, these query-responses repeated indefinitely, might occur. Such conversations are especially useful to human astronauts in maintaining a social connection with family and friends.

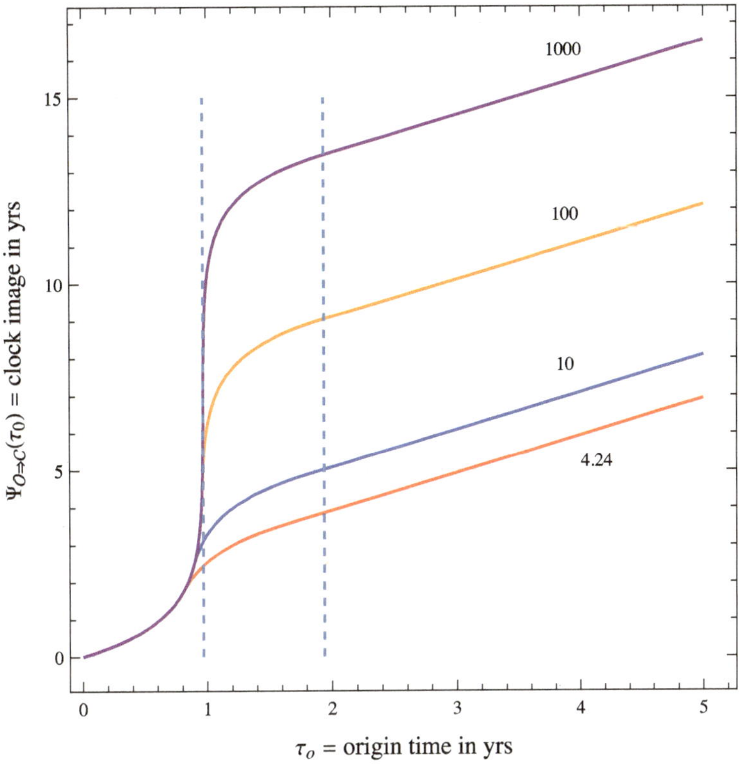

FIG. 8

A plot of the clock image $\Psi_{O\Rightarrow C}(\tau_o)$ in years for photon transfer from origin to spacecraft, for a canonical 1-g mission. The *vertical dashed lines* mark the event horizon t_H and twice the event horizon $2t_H$, which are, respectively, upper bounds on the first half and entirety of the spacecraft cruise phase. Beyond $\sim 2t_H$ the clock image is linear with unit slope because the spacecraft is at rest at the destination. Both horizontal and vertical scales are linear (rather than logarithmic), reflecting the relatively short METT experienced by the spacecraft as well as the relatively short interval before the event horizon is reached at the origin (beyond that, transfer has to await the second half of the cruise phase or following landing). In practical terms, a reasonable delay is achieved for a relatively short time interval following launch. The different curves show the clock image function for different distances $D \in \{4.24, 10, 100, 1000\}$ ly to the destination, with an expansion in METT for the longer distances.

4.6.1 Indefinite constant self-acceleration mission

When the self-acceleration continues indefinitely, there are only two query-response latencies of interest, $\mathfrak{L}_{O\Leftrightarrow C}$ and $\mathfrak{L}_{C\Leftrightarrow O}$. The values are determined in Section 5.6 and illustrated in Fig. 10 for a 1-g mission. Because either the response or the query has to propagate in the same direction as the spacecraft motion, in both cases, the traveler's time of a query is limited by the time horizon and latency increases dramatically as that horizon is approached.

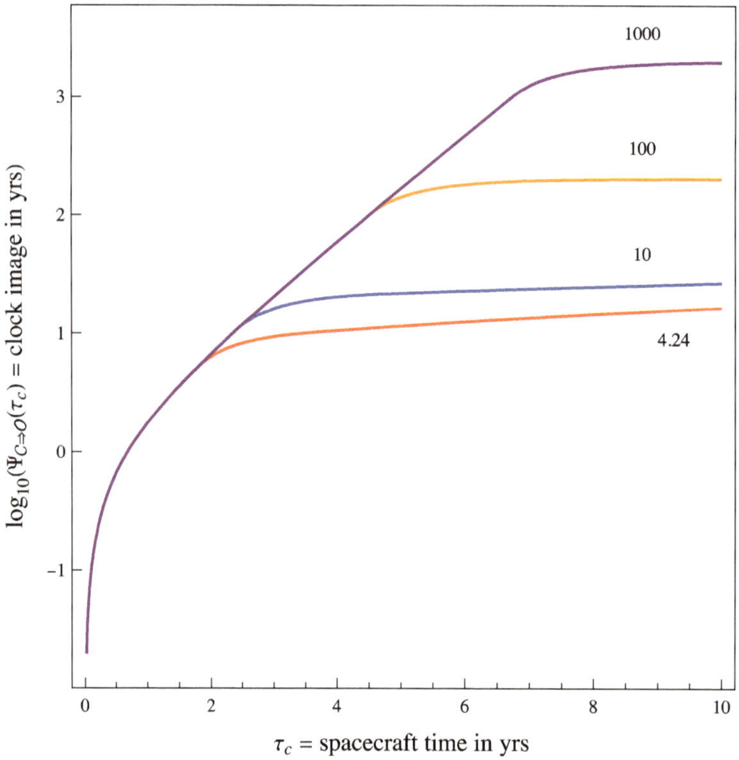

FIG. 9

The clock image $\Psi_{\mathcal{C}\Rightarrow\mathcal{O}}(\tau_c)$ for a photon transfer from a spacecraft to origin. All the conventions of Fig. 8 are repeated, except the direction of photon transfer is reversed. The image axis is now logarithmic (in contrast to Fig. 8) because photon transfers from the spacecraft now persist over a much longer period (reflecting MECT ≫METT).

4.6.2 Canonical mission

For the canonical mission, which adds the possibility of communicating with the destination, there are six distinctive query-response latencies,

$$\{\mathfrak{L}_{\mathcal{O}\Leftrightarrow\mathcal{C}}, \mathfrak{L}_{\mathcal{C}\Leftrightarrow\mathcal{O}}, \mathfrak{L}_{\mathcal{C}\Leftrightarrow\mathcal{D}}, \mathfrak{L}_{\mathcal{D}\Leftrightarrow\mathcal{C}}, \mathfrak{L}_{\mathcal{O}\Leftrightarrow\mathcal{D}}, \mathfrak{L}_{\mathcal{D}\Leftrightarrow\mathcal{O}}\},$$

but only $\{\mathfrak{L}_{\mathcal{O}\Leftrightarrow\mathcal{C}}, \mathfrak{L}_{\mathcal{C}\Leftrightarrow\mathcal{O}}\}$ are quantified here, with the photon trajectories illustrated in Figs. 11 and 12. There are also two trivial cases

$$\mathfrak{L}_{\mathcal{O}\Leftrightarrow\mathcal{D}}[\tau] = \mathfrak{L}_{\mathcal{D}\Leftrightarrow\mathcal{O}}[\tau] = 2D/c$$

for any traveler's time τ, since $\mathcal{O}$ and $\mathcal{D}$ are both at rest in S.

$\mathfrak{L}_{\mathcal{O}\Leftrightarrow\mathcal{C}}$ is plotted in Fig. 13 and $\mathfrak{L}_{\mathcal{C}\Leftrightarrow\mathcal{O}}$ in Fig. 14. There is a rapid increase in the latency until the point at which the spacecraft begins to slow during the deceleration phase of the mission. In the case of $\mathfrak{L}_{\mathcal{C}\Leftrightarrow\mathcal{O}}$,

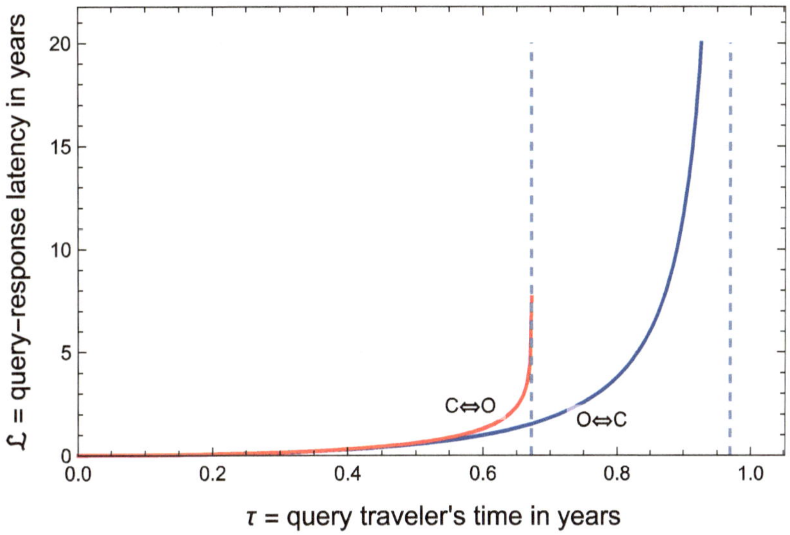

FIG. 10

The query-response latencies $\mathfrak{L}_{O\Leftrightarrow C}$ and $\mathfrak{L}_{C\Leftrightarrow O}$ are compared for an indefinite self-acceleration mission with $\alpha = g$. The opportunity for obtaining a response to a query originating at the origin is limited to the time horizon $\tau_o < t_H$ (shown as the *second dashed line*), and queries from the spacecraft are more severely limited to $\tau_C < t_H \cdot \log 2$ (also shown as the *first dashed line*). The latter restriction allows a query to reach the origin soon enough that the response is generated prior to t_H.

that rapid increase actually occurs prior to the event horizon, since the query, following a downlink propagation delay, has to reach $\mathcal{O}$ prior to the event horizon.

Save for a period following launch in the order of days or months, the conclusion is that two-way communication becomes at best cumbersome due to large response latencies (in the order of years, decades, or centuries, except for missions to relatively nearby targets). Autonomous control of the spacecraft is a must for most of the mission. The implications are profound for a human crew and for humanity back at the origin, because for most of the mission both must rely primarily on one-way streaming of crew status and current events (see Section 4.5) rather than a conversation.

4.7 Time warping in streaming communication

The query-response latency captures one common type of communication protocol where one party has a question or comment and the other party responds. Our conclusion is that the latency is impractically large except for communication with the origin for a short period after launch or with the destination for a short period near landing. *Streaming* is a second mode of communication, in which either spacecraft or origin or destination transmits information continuously with no expectation of a response.

Example 9: The origin can transmit a continuous stream of sports programming to entertain human astronauts (see Section 4.7) or the spacecraft can transmit a webcam image of activity onboard the spacecraft back to the origin.

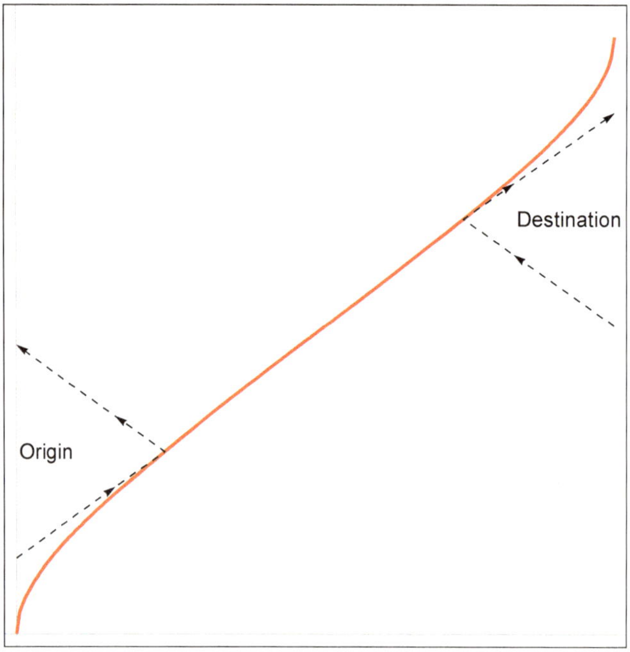

FIG. 11

An illustration of the photon trajectories for two-way query-response communication for a canonical 1-g mission with linear axes. Shown are queries originating at the origin and destination, and in each case a response from the spacecraft. The query-response latencies $\mathfrak{L}_{O\Leftrightarrow C}$ and $\mathfrak{L}_{D\Leftrightarrow C}$ are generally very large (multiple years) unless the query is generated shortly after launch or shortly before landing, respectively. The query and the response contribute equally to this latency.

Of course, achieving sufficiently high data rates to support reasonable program fidelity is assumed. (Low data rates may limit the fidelity of the streamed material. To improve that fidelity there is the option to stretch the transmission out in time and then transmit it at a lower data rate.) Subject to this limitation, streaming is more practical than query-response communication during the bulk of the cruise phase, in that the inevitable propagation delay should not disturb the quality of the viewing experience at the time of reception.

In streaming, it is natural to consider a continuous-time transmit and receive signal waveform $s[\tau]$ representing the signal intensity (average rate of photon emissions or detections) versus time. In one common form of modulation, the underlying data is represented by a variation of that intensity. (There are other less energy-efficient modulation techniques that modulate phase and/or wavelength, typically in the context of heterodyne as opposed to photon-counting detection.) The effect of accelerated motion is a nonlinear "warp" in the timebase τ of the waveform $s[\tau]$ between the transmitter and receiver.

When streaming to or from the spacecraft, it is evident from Fig. 4 (and keeping in mind that coordinate time is greatly expanded relative to the spacecraft traveler's time) that there is a significant expansion or contraction of a streaming program's duration. In the case of streaming to or from the

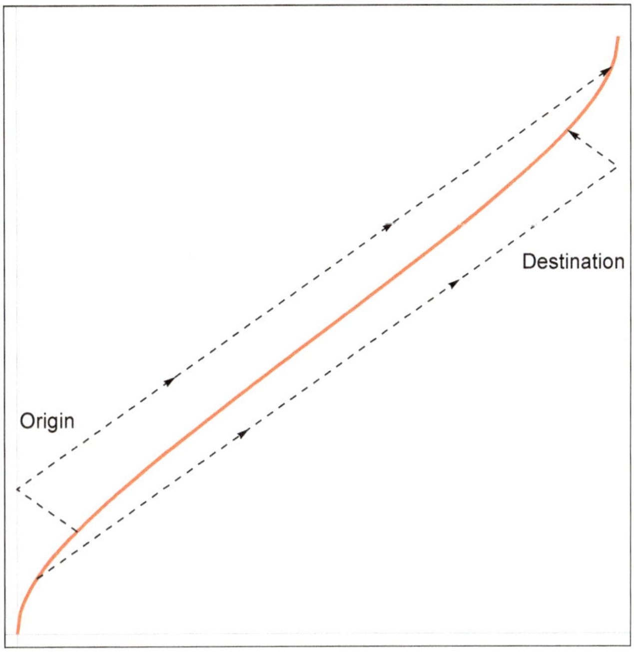

FIG. 12

A repeat of Fig. 11 for queries from the spacecraft to the origin and to the destination, also with linear axes. Again, the latencies $\mathfrak{L}_{\mathcal{C}\Leftrightarrow\mathcal{O}}$ and $\mathfrak{L}_{\mathcal{C}\Leftrightarrow\mathcal{D}}$ are generally very large (multiple years) unless the query is generated shortly after launch or shortly before landing, respectively. The photon propagation in the direction of the spacecraft trajectory contributes disproportionally to latency in both cases.

origin (the only case considered here), a streaming program's duration is expanded from transmitter to receiver in both directions:

- From origin to spacecraft this is due to the time horizon limitation (which is less than the METT) on total transmit time.
- From spacecraft to origin this is due to the expansion in coordinate time relative to spacecraft traveler's time (since MECT $\gg$ METT).

In both cases, if the program is streamed in "real time" at the transmitter, it is not only delayed (due to propagation delay) but also has an expanded duration at the receiver. If desired it can be stored at the receiver and subsequently played back at a higher speed (to return to real time). This time expansion is due to very different physical effects in the two directions. For $\mathcal{O} \Rightarrow \mathcal{C}$ this is due to the rapidly increasing propagation delay experienced by photons as the event horizon is approached (this is a classical effect). For $\mathcal{C} \Rightarrow \mathcal{O}$ this is a manifestation of the relatively slow evolution of the traveler's time τ_c relative to the coordinate clock t_o, which is a manifestation of relativistic time dilation (this is directly related to the observation that METT $\ll$ MECT).

For the streaming of programs of finite length (hours rather than years), this motion-induced time warp has minimal variation within the program duration. The timebase expansion can then be

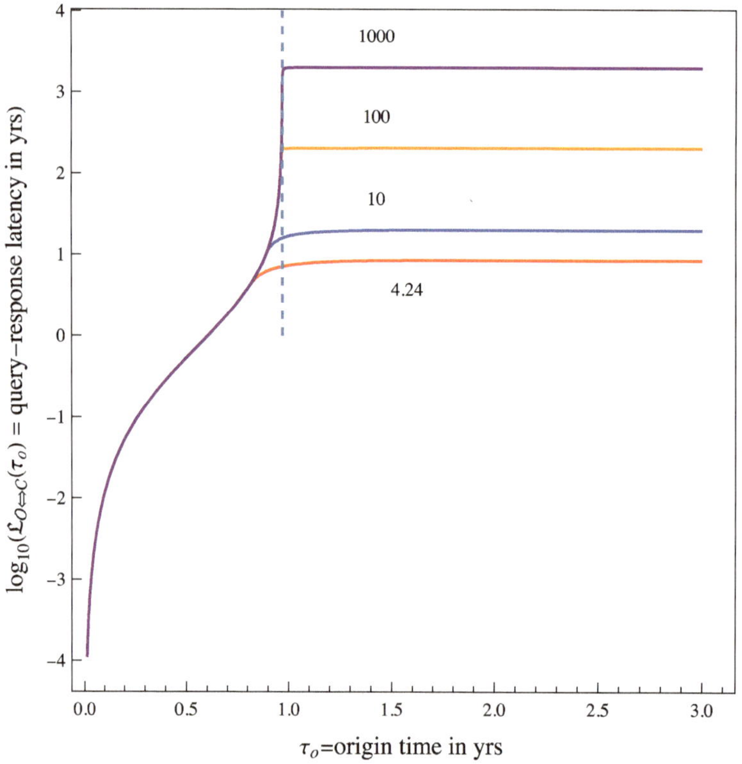

FIG. 13

A plot of the query-response latency $\mathfrak{L}_{O\Leftrightarrow C}$ in years for queries sent from the origin to the spacecraft for a canonical 1-g mission. The photon trajectories were illustrated in Fig. 11. The horizontal axis is the time after launch ($\tau_o = 0$) that a query to the spacecraft is originated at the origin, and the vertical axis is the resulting query-response latency. Only a very small fraction of the total MECT is plotted, especially at the longer distances. The vertical axis is logarithmic, reflecting the very large $\mathfrak{L}_{O\Leftrightarrow C}$ following the event horizon t_H (shown as a *vertical dashed line*). Queries sent following t_H result in much greater latency since the query does not reach the spacecraft until the second half of its cruise phase. The different curves show $\mathfrak{L}_{O\Leftrightarrow C}$ for different distances $D \in \{4.24, 10, 100, 1000\}$ ly to the destination, with longer distances resulting in greater $\mathfrak{L}_{O\Leftrightarrow C}$ beyond the event horizon due to the longer duration of acceleration (as opposed to deceleration).

approximated by the derivative of the receiver local time with respect to the transmitter local time referenced to a specific traveler's time τ_c at the spacecraft, and assuming that the result is constant over time intervals of interest. This derivative is evaluated in Appendix B.4 for the simpler indefinite self-acceleration case, and found to be:

$$\left.\frac{d\tau_c}{d\tau_o}\right|_{O\Rightarrow C} = \left.\frac{d\tau_o}{d\tau_c}\right|_{C\Rightarrow O} = \frac{\Psi_{C\Rightarrow O}[\tau_c]}{\Omega_{O\Rightarrow C}[\tau_c]} = e^{\tau_c/t_H}. \tag{8}$$

The timebase is expanded (rather than contracted) and is identical in the two directions, when referenced to the same instant from the spacecraft perspective, in spite of the distinctive physical processes

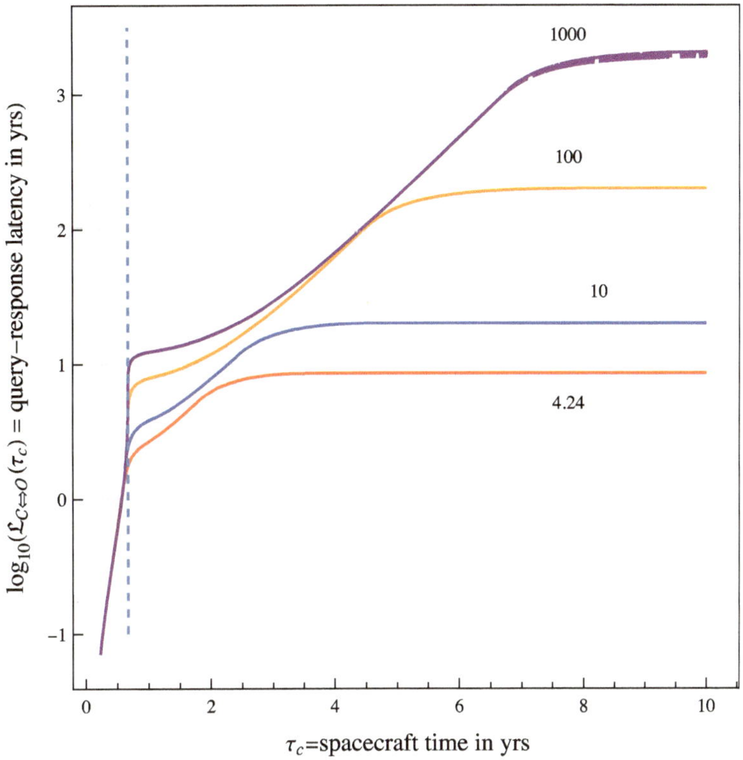

FIG. 14

A plot of the query-response latency $\mathfrak{L}_{C\Leftrightarrow O}$ in years for queries sent from the spacecraft to the origin for a canonical 1-g mission. The photon trajectories were illustrated in Fig. 12. All the conventions of Fig. 13 are repeated, except the direction of photon transfer is reversed. Thus the horizontal axis is the traveler's time τ_c at which the spacecraft originates the query, and the vertical axis is the logarithm of the resulting query-response latency as measured by that same traveler's time. The *vertical dashed line* is at $\tau_c = t_H \cdot \log 2$, which is the τ_c for which a photon emitted from the spacecraft reaches the origin at the event horizon $\tau_o = t_H$ (see Eq. 17b). This is the point at which responses begin to experience much larger delays (see Fig. 8).

at work. It increases exponentially with the spacecraft traveler's time τ_c due to the continuous increase in speed under constant acceleration.

Example 10: The origin $\mathcal{O}$ streams a football match to the spacecraft $\mathcal{C}$, and the spacecraft streams back a webcam video of the astronauts watching the match so that the controllers at the origin can gain vicarious enjoyment from watching the astronauts' reaction to the goals.

We are interested in the timing of this interaction. The appropriate timebases are the local clocks at $\mathcal{O}$ and $\mathcal{C}$ that govern the streaming timebases in the respective directions. Assume that these clocks are initialized so that MECT $= t_o$ at $\mathcal{O}$ and METT $= \tau_c$ at $\mathcal{C}$.

Consider the case where the match video arrives at $\mathcal{C}$ following 2 years of 1-g self-acceleration, or $\tau_c = 2$ years, and the webcam video is simultaneously streamed back. What are the times of transmission and reception of these streams at the origin? These times are revealed by the appropriate clock images, which are derived later in Section 5.6, with the result:

$$\mathcal{O} \Rightarrow \mathcal{C}:\ t_o = \Omega_{\mathcal{O}\Rightarrow\mathcal{C}}[\tau_c = 2\,\text{years}] = 0.846\,\text{years} = 309\,\text{days}, \tag{9a}$$

$$\mathcal{C} \Rightarrow \mathcal{O}:\ t_o = \Psi_{\mathcal{C}\Rightarrow\mathcal{O}}[\tau_c = 2\,\text{years}] = 6.66\,\text{years}. \tag{9b}$$

Although both directions of streaming are concurrent at the spacecraft, at the origin the $\mathcal{C} \Rightarrow \mathcal{O}$ webcast is delayed by 5.8 years relative to the $\mathcal{O} \Rightarrow \mathcal{C}$ match. The controllers at the origin have to wait multiple years to observe the astronauts' reaction.

What is the timebase expansion of the two streaming programs? Applying Eq. (8), the expansion in $\mathcal{O} \Rightarrow \mathcal{C}$ happens to be the same as the expansion $\mathcal{C} \Rightarrow \mathcal{O}$. (Again this assumes the spacecraft's reception of the match and the spacecraft's transmission of the webcast are simultaneous.) In both cases this expansion is a factor of 7.87. (The ratio of Eqs. (9b)–(9a) also evaluates to 7.87, which is also anticipated in Eq. (8).) Thus a 1-h football match would be expanded to 7.87 hours at the receiver. If the match video is played back as it is received, it would appear to be running in slow motion. To make the presentation more natural for the astronauts, the streamed match could be stored and subsequently played back at its original speed (which is 7.87 times faster than it was received). Identical logic applies independently to the spacecraft's webcast as it is received by the controllers.

Similarly, in streaming to and from the destination, it can be shown that the timebase contracts (rather than expands), is the reciprocal (decreases exponentially), and is identical in the two directions $\mathcal{C} \Rightarrow \mathcal{D}$ and $\mathcal{D} \Rightarrow \mathcal{C}$. Although these symmetries generalize to an arbitrary spacecraft trajectory, they are attributable to $\mathcal{O}$ and $\mathcal{D}$ both being at rest in S and do not apply when they are in motion.

4.8 Communication between two spacecraft

We can imagine two spacecraft with the same acceleration profile, but launched at different times, for example, in an alien-world colonization scenario. In that case it is interesting to consider the possibilities for communication between this pair of spacecraft.

Suppose that there are two spacecraft $\mathcal{C}1$ and $\mathcal{C}2$ launched in that order, so the ordering of spacecraft versus distance is

$$\mathcal{O} \rightarrow \mathcal{C}2 \rightarrow \mathcal{C}1 \rightarrow \mathcal{D}.$$

Thus, during their respective cruise phases, their trajectories are illustrated in Fig. 15. The traveler's times aboard the two spacecraft initialized to $\tau_{c1} = \tau_{c2} = 0$ at their respective launch times, so that they measure the METT for each spacecraft.

Our immediate concern is exploration of the photon exchange between $\{\mathcal{C}2, \mathcal{C}1\}$ in both directions. As illustrated in Fig. 15, photon exchange $\mathcal{C}2 \Rightarrow \mathcal{C}1$ is feasible as long as the launch time of $\mathcal{C}2$ falls within $0 < t_o < t_H$ (as it does in this case). However, the photon emission time is limited by the event horizon of $\mathcal{C}1$, after which there is a communication blackout.

The clock images for communication in both directions are illustrated in Figs. 16 and 17 for a $\mathcal{C}2$ launch coordinate time $t_o = \epsilon t_H$ where $0 < \epsilon < 1$ (see Appendix B.3). The query-response latencies for two-way communication between $\mathcal{C}2$ and $\mathcal{C}1$ are illustrated in Figs. 18 and 19.

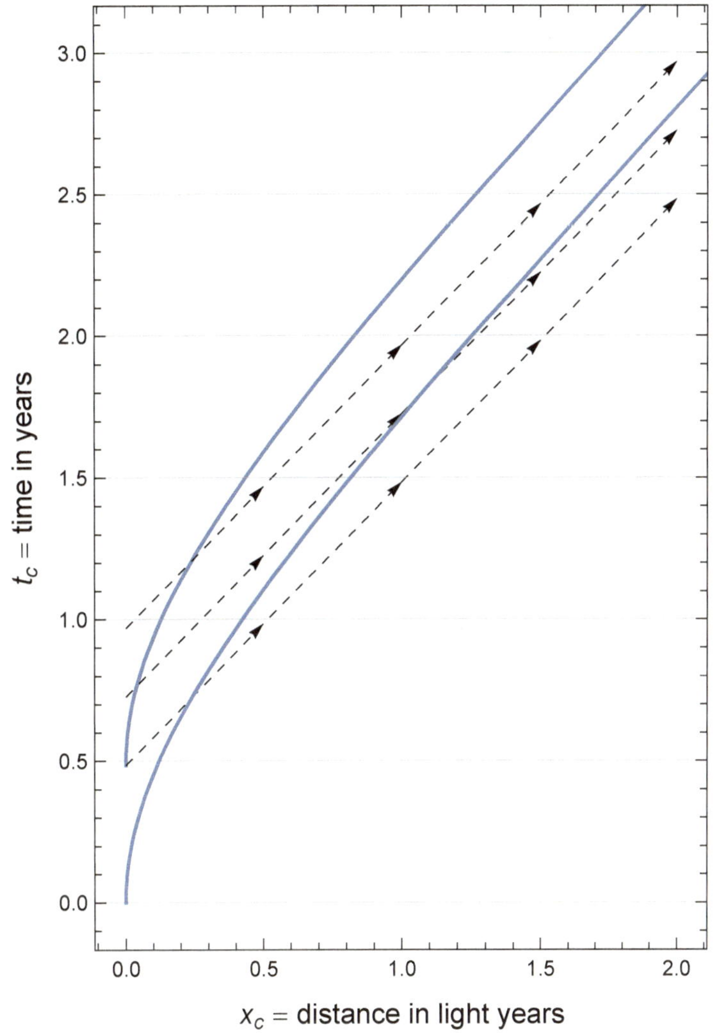

FIG. 15

A plot of the trajectories of two spacecraft launched from the origin $\mathcal{O}$ at different times, each with indefinite 1-g self-acceleration. One spacecraft (which is traveler $\mathcal{C}1$) is launched at $t_o = 0$ and the other (which is traveler $\mathcal{C}2$) is launched at $t_o = 0.5 \cdot t_H$. Also shown as *dashed lines* are the trajectories of a photon emitted from the origin at three coordinate times $t_o \in \{0.5, 0.75, 1\} \cdot t_H$ in the $+x$-direction. The *shaded area* represents the ensemble of photon emission times from $\mathcal{C}2$ that intersect the trajectory of $\mathcal{C}1$ (and hence can be detected by $\mathcal{C}1$). This represents a finite interval beginning at the instant of launch of $\mathcal{C}2$.

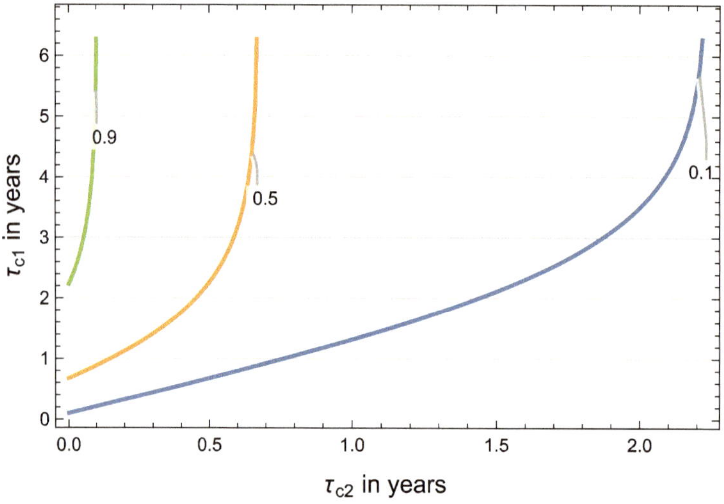

FIG. 16

The clock image $\Psi_{C2 \Rightarrow C1}$ for photon exchange from $C2$ to $C1$ is plotted. Two spacecraft $C1$ and $C2$ are launched in that order, so that photon propagation is in the same direction as the spacecraft trajectories. Both spacecraft are assumed to have indefinite 1-g acceleration, and the traveler's times $\{\tau_{c1}, \tau_{c2}\}$ each measure METT for their respective spacecraft ($\tau_{c1} = \tau_{c2} = 0$ at launch). The different curves represent $\epsilon \in \{0.1, 0.5, 0.9\}$, where the later launch is delayed by $\epsilon\ t_H$ relative to the earlier launch. The time τ_{c1} for photon detection by $C1$ occurs increasingly late in its trajectory as ϵ increases. There is a limit $-t_H \cdot \log \epsilon$ on the emission time τ_{c2} that falls within the time horizon of $C1$. The image $\Psi_{C2 \Rightarrow C1}$ is unbounded toward the end of that interval.

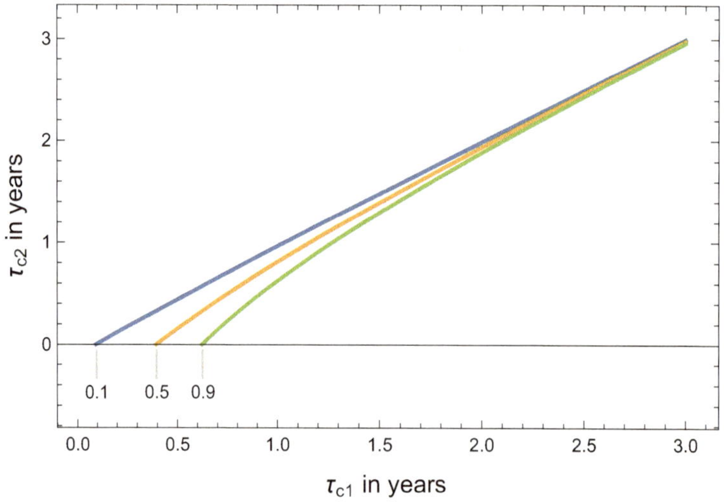

FIG. 17

Fig. 16 is repeated for photon exchange from an earlier-launched $C1$ spacecraft to a later-launched spacecraft $C2$, with photon propagation in the opposite direction to the spacecraft trajectories. The clock image is $\Psi_{C1 \Rightarrow C2}$ and for this case a communication blackout never occurs. The earliest photon emission time τ_{c1} by $C1$ is $t_H \cdot \log(1 + \epsilon)$, since earlier emissions would arrive prior to the launch of $C2$ (recall that $C2$ is launched later than $C1$). Later in $C1$'s trajectory, ϵ has minimal impact on $\Psi_{C1 \Rightarrow C2}$.

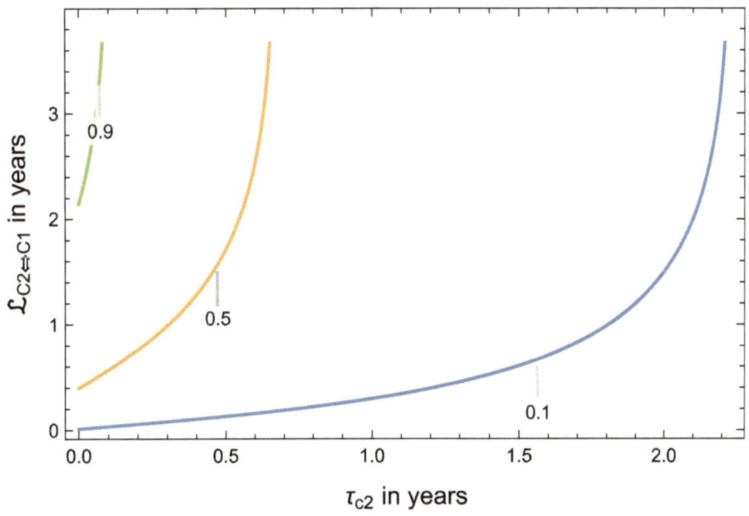

FIG. 18

The query-response latency $\mathfrak{L}_{C2\Leftrightarrow C1}$ for a query initiated by the later-launched rocket $C2$ to the earlier-launched rocket $C1$ is plotted for the same assumptions as in Fig. 16. Queries can be initiated by $C2$ immediately following launch, but receiving a response is only possible for an interval $0 < \tau_{c2} < -t_H \cdot \log \epsilon$ following launch. Toward the end of that interval $\mathfrak{L}_{C2\Leftrightarrow C1}$ becomes unbounded.

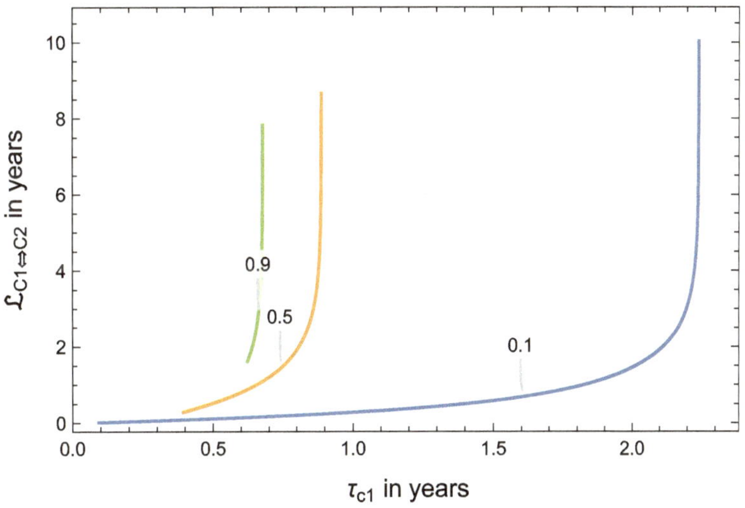

FIG. 19

The query-response latency $\mathfrak{L}_{C1\Leftrightarrow C2}$ for a query initiated by the earlier-launched spacecraft $C1$. This is a repeat of Fig. 18 for a query in the opposite direction. Both the earliest and latest such query are constrained. The earliest query arrives at $C2$ at the instant of its launch. There is also a latest such query time $t_H \cdot \log(\epsilon + 1/\epsilon)$ that allows for a response from $C2$ to reach $C1$. This is because that response must be initiated early enough to fall within the time horizon of $C1$. As the time horizon is reached, $\mathfrak{L}_{C1\Leftrightarrow C2}$ grows without bound.

4.9 Role of acceleration magnitude and duration

The canonical mission of Fig. 1 makes the assumption that acceleration magnitude is time-invariant during the entire travel time from launch to landing. Having developed a complete picture of one-way image and two-way query-response latency, two modifications that can be made in the interest of easing the restrictions on communication can be considered. One is reducing the acceleration and deceleration time durations, thereby introducing an intermediate "ballistic" interval during which the spacecraft's speed is constant. The second is reducing the acceleration magnitude α (thereby increasing t_H). Neither will make a meaningful change in MECT, and hence neither change would have significant impact on the origin's awareness of landing, nor on response latencies. However, these modifications would have a significant effect on one-way uplink communication to the spacecraft by lengthening the event horizon t_H (allowing longer communication with the origin during the acceleration phase) and/or by earlier termination of the acceleration phase (allowing uplink signals transmitted past the event horizon to be received earlier). Employing a combination of larger t_H and a shorter acceleration phase could allow an extension of the event horizon past the acceleration phase altogether.

A significant impact of either of these modifications would be in increasing the METT, with the resulting adverse implications for resources aboard the spacecraft and the required lifetime of equipment and crew. The greater mass associated with greater resources would be compensated by lower mass devoted to spacecraft propulsion. The human crew would devote more of their lifetime to the cruise phase (and commensurately less to the postlanding) experience.

5 Analysis

Relativistic effects can generally be neglected in the case of spacecraft missions within our Solar System, at least when chemical spacecraft propulsion is employed and the speeds are relatively low compared to light-speed c. This implies that the Newtonian theory of gravity and a single universal measure of coordinate time in the predication of spacecraft trajectories and communication latencies is suitably accurate. This approach has to be abandoned for interstellar missions, where the full General Theory of Relativity (GR) [2] could be adopted. In this chapter, the "empty-space" approximation is utilized, in which the spacecraft trajectory is determined entirely by its own propulsion, neglecting the effects of gravity from astronomical bodies, including the origin and destination. This is the standard approach for development of the relativistic spacecraft equation [3, 4], and relies on the simpler (and more analytically tractable) Special Theory of Relativity (SR).

There are numerous resources for studying SR, including some excellent textbooks [5–7]. We follow an alternative approach and notation to the development of SR originating in Ref. [1], which (it was argued) is more appropriate for engineering applications such as interstellar travel (as opposed to scientific applications). In particular, the perspective of the traveler (accelerated observer) was adopted in place of the usual denizen (inertial) observer universally adopted in physics, starting with Einstein's initial formulation. This alternative development greatly simplifies the derivation of trajectories involving acceleration, and is consistent with the universal practice of relying on clocks carried by spacecraft. This approach was inspired in part by earlier pedagogical innovations by Fraundorf [8–10] and a new derivation of the relativistic spacecraft equation by Walter [11, 12]. The notion of a clock image, which forms the basis of the communication latencies developed here, was introduced in Ref. [1].

Most relativistic spacecraft derivations assume that the propulsion is based on a fixed-rate consumption of a propellant carried by the spacecraft, in which case the force exerted on the spacecraft will be constant. As the mass of the spacecraft decreases, the self-acceleration thereby increases. In contrast, the canonical mission considered here assumes that the self-acceleration of the spacecraft is fixed. This would require that the rate of fuel consumption start out high to compensate for the higher initial fuel mass, and decrease during the mission. Alternative versions of the spacecraft equation are derived and compared in Ref. [1]. Here we do not take propulsion challenges into account, but rather make the "empty-space constant self-acceleration" simplifying assumption in order to focus our efforts on understanding the communication timing issues in a simple context. Our methods readily extend to more complicated mission scenarios.

5.1 Map and two-clock metaphor

A map and two-clock metaphor [1] forms the basis for the following. The kinematic variables used in this metaphor are summarized in Table 2.

A coordinate system to describe the trajectory of a moving body like a spacecraft $\mathcal{C}$ is defined relative to an inertial frame S, which defines a set of spatial coordinates. Relative to S, any observer $\mathcal{R}$ with fixed spatial coordinates (said to be *at rest*) experiences no acceleration. The coordinates defined by S are called a *map*, in analogy to terrestrial travel, and the position of an observer is called the *map-position*. For the general case a map-position is three-dimensional, but for simplicity here we specialize to the 1D (rectilinear) case. Thus there is a single scalar spatial coordinate, which is labeled r or x. Anybody such as $\mathcal{C}$ not permanently at rest is called a traveler.

There are two types of clocks of interest. The traveler's clock (carried along with $\mathcal{C}$) is the clock of interest in the context of interaction (such as communication) among two or more travelers, and measures traveler's time τ (as was described in Section 1). Any clock at rest in S is called a coordinate clock, because it measures *coordinate time* denoted by t. Traveler and coordinate clocks advance at different rates, excepting the degenerate case of a traveler temporarily or permanently at rest relative to S. Clock synchronization, the issue of initialization of all clocks, was discussed in Section 3.3.

Kinematic variables quantify the motion of travelers, and for our purposes constitute map-position as well as the map-speed and map-acceleration measured at the same map-position. SR offers a complete model of kinematics, for accelerated bodies as well as inertial, in the absence of gravitational influence. Since map-speed is the derivative of map-position, and map-acceleration is the derivative

Table 2 Kinematic variables for a 1D map and two-clock metaphor.

Variable	Traveler's version	Coordinate version
Map-position	r	x
Time	τ	t
Time-speed		$\gamma = t' = \dfrac{dt}{d\tau}$
Map-speed	$w = r' = \dfrac{dr}{d\tau}$	$u = x' = \dfrac{dx}{dt}$
Self-acceleration	α	
Map-acceleration	$w' = \dfrac{dw}{d\tau}$	$u' = \dfrac{du}{dt}$

of map-speed, both relative to an appropriate definition of time, there are two distinct sets of kinematic variables. The traveler's set, measured relative to τ, is denoted by $\{\tau, r, w, w'\}$, where w' denotes the derivative of w with respect to τ. Similarly the coordinate set $\{t, x, u, u'\}$ is measured relative to t.

In addition to $\{w', u'\}$, there is a third measure of acceleration α measured by an accelerometer carried aboard the spacecraft, and called the *self-acceleration*. Any accelerometer must implicitly incorporate a measurement of time, and the appropriate clock for this purpose is the traveler's clock (which is carried along with that accelerometer as part of the traveler's payload). Thus α is unambiguously measured relative to traveler's time τ (there is no coordinate-time version). In general, the three forms of acceleration $\{w', u', \alpha\}$ possess different numerical values at any instant of time. The coordinate map-acceleration u' is of no concern in the following.

Finally, the *time-speed* γ (called the *Lorentz factor*) measures the rate of advance of a coordinate clock t relative to a traveler's clock τ situated at the same position.

5.2 Time dependence of kinematic variables

Since there are two distinct time measurements of interest, to keep things straight we exercise care in expressing the time dependence of kinematic variables. There are two kinematic variables that are unaffected by the choice of S, and depend only on τ and not t: These are τ itself and self-acceleration $\alpha[\tau]$. The other kinematic variables not only have a coordinate-time version, but more fundamentally these variables are affected by the choice of S. Since it is frame-invariant, τ is preferred as the fundamental timebase for accelerated motion, and the traveler's time kinematic variables are written as $\{\tau, \gamma[\tau], r[\tau], w[\tau], w'[\tau]\}$. This kinematic description is also practical because a traveler is assumed to carry a clock measuring τ, and the traveler is normally navigationally aware of its own position $r[\tau]$.

This is so simple and clean, it is reasonable to ask "in the communications context, what purpose is served by a coordinate-time description of kinematics?" A fundamental postulate of SR is that the speed of light c is invariant to the choice of inertial frame S. The interaction among different travelers is based on photon exchange, and the propagation delays for photons are defined in terms inertial frame S and coordinate time t. Thus, for us to quantify photon-transfer interactions, for each traveler emitting or detecting a photon we must know its traveler's time τ and map-position $r[\tau]$ at that instant, but also the associated coordinate time at that map-position and instant of time, which we write as $t[\tau]$. It would be unusual for there to be an actual physical clock at rest at that map-position, so typically this coordinate clock is *notional* rather than physical. The entire set of kinematic variables can thus be written as $\{t[\tau], x[t[\tau]], u[t[\tau]], u'[t[\tau]]\}$. For purposes of photon transfer timing calculations (the focus of this chapter), the only coordinate variable needed is $t[\tau]$, since map-position is conveniently specified by $r[\tau]$.

The calculations establishing all the useful kinematic variables proceed in the following sequence:

$$\alpha[\tau] \to w'[\tau] \to w[\tau] \to r[\tau], \tag{10a}$$

$$w[\tau] \Rightarrow \gamma[\tau] \to t[\tau]. \tag{10b}$$

Each "$\to$" represents the solution to a first-order differential equation (DE) given in Table 2 (with the exception of $\alpha[\tau] \to w'[\tau]$, which is noted below), and the single "$\Rightarrow$" is a functional relation also noted below. All kinematic variables follow from knowledge of $\alpha[\tau]$, which is the input. The final outcome $\{r[\tau], t[\tau]\}$ is everything needed for the photon transfer propagation delay calculations that enter into clock images and the resulting query-response latencies.

5.3 Kinematic variable solutions

The role of time-speed $\gamma[\tau]$ becomes evident when the two versions of map-speed are compared. Applying the chain rule of differentiation:

$$w[\tau] = r'[\tau] = x'[t[\tau]] \cdot t'[\tau] = u[t[\tau]] \cdot \gamma[\tau]. \tag{11}$$

Thus $\gamma[\tau]$ directly relates the two map-speeds. The $w[\tau] \Rightarrow \gamma[\tau]$ relation in Eq. (10b) is [1]

$$\gamma[\tau] = \sqrt{1 + w[\tau]^2/c^2}. \tag{12}$$

If S is changed for any reason, both $\gamma[\tau]$ and $w[\tau]$ are affected as a result, but their frame-invariant relationship in Eq. (12) is preserved.

Relation equation (12) has profound implications for interstellar travelers (analogous to $E = mc^2$ for energy and propulsion) because it directly relates motion and time. In particular, since for a moving body $1 < \gamma[\tau]$, it follows that τ always advances more slowly than $t[\tau]$, and the larger the map-speed, the larger this *time-dilation* effect. A natural question is "what is the asymmetry that causes this disparity to always occur in the same direction." One simple answer is that a measurement of the traveler kinematic variables expressed in terms of traveler's time can be accomplished using the single traveler's clock carried by the traveler, but measuring such variables expressed in terms of coordinate time requires two or more coordinate clocks.

Eqs. (12), (11) together imply that ($|u[t[\tau]]| < c$), leading to the common perception that objects cannot travel faster than light. This is at best an incomplete statement when applied to interstellar travel, because spacecraft self-aware of their own motion employing their own clock as the basis of observations see no such limitation on speed (since $|w| \gg c$ is possible). The spacecraft's perception of superluminance (travel faster than light-speed) is attributable to its traveler's clock advancing more slowly. There is no inconsistency since a traveler's clock is incapable of measuring photon speed. Since $|u| < c$, a traveler's clock as observed in S necessarily travels slower than a photon, and thus could measure either the time of a photon's emission or the time of that same photon's detection, but never both.

The $\alpha[\tau] \rightarrow w'[\tau]$ DE in Eq. (10a) is given by [1]

$$w'[\tau] = \gamma[\tau] \cdot \alpha[\tau]. \tag{13}$$

After substituting for $\gamma[\tau]$ from Eq. (12), this becomes a first-order DE in $w[\tau]$. This is another frame-invariant relation; that is, both $w[\tau]$ and $\gamma[\tau]$ are affected by the choice of S, but relationship Eq. (13) is not. The self-acceleration $\alpha[\tau]$ is of practical interest for two important reasons:

- Acceleration α relates directly to the force imparted by the propulsion system [11], if propulsion is the cause of the acceleration. (Another possibility is gravity, but that effect is neglected in SR.)
- Acceleration α is that experienced by onboard equipment and material, including biological material, both of which are normally sensitive to and affected by the acceleration magnitude. In numerical examples a time-invariant 1-g self-acceleration magnitude is assumed, because this is consistent with that normally experienced by humans and other biological material originating from Earth.

Since again $1 < \gamma[\tau]$, what follows from Eq. (13) is the profound observation that $|w'[\tau]| > |\alpha[\tau]|$, which is called an *acceleration boost*. That map-acceleration is always greater than self-acceleration is great news for interstellar travelers. Although biological organisms (like humans) have a limited tolerance for high self-acceleration, that intolerance need not limit the map-acceleration. Even better news for interstellar travelers is a positive feedback effect: As $w[\tau]$ grows so does $\gamma[\tau]$, and hence acceleration boost magnitude is amplified without limit for as long as self-acceleration (in the direction of travel) persists. This makes it advantageous to maintain self-acceleration as long as possible (as in the canonical mission assumption, see Section 4.9), because longer self-acceleration reduces the METT (while its effect on MECT is minimal) and thus reduces the resources (food, water, oxygen, etc.) required on the spacecraft, as well as biological aging.

5.4 Trajectory with indefinite self-acceleration

The coordinate-time spacecraft trajectory $\{r_c[\tau_c], t_c[\tau_c]\}$ for constant acceleration and time horizon t_H (see Appendix B.1) is

$$t_c[\tau_c] = t_H \cdot \sinh(\tau_c/t_H), \tag{14a}$$

$$r_c[\tau_c] = c\, t_H(\cosh(\tau_c/t_H) - 1). \tag{14b}$$

This is consistent with the textbook *hyperbolic trajectory* for a uniformly accelerated body (see Chapter 14 of [7]), although here it has been derived in an unconventional way.

Example 11: Let the units of time be year and the units of distance be light years, and determine the kinematic variables at $\tau_c = 2$. The coordinate map-speed is $u = 0.968$ (96.8% of c), the traveler's map-speed is $w = 3.87$ (387% of c), and the time-speed is $\gamma = 4.0$. While the traveler's self-acceleration is $\alpha = g = 1.03$, the traveler's map acceleration is $w' = 4.13$ ($\gamma = 4.0$ times the self-acceleration α).

5.5 Event horizon

Before verifying the event horizon analytically, the interpretation of the phrase "trajectory of a photon" should be clarified. As $u[t[\tau]] \to c$ we find that $\gamma[\tau] \to \infty$, which implies that $\tau \to 0$. Thus from the perspective of a photon (if indeed it had its own perspective!) we have $\tau \equiv 0$. A photon trajectory cannot be parameterized by τ (as in the case of nonzero mass particles). Rather, a photon trajectory $\{x[t], t\}$ must omit τ altogether. A fundamental postulate of SR is that a photon's position versus coordinate time obeys the *range equation*:

$$t = t_0 + x[t]/c \tag{15}$$

for a photon that is emitted from position $x[0] = 0$ at time $t = t_0$. This expresses the simple concept that the photon travels at speed c from the perspective of observers at rest in inertial frame S. Photon trajectories in previous plots have this special interpretation.

Analytically verifying the time horizon phenomenon follows from comparing two coordinate times:

Spacecraft. When the spacecraft travel time is τ_c, its distance reached is $r_c[\tau_c]$ and the coordinate time at which it reaches this distance is $t_c[\tau_c]$.

Photon. Based on Eq. (15), the coordinate time that a photon emitted from the origin at coordinate time $t_c[\tau_c] = t_H$ arrives at the same distance $r_c[\tau_c]$ is $(t_H + r_c[\tau_c]/c)$.

Substituting from Eq. (14), the difference between these two times:

$$(t_H + r_c[\tau_c]/c) - t_c[\tau_c] = t_H \cdot e^{-\tau_c/t_H} > 0 \tag{16}$$

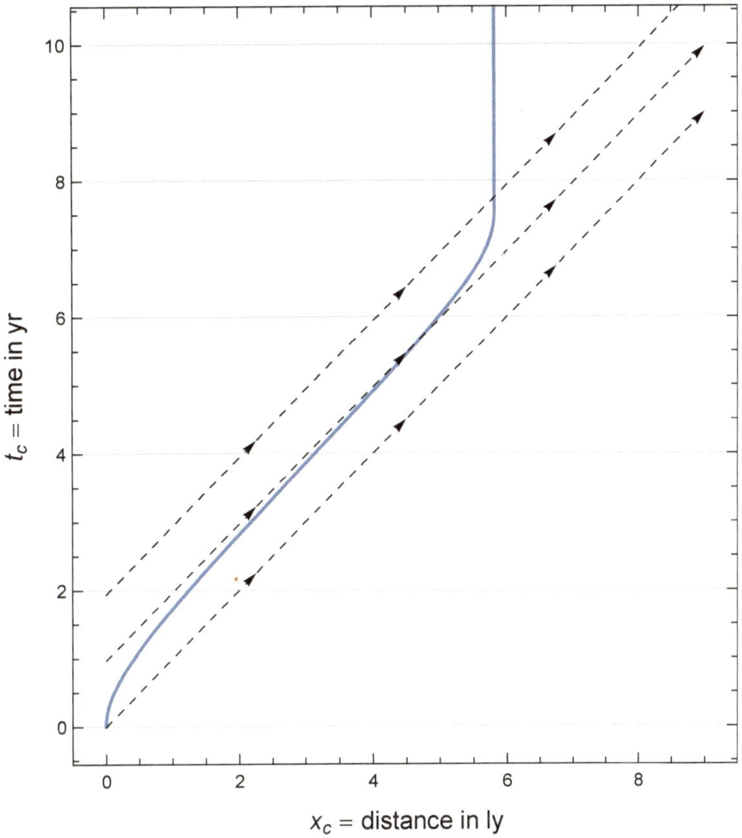

FIG. 20

Illustration of the trajectory of a canonical launch-to-landing 1-*g* mission as defined in Fig. 21. The origin to destination distance D is chosen so that the total METT at landing is $\tau_c = 4$ years (see Section 6). This is a repeat of Fig. 3 with the addition of a deceleration cruise phase. The *dashed lines* show the trajectories of photons emitted from the origin at $t_o \in \{0,\ t_H,\ 2t_H\}$. In this case the spacecraft can detect photon emissions during the interval $\sim t_H < t_o < \sim 2t_H$ during the deceleration, but emissions for $\sim 2t_H < t_o$ can only be detected following landing at the destination. A photon that is emitted from the origin at $t_o = 2t_H$ is detected shortly following landing, so that MECT $\sim (D/c + 2t_H)$.

is positive and approaches zero exponentially as $\tau_c \to \infty$. Thus the photon and accelerated spacecraft trajectories approach one another (but never intersect). The event horizon no longer applies when the spacecraft terminates its acceleration and/or starts decelerating, as illustrated in Fig. 20.

5.6 Clock images and query-response latencies

There are two clock image functions (see Section 4.5) of interest, $\Psi_{\mathcal{O}\Rightarrow\mathcal{C}}$ and $\Psi_{\mathcal{C}\Rightarrow\mathcal{O}}$. For trajectory Eq. (14) (see Appendix B.2):

$$\tau_c = \Psi_{O \Rightarrow C}[\tau_o] = -t_H \cdot \log(1 - \tau_o/t_H), \quad 0 \le \tau_o < t_H, \tag{17a}$$

$$\tau_o = \Psi_{C \Rightarrow O}[\tau_c] = t_H(e^{\tau_c/t_H} - 1), \quad 0 \le \tau_c. \tag{17b}$$

The domain of $\Psi_{O \Rightarrow C}$ is restricted to the interval between launch and event horizon as expected, whereas the domain of $\Psi_{C \Rightarrow O}$ is only restricted to postlaunch, although it grows exponentially. The corresponding inverse clock image functions are

$$\tau_o = \Omega_{O \Rightarrow C}[\tau_c] = t_H \cdot (1 - e^{-\tau_c/t_H}), \quad 0 \le \tau_c, \tag{18a}$$

$$\tau_c = \Omega_{C \Rightarrow O}[t_o] = t_H \cdot \log(1 + \tau_o/t_H), \quad 0 \le \tau_o. \tag{18b}$$

The resulting query-response latencies are (see Appendix B.2)

$$\mathfrak{L}_{O \Leftrightarrow C}[\tau_o] = \frac{\tau_o^2}{t_H - \tau_o}, \quad 0 \le \tau_o < t_H, \tag{19a}$$

$$\mathfrak{L}_{C \Leftrightarrow O}[\tau_c] = -t_H \cdot \log(2 - e^{\tau_c/t_H}) - \tau_c, \quad 0 \le \tau_c < t_H \cdot \log 2. \tag{19b}$$

The range of $\mathfrak{L}_{C \Leftrightarrow O}$ is more restricted (since $\log 2 \approx 0.69$) because the query has to reach the origin early enough that the origin's response falls within the time horizon of the spacecraft.

5.7 Other motion effects

Closely related to the warping phenomenon is red Doppler shift, which becomes substantial at relativistic speeds (not considered further here). This (as well as relativistic aberration) affects the directivity of transmit and/or receive apertures, detectors, etc. On longer missions multiple receivers are needed to deal with a wide variation in received wavelength, or alternatively the transmitter has to employ multiple compensatory wavelengths to hold the receive wavelength in a narrower range.

Due to both red shift and timebase expansion, the signal power decreases substantially from transmitter to receiver. Power and energy are conspicuously not conserved when they are compared with respect to two distinct frames of reference, the origin and the spacecraft. Reduced signal power substantially reduces the data rate and data volume that can be achieved, and this effect increases as long as the constant self-acceleration continues.

6 Canonical mission trajectory

The kinematic variables in the specific context of the canonical mission scenario (defined in Section 2.2) are illustrated in Fig. 21. Although the trajectory of a spacecraft $C3$ on a canonical mission could be calculated directly from the modeling DEs given by SR, it is easier to recognize the symmetry of canonical self-acceleration. In particular, the trajectory for the second half of the cruise phase is related to the trajectory in the first half (see Section 5.4) with time running backward, from maximum to zero map-speed (rather than the other way around). The second half of the $C3$ trajectory is thus the mirror image of the first half of the $C1$ trajectory.

To determine that mirror image, the two times $\{\tau_h, t_h\}$ at the halfway point ($r_c[t_h] = D/2$) are needed. These are determined from Eq. (14) to be

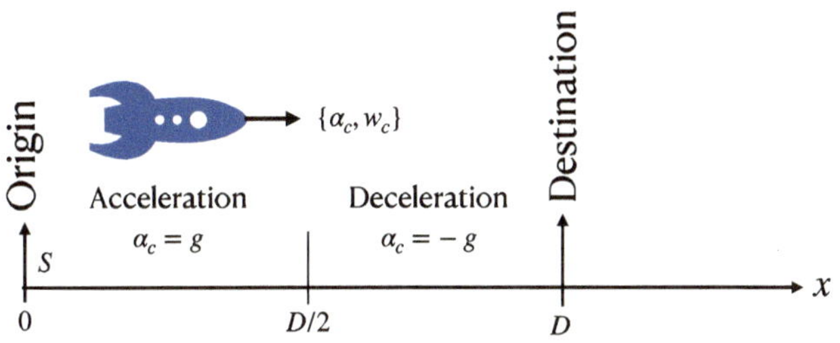

FIG. 21

A spacetime coordinate system and the kinematic variables describing the trajectory of a spacecraft relative to an inertial frame S with coordinates defined in Fig. 2. Added variables are the self-acceleration α_c and traveler's map-speed w_c, which are both implicitly a function of traveler's time τ_c. For the canonical 1-g launch-landing mission of Fig. 1, the acceleration has only two constant values of acceleration, $\alpha_c = \pm g$.

$$\tau_h = t_H \cdot \cosh^{-1}\left(1 + \frac{D}{2c \cdot t_H}\right) \quad \text{and} \quad t_h = \frac{D}{2c}\sqrt{1 + \frac{4c \cdot t_H}{D}}. \tag{20}$$

It can be verified that $t_h \to (t_H + D/2c)$ as $D \to \infty$, or in other words, t_h at longer distances approaches the coordinate time at which a photon emitted from the origin at $t_o = t_H$ reaches the halfway point.

The second half of the canonical trajectory equals the first half reversed in time, but in addition displaced in both coordinate time (by t_h) as well as in map-position (by $D/2$). Taking these three factors into account, the canonical trajectory becomes

$$\{t_{c3}[\tau_{c3}], r_{c3}[\tau_{c3}]\} = \begin{cases} \{\tau_{c3}, 0\}, & \tau_{c3} < 0 \\ \{t_{c1}[\tau_{c3}], r_{c1}[\tau_{c3}]\}, & 0 \le \tau_{c3} \le \tau_h \\ \{2t_h - t_{c1}[2\tau_h - \tau_{c3}], D - r_{c1}[2\tau_h - \tau_{c3}]\}, & \tau_h < \tau_{c3} \le 2\tau_h \\ \{2t_h + \tau_{c3} - 2\tau_h, D\}, & 2\tau_h < \tau_{c3}. \end{cases} \tag{21}$$

7 Conclusions

The timing relationship (characterized by the clock image functions, in terms of local traveler's clocks) between photon emission and detections as measured at the origin of launch, and one or two accelerating spacecraft, and a landing destination has been characterized by drawing on SR. The methods employed apply generally to accelerated bodies performing photon transfer in their communications. Here they have been applied in two simple cases of one or two spacecraft experiencing indefinite fixed self-acceleration and a canonical launch-landing mission. From these results, the query-response latencies are inferred directly through the composition of two appropriate clock image functions.

The overall conclusion is that these round-trip latencies in both directions are a significant issue for interstellar travel, and in fact such two-way communication is for all practical purposes very cumbersome (due to large latencies) except in the vicinity of the origin or the destination. Interstellar spacecraft and their crews must accept highly autonomous operations, and abandon notions of maintaining operational and social interactions with those at the origin or destination throughout the mission, with the exception of a short period following launch or prior to landing. Since the root cause is the large propagation distances, this general conclusion applies to many other types of interstellar missions, largely irrespective of spacecraft speeds.

There are other issues in intertraveler communications arising out of relativistic effects that have not been considered here. These include aperture and antennae beam patterns as affected by a significant red shift in wavelength as well as relativistic aberration. Gravitational effects, which have also been neglected, are a factor in the vicinity of origin and destination.

Acknowledgments

The research represented in this tutorial review was supported by a grant from the Breakthrough Foundation and its Breakthrough StarShot program, NASA grants NIAC Phase I DEEP-IN—2015 NNX15AL91G and NASA NIAC Phase II DEIS—2016 NNX16AL32G and the NASA California Space Grant NASA NNX10AT93H, the Emmett and Gladys W. Technology Fund, and the Limitless Space Institute.

Appendix A Nomenclature

S	A one-dimensional (rectilinear) inertial frame with map-position (spatial) and coordinate time $\{x, t\}$
c	Speed of light in a vacuum as observed by any two synchronized coordinate-time clocks in an arbitrary inertial frame S
$\{\mathcal{A}, \mathcal{B}\}$	Accelerated travelers observing traveler's times $\{\tau_a, \tau_b\}$ (in a degenerate case, may also be at rest)
$\mathcal{O}$	Location of the origin of a spacecraft launch, assumed to be at rest in S
$\mathcal{C}$	A spacecraft, in motion and accelerating during its cruise phase
$\mathcal{D}$	Location of the destination for a spacecraft, assumed to be at rest in S
D	The fixed distance from $\mathcal{O}$ to $\mathcal{D}$
MET	Mission-elapsed time (initialized at the instant of launch)
τ_a	Traveler's time measured by a clock carried by traveler $\mathcal{A}$
METT	The MET as measured by the spacecraft traveler's clock
$t_a[\tau_a]$	Coordinate time measured by a (typically notional) clock $\mathcal{A}$ at rest in an inertial frame S at the instantaneous map-position of $\mathcal{A}$ at its traveler's time τ_a
MECT	The MET as measured by a (typically notional) coordinate clock at rest at the instantaneous map-position of the spacecraft
$\pm\alpha$	A spacecraft traveler's self-acceleration ($0 < \alpha$); if positive, this is an acceleration, and if negative, this is a deceleration
t_H	The time horizon of an accelerating spacecraft, equal to α/c
g	Acceleration equal to that experienced at the Earth's surface due to gravity
$\Psi_{\mathcal{A}\Rightarrow\mathcal{B}}[\tau_a]$	Clock image ($\mathcal{B}$'s traveler's arrival time) in photon exchange for a photon propagating in the direction $\mathcal{A}\Rightarrow\mathcal{B}$ that was emitted by $\mathcal{A}$ at its traveler's time τ_a
$\Omega_{\mathcal{A}\Rightarrow\mathcal{B}}[\tau_b]$	Inverse clock image ($\mathcal{A}$'s traveler's photon emission time) in photon exchange for a photon propagating in the direction $\mathcal{A}\Rightarrow\mathcal{B}$ that arrives at $\mathcal{B}$'s traveler's time τ_b
$\mathfrak{L}_{\mathcal{A}\Leftrightarrow\mathcal{B}}[\tau_a]$	Query-response latency (elapsed traveler's time) for queries originating with $\mathcal{A}$ traveler's time τ_a, directed at $\mathcal{B}$ with an immediate response back to $\mathcal{A}$
$x_a[t_a[\tau_a]]$	Map-position of $\mathcal{A}$ at coordinate time $t_a[\tau_a]$ corresponding to traveler's time τ_a
$r_a[\tau_a]$	Map-position of $\mathcal{A}$ at traveler's time τ_a, equal to $x_a[t_a[\tau_a]]$

$\gamma_a[\tau_a]$	Time-speed (Lorentz factor) for $\mathcal{A}$ at traveler's time τ_a, equal to the derivative of $t_a[\tau_a]$ with respect to τ_a
$w_a[\tau_a]$	Map-speed of $\mathcal{A}$ as measured using map-position $r_a[\tau_a]$ in conjunction with traveler's clock τ_a
$u_a[t[\tau]]$	Map-speed of $\mathcal{A}$ as measured using map-position $x_a[t_a[\tau_a]]$ in conjunction with coordinate clock $t_a[\tau_a]$, both measured at traveler's time τ_a
$w'_a[\tau_a]$	Map-acceleration as measured using map-speed $w_a[\tau_a]$ in conjunction with traveler's clock τ_a
$\{\mathcal{C}1,\mathcal{C}2\}$	Two spacecraft launched at different times, with $\mathcal{C}1$ launched prior to $\mathcal{C}2$
ϵ	For a second spacecraft $\mathcal{C}2$, the delay in its launch relative to a first spacecraft $\mathcal{C}1$, expressed as a fraction of the time horizon t_H (so $0 < \epsilon < 1$)

Appendix B Analytical results for indefinite acceleration

The assumption of indefinite fixed-magnitude self-acceleration yields straightforward analytical results, which are now derived.

B.1 Spacecraft trajectory

The trajectory of the spacecraft of Eq. (14) is now derived. Solving differential equation (13) with initial condition $w_c[0] = 0$ and substituting the resulting $w_c[\tau_c]$ into Eq. (12) yields

$$w_c[\tau_c] = c \cdot \sinh(\tau_c/t_H), \tag{B.1a}$$

$$\gamma_c[\tau_c] = \cosh(\tau_c/t_H), \tag{B.1b}$$

Note that c as a subscript in Eq. (B.3) is distinct from c as a physical constant (the speed of light). Solving the differential equation $t'[\tau_c] = \gamma_c[\tau_c]$, substituting Eq. (B.1b) and with initial condition $\tau_c[0] = 0$, yields Eq. (14a).

The map-position can be recast as

$$r'_c[\tau_c] = x'_c[t_c[\tau_c]] \cdot \gamma_c[\tau_c] = u_c[t_c[\tau_c]] \cdot \gamma_c[\tau_c] = w_c[\tau_c], \tag{B.2}$$

from Eq. (11). The $r_c[\tau_c]$ in Eq. (14b) is the solution to differential equation (B.2) after substituting from Eq. (B.1a) with initial condition $r_c[0] = 0$. The canonical mission trajectory is readily inferred from this simpler mission (see Section 6).

B.2 Clock images and query-response latency

Consider the trajectory of a photon emitted from position $\{0, t_0\}$ in direction $\mathcal{O} \Rightarrow \mathcal{C}$ and intersecting the spacecraft trajectory at $\{r_c[\tau_c], t_c[\tau_c]\}$, and also a photon emitted in direction $\mathcal{C} \Rightarrow \mathcal{O}$ from the spacecraft position $\{r_c[\tau_c], t_c[\tau_c]\}$ that intersects origin $\{0, t_0\}$. Then the respective photon trajectories, based on range equation Eq. (15), are

$$\mathcal{O} \Rightarrow \mathcal{C}: \quad t_c[\tau_c] = t_o + r_c[\tau_c]/c, \tag{B.3a}$$

$$\mathcal{C} \Rightarrow \mathcal{O}: \quad t_o = t_c[\tau_c] + r_c[\tau_c]/c. \tag{B.3b}$$

The clock images can be determined numerically by calculating t_o for a set of different values of τ_c in Eq. (B.3) (and equivalent relations for the destination). This yields a list of $\{\tau_o, \tau_c, \tau_d\}$ triplets that directly sample the relevant clock images. This approach was used to generate Figs. 8, 9, 13, and 14.

For simple trajectory Eq. (14), analytical results are easily obtained, and these provide additional insight. The clock images in Eq. (17) are the solutions (with the trajectory variables substituted from Eq. 14) to Eq. (B.3a) for τ_c and to Eq. (B.3b) for $t_o \equiv \tau_o$. Similarly, the inverse clock images in Eqs. (18a), (18b) are the solutions to Eq. (B.3a) for $t_o \equiv \tau_o$ and to Eq. (B.3b) for τ_c. The query-response latencies associated with the indefinite self-acceleration trajectory of Eq. (19) follow directly from Eq. (17) substituted into Eq. (7).

B.3 Communication between two accelerating spacecraft

Consider two spacecraft, each with indefinite acceleration, as shown in Fig. 15, which permits communication between the spacecraft in both directions under some conditions. Analytical results for clock images are readily obtained as an extension of the range equation technique used in Section B.2.

The two spacecraft are labeled $\mathcal{C}1$ (the earliest launch) and $\mathcal{C}2$ (the later launch). It is assumed that $\mathcal{C}2$ is launched at coordinate time $t_o = \epsilon t_H$, and as well that the traveler's clock for $\mathcal{C}2$ is initialized to $\tau_{c2} = 0$ at that instant of launch. We must have $0 < \epsilon < 1$, or else all photon emissions from $\mathcal{C}2$ violate the event horizon of $\mathcal{C}1$ and thus will never be detected. In that case, the trajectory of $\mathcal{C}2$ is identical to that of $\mathcal{C}1$, except that it is shifted up by ϵt_H in the time direction. Analogously to Eq. (B.3), the three points at which the trajectory of a photon emitted from $\mathcal{O}$ and traveling in the $+x$-direction intersects the trajectories of $\mathcal{C}2$ and $\mathcal{C}1$ are then given by

$$t_o = t_{c1}[\tau_{c1}] - r_{c1}[\tau_{c1}]/c = \epsilon t_H + t_{c2}[\tau_{c2}] - r_{c2}[\tau_{c2}]/c. \quad \text{(B.4)}$$

Substituting for $\{r_c, t_c\}$ from Eq. (14a) and solving for each traveler's time in terms of the other yields the image function for photon exchanges in the direction $\mathcal{C}2 \Rightarrow \mathcal{C}1$:

$$\Psi_{\mathcal{C}2\Rightarrow\mathcal{C}1}[\tau_{c2}] = -t_H \cdot \log\left(e^{-\tau_{c2}/t_H} - \epsilon\right), \quad 0 \le \tau_{c2} < -t_H \cdot \log \epsilon, \quad \text{(B.5a)}$$

$$\Omega_{\mathcal{C}2\Rightarrow\mathcal{C}1}[\tau_{c1}] = -t_H \cdot \log\left(e^{-\tau_{c1}/t_H} + \epsilon\right), \quad 0 \le \tau_{c1}. \quad \text{(B.5b)}$$

There is an upper bound on the time over which $\mathcal{C}2$ can communicate with $\mathcal{C}1$; namely, only as long as $\mathcal{C}2$ remains within the time horizon of $\mathcal{C}1$.

Likewise, a photon emitted in direction $\mathcal{C}1 \Rightarrow \mathcal{C}2$ from either $\mathcal{C}1$ or $\mathcal{C}2$ and reaching $\mathcal{O}$ at the same t_o must satisfy the range equations:

$$t_o = t_{c1}[\tau_{c1}] + r_{c1}[\tau_{c1}]/c = \epsilon t_H + t_{c2}[\tau_{c2}] + r_{c2}[\tau_{c2}]/c. \quad \text{(B.6)}$$

Solving for the clock image functions in this direction:

$$\Psi_{\mathcal{C}1\Rightarrow\mathcal{C}2}[\tau_{c1}] = t_H \cdot \log\left(e^{\tau_{c1}/t_H} - \epsilon\right), \quad t_H \cdot \log(1+\epsilon) \le \tau_{c1}, \quad \text{(B.7a)}$$

$$\Omega_{\mathcal{C}1\Rightarrow\mathcal{C}2}[\tau_{c2}] = t_H \cdot \log\left(e^{\tau_{c2}/t_H} + \epsilon\right), \quad 0 \le \tau_{c2}. \quad \text{(B.7b)}$$

There is a lower limit on τ_{c1} since earlier emissions would arrive prior to launch.

The query-response latencies follow from substituting Eqs. (B.5), (B.7) into Eq. (7):

$$\mathfrak{L}_{\mathcal{C}2\Leftrightarrow\mathcal{C}1}[\tau_{c2}] = t_H \cdot \log\left(\frac{1}{e^{-\tau_{c2}/t_H} - \epsilon} - \epsilon\right) - \tau_{c2}, \; 0 \le \tau_{c2} < -t_H \cdot \log \epsilon, \quad \text{(B.8a)}$$

$$\mathfrak{L}_{\mathcal{C}1\Leftrightarrow\mathcal{C}2}[\tau_{c1}] = -t_H \cdot \log\left(\frac{1}{e^{\tau_{c1}/t_H} - \epsilon} - \epsilon\right) - \tau_{c1}, \; t_H \cdot \log(1+\epsilon) \le \tau_{c1} < t_H \cdot \log\left(\epsilon + \frac{1}{\epsilon}\right). \quad \text{(B.8b)}$$

The upper bound on τ_{c1} in $\mathfrak{L}_{C1\Leftrightarrow C2}$ follows because the arrival time of a query at $C2$ must fall within the allowed emission time given in Eq. (B.5a). The upper bound as referred to τ_{c1} is found by applying the inverse clock image in Eq. (B.7b):

$$\Omega_{C1\Rightarrow C2}(-t_H \cdot \log\epsilon) = t_H \cdot \log\left(\epsilon + \frac{1}{\epsilon}\right). \tag{B.9}$$

The resulting $\mathfrak{L}_{C2\Leftrightarrow C1}$ and $\mathfrak{L}_{C1\Leftrightarrow C2}$ are plotted in Figs. 18 and 19.

B.4 Time warping derivatives

The derivatives in Eq. (8) can be derived analytically by taking the total derivative of the range equations [1]. Alternatively they can be expressed in terms of clock images as

$$\left.\frac{\mathrm{d}\tau_c}{\mathrm{d}\tau_o}\right|_{O\Rightarrow C} = \left(\frac{\mathrm{d}}{\mathrm{d}\tau_o}\Psi_{O\Rightarrow C}[\tau_o]\right)_{\tau_0=\Omega_{O\Rightarrow C}[\tau_c]}, \tag{B.10a}$$

$$\left.\frac{\mathrm{d}\tau_o}{\mathrm{d}\tau_c}\right|_{C\Rightarrow O} = \frac{\mathrm{d}}{\mathrm{d}\tau_c}\Psi_{C\Rightarrow O}[\tau_c]. \tag{B.10b}$$

Both derivatives are evaluated at the same spacecraft traveler's time τ_c. Substituting from Eqs. (17), Eq. (18), both derivatives evaluate to the same value: that given in Eq. (8). In addition, the ratio of clock images in Eq. (8) evaluates to that same value when substituted from Eqs. (17b), (18a).

References

[1] D.G. Messerschmitt, Relativistic timekeeping, motion, and gravity in distributed systems, Proc. IEEE 105 (8) (2017) 1511.
[2] M.P. Hobson, G.P. Efstathiou, A.N. Lasenby, General Relativity: An Introduction for Physicists, Cambridge University Press, 2006.
[3] W. Bade, Relativistic rocket theory, Am. J. Phys. 21 (4) (1953) 310–312.
[4] R. Forward, A transparent derivation of the relativisitic rocket equation, in: 31st Joint Propulsion Conference and Exhibit, 1995, p. 3060.
[5] A.P. French, Special Relativity, CRC Press, 1968.
[6] R. Resnick, Introduction to Special Relativity, Wiley, 2007.
[7] W. Rindler, Introduction to Special Relativity, Clarendon Press, 1991.
[8] P. Fraundorf, A one-map two-clock approach to teaching relativity in introductory physics, 1996. arXiv preprint physics/9611011.
[9] P. Fraundorf, Some minimally-variant map-based rules of motion at any speed, 1998. arXiv preprint physics/9704018.
[10] P. Fraundorf, One-map two-clock relativity for introductory physics classes, in: APS April Meeting Abstracts, vol. 1, 1997, p. 1110.
[11] U. Walter, Relativistic rocket and space flight, Acta Astronaut. 59 (6) (2006) 453–461.
[12] S. Westmoreland, A note on relativistic rocketry, Acta Astronaut. 67 (9–10) (2010) 1248–1251.

CHAPTER

Prospects for human hibernation

6

John Bradford

SpaceWorks Enterprises, Inc., Atlanta, GA, United States

Chapter outline

Interstellar Travel. https://doi.org/10.1016/B978-0-323-91280-8.00005-8

1 Introduction

1.1 Challenges of human interstellar travel

While interstellar transit faces numerous technical obstacles, ensuring the survivability of the crew and passengers presents an even more daunting series of challenges. Any human spaceflight mission involving extended periods of time in microgravity environments faces a myriad of medical and health issues. In the context of interstellar travel, the extremely long mission durations bring numerous additional issues that must be solved before we can advance beyond plans for intersolar missions.

When we consider potential travel to our nearest star systems (e.g., Alpha Centauri, Epsilon Eridani, Tau Ceti, etc.), mission durations for the crew are likely to be in the order of 40–50 years at subluminal velocities. Fortunately, this duration corresponds with time periods over which most individuals are physically strong, have excellent mental faculties, and have obtained a minimum required knowledge set to either support complex engineering operations, conduct science experiments, or solve difficult problems.

While near lightspeed travel would solve many of these challenges, if we temper our expectations regarding future technological capabilities to travel at significantly lower sublight speeds, we can begin to establish a framework in which we can discuss the human support systems and requirements that will be needed. These range from the need for fully regenerative and redundant life support systems that provide nutrition, oxygen, water, and medicine, to artificially-induced gravity for reproduction, and the ability to mitigate and/or repair the damage due to cosmic radiation exposure. Furthermore, we must consider that any crew member or passenger would spend the vast majority of their "useful life" in transit. Either we will need to make life during transit meaningful and purposeful, which will require even larger spacecraft with more resources, or we will need to develop ways to effectively shorten the transit duration. Both solutions will increase the energy requirements for transit, in whatever form it is provided.

1.2 Rationale for hibernation

Enabling interstellar travel is arguably one of the most difficult problems ever conceived, the achievement of which would undoubtedly represent one of the greatest feats in human history. The challenges are extremely diverse and range from engineering to human factors, and even affordability. Committing to such an endeavor will surely test our commitment and resolve to be a spacefaring species.

Hibernation (or stasis or suspended animation, however it may be referred to) has been featured as a major plot element or technological capability in almost every science fiction work since the earliest days of the genre. From Arthur C. Clarke's *2001: A Space Odyssey*, to Gene Roddenberry's *Star Trek* series, and 2014's *Interstellar*, various implementations have been portrayed to the audience or reader, ranging from rapid, whole-body freezing/unfreezing, submersion in liquid-filled tanks, to simple gas-filled sleep chambers or pods. While some of these implementations are specious at best, in general, the concept has been rooted in science with an eye toward mimicry of animals. Hibernation in nature is most often initiated due to a scarcity of resources (i.e., food). The associated hypometabolic state—often referred to as stasis—also conveys a lack of available oxygen. Both conditions are extremely desirable for human space travel where consumables necessary for sustaining life are limited.

For intersolar travel with transit times in the order of weeks or months, placing the crew in a hibernation or stasis state has been researched and shown to provide a number of benefits to both crew health and spacecraft systems [1,2]. At a minimum, there is a significant reduction in consumables due to the reduced levels of activity. Benefits with regard to crew health have been proposed that range from reduced muscular atrophy and rates of bone loss to radioprotective effects and reduced intracranial pressure. Spacecraft system-level benefits have been demonstrated that range from reduced habitat power requirements to elimination of many ancillary systems, and smaller pressurized structural volumes—both of which can significantly lower overall system mass. There are even potential psychosocial benefits for the crew due to reduced levels of interpersonal interactions during high-stress, low activity periods of the mission. All these benefits can ultimately be used to enable larger crew complements and increased system redundancy, permitting a range of options for artificial gravity configurations and/or enabling faster transit times via higher available delta-Vs.

In the context of interstellar travel, the driving need for stasis becomes "life extension" to support the significantly longer transit durations coupled with the relatively short human lifespan. The key objectives and benefits noted for intersolar travel scenarios, while still beneficial, generally become secondary benefits to the mission. The primary question, and challenge, now becomes how to utilize hibernation, in some form, to tackle the challenges of a multidecade spaceflight mission. Fortunately, while life extension is not the driving impetus in nature for hibernation, it is generally established that animals that hibernate live longer than their nonhibernating counterparts. It is also well researched and studied that caloric restriction is one of the most effective and demonstrated approaches to achieving longer lifespans. Therefore, we can broadly state that, in theory, reduced metabolic activity could be utilized to extend the human lifespan, in addition to providing several more mission-level benefits.

Lastly, a 50-year transit period will be extremely difficult both socially and emotionally for the crew members. Interpersonal conflicts are likely to amplify as the mission progresses. During operations on the ISS, interpersonal conflicts among the ISS astronaut core have been relatively minor. The frequency of crew rotations and introduction of new crew members every few months has likely helped to achieve this. Significant tensions do occasionally arise between the crew and mission control personnel on the ground. Interestingly, these conflicts are thought to lessen the chance and degree of any interpersonal conflicts among onboard crew members by creating a common "foe" that builds comradery. For deep-space missions or interstellar travel, interactions with this perceived "foe" will be reduced substantially due to the communications time delays, and chances of interpersonal conflicts are likely to increase without the regular crew rotations. Results from the 2010–11 Russian-ESA-Chinese Mars500 experiment with its crew held in isolation for 520 days also indicated interrupted sleeping patterns, depression, lethargy, and even willful isolationism [3]. With larger crew and passenger complements, these interpersonal dynamics will become less pronounced, but this will place even greater demands on the technology required to facilitate an interstellar mission due to the larger spacecraft required. The introduction of hibernation permits the crew to undergo long periods of inactivity and can be used to mitigate the psychological challenges of the mission. This will also lessen the technological demands by permitting smaller passenger complements and living spaces.

2 Overview

2.1 Limitations of the human lifespan

In the early 21st century, humans have an average life expectancy of about 76 years for males and 81 years for females [4]. While some countries and/or regions of the world have slightly higher life expectancies due to lifestyle, environment, and dietary choices, this only tends to add another 2–3 years. As Fig. 1 indicates, human life expectancy has improved over time; this is, however, almost entirely attributable to reduced infant mortality rates rather than people actually living longer. As an upper limit or bound, the oldest officially documented person was a woman in France who lived for 122 years [6].

It is estimated that, in the future, the average life expectancy will continue to increase due to improvements in healthcare treatments, preventative medicine, and understanding of optimal diet/lifestyle decisions. These are relatively modest gains though, which translate to about 0.1 years gained at birth per year in time.

Based on recent medical research and computer simulations, a number of scientists believe that there is a theoretical limit to the human lifespan of 120–150 years [7]. It is believed that, at this point, the body is operating in a greatly diminished state and cannot fight off basic illnesses. The predictive modeling for these forecasts assumed best case scenarios but did not account for any additional drug therapies or advancements. Other groups believe the human body can continue indefinitely if managed properly, approaching the issue of aging as a disease that must be cured [8,9].

2.2 Metabolic reduction in nature

If we broadly define stasis as a reduction in metabolic and physical activity for a period of time, there are three types found in nature: hibernation, torpor, and aestivation. Hibernation typically occurs over periods of months and in cold climates, with the body undergoing fairly significant reductions in

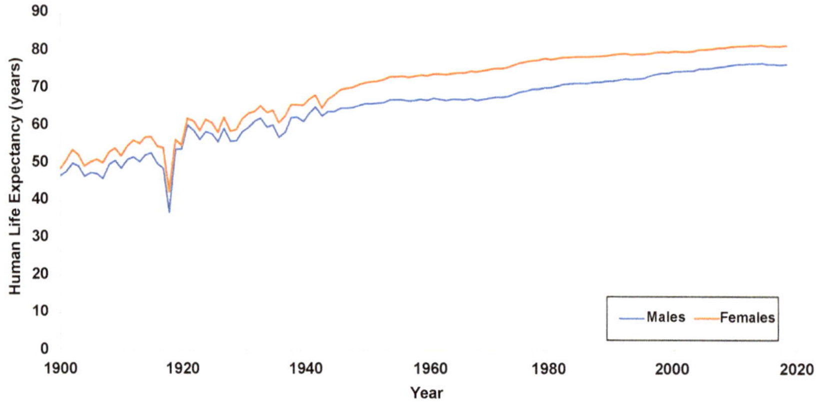

FIG. 1

Human life expectancy trends (United States Mortality Rates from 1900 to 2018) [5].

physiological state (heart rate, blood pressure, core temperature). Torpor generally occurs over a shorter period of time (days or weeks) and with smaller changes in physiological state. Aestivation is hibernation undergone during warmer climate conditions. Within these types of stasis, species are labeled as being either obligate or facultative. Facultative hibernators initiate their dormancy based on external factors (e.g., scarcity of food, extreme temperature), while obligates are triggered spontaneously and typically on an annual basis.

Hibernating animals can be found throughout nature. Despite this, we still know very little about the how and why of hibernation. For instance, black bears can maintain significant muscle mass despite months of disuse, and squirrels' blood does not freeze despite the extreme temperature reduction they experience [10,11]. Fat-tailed dwarf lemurs are believed to enter REM-sleep exclusively while in torpor, whereas other hibernators will only experience non-REM sleep [12]. These amazing animals may hold the key to enabling and perfecting human hibernation/torpor capabilities, whether it is induced artificially or can be activated naturally.

During stasis periods, heart rates can decrease from hundreds of beats per minute to just a few beats. Metabolic rate is a mere fraction of that when active, resulting in just a few breaths per minute being required to sustain the body. Regardless of the mammalian order, most species can maintain this state for months. It is worth noting that some animals do awaken during these stasis periods to briefly forage for food, if conditions are right, or even to sleep. Table 1 provides a summary of some mammals that undergo hibernation or torpor during their lifecycle.

2.3 Evidence for hibernation in hominins

As we look across the animal kingdom, we see that the ability to hibernate crosses multiple species, large and small, and extends beyond mammals. In many cases, the initiation is tied to changes in the local environment (e.g., temperature reduction, food scarcity) or other external stressors. Scientists

Table 1 Hibernating mammalian species.

Species	Order	Type	Avg. duration (months)	Core temperature (°C) (normal/inactive)	Heart rate (BPM) (normal/inactive)
Black bear	Carnivora	Obligate	3–8	37.8/31.1	45/8
Arctic ground squirrel	Rodentia	Obligate	Up to 9	37.2/−2.8	300/4
Marmot	Rodentia	Obligate	Up to 8	36.1/5	180/30
Black tailed prairie dog	Rodentia	Facultative	*Days/weeks*	37.2/19	–
Groundhog	Rodentia	Obligate	Up to 6	37.2/2	80/10
Edible dormouse	Rodentia	Obligate	Up to 11	37.2/3	350/4
Fat-tailed dwarf lemur	Primate	Obligate	Up to 7	36.7/25	180/8

continue to research the basic mechanisms that trigger induction and the internal mechanisms that support the extreme conditions of caloric restriction, oxygen deprivation, waste production management, brain synapses, etc. that are required to permit this state.

Regardless of the why or how, the end result yielding a significant reduction in metabolic activity and energy expenditures is the common effect. As there are not any specialized organs required for hibernation, and considering the broad ranges of species capable of entering hibernation states, the question is often asked if any human ancestors may have also had this ability. One viewpoint held by several scientists is that humans likely could enter a hibernation state during times of limited resources or extreme weather conditions.

A recent study of skeletal remains from the Sima de los Huesos in Atapuerca, Spain, has produced some of the first compelling evidence of early hominins who appear to have been undergoing annual hibernation or torpor periods [13]. Scans of skeletal remains from the cave revealed multiple indicators of changes in bone structure and density. These changes were consistent with having undergone repeat cycles of disuse or offloading. Dating of these remains indicated that the hominins lived during a period of extreme glaciation. This would give some further credence to why the group would have entered hibernation during certain times of the year.

Another potential indicator of some innate ability in early humans may still be presenting itself during cases of modern hypothermia. There is a documented phenomenon and well-known characteristic of hypothermia victims to tunnel, or "hide-and-die," after prolonged periods of exposure and as their body begins to effectively shut down [14]. Formally referred to as "terminal burrowing," the bodies of people who have succumbed to the cold are often found located in small, enclosed spaces (e.g., closets, behind furniture, etc.). As all known mammalian species create burrows, or dens, or locate a small, confined area to hibernate in, this tendency in humans may indicate the existence of some intrinsic, autonomic response that has long been dormant but is being re-triggered as the body cools and metabolism decreases.

Lastly, there are some well-documented cases in modern times of human's ability to survive extreme conditions of hypothermia. These cases have typically involved people who have become lost outside in extremely cold weather or who have broken through an ice-covered lake and fallen into freezing water. In 1999, Dr. Anna Bagenholm was revived after her heart was stopped for 3 hours after being submerged under the ice while skiing. Her body temperature quickly dropped to 13.7°C and entered a torpor state, enabling her to survive the accident [15]. There is also the case of Erika Norby in 2002, who, at a year old, was revived after her heart stopped beating for over 2 hours. An accident left her exposed to −20°C weather conditions and her core temperature had dropped to 17°C [16]. There was also the miraculous survival of a 16-year-old boy who stowed away in the wheel hub of a 747 traveling to Hawaii. He survived freezing temperatures and very low oxygen levels occurring at the 12 km flight altitude for several hours and recovered with no medical complications. Doctors speculate that his body quickly entered a hibernation state due to the rapid temperature drop, thus permitting him to survive at the minimal oxygen levels [17].

While all these occurrences are rare, they do provide some additional evidence that the human body is fairly robust and that under the right, narrow set of conditions, an opportunity exists under which we can survive these extreme conditions of stasis. If this is accurate, we will just need to understand how to reactivate these abilities and quantify the duration over which it can be sustained for spaceflight.

3 Maintaining the human body

The human body is an extremely complex organism requiring energy and resources from multiple sources to maintain basic functions and internal homeostasis. If we look at the minimum areas required to sustain the body and that are impacted in hibernation or stasis states, we can consider hydration, caloric intake, and oxygen.

For discussion purposes, we will assume any spaceflight travelers are within the standard 95th percentile anatomy model for males and females. For males, this corresponds to a body mass of 98.5 kg and height of 1.9 m. For females, this corresponds to 61.7 kg and height of 1.65 m [18].

3.1 Hydration

Hydration is essential to the body and, ideally, provided in the form of pure water. It is consumed, processed, excreted, and/or released to manage everything ranging from maintaining the health and functioning of our cells, to temperature control via sweating, to aiding in digestion and waste elimination. On average, a human male nominally requires about 3.7 L of water per day (~3.7 kg), while women require 2.7 L (~2.7 kg) [19]. If adequate hydration levels are not maintained, metabolic processes are quickly disrupted, leading to feelings of dizziness, confusion, and fatigue.

The required levels of hydration for the body are reduced in a low metabolic state, primarily due to lower rates of breathing and the reduced demands on the thermoregulatory system to manage variations in body temperature. While a reduction in the body's demand for water is beneficial for spaceflight, our demonstrated ability to design systems with very high recovery efficiencies (and the apparent abundance of water throughout the solar system) makes this issue one of the least concerning areas for achieving long-duration spaceflight.

3.2 Oxygen demand

Similar to hydration, the human body requires oxygen to breathe. The body will consume about 400 L of oxygen (O_2) per day, with demand increasing significantly with any strenuous activity [18]. Unlike hydration or nutrition, however, the human body requires a continuous intake of oxygen. The body can survive weeks without food and days without water. Under normal metabolic conditions, permanent brain damage can start to occur in a matter of minutes due to hypoxia (oxygen deprivation), and within approximately 10 minutes, death will occur.

3.3 Caloric intake

In the sedentary state of resting or sleeping, the human body requires a minimum amount of energy to support the basic functions of breathing, circulation, digestion, and cellular activity. This energy requirement, known as the basal metabolic rate (BMR), is based on our metabolism and varies depending on body mass, size, and percentage of fat tissue [20]. One additional factor affecting an individual's BMR is their age, with our metabolism tending to reduce over time.

Using the Mifflin St. Jeor equation for estimation of daily required caloric intake, we can estimate the reduction in nutritional demands by the body at varying levels of metabolic reduction [20]:

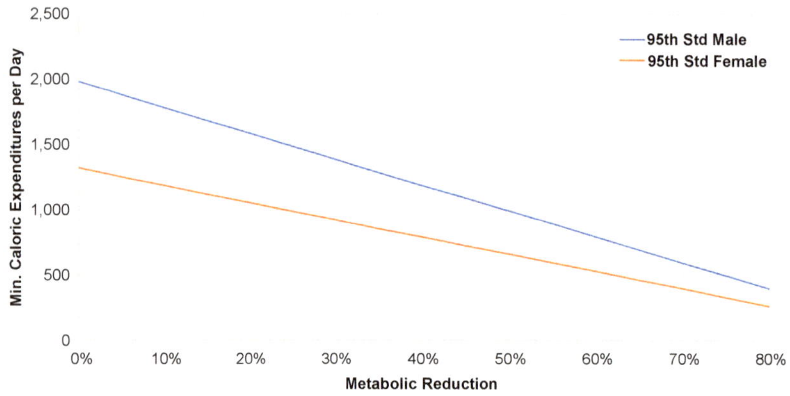

FIG. 2

Standard male and female minimum daily caloric expenditures vs metabolic rate reduction.

$$\text{Body Heat Production at Rest} = \left(\frac{10.0\,\text{body mass}}{1\,\text{kg}} + \frac{6.25\,\text{height}}{1\,\text{cm}} - \frac{5\,\text{age}}{1\,\text{year}} + s\right)\frac{\text{kcal}}{\text{day}}$$

where $s = +5$ for males and -161 for females.

For the standard male and female models and with no metabolic reduction, the Mifflin St. Jeor equation yields values of ~1980 and ~1320 kcal/day, respectively. If we consider a reduction in metabolic activity of 70%–80%, it is estimated daily caloric demand will decrease to only 495 and 320 kcal/day for males and females, respectively (Fig. 2). For spaceflight, this metabolic reduction would translate to more than a threefold reduction in the mass of nutrition required.

4 Approaches to human stasis

There are currently no known methods to definitively extend human life or place the human body into a permanent stasis condition. This stands in sharp contrast to the prominence of stasis capabilities in nearly every science fiction work involving spaceflight and being used as a plot device to easily enable or explain the ability to conduct interstellar travel. Fortunately, while there are likely a variety of interesting terrestrial uses for stasis, the dawn of the space age has created new demands and potentially will be the driving impetus to enable this capability.

We can frame a discussion on stasis approaches and associated technologies into two methods, that of "suspended animation" and "hypothermic states." Suspended animation results in a near cessation of all cellular activity, whereas hypothermic approaches involve undergoing significant reductions in metabolic activity, but still yield relatively active physiological activity. The technologies required for implementation of either approach will be radically different, but we could utilize either approach for interstellar travel.

4.1 Suspended animation

Current approaches to achieving a suspended animation have been focused on either cryogenics or chemically induced states. If ever achievable, suspended animation would obviously be the preferred method for interstellar travel. Presumably, almost no aging would occur and the need to provide oxygen, hydration, nutrition, etc. during transit could be eliminated. As with any critical ship functions, the preservation systems would still require extremely high levels of redundancy, power, and potentially processing/maintenance of any chemicals or fluids used to maintain the state.

In the area of cryonics, successful implementation has not been achieved primarily due to the lack of a viable process to reverse the state. Despite this obviously critical aspect, there are several commercial companies that have offered these services to individuals (i.e., Alcor, Cryonics Institute, KrioRus). A few hundred people around the world are currently being held in this state indefinitely and they are all considered deceased for legal purposes.

The overall cryonics process involves storing the body (or head) in a liquid nitrogen tank after it has been slowly cooled to −195°C. A cryoprotectant fluid is injected into the blood stream prior to storage in order to prevent any crystallization. While these fluids prevent the formation of ice, which can destroy the cells, they are also increasingly toxic as their concentration in the body (or organ) is increased [21]. While there is continued research in the pursuit of new cryoprotectant chemicals, primarily to support organ transplantation and longer timeframes for donor matching, there does not appear to be any new or recent developmental progress that would benefit anyone currently in cryostasis.

In contrast to cryonics, cryogenics involves the study and application of cold temperature therapeutically. The focus and goal is not to freeze the body, organs, or tissues, but to merely manipulate the metabolic process and cellular demand for oxygen. Medical research in this area has led to a procedure that entails the full cessation (and resumption) of functioning in all vital body organs [22]. This life-saving measure is still experimental and can only be attempted after other measures have been exhausted. Unfortunately, it is only useful over short time periods, measuring in a few hours. While the process is seemingly miraculous, it is dependent on lingering oxygen in the body to prevent any permanent ischemic damage. Eventually, any oxygen is exhausted, and resuscitation must be performed to prevent permanent damage to the brain. This technique does not appear to have a path forward toward longer durations or use in less extreme conditions.

In 2010, an alternative approach to achieving suspended animation was demonstrated. Using hydrogen sulfide gas (H_2S), mice were placed in a suspended animation state by scientist Dr. Mark Roth and successfully reanimated 6 hours later [23]. While a poisonous gas, when administered in small doses the H_2S appears to bond to oxygen receptors to temporarily reduce the demand for oxygen, resulting in a significant reduction in metabolic rate. There was a brief period of optimism that this noncryogenic, chemically induced state represented a major breakthrough. The effects demonstrated in mice, however, could not be duplicated when applied to larger animals.

4.2 Hypothermic states

Humans are endothermic and our body temperature, which is correlated with our metabolic rate, is managed by the thermoregulatory system. A large amount of the food and energy we consume is converted to energy that is used to maintain our body in homeostasis at a nominal core temperature between 36.5°C and 37.5°C. As our external environment changes, our body is constantly adjusting to maintain this nominal temperature.

The thermoregulatory system of hibernating animals manages their core temperature and metabolism based on various innate or external triggers and can result in variations to their metabolic activity level and body temperature. In humans, as our body temperature begins to fluctuate, our thermoregulatory system fights to maintain it at its nominal temperature. To achieve this, the body has various tools at its disposal that include sweating, shivering, vasoconstriction/dilation, and metabolic changes driven by the thyroid.

Through different mechanisms (conduction, convective cooling, etc.) that result in thermal energy imbalances, the body's temperature may become lower than the nominal condition, leading to a state called hypothermia. As the core temperature is reduced, metabolic activity follows, with an estimated 5%–7% reduction for every one degree temperature change [24]. The effect of this metabolic reduction on the body is lower oxygen demand, energy expenditures, heart rate, and cellular activity.

There are limits to the extent the body's core temperature can be safely reduced. The hypothermic state is generally classified into three different levels: mild, moderate, and severe [25]. Mild hypothermia is regarded as a core temperature within the range 32–35°C and is accompanied by shivering, fatigue, and declining cognitive abilities. Moderate hypothermia is typically when the core temperature has dropped to 28–32°C. In this range, an individual will stop shivering, become lethargic, and undergo a phenomenon called "paradoxical undressing." With severe hypothermia, characterized as a core temperature below 28°C, the body will become unresponsive, with declining heart rate, blood pressure, and breathing. At this stage, the body is beginning cardiorespiratory failure.

In clinical settings, rewarming of patients from unintentional hypothermia has a fairly high success rate and has been demonstrated in cases of severe hypothermia with core temperatures as low as 14°C and prolonged cardiac arrest [15]. Rewarming techniques and the extent of the intervention for recovery vary widely, ranging from the application of simple heat pads to extracorporeal blood warming.

5 Achieving limited-duration stasis

The concept of hypothermia in medical applications has been theorized since the Greek physician Hippocrates, with in-depth scientific research into its effectiveness and utilization starting in the 1950s [26]. During this time, multiple studies have indicated the ability of mild hypothermia to assist in surgeries by serving as a general neuroprotectant from incidents that resulted in reduced blood flow to the brain. Additional studies have shown benefits in reducing long-term damage to the brain after incidents of traumatic brain injury, stroke, and cardiac arrest [27]. The practice and procedure, referred to as therapeutic hypothermia or targeted temperature management (TTM), has become a valuable tool for care that is utilized by all major healthcare centers and, more recently, by first responders around the world.

It has been proposed that this procedure could be adapted to serve as a starting point for achieving limited-duration stasis capabilities for human exploration missions to Mars [2].

5.1 Therapeutic hypothermia background

In clinical settings, a torpor state can be artificially achieved by inducing mild hypothermia in humans during a procedure known as therapeutic hypothermia (TH). This medical treatment helps to reduce the risk of ischemic injury to tissue following a period of insufficient blood flow by lowering a patient's body temperature and metabolic rate [27]. As the patient's metabolism is reduced, so is their demand for oxygen during a time in which the body is also struggling to provide this oxygen.

Once a determination is made for the application of TH, the injured patient is placed under heavy sedation (if not already) using an anesthetic such as fentanyl or propofol. Sedation is required to reduce overall physical movement by the patient, suppress the shivering response, and relieve any pain in order to make the patient comfortable. More recent research has evaluated the use of dexmedetomidine (dexmed) and benzodiazepines for core body cooling [28]. These hypnotic sedatives are nonnarcotic and may be safer options for longer duration periods (among other potential benefits).

Regional or whole-body active cooling can be accomplished through a variety of methods. This includes external ice packs, cold saline injection, and intranasal catheters that emit a cooling gas in the nasopharynx. The recommended core body temperature is between 32°C and 35°C, which can be achieved through cooling at a rate of about 1–2°C per hour. Medical studies have attempted to identify an ideal target temperature, but results have generally been inconclusive. Once the target temperature is reached, the patient is maintained in this unconscious, inactive state for a period of 24–72 hours. After the treatment period is over and their medical status has improved, the body is gradually rewarmed at rate of 0.5°C per hour and administration of any sedatives is stopped.

It is noteworthy that there have been some medical studies evaluating extended-duration application of TH over periods of 10–14 days [29]. The outcomes appeared to be no different for patients that were held in TH over an extended period when compared to the standard 24–72 hour treatment periods. The end result being that there was no medical benefit for maintaining patients in these longer states, so shorter treatment periods have continued as the standard for critical care.

5.2 Adaptations for human spaceflight

For space travel, the model of therapeutic hypothermia may serve as a point-of-departure for providing an artificially-induced torpor state in humans during a mission. Unlike the clinical application of TH, this process will be performed on healthy individuals, and a number of complications and factors can be eliminated. There are, however, still a number of advancements that will be required to enable regular and routine use of any stasis capability in the spaceflight environment.

The true measure of our mastery of this process lies in how naturally we can induce this state. Hibernating animals are able to trigger a reduction in their metabolism, a process scientists are still trying to fully understand. It is this metabolic reduction that leads to the lowering of their core temperature and stasis condition. This is in contrast to the process of forced cooling used in TH that results in the metabolic reduction. To support this artificial process in humans, sedation is required to effectively disable the body's rewarming defenses (e.g., shivering).

Fig. 3 provides a vision for what the implementation of a torpor-inducing process would look like for a space mission [2]. Some key features include installation of an implantable, subdermal venous access device and a percutaneous endoscopic gastrostomy (PEG) tube, external sensors for basic health vitals monitoring, and application of neuromuscular electrical stimulation (NMES) [29,30].

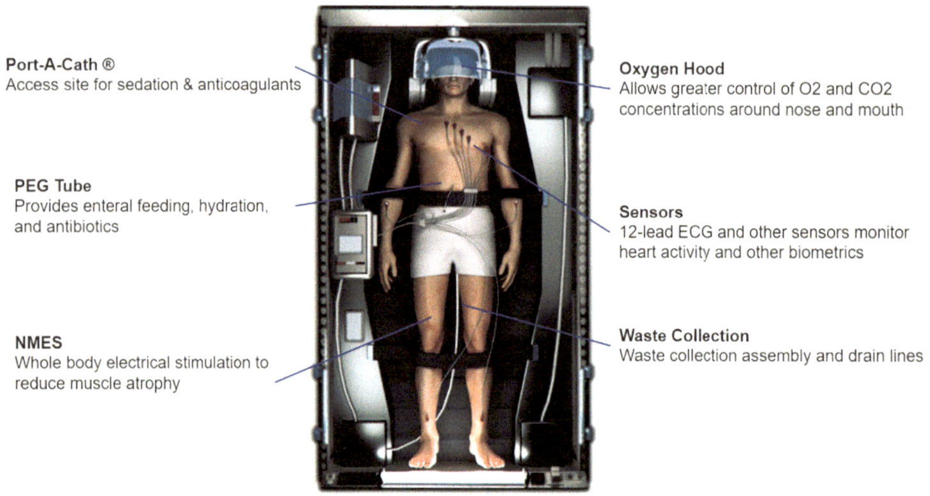

FIG. 3

Implementation of limited-duration stasis in spaceflight environment.

Credit: SpaceWorks.

The venous access is used for longer-term delivery of intravenous medications and has been demonstrated to be safe for use over long periods. Its location under the skin decreases the risks from injection and allows them to remain installed for months or years. The PEG tube is for providing nutrition in the form of a liquid directly into the stomach, bypassing the throat and esophagus during the inactive state. These devices are in widespread clinical use, are very safe, and permit an individual to resume normal eating at any time [30].

As the body enters a low metabolic state, it is expected that the rates of muscle atrophy and bone loss common in spaceflight will be reduced. Bone demineralization has been countered successfully through use of bisphosphonates [31,32]. Muscle disuse over extended periods of time, however, is likely to result in some level of atrophy. To counter this effect, an automated system is used to provide whole-body muscle stimulation by applying small electrical impulses at key locations on the body. This technique for maintaining muscle mass, known as neuromuscular electrical stimulation (NMES), has undergone limited demonstration in space and shown to be effective [33].

For spaceflight, cooling methods such as saline injection or use of gel pads are not conducive, and less intrusive techniques are desired. To induce and/or maintain a target core temperature, simplified options for space travel include ambient cooling of the cabin and use of a nasopharyngeal evaporative cooling system. With the majority of heat transfer in humans occurring through the hands and feet, a simple cooling system contacting the glabrous skin surfaces could also be used.

A goal of medical experts and researchers is to be able to naturally trigger metabolic reductions in humans. Current interest in achieving this is through the manipulation of the body's adenosine receptors through the use of new pharmaceuticals. Alternatively, if early hominins did possess the ability to hibernate, it may be possible to reactivate expression of these dormant genes.

The longer periods of administration for TH that have been conducted provide confidence that a modest (artificial) torpor capability can ultimately be achieved in humans for periods lasting up to 2 weeks. If this can be perfected for spaceflight, it could be used extensively to support intersolar transit. Spacefaring crews would undergo repeated torpor bouts consisting of 10–14 day periods of inactivity followed by brief periods of activity/nontorpor.

5.3 Associated medical challenges

There are commonly encountered medical risks associated with any procedure, and inducing hypothermia is no different. Typical complications that can arise involve infection, anticoagulation, bradycardia, and hypotension.

Infection is a concern whenever the skin tissue is broken (e.g., IV access) and can be mitigated through stringent site monitoring. Extended periods of inactivity can lead to the formation of pressure ulcers and blood clots, as is often seen in spinal cord injury patients. This is expected to be less of a concern in space compared to Earth, with the microgravity environment reducing high-pressure contact levels and locations on the body. There are also common pharmaceuticals that can be used to reduce blood clot formation (e.g., heparin). Body cooling itself actually has an anticoagulant effect on blood [34]. The combination of being in a hypothermic state while in microgravity may balance out these two conditions.

Hypothermia also impacts the cardiac pacemaker cells that trigger heartbeats and therefore may result in bradycardia (reduced heart rate) or hypotension (low blood pressure). At mild or moderate hypothermic conditions (e.g., 30–32°C) a lowered heart rate and blood pressure are of less concern. These are conditions that can easily be monitored continuously and counteracted if needed through rewarming, drug administration, or artificial heart stimulation.

6 Recent research and advancements

A number of potential clinical and terrestrial applications for the use of hypothermia are permitting rapid advancements to improve our knowledge and understanding of this process. Multiple approaches are being pursued that may prove valuable individually or collectively.

6.1 Hypothalamus manipulation

The hypothalamus portion of the brain controls metabolic processes and the thermoregulatory system of the body. It has been proposed that it could be artificially manipulated to reduce metabolic rate. This approach closely resembles spontaneous or natural torpor in animals and contrasts with methods that use pharmaceuticals and physical cooling to lower the body's temperature.

Recently, two independent research teams identified cells in a region of the hypothalamus that, when they were stimulated, appeared to induce a torpor state in laboratory mice [35]. When this occurred, the mice showed signs of reduced heart rate and breathing, with their core temperature lowering and stabilizing at about 10°C lower than their normal core temperature. While the specific neurons identified by each team were different, they were located in a similar region of the hypothalamus and might interact with each other.

6.2 Adenosine receptors

Researchers have been pursuing methods to trigger the hibernation state in both natural hibernators and in nonhibernating species. This has led to a focus on adenosine receptors with a specific interest in the A1 receptor, which controls sleep.

Agonists for the A1 adenosine receptor (A1AR) have been used in experiments since 2010. The synthetic adenosine receptor agonist N6-cyclohexyl adenosine (or CHA) is a pharmacological agent that has shown the ability to spontaneously initiate a torpor state with the associated decrease in metabolic rate in rats [36]. In addition, through the use of initial bolus injection of CHA and then continuous drip administration, this drug has been shown to keep laboratory rats in an induced hibernation state for over 24 hours [36].

A recent study evaluated the impact of utilizing CHA for torpor induction on a large animal model [37]. During the experiment, the animals' core body temperature was lowered and remained steady between 30°C and 32°C. Additionally, no shiver response was exhibited by the animal. Parallel research studies have also examined the use of 8-sulfophenyl theophylline (or 8-SPT), an adenosine receptor agonist. Since 8-SPT does not cross the blood-brain barrier, it does not mix with and impact the effects of CHA. This drug was shown to successfully relieve bradycardia and hypotension in animals undergoing torpor.

7 Long-duration stasis for interstellar travel

At this time, there are no viable approaches to achieve the long-duration stasis capabilities likely needed for interstellar missions. Based on current medical capabilities, recent technological progress in certain areas, and some anecdotal evidence, there are some potential options for achieving stasis capabilities that could advance sufficiently in the future to be useful for interstellar missions. These include achieving profound hypothermia, a combined stasis and longevity approach, or human adaptations through enhanced biology.

7.1 Profound hypothermia

A core body temperature below 20°C is defined as profound hypothermia and, in this condition, there are no detectable vital signs or brain electrocerebral activity. This state is induced in some rare clinical treatments or as a last resort for victims of extreme trauma, a procedure known as emergency preservation and resuscitation (EPR) [22]. Maintaining this condition requires a complete cardiopulmonary bypass (CPB) due to heart arrhythmias and cardiac arrest that can develop below core temperatures of 28°C [38].

While this is an invasive approach, it is likely a more plausible solution toward achieving deep stasis compared to cryonics. One of the main questions lies in how long the body and brain can be held in this state and safely recovered. From the late 1990s to early 2020s, sustained durations of moderate hypothermia on the order of days are achievable with generally positive outcomes. Periods of profound hypothermia has been achieved in the order of hours. Longer durations are associated with increased morbidity in unhealthy patients. Aside from some potential medical complications, such as an increased risk of blood clot formations, infection, and pneumonia, it may be possible to safely extend this condition to significantly longer periods in the future.

At these core temperatures, the metabolic rate is reduced substantially, and the daily caloric demands of the body are extremely low. While the body will still require some nutritional intake, in a profound hypothermic state, any gut motility will be nonexistent and even enteral feeding will not be an option. One potential solution would be to periodically rewarm the individual to a mild hypothermic state that permits replenishment of nutritional requirements through enteral feeding.

7.2 Combined stasis and longevity

There are several entities that are currently pursuing advancements in human longevity. These entities range from medical research centers to commercial companies and nonprofit foundations. Areas being targeted and investigated that may yield life extension include dietary improvements, new pharmaceuticals, stem cell therapies, tissue regeneration, organ replacement, and gene therapy.

Recent advances have not translated into any benefits for humans. Progress through 2020 has been limited to extending the life of worms, mice, and nematodes, with up to a fivefold increase in some cases. At this time, the only recognized method to likely offer longer human lifespans is through caloric restriction with intermittent fasting [39].

In contrast to profound hypothermia, it may be possible to combine stasis approaches at mild or moderate hypothermic conditions with any future longevity advances. If we assume the human lifespan could be extended to approx. 150 years, we could travel interstellar distances by undergoing 2 or 3 week stasis sleep periods. These periods of inactivity could be interspersed with active periods spanning a week, resulting in a 4 or 5 year timeframe that would, in effect, be perceived as only a year. Given the isolation of the passengers from the ever-changing world on Earth, the perception is even easier to achieve. Ultimately, a 50 year trip to another star may only feel like 10–12 years have actually elapsed.

While these are still long durations for space travel by any measure, they would permit an individual to travel and visit a different star system within their lifetime. Additionally, they could depart at an age where a conscious decision could be made to undertake the journey and an individual would still have a significant portion of their life remaining upon arrival.

7.3 Enhanced biology

With the emergence of gene-editing capabilities such as CRISPR-Cas9 technology, the possibility of altering the human genome to either extend our lifespan or enable natural hibernation capabilities in humans is increased. This approach could address multiple issues that currently have to be mitigated or managed with synthetic induction of hibernation. This may include the ability for natural or on-demand metabolic reduction without sedation, preservation of muscle mass during long periods of disuse, rapid awakening, an improved ability to store energy reserves, and significantly deeper metabolic reductions with the ability to maintain homeostasis.

If we assume hibernation capabilities similar to other hibernating mammals could be feasible, uninterrupted stasis periods of 3–9 months could be achieved without requiring a number of medical systems. Unlike artificially induced states where forced body cooling is used to lower the metabolism, natural hibernation begins with a downregulation of the metabolism, which eventually leads to a lowering of the body's core temperature. This process could eliminate the need for cooling systems, like gel pads and intranasal catheters, as well as any requirement for enteral feeding systems like the PEG tube.

Editing of human DNA in this manner does raise serious ethical concerns and questions. Encouragingly, there is some indication that gene editing may not be required to enable natural metabolic suppression, as "no unique factors or genes or neuro-endocrine properties have been identified for this state" [40]. In this case, the challenge becomes understanding how to trigger the dormant gene expression necessary for hibernation.

8 Summary

There are a number of pathways and approaches that could be pursued to achieve human stasis capabilities in support of interstellar travel. The requirements and utility of these will be highly dependent on the propulsion capabilities available for the mission. In the scenario of breakthrough physics (e.g., wormholes) that provide rapid or instantaneous travel between star systems, there will obviously be little need for the crew and passengers to undergo hibernation. If we assume a much more conservative propulsion technology that results in a very long-duration spaceflight necessitating "generational ships," a form of simple hibernation may be valuable to simply conserve resources during the journey. If we end up between these two extremes, then an advanced form of hibernation or stasis should prove extremely valuable and possibly one of the key enabling technologies for interstellar travel.

Fortunately, we can pursue the necessary medical technologies and health advancements needed for interstellar travel in parallel. Limited forms of hibernation capabilities, ranging from inactive periods of a few weeks to months, have been demonstrated via simulation to be extremely beneficial for intersolar travel. The system-level impacts of this capability on missions to Mars or Ceres have proven to be more impactful and important than the propulsive technologies used for these missions. The recent global surge in the development of new spaceflight systems and human exploration plans will likely accelerate advancements and yield a variety of new approaches on how to solve some of the challenges of human torpor.

Occurring in parallel with space-focused technology development, renewed interest and investment targeted at improving human longevity is occurring and will be extremely valuable. New approaches and understanding of key contributors to human health will ultimately accelerate increases in the expected human life expectancy. Combining hibernation with these improved human longevity advancements will enable new approaches to interstellar travel that may permit the use of more conservative technological advancements as well as significant mass and cost reductions.

Prospects for achieving human stasis for spaceflight have never been more encouraging. Increased interest and investment from both private and government sources continues to address the myriad of questions and challenges. Meanwhile, as we look toward past and current times, recent archeological discoveries and seeming medical miracles increasingly indicate the potential that humans may indeed possess an innate capability for hibernation.

References

[1] J. Bradford, M. Schaffer, D. Talk, Torpor Inducing Transfer Habitat for Human Stasis to Mars, Phase I Final Report, NASA NIAC Grant No. NNX13AP82G, 2014.

[2] J. Bradford, B. Merrel, M. Schaffer, C. Williams, D. Talk, Advancing Torpor Inducing Transfer Habitat for Human Stasis to Mars, Phase II Final Report, NASA NIAC Grant No. NNX16AL30G, 2018.

[3] M. Basner, D. Dinges, et al., Mars 520-d mission simulation reveals protracted crew hypokinesis and alterations of sleep duration and timing, Proc. Natl. Acad. Sci. 110 (7) (2013) 2635–2640.

[4] L. Medina, S. Sabo, J. Vespa, Living Longer: Historical and Projected Life Expectancy in the United States, 1960 to 2060, 2020. P25-1145.
[5] Mortality Trends in the United States, 1900–2018. https://www.cdc.gov/nchs/data-visualization/mortality-trends/. (Accessed 20 January 2024).
[6] J. Robine, M. Allard, F. Herrmann, The real facts supporting Jeanne Calment as the oldest ever human, J. Gerontol. Ser. A 74 (2019).
[7] A. Fox, Study suggests 150 years may be the human lifespan's upper limit, Smithsonian Mag. (2021).
[8] L.R. Belzile, A.C. Davison, H. Rootzén, D. Zholud, Human mortality at extreme age, R. Soc. Open Sci. 8 (9) (2021) 1–15.
[9] Z. Istvan, Death could soon become a curable disease, in: The Future of Everything, 2019.
[10] M. Berg von Linde, et al., Insights from the den: how hibernating bears may help us understand and treat human disease, Clin. Transl. Sci. 8 (2015).
[11] S. Cooper, S. Lloyd, A. Koch, Temperature effects on the activity, shape, and storage, of platelets from 13-lined ground squirrels, J. Comp. Physiol. B Biochem. Syst. Environ. Physiol. 187 (5–6) (2017) 815–825.
[12] M. Blanco, K. Dausmann, S. Faherty, et al., Hibernation in a primate: does sleep occur? R. Soc. Open Sci. 3 (8) (2016) 160282.
[13] A. Bartsiokas, J. Arsuaga, Hibernation in hominins from Atapuerca, Spain half a million years ago, L'Anthropologie 124 (5) (2020).
[14] M.A. Rothschild, V. Schneider, 'Terminal burrowing behaviour'—a phenomenon of lethal hypothermia, Int. J. Legal Med. (1995) 250–256.
[15] M. Gilbert, R. Busund, et al., Resuscitation from accidental hypothermia of 13.7 degrees C with circulatory arrest, Lancet 355 (2000) 375–376.
[16] E. Stolte, Saskatchewan tragedy strikes painful chord for city mom, Edmont. J. (2008) (February 6).
[17] Hawaiian airlines stowaway survives 5-hour flight in wheel well; teen's body went into 'hibernation', Medical Daily (2014).
[18] NASA-STD-3001, Space Flight Human-System Standard, vol. 1, Crew Health.
[19] B. Popkin, et al., Water, hydration, and health, Nutr. Rev. 68 (8) (2010) 439–458.
[20] D. Frankenfield, L. Roth-Yousey, C. Compher, Comparison of predictive equations for resting metabolic rate in healthy, nonobese and obese adults: a systematic review, J. Am. Diet. Assoc. 105 (5) (2005) 775–789.
[21] B. Best, Cryoprotectant toxicity: facts, issues, and questions, Rejuvenation Res. 18 (5) (2015) 422–436.
[22] S.A. Tisherman, Emergency preservation and resuscitation for cardiac arrest from trauma, Ann. N. Y. Acad. Sci. 1509 (1) (2021) 5–11.
[23] E. Blackstone, M. Morrison, M. Roth, H2S induces a suspended animation-like state in mice, Science 308 (5721) (2005).
[24] A.M. Omairi, S. Pandey, Targeted Hypothermia Temperature Management, StatPearls, 2021.
[25] H. Duong, G. Patel, Hypothermia, StatPearls Publishing, 2022.
[26] H. Guly, History of accidental hypothermia, Resuscitation 82 (1) (2011) 122–125.
[27] K.H. Polderman, Application of therapeutic hypothermia in the ICU, Intensive Care Med. 30 (2004) 556–575.
[28] D. Hostler, W. Northington, C. Callaway, High-dose diazepam facilitates core cooling during cold saline infusion in healthy volunteers, Appl. Physiol. Nutr. Metab. 34 (2009) 582–586.
[29] J.-Y. Jiang, M.-K. Yu, C. Zhu, Effect of long-term mild hypothermia therapy in patients with severe traumatic brain injury, J. Neurosurg. 93 (2000) 546–549.
[30] J. Anthony, K. Kapil, B. Hawthorne, Long-term outcome of percutaneous endoscopic gastrostomy feeding in patients with dysphagic stroke, Age Aging 27 (6) (1998) 671–676.
[31] A. Okada, T. Matsumoto, et al., Bisphosphonate use may reduce the risk of urolithiasis in astronauts on long-term spaceflights, JBMR Plus 6 (1) (2021) 1–10.
[32] A. Leblanc, T. Matsumoto, et al., Bisphosphonates as a supplement to exercise to protect bone during long-duration spaceflight, Osteoporos. Int. 24 (7) (2013) 2105–2114.

[33] N. Shiba, H. Matsuse, Y. Takano, et al., Electrically stimulated antagonist muscle contraction increased muscle mass and bone mineral density of one astronaut—initial verification on the International Space Station, PLoS One 10 (8) (2015) 1–14.
[34] J. Ruzickaa, M. Stengla, L. Boleka, J. Benesa, M. Matejovicb, A. Krouzeckyb, Hypothermic anticoagulation: testing individual responses to graded severe hypothermia with thromboelastography, Blood Coagul. Fibrinolysis 23 (2012) 285–289.
[35] T.M. Takahashi, G.A. Sunagawa, S. Soya, et al., A discrete neuronal circuit induces a hibernation-like state in rodents, Nature 583 (7814) (2020) 109–114.
[36] T. Jinka, V. Combs, K. Drew, Translating drug-induced hibernation to therapeutic hypothermia, ACS Chem. Neurosci. 6 (6) (2015) 899–904.
[37] K. Drew, Thermolytic efficacy and pharmacodynamics of CHA/8SPT in rats and pigs, in: ESA-Topical Team Workshop on Hibernation and Torpor, Innsbruck, Austria, 2018. February 10.
[38] M. Sarkar, V. Prabhu, Basics of cardiopulmonary bypass, Indian J. Anaesth. 61 (9) (2017) 760–767.
[39] D.S. Hwangbo, H.Y. Lee, L. Abozaid, K.J. Min, Mechanisms of lifespan regulation by calorie restriction and intermittent fasting in model organisms, Nutrients 12 (4) (2020) 1–24.
[40] A. Choukèr, J. Bereiter-Hahn, D. Singer, et al., Hibernating astronauts—science or fiction? Pflug. Arch Eur. J. Physiol. 471 (2019) 819–828.

CHAPTER

Ship governance and culture

7

James E. Dunstan*
Mobius Legal Group, PLLC, Fredericksburg, VA, United States

Chapter outline

Captain's Log, Day 215 of the journey of the interstellar ship Nellie Bly. We're through the "shake-out" portion of the flight and have worked out the major kinks of the ship. Systems are all nominal, and morale onboard is high. We're enjoying our great adventure in colonizing a new world.

*James E. Dunstan is the founder of Mobius Legal Group, PLLC, and serves as General Counsel to TechFreedom. Jim has over 40 years of private legal practice centered on high technology, telecommunications, computer games, and outer space law. Jim was the 1978 Harry S. Truman Scholar for the state of Arizona, is an honors graduate of Claremont McKenna College and the Georgetown University Law Center. At Georgetown, he founded the Georgetown Space Law Group, the first US-based law student organization focused on the law of outer space.

Interstellar Travel. https://doi.org/10.1016/B978-0-323-91280-8.00007-1

1 Introduction and key assumptions

This chapter examines governance models for an interstellar journey on a multigenerational ship. "Governance" involves many complex issues, ranging from the form of government adopted, to issues such as property, economics, sociology, social psychology, and culture. This chapter will touch on all of these, but a full "deep dive" on each subject is beyond the space limits of this work.

As difficult as it may be, getting the right governance model to fit the particular needs of a society, even a limited one such as the occupants of an interstellar ship, has a huge impact on the quality of life of those people. In "Why Nations Fail: The Origins of Power, Prosperity, and Poverty," D. Acemoglu and J. Robinson [1] analyze a number of historical societies, their governments, and why even neighboring cities can have radically different qualities of life based on how they are governed.

> The reason that Nogales, Arizona, is much richer than Nogales, Sonora, is simple; it is because of the very different institutions on the two sides of the border, which create very different incentives for the inhabitants of Nogales, Arizona, versus Nogales, Sonora. The United States is also far richer today than either Mexico or Peru because of the way its institutions, both economic and political, shape the incentives of businesses, individuals, and politicians. Each society functions with a set of economic and political rules created and enforced by the state and the citizens collectively. Economic institutions shape economic incentives: the incentives to become educated, to save and invest, to innovate and adopt new technologies, and so on. It is the political process that determines what economic institutions people live under, and it is the political institutions that determine how the process works (p. 42).

Some key assumptions are necessary to provide boundaries for this work. Without them, there would be just too many variables to be able to reach any usable conclusions about governance models. Those key assumptions are:

(1) The ship is not capable of faster than light travel, meaning that it will take many years to reach its destination. For the sake of simplicity, and to allow for a longer-term analysis of governance models, I have assumed a minimum of a 100-year journey.
(2) The total occupancy of the ship is greater than 100 but less than 1000. For simplicity, I've chosen an occupancy of 500.
(3) The occupants on the ship will be awake for the duration of the journey.
(4) The occupants will be multicultural and diverse. Obviously, given the 100-year journey, multiple generations will exist during the transit.
(5) The ship has a specific destination and is not just wandering through space.
(6) Each of the original occupants has chosen to go on the ship (no conscription).
(7) The ship is not an "ark" transporting the few survivors left from a dying Earth. This assumption is necessary because otherwise the choice of governance model boils down to "anything" versus "death," making the analysis useless.
(8) The occupants will resemble current 21st century humans, to the extent that they will have a life expectancy of around 80 years, with variable declines in both physical and mental capabilities depending on the individual as they approach their end of life.
(9) Cloning has not become an acceptable means of either procreation or population control on Earth, or for "replacement parts" for the living.

(10) The ship itself will have some form of gravity such that "normal" human activities are possible, and the occupants won't have to undergo the grueling exercise regimen required of today's astronauts for long duration missions in space.

(11) No advances have been made in quantum (instantaneous) communications so that the ship could be governed remotely by some other entity. The ship will be governed internally only; at least once it leaves effective communications range by radio.

Beyond these, no further assumptions are made. That is not to say that other key inputs into the governance model are not present, but rather, each of these key inputs is described below to demonstrate how they translate into governance models. It would be far too easy and limiting to just declare, as an example, that the ship is built by the United Nations and subject to the full panoply of UN treaties, conventions, and other agreements as decided by a government installed by the United Nations.

2 Output goals: Safety, efficiency, and freedom

In choosing the optimal governance model for an interstellar ship, there are three main output goals: safety, efficiency, and freedom. How these goals are prioritized and balanced has a significant impact on the form of government adopted.

A safety-driven governance model prioritizes the safety of the occupants above everything else. In its truest sense, all decisions are made with safety in mind, and anything that could endanger the safety of the occupants is not allowed. Safety can be defined both as physical safety as well as the psychological wellbeing of the occupants. The term itself, however, can easily be hijacked (and potentially be abused as it has been here on Earth) if it is expanded to include concepts of cultural safety (e.g., limiting speech that might upset another occupant or challenges the governance model itself).

An efficiency-driven governance model prioritizes the ability of the government to make decisions quickly and efficiently. At one level, this allows for quick reactions to exigent circumstances, which might be necessary (e.g., issuing an order to stop taking showers because of a short-term problem with the closed-loop water filtration system). Favoring efficiency, however, can also act to limit discourse and discussion on an issue impacting all the inhabitants of the ship, especially if the decision will have a long-term impact on ship culture or moral.

A freedom-driven governance model prioritizes the individual rights and liberties of the occupants. This is the model most Westerners think of as the natural model that will be applied to an interstellar ship with slight modifications to accommodate the unforgiving environment of space. Most academic work on space governance models has focused on the issue of individual freedoms, and rightly so. In Robinson and White's [2] "Declaration of First Principles," the authors noted the need to protect individual rights and liberties:

> Appropriate constraints upon and limitations of authority shall be defined so as to protect the personal freedom of each individual, such as the right to reasonable privacy, freedom from self-incrimination, freedom from unreasonable intrusion, search and seizure, and freedom from cruel and unusual treatment (Article I (A)).

In 1991, I challenged many of the assumptions of these freedoms in "From Flag Burnings to Bearing Arms to States' Rights: Will the Bill of Rights Survive a Trip to the Moon?" [3] My argument was that the very concepts contained in the US Bill of Rights might not be compatible to outer space governance—in that instance, a lunar base. For example, the burning of an American flag is considered protected speech, and the US Supreme Court struck down a Texas law making flag burnings illegal in Texas vs Johnson [4]. Yet in a closed-loop environmental control system, which would be found on a lunar base or interstellar ship, the safety of the entire population of the base or ship could be jeopardized in the name of free speech when an item is ignited in protest, thereby justifying such a limit on speech rights. Similarly, what we might term as cruel and unusual treatment under the Eighth Amendment [5] will probably have to be modified in space. Everything from inmate cell size to capital punishment will test the law and patience of an interstellar ship government if it is faced with an occupant who violates the laws of the ship, especially in violent ways.

What, then, are the key freedoms that will be put to the test on an interstellar ship? The US Bill of Rights [6] gives a list of "inalienable" or "natural" rights that a government should not control:

(1) Freedom of speech and press (First Amendment)
(2) Freedom of religion (First Amendment)
(3) Freedom to petition the government with grievances (First Amendment)
(4) Right to bear arms (Second Amendment) (most likely the first natural right to be jettisoned, or at least the current reading of that right under Supreme Court precedent)
(5) Right to refuse to quarter soldiers for the government (Third Amendment) (this implicates property rights, which is a whole issue unto itself beyond the scope of this chapter)
(6) Right to be safe from unreasonable search and seizure (Fourth Amendment)
(7) Right to grand jury indictment (Fifth Amendment)
(8) Right to speedy trial and criminal due process (Sixth Amendment)
(9) Right to jury trial (Seventh Amendment)
(10) Freedom from cruel and unusual treatment (Eighth Amendment)

The one freedom not listed above and not explicitly set forth in the Bill of Rights may turn out to be the most important, and most hotly debated, in the context of an interstellar ship: the right to procreate. In the United States, we've debated for decades one portion of that freedom in the form of abortion and the Supreme Court's decision in Roe vs Wade [7]. Population control, in both directions, will be necessary onboard the interstellar ship. A birth boom could overtax the resources of the ship. The failure to maintain a next generation of occupants would similarly jeopardize the voyage because a lack of necessary crew would threaten the ability to maintain the ship in working order. We can, therefore, envision that the biggest freedom that will be forfeited by the occupants of the ship—or at least highly controlled by the government—will be the right to choose to have or not have children. I can envision the issuance of strict birth control measures to stop societal "unwanted" pregnancies, up to and possibly including forced abortions. I can also envision forced pregnancies, where the government orders couples or single women to have a child because the population has begun to dip because of natural aging, accident, or disease among the ship's contingent. One can shudder at the dystopian eugenics discussion that inevitably will occur: "We think that X and Y should have a child because of the strong DNA in the offspring that will result." Arranged marriages, which have a significant role in several modern cultures, including many traditional African societies, royal families, some Muslims, the Amish and other Anabaptist groups, and members of the Unification Church, may play a significant role in providing generational population control. If that becomes part of the ship's culture, it will bring into question and/or conflict other cultural norms and potentially some of the religious beliefs of inhabitants.

The second freedom/right not enumerated in the Bill of Rights is healthcare. The debate we have on Earth about whether universal healthcare is a right or a privilege will come into sharp focus on the interstellar ship. Many US courts have held that a human has the right to extraordinary efforts to maintain life, even overruling the wishes of the family in some instances [8]. The occupants of the interstellar ship may experience the reverse—the much-hyped "death panels" some predicted after passage of the Affordable Care Act in the United States [9]. On an interstellar ship, that debate will be very real. How precious is a single life when compared to the resources necessary to keep it alive? Will the ship's medical crew be expected to do everything in their power to keep a sick occupant alive, even if to do so is highly costly in terms of finite resources (medical supplies, blood, etc.)? Will euthanasia become an accepted, or at least, acceptable alternative to maintaining an individual's life? Will there be discussions of senicide—the idea that the elderly should be euthanized once their ability to contribute to the ship's complement is approaching zero? While tales of putting the elderly out on ice floes to die exist in many cultures, it is unclear whether such practices were prevalent in ancient times. But the ability of the elderly to contribute to the overall welfare of the ship's contingent will undoubtedly rekindle this discussion. It may even become a custom to return one's elements to the collective chemical mix onboard a starship. The very freedom to age will have to be debated by the inhabitants. With that backdrop, I will turn to the types of governance models that might be applicable to an interstellar ship.

3 Basic governance models

There are several historical governance models that could provide the foundations for an interstellar ship government. Each is described in Table 1 within the context of our three main output goals, with a score of 5 representing the highest coefficient and 1 being the lowest coefficient. Each will be discussed in more detail below.

3.1 Military ships

This model is an excellent analog to the extent that the physical environment of a ship most closely resembles an interstellar vessel. Indeed, to date, virtually all space missions have adopted something akin to a Navy ship model. There is a "commander" of each ISS crew and US law and policy gave Space Shuttle commanders absolute authority over their missions. 14C.F.R. § 1214.702 [10] states, for example:

Table 1 Basic governance models.

Name	Description/example	Safety	Efficiency	Freedom
Military ships	Modeled after large navy ships	5	5	1
Antarctic stations	Modeled after existing research institutes in Antarctica	4	4	2
Military base	A self-contained base with adjacent civilian communities	4	4	3
Company town	Mining towns of the 19th century	2	5	2
Private compacts	Jamestown and Plymouth examples	2	2	4
Democracies	Western democratic systems, United States	3	2	5
The commune	Small breakaway settlements	1	1	3

During all flight phases, the NASA Commander shall have the absolute authority to take whatever action is in his/her discretion necessary to: 1) Enhance order and discipline; 2) Provide for the safety and well-being of all personnel on board; and 3) Provide for the protection of the spacecraft and payloads. The NASA Commander shall have authority, throughout the mission, to use any reasonable and necessary means, including the use of physical force, to achieve this end. The authority of the NASA Commander extends to any and all personnel on board the spacecraft including Federal officers and employees and all other persons whether or not they are U.S. nationals.

This form of governance is a top-down authoritarian form of government in which efficiency, order, and crew safety are paramount and individual freedoms take a far back seat. From that standpoint, it is a highly effective form of government and may be the best model in this study that can achieve the requisite level of safety and efficiency to get the ship to its intended destination. There are three major differences between a navy vessel and the interstellar spacecraft, however, which make this model highly problematic:

(1) Everyone onboard the navy vessel is a military member and part of the crew. It is highly unlikely that the interstellar spacecraft will contain only military personnel, trained to respond favorably to the chain of command and military discipline that is vital to the functioning of the navy ship. It is not even clear that all of the inhabitants of the spacecraft would be crew, or whether the occupants might be a mix of crew and passengers who are onboard the ship merely to travel to the ultimate destination planet, although it is highly likely that any "passengers" will have a role in the ship upkeep and basic maintenance, and not just be part of a multigenerational cruise.

(2) Naval deployments nominally last months and rarely, if ever, go beyond a year in duration before the crew goes back to their home port to be returned to a more normal life. However, for the occupants of the interstellar spacecraft, their very world is inside that spacecraft, and in the scenario posited here, it is only their progeny that will ever see the final destination planet. There is no returning "home" to see loved ones and eventually leaving military service to go back to the rest of the world.

(3) Finally, and harkening back to the prior difference, there's never going to be a shore leave, where sailors can blow off steam and exercise a degree of freedom that isn't allowed onboard a navy vessel. Freedoms that were lost while onboard the navy ship can be recaptured or rekindled on those much-needed breaks, as sailors abide with local laws and customs.

Conclusion: It is hard to imagine that a governance model built strictly on a military naval vessel system can survive multiple generations. Occupants simply will not be willing to sign up for a government that so limits individual liberties. Even if occupants might agree at the outset to subject themselves to such a regime, over time, the human urges for freedom will manifest themselves somehow. We saw that with the so-called "Skylab Mutiny," [11] where the crew of Skylab 4 simply turned off their radios and took a day off in protest against the micromanagement of their time onboard the spacecraft; and that was a mission of merely 84 days.

Captain's Log, Day 1215 of the journey of the interstellar ship Nellie Bly. Over three years into our voyage, the ship continues to perform admirably. Our 3D printers are kept busy cranking out replacement pieces for key components that are beginning to show signs of wear. I wish it were just as easy to replace the passengers and crew on the ship, who are showing increasing signs of psychological

> strain. We've got to find a better balance between work and leisure and must find a way to let people 'blow off steam' in a more constructive way. I find myself having to judge dozens of cases of fights and assaults, often between passengers and crew, the former not appreciating that the crew must be rewarded for their Herculean efforts to keep the ship going.

3.2 Antarctic stations

We study this model because we've had a near-constant series of scientific research stations in Antarctica since the 1950s, and military installations are prohibited by treaty. There are now close to 70 bases on the continent, either seasonal or permanent, which are mostly set up by individual countries, each bringing their own governance models to the bases. As such, there is no single "Antarctic governance model." Each base must adhere to the Antarctica Treaty, and the Convention for the Conservation of Antarctic Marine Living Resources. Beyond that, the internal workings of the individual bases are decided by the countries that establish the base.

The United States bases McMurdo, Amundsen-Scott South Pole, and Palmer are subject to US law, including its criminal code. The bases are managed by the National Science Foundation, but with input from the Department of Defense, Department of Homeland Security, Department of the Interior, and the Department of State, which makes governance complicated.

Being at an Antarctic base [12] is certainly no cake walk. Those who make it there are expected to work and work hard:

> In Antarctica, the work week is 54 hours, nine hours per day, Monday through Saturday. Aboard the research vessels, the work week is 84 hours, 12 hours per day, Monday through Sunday. At times, everyone may be expected to work more hours, assist others in the performance of their duties, and/or assume community-related job responsibilities such as washing dishes or cleaning the bathrooms. Due to the challenges of working in Antarctica, no guarantee can be made regarding the duties, location or duration of work. The objective is to support science, maintain the station, and ensure the well-being of all station personnel.

The lack of effective local governance over even minor criminal matters was highlighted by a fight that occurred at the McMurdo base in 1996 [13]. What was a minor altercation between two cooks required a visit by an FBI agent who was flown in to take the alleged perpetrator into custody and remove him from the base, where he had been held under house arrest after the incident. "Parachuting in" law enforcement on an interstellar ship obviously is not a workable solution.

What makes an Antarctic Base slightly different from the military vessel model is not only the overall mission (military versus scientific), but also the fundamental difference in the cultural hierarchy. Thus, we can learn much from these dynamics. As Ragosta and Reynolds [14] put it:

> A winter base in Antarctica is a unique world, where the cook often has greater prestige than the officer-in-charge and the radio operator can have more influence than an established scientist. The traditional hierarchical structure of the military, and of government as whole, breaks down. This was a controversial and embarrassing realization for the Navy. Flexible authority and sharing of tasks among everyone are vital. This can run against the grain of highly specialized scientists and career military officers.

Conclusion: Although seemingly independent, Antarctic bases are closely tied back to their establishing countries both in terms of constant resupply, but also in the forms of government they use. Such close ties are not possible for an interstellar ship. The fact that external law enforcement had to be flown in to extract someone after a simple assault shows the limits of the multiple overlapping jurisdictions of the various US bureaucracies that oversee US stations there, and clearly is not feasible for an interstellar ship.

Captain's Log, Day 4257. I've had to order the cantina closed again. Another fight broke out, Jupiter only knows what caused this fracas. I did find out that someone has been running a clandestine still and has been spiking the 2% beer we brew. I've tasked our recreation committee with coming up with better outlets for our energies. Past attempts at sports leagues have failed, with many games devolving into heated arguments or worse.

3.3 Military base

The military base model differs significantly from both the military vessel and Antarctic station in significant respects. Except for the most secure military bases, civilians are often present in large numbers on a military base, sometimes outnumbering members of the military. With on-base housing for spouses and families, the military base has more of the hallmarks of towns than either military vessels or Antarctic stations, whose populations are less heterogeneous, and, thus, a military base is more analogous to the population aboard an interstellar ship. There are certainly greater individual freedoms to be found on a military base than in the prior two examples, while the military mission and other aspects of life on base are still conducted under a chain of command. The overall area encompassing the base, depending on location, history, and other factors, may also house "military towns," municipalities that grew up around a military base. Especially in highly rural areas, the symbiosis between the actual base and military town is essential, which is why military base closures are so contentious because of the damage the closure has on the surrounding community. Minot, North Dakota is a prime example of such a military town, as Minot Air Force Base employs more people than the local government and 15 top businesses combined.

The main difference between the military base-plus-town model and the interstellar ship, however, is the clear demarcation between the military and civilian areas in the former. Everyone who has ever served on a remote base knows that the rules inside the gate and outside the gate differ, and rarely is there any confusion as to which or whose rules apply. These demarcations will be much more difficult onboard an interstellar ship.

Conclusion: Taken as a whole, the base-plus-town model may prove to be an attractive overall approach to governing an interstellar ship. The operations of the ship itself may be governed by a military-style model, while other aspects of community life, including a private-sector economy, could be governed by what we Westerners have come to know as democracy. The challenge, as noted above, is defining the jurisdictional boundaries of two very different governance models and making clear where, and when, each set of rules applies.

Joint Statement of the Captain and Governor, Day 2367 of the journey of the interstellar ship Nellie Bly. "We are instituting new measures to remind citizens of their duties to respect the laws onboard this ship. Henceforth, we will be changing the lighting in all military sectors to a blue tint. Citizens and military personnel are reminded that in these sectors, military law will be enforced, whether or not you are a member of the military crew, and any violations will be enforced under the military code of conduct and not under the civil laws as they have been voted upon during the voyage."

3.4 Company town

During the 19th century in America, for example, in remote locations such as railroad construction sites, lumber camps, or mining operations, the nongovernmental entity engaged in the predominant business sometimes developed a company town, where it owned all the buildings and businesses. The history of the company town is mixed. In some situations, company towns developed out of a paternalistic effort to create a utopian worker's village. Churches, schools, libraries, and other amenities were constructed to foster healthy communities and productive workers. Saloons or other places or services believed to be negative influences were limited, heavily regulated, or prohibited outright. In other cases, the company's motivations were less ideal. The remoteness prevented workers from leaving for other jobs or to buy from other, independent merchants. Without external competition, housing costs and groceries in company towns could become exorbitant, and the workers built up large debts that they were required to pay off before leaving, creating the functional equivalent of slavery. As a corollary, my science fiction novel *Atlantis Ascending* uses the backdrop of a company town building the first large-scale space solar power station.

Indeed, the first national strike in US history was catalyzed by the poor living conditions in the Pullman company town in 1894. When wages were cut but housing rents in company-owned housing remained at levels above the surrounding areas, workers began a strike that ultimately involved over 150,000 workers in 27 states. President Grover Cleveland ordered federal troops to crush the strike and dozens were killed in violent clashes.

The company town has all the efficiency of the military vessel, but with its main goal of maximizing the economic output of its citizens, safety is often compromised. As the Pullman strike of 1894 also demonstrates, the efficiency goal often results in the loss of fundamental rights and freedoms of workers.

This form of government warrants analysis because the company town model may be the default regime in the event that the interstellar ship is built by a private company or a consortium of companies. Today, we are witnessing what many see as the rise of new company towns when giant corporations like Apple, Amazon, Microsoft, and others build "campuses" that perform many of the functions of a company town. Having successfully built such enclaves, why wouldn't a rich entrepreneur decide to build an interstellar ship and govern it as a company town, thinking that she can make it more of a Shaker utopia than a Pullman disaster? For this reason, we must consider, and study, the history of the company town.

Conclusion: The efficiencies inherent when a company town is built can't be ignored, and the fact that one organization is building the "campus" that becomes the interstellar ship lends itself to considering this governance model. The shortfall of this governance regime, however, is that there is no true motivation on the part of the creator to maximize personal liberty, freedom, or even safety, which are often sacrificed for economic output. The inhabitants of the interstellar ship would, like occupants of a company town, be at the mercy of the government to determine whether the environment will be Utopia, or slavery.

Chief Commissar's Log, Day 10,923. Had to call in the goons again to quash another strike, this time in the janitorial section. The ship must be kept clean, and these people simply don't want to work. It's especially annoying that the younger people, those born on this ship, refuse to accept the positions the Council assigns to them. Not everyone can be the next Commissar, or foreman, or other position requiring special expertise. Rationing food seems to be the only way to keep people in line, but how far can that go? We're fortunate none of the hunger strikes have actually resulted in any deaths.

3.5 Private compacts

When we think of great exploration and migration, we often look at private expeditions. How did they govern themselves? What type of social contracts did they enter into? The goal of this section is not to discuss the philosophy of social contracts in the abstract as exemplified by Hobbes [15], Locke [16], or Rousseau [17], but rather to discuss whether, from a practical perspective, a group of individuals boarding a vessel where they would live the remainder of their lives could collectively come up with a social contract that would establish their rights and responsibilities. Certainly, the approaches of Hobbes, Locke, and Rousseau would be debated and applied, but only if, ab initio, we conclude that a form of government could be created outside of the Earthbound governments that existed at the time the ship is built.

Although now rebutted as fictitious, it is said that Ernest Shackleton placed the following ad for his 1907 expedition to Antarctica: "Men wanted for hazardous journey. Low wages, bitter cold, long hours of complete darkness. Safe return doubtful. Honour and recognition in event of success." This, in a way, is a compact—a private agreement between individuals. What Shackleton was saying to his prospective crew was, essentially, that he was in charge, sign on at high risk, but there's a possibility of great reward if the expedition was successful. His plans were well covered by the press, even if that advertisement never ran.

More familiar to most people are the two famous early expeditions to establish colonies in the Americas, Jamestown (1607) and Plymouth (1620). The Jamestown colony famously failed, while the Plymouth colony survived, still with many deaths, and brought us the tradition of Thanksgiving. Many factors played into the different fates of the colonies. Jamestown was effectively a company town because it was owned by the England-controlled Virginia Company. Many historical reports cite the fact that many of the Jamestown colonists were duped into believing that there were riches to be made early on, and many of the mostly single males refused to work to build the colony's defenses, instead searching for gold and other precious metals that they had been promised were there for the picking. There is also a sense that the Jamestown leaders thought that they would be able to enslave the native population to do the hard work for them. Acemoglu and Robinson put it this way:

> It took the Virginia Company twelve years to learn its first lesson that what had worked for the Spanish in Mexico and in Central and South America would not work in the north. The rest of the seventeenth century saw a long series of struggles over the second lesson: that the only option for an economically viable colony was to create institutions that gave the colonists incentives to invest and to work hard.

The Plymouth colony was populated partly by families of religious dissenters. Only 41 of the 102 passengers were true Pilgrims. They wanted to flee England and Europe in general in search of a place to practice their religion. Because the *Mayflower* landed outside of the jurisdiction of the Virginia colony (where they were supposed to land), the passengers decided that they no longer were required to abide by the contract they signed with the Virginia Company, and instead entered into what we now know as the Mayflower Compact.

The Mayflower Compact is a short document, but it contains core declarations of rights and responsibilities that governed the colony:

> [C]ovenant and combine ourselves together into a civil Body Politick, for our better Ordering and Preservation, and Furtherance of the Ends aforesaid: And by Virtue hereof do enact, constitute, and frame, such just and equal Laws, Ordinances, Acts, Constitutions, and Officers, from time to time, as shall be thought most meet and convenient for the general Good of the Colony; unto which we promise all due Submission and Obedience.

Mayflower colonists certainly knew about the failure of the Jamestown settlement, and it was their declaration of a "civil Body Politick" that acted as a pledge that they would act for betterment of the colony, including agreeing to obey the subsequent laws enacted. While the Plymouth colony evidenced an early form of democratic self-governance, it should be noted that the Pilgrims nonetheless granted significant power to Myles Standish, an English military officer hired to run the defense of the colony, which included the design and construction of a large wall surrounding the settlement, which must have consumed an extensive amount of the colonists' time.

Is such a new crafting of a government through private compact, completely independent of all other governments, even possible for an interstellar ship? I ask this because attempts to create new governments on Earth, away from existing countries, have mostly failed. From Sealand (a former oil drilling tower in the North Sea, which declared itself a sovereign nation in 1967) to the Seasteading Institute (a movement created by a number of Silicon Valley entrepreneurs), attempts to establish new nations apart from the existing countries on Earth have been mostly unsuccessful, for one critical reason: existing countries won't willingly give up the ability to tax an individual just because that person wants to move away unless there is a reciprocal taxing agreement between the country of egress and the country of ingress. There is a second big hurdle to just jettisoning traditional notions of government and writing one, cutting from whole cloth. Article 73 of the United Nations Charter speaks to the issue of the administration of "territories whose peoples have not yet attained a full measure of self-government" and assumes that any new nation would come into the fold of the UN treaty regime. Thus, while there can be new governments and even new countries established, the assumption is that such new nations would not deviate from established international norms and laws. Would the United Nations, as a body, even recognize any government on the interstellar ship if that government wasn't specifically sanctioned by the UN? Taiwan, for example, is not recognized as a separate country by the UN, based on China's position on the Security Council. Of course, what would happen if the interstellar ship simply thumbed its nose at the UN? Would the UN try to stop the ship from leaving orbit because it had adopted a private compact viewed as being at odds with the UN Charter? Would we see a revolt ala Heinlein's *The Moon is a Harsh Mistress* [18] or Allen Steele's *Lunar Descent* [19]?

Conclusion: It is convenient to think that the interstellar ship's inhabitants would be free to enter into a private compact to govern the ship. The reality might be far different. Whatever country or countries built the ship would certainly want a say in the ship's governance structure, and the UN itself might want certain assurances before "recognizing" the ship's government.

> Captain's Log, Day 12,228. The people have agreed to the 22nd Charter rewrite, and thus we've dodged yet another mutiny. The things these people want to argue about! If they spent half as much time actually doing their work than arguing about how to allocate resources, there'd be so much more to go around for everyone! The latest division between communal crops and personal crops appears to appease the dissenters, with a larger percentage to be kept by individuals, but what happens if any of those individuals have a crop failure? You'll bet they'll be back in my office complaining that they need a larger share of the communal percentage.

3.6 Democracies

Many in the West might simply assume that the government of the interstellar ship would be some form of democracy, since that governmental model has survived, at least somewhat successfully, for nearly 250 years. The Cambridge Online Dictionary [20] defines "democracy" as "the belief in freedom and equality between people, or a system of government based on this belief, in which power is either held by elected representatives or directly by the people themselves." A corollary to democracy is "republic," which the Cambridge Online Dictionary defines as "a country without a king or queen, usually governed by elected representatives of the people and a president."

Of course, democracy and republic are often in the eye of the beholder. Historically, there are many countries who have adopted "Democratic" or "Republic" as part of their names, yet no one would rationally argue that their forms of government met the definition of democracy. The Democratic People's Republic of Korea, German Democratic Republic (East Germany), and the People's Republic of China all fall within this category of nondemocratic democracies—effectively, democracies in name only. While these nations may be outliers in terms of lying about having democratic governments or being republics, the very notion of what it means to be in a democracy is hotly debated.

For purposes of this chapter, in order to qualify as a democracy, an interstellar ship's government must evidence three characteristics:

(1) The government must recognize certain rights and liberties of the people that cannot be overridden by government fiat;
(2) The government must be elected by the people on a regular basis through free and open elections; and
(3) There is a mechanism by which the people have a means to "petition" the government with their grievances, and have some neutral arbiter (e.g., a court system) decide these matters.

Beyond that, the type of democracy (number of branches, types of legislative bodies, etc.) is far less important.

It is the agreement as to the rights and liberties of people where the discussion of democratic rule will be most hotly debated if the interstellar ship wishes to pursue democracy as its form of government. Earthbound countries differ widely in terms of what those fundamental rights are, and how they can be exercised. I discussed some of these fundamental rights above. Of greatest importance in this discussion may well be the notion of what Americans call "the pursuit of happiness." The term was first coined by philosopher John Locke [21] in his book *An Essay Concerning Human Understanding*. The term took on a more concrete meaning when Thomas Jefferson used it as part of his three inalienable rights: "Life, liberty and the pursuit of happiness" in the Declaration of Independence [22]. There is much debate as to what exactly Jefferson meant, whether "pursuit of happiness" was a synonym for wealth or property, or whether it encompassed much broader concepts of individual, as opposed to societal, wellbeing. George Mason [23], who wrote the Virginia Declaration of Rights at virtually the same time as Jefferson (a Virginian himself) was drafting the Declaration, said this about inalienable rights:

> That all men are by nature equally free and independent and have certain inherent rights, of which, when they enter into a state of society, they cannot, by any compact, deprive or divest their posterity; namely, the enjoyment of life and liberty, *with the means of acquiring and possessing property, and pursuing and obtaining happiness and safety* [emphasis added].

While many countries and other institutions have adopted tripartite mottos for inalienable rights, the UN Declaration of Human Rights, Article 3 [24] uses a very different term: "Everyone has the right to life, liberty and the security of person." Totalitarian and other nondemocratic regimes often use "security of person" as an excuse to limit other liberties. More fundamentally, the term "pursuit" is a verb evidencing a very individual process by which a person quests for, and potentially receives, "happiness." By comparison, "security" is a noun, defined by the Cambridge Online Dictionary as "protection of a person, building, organization, or country against threats such as crime or attacks by foreign countries." It is not a right an individual has, but, rather, a responsibility of a state to protect its citizens.

In essence then, the debate over whether democracy should spread out into the cosmos comes down to whether one believes that the individual quest for happiness should be paramount, or whether the security of person is more important. There is a fundamental tension between the two. Should people be free to make decisions that ultimately may kill them (e.g., by deciding to try and climb Mount Everest) because it makes them happy, or should the government step in to stop such decision making, even if those decision impact only the single individual?, This goes to the heart of the intersection of two of the three assumed goals of governance: safety, and freedom. Underlying this dichotomy is the question of whether an individual's happiness is to be measured, at least in part, based on the notion that happiness includes a component of wealth and property, and whether that is a fundamental right at all. In some sense then, the issue of the pros and cons of democracy has become a referendum on what we know as capitalism.

How this is playing out in the debate over governments in outer space is fascinating. Take the Arab Republic of Egypt, for example. Freedom House [25] rated Egypt as "not free" and gave Egypt a "Political Rights" score of 6 out of 40 and a "Civil Liberties" score of 12 out of 60, with a total score of 18 out of 100. Yet the Egyptian delegation presented a proposal, on June 2, 2021, to the UN's Committee on the Peaceful Uses of Outer Space (COPUOS) calling for "a new era for human civilization." The document argues that the entire underpinning of what most would call Western Democracy (and the "pursuit of happiness") cannot be allowed to migrate to outer space:

> The result of [the development of civilization on Earth] has shaped our collective governing culture to become a culture of conflict, resulting in many wars, struggles, and catastrophes that have plagued

> humanity. It becomes evident [...] that our current human civilization with all its negative attributes cannot be the basis for a new human civilization in outer space. Therefore, we as humans are required to lay down the building blocks of a new human civilization in outer space, that takes in consideration all the above-mentioned resources and potentials, as well as the possible dangers and threats. We will call this the second human civilization or the human space civilization. Having also witnessed all the negative and destructive attributes of our current human civilization, it becomes our duty to not transfer these negative attributes to any new human civilization in space.

Similarly, even in the United States, the thought of exporting our democratic principles to outer space, including capitalism, is under challenge. In October 2020, NASA's Planetary Science and Astrobiology Decadal Survey [26] committee received a manifesto from its Equity, Diversity and Inclusion Working Group (EDIWG) entitled "Ethical Exploration and the Role of Planetary Protection in Disrupting Colonial Practices." Its core thesis is that the exploration and development of Western civilization, and the establishment of democratic norms, is evil.

> Violent colonial practices and structures—genocide, land appropriation, resource extraction, environmental devastation, and more—have governed exploration of Earth, and if not actively dismantled, will define the methodologies and mindset we carry forward into space exploration. It is critical that ethics and anticolonial practices are a central consideration of planetary protection. We must actively work to prevent capitalist extraction on other worlds, respect and preserve their environmental systems, and acknowledge the sovereignty and interconnectivity of all life.

In theory, one can divorce the political systems of democracy from the economic system of capitalism. It is theoretically possible to have a democracy without relying on capitalism to fund it. In practice, however, few, if any, democracies have survived long without an undergirding economic system that is reliant on property rights and which rewards entrepreneurial behavior and innovation. The question, to be taken up below, is whether such a system can work for the interstellar ship, since few, if any, countries can claim to be true democracies but boast an economy free of all remnants of capitalism.

Conclusion: Winston Churchill's [27] take on democracy is often misquoted. What he actually said in 1947 was this:

> [I]t has been said that democracy is the worst form of Government except all those other forms that have been tried from time to time; but there is the broad feeling in our country that the people should rule, and that public opinion expressed by all constitutional means, should shape, guide, and control the actions of Ministers who are their servants and not their masters.

With all its faults, the notion that the people should have a say in the decisions of the government is a compelling model that may ultimately prevail.

> Prime Minister's report, Day 17,345. The Ship Council has informed me of their intention to put to the people a Vote of No Confidence of my administration. This will constitute the 35th such vote over the past three years, roughly one such election every Earth month. With 17 official parties now

represented on this ship, it is becoming increasingly difficult to piece together a government that can survive constant calls for votes of no confidence. I fear we are nearing a time where our democracy cannot survive and may be replaced, as some have called for, by a far more dangerous authoritarian or military regime.

3.7 The commune

The last general model I address is that of the true commune, a place where everyone lives and works together, but each individual lacks any ownership of property. The Cambridge Online Dictionary defines a "commune" as: "a group of families or single people who live and work together sharing possessions and responsibilities." I distinguish commune from "communism," which the same dictionary defines as "the belief in a society without different social classes in which the methods of production are owned and controlled by all its members and everyone works as much as they can and receives what they need." I also distinguish both generic terms from 20th century communism as a form of government, since no such government came close to achieving this ideal.

At first blush, a commune would appear to be a viable model for the interstellar ship, since it is highly likely that there will be no real property rights (e.g., inhabitants likely will not own their lodgings), and everyone will work together for the common good. Communes, thus, are more of an economic system than a government, since who actually makes decisions in a commune can be as diverse as in a free market society. The commune needs to be studied mainly because it is in this form of social grouping that significant free rider problems historically appear, some with devastating impacts on the viability of the community. I highlighted the problems with the Jamestown expedition above. But even the Plymouth colony, with its Mayflower Compact, was not immune to the free rider; it appears that, at first, the colony operated as a commune, where all crops raised were shared equally, regardless of who did the work. Tom Bethell of the Hoover Institute describes the double problem faced by the Plymouth Colony of heavy taxation by their investors, as well as the free rider problem:

> By the spring of 1623, the population of Plymouth can have been no larger than 150. But the colony was still barely able to feed itself, and little cargo was returning for the investors in England. On one occasion newcomers found that there was no bread at all, only fish or a piece of lobster and water. "So they began to think how they might raise as much corn as they could, and obtain a better crop than they had done, that they might not still thus languish in misery," [William] Bradford wrote in his key passage on property [in *Of Plymouth Plantation*].

> Having tried what Bradford called the "common course and condition"—the communal stewardship of the land demanded of them by their investors—Bradford reports that the community was afflicted by an unwillingness to work, by confusion and discontent, by a loss of mutual respect, and by a prevailing sense of slavery and injustice. And this among "godly and sober men." In short, the experiment was a failure that was endangering the health of the colony.

> On the one hand, it is true that the colonists did feel "exploited" by the investors because they were eventually expected to surrender to them an undue portion of the wealth they were trying to create. It is as though they felt that they were being "taxed" too highly by their investors—at a 50 percent rate, in fact.

> But there was another problem, separate from the "tax" burden. Bradford's comments make it clear that common ownership demoralized the community far more than the tax. It was not Pilgrims laboring for investors that caused so much distress but Pilgrims laboring for other Pilgrims. Common property gave rise to internecine conflicts that were much more serious than the transatlantic ones. The industrious (in Plymouth) were forced to subsidize the slackers (in Plymouth). The strong "had no more in division of victuals and clothes" than the weak. The older men felt it disrespectful to be "equalized in labours" with the younger men.

> This suggests that a form of communism was practiced at Plymouth in 1621 and 1622. No doubt this equalization of tasks was thought (at first) the only fair way to solve the problem of who should do what work in a community where there was to be no individual property: If everyone were to end up with an equal share of the property at the end of seven years, everyone should presumably do the same work throughout those seven years. The problem that inevitably arose was the formidable one of policing this division of labor: How to deal with those who did not pull their weight?

> The Pilgrims had encountered the free-rider problem. Under the arrangement of communal property one might reasonably suspect that any additional effort might merely substitute for the lack of industry of others. And these "others" might well be able-bodied, too, but content to take advantage of the communal ownership by contributing less than their fair share. As we shall see, it is difficult to solve this problem without dividing property into individual or family-sized units. And this was the course of action that William Bradford wisely took.

A similar fate has befallen both small communes ("intentional communities") as well as the modern communist state. Collective farms in China are a prime example. Richard Baum, PhD, explains this fundamental problem in "Collective Farming in Communist China: The Free Rider Problem."

> With as many as 500 or 600 able-bodied peasants in a single collective, personal responsibility was very difficult to assess and monitor let alone to assign to individuals or families. Hence, the incentive to work diligently was correspondingly diminished for each individual. The strength of that disincentive was being directly proportional to the size of the collective: The bigger the collective unit, the weaker the incentive to work hard.

> Under this system, people tended to just go through the motions in their fieldwork. Content to free-ride on the efforts of the others, they paid scant attention to the quality or efficiency of the work performed.

Largely as a result of these two factors—resistance to collectivization by rich peasants and a widespread free-rider problem—the collectivization of the mid-1950s failed to yield the anticipated farm output. Agricultural production limped, rather than leapt forward, with growth rates barely reaching 2 or 3 percent annually, hardly enough to keep pace with China's rapidly burgeoning population.

Despite clear evidence that free-riders were undermining the effectiveness of China's collective farms, Mao steadfastly denied that there was any such problem. In his belief system, the worker-peasant masses could be educated to understand the benefits of working for the good of the collective, rather than for themselves. "Serve the people," was Mao's oft-repeated mantra.

Is this the fate of an interstellar ship, were it to adopt true communism? One can argue that there is sufficient motivation to serve the people to survive in the harsh environment of outer space. Others might argue that the type of person deciding to go on such a voyage would have to be highly self-motivated such that they would not succumb to the enticement of free-riderism. That might, and I stress "might", be true in the macro sense, but will it be true in every case? What do you do with that person who simply doesn't want to work or, as was seen in China's collective farms, just goes through the motions? Does the commune leadership simply cut off future food and other resources from someone who doesn't pull his/her weight? Fundamentally, will large numbers of people willingly set sail on a ship betting that somehow they've changed human nature sufficiently to rid the group of the free rider problem?

Conclusion: The commune model is both attractive and dangerous. While free rider problems exist in any government system, in a commune, the sharing of all outputs makes the problem more evident, and if there is no "upside" incentive to work harder; even the most "sober" individual would eventually think twice about putting in extra effort when, at the end of the day, she is still getting the exact same rations as someone who puts forth minimum effort. Can this rush to mediocrity be combatted?

3.8 Are there models we've missed?

Are there other governance models that we've missed with this analysis? Certainly, there could be hybrids of the models above that would look different from its progenitor systems. Running through all of those hybrids is beyond the scope of this chapter. It has also been suggested that the interstellar ship could be governed via a technocracy, defined as "the government or control of society or industry by an elite of technical experts." Those who suggest a technocracy presumably believe that engineers and technicians would be in a better position to govern than Earth governments, which tend to be led by lawyers and political scientists, steeped in law and politics. The problem with this approach, however, is twofold. First, it focuses on the resumes of individuals in power, rather than on institutions established to govern. Second, the track record of governments run by engineers and technocrats is no better than governments run by lawyers and politicians, and some could argue far worse. Soviet Leader Leonid Brezhnev held degrees in land management (agriculture) and metallurgy, yet oversaw massive economic stagnation, and ultimately the loss of the Cold War to the United States. Current Chinese President Xi Jinping has a degree in chemical engineering (certainly a technology critical to an interstellar ship), and the Chinese Communist Party has a history of advancing engineers up through its ranks. The current First Secretary of the Communist Party of Cuba, Miguel Diaz-Canel, has a degree in electrical engineering. Are these the types of technocrats that one would envision governing an interstellar ship?

By definition, the government of the interstellar ship will be populated by engineers, technicians, and other technocrats, since the vast majority of the ship's complement will come from highly technical backgrounds. They will most likely be "elites" as well; one can hardly imagine the crew and overall population of an interstellar ship to be of below-average intelligence or achievement. Saying that the government is a "technocracy," therefore, doesn't define the system so much as highlight the individuals who will serve in whatever governance model is chosen.

Second, can technology itself be harnessed as a form of government? Can we envision a form of artificial intelligence (AI) that could emerge as the equivalent of Plato's "philosopher king," putting an end to all the troubles of the state? Certainly, the systems onboard the ship will be complex and will govern themselves in most instances. A properly engineered ship will not allow human error to allow all its oxygen to be vented out into space, for example. AI systems can govern food production and many other aspects of daily life on the ship, without need of human intervention. But can AI systems form the basis of an actual government, and would humans agree to be governed by a set of algorithms, disassociated from feelings, instincts, morals, and all the other traits that make us human? I believe the likelihood is extremely small that 500 people would willingly sign over their fates to a machine without at least one human in the loop.

4 The role of innovation and its rewards

At the outset of the chapter, I said that there should be three goals that drive the governance model for an interstellar ship: safety, efficiency, and freedom. As we study different governance models, especially coming from the perspective of Western democracies, we often forget a corollary to efficiency: innovation. We take it for granted. We forget that the American founding fathers found innovation to be so important that it appears in Article I, Clause 8 [28] of the US Constitution: "[Congress shall have the power] To promote the progress of science and useful arts, by securing for limited times to authors and inventors the exclusive right to their respective writings and discoveries." It can be argued that placing a reward for innovation in the Constitution merely codified centuries of English common law, and followed suit from other European country patent laws, such as the Venetian Patent Statute of 1474, which appears to be the first codification of the notion that an inventor should be rewarded for his/her efforts with an exclusive right to use the invention. Such arguments miss the point that, at least to the American Founding Fathers, being rewarded for innovation is a fundamental right, and part and parcel of the notion of "the pursuit of happiness."

The "Patent Clause," as it is known, was so uncontroversial that it is mentioned only once in the Federalist Papers. In Federalist Paper No. 43, James Madison [29] writes: "The utility of this power will scarcely be questioned. The copyright of authors has been solemnly adjudged, in Great Britain, to be a right of common law. The right to useful inventions seems with equal reason to belong to the inventors."

President Washington [30], in the first State of the Union address (which consisted of a mere 1096 words), said this about the need for Congress to quickly pass a comprehensive patent statute:

> The advancement of agriculture, commerce, and manufactures by all proper means will not, I trust, need recommendation; but I can not forbear intimating to you the expediency of giving effectual encouragement as well to the introduction of new and useful inventions from abroad as to the exertions of skill and genius in producing them at home, and of facilitating the intercourse between the distant parts of our country by a due attention to the post-office and post-roads.

Congress responded quickly; the Patent Act of 1790 was just the third law passed by the newly formed Congress of the United States.

With all of its natural resources and other qualities, the great success of the United States, especially in the 20th century, can be attributed in no small measure to the intellectual property created here. This includes the nearly 11 million patents issued since 1790 (currently running at over 300,000 new patents issued every year), as well as trademarks, copyright, and trade secrets developed.

I raise the issue of innovation, and specifically rewards for innovation, in this chapter for several reasons. First, there is little doubt that there will be a great need for smart people on the interstellar ship, finding creative solutions to problems as they develop. One can imagine a thousand scenarios similar to what happened with the Apollo 13 mission, where an entire team of engineers on the ground had to solve dozens of problems to bring the astronauts home safely. Such "crisis engineering" alone, however, can't be counted on to fuel progress onboard the interstellar ship on a daily basis. Creating innovative solutions to mundane problems onboard can go a long way to making life better and easier for the inhabitants. But can we count on everyone on board to go the extra mile to innovate if there is no reward for innovation? Will a "good job!" and a pat on the back encourage a young maintenance worker to suggest to a higher-up a way to squeeze a half of a percent more efficiency out of the water filtration system?

If Earthbound society provides any lessons, it is that without some reward, most people will not take on the risks of innovating. Without some profit, and I use that term in its broadest form, many people will not sacrifice for that innovation. Elon Musk has been criticized for saying that "nobody ever changed the world on 40 hours a week." Those critics missed the true point—it takes a lot of very hard work to create new things. And it takes being willing to risk failure. Will the interstellar ship have such innovation incentives built into its government, its culture? How can this be accomplished by the ship government? Is it possible without any private property rights attached to the innovation?

Some may argue that we don't need to worry about this issue, that the nature of the inhabitants of the interstellar ship themselves will be such that innovation will course through their very veins. While that might be the case for the initial inhabitants (I have my doubts), what about future generations? Will the innovation and self-motivation gene replicate across generations to assure that no future generations of inhabitants will become free riders?

Finally, some may ask why I chose to hold this discussion of innovation incentives until near the end of this chapter. Why introduce a new "character" into this story so late? Unfortunately, at least in my mind, to understand the importance of this variable, one first needs to examine the different governance models and understand them (and their various failure modes). Without understanding the nature of the free rider problem, and how it has cropped up again and again in various forms of societal association, one will assume that innovation incentives will always (or nearly always) counteract the fact that humans can be lazy. In Western cultures, progress, in the form of increased technology resulting in ever-increasing standards of living, has exceeded the drag put on society by the free riders. In short, a productive, advancing society can tolerate a certain percentage of free ridership (or other forms of poor relative productivity owing to a variety of factors) so long as the net production remains positive. It is precisely those innovation incentives that have kept us going. A society that does not have a mechanism to reward innovation, on the other hand, is in constant danger of losing momentum, suffering from stagnation, and ultimately, collapse. How an interstellar ship deals with this issue may prove to be the most important governance issue it faces.

5 Ultimate conclusions

Consideration of what laws and form of government should be applied to societies in space is not a new endeavor. In addition to the other sources cited in this chapter, this work draws from space lawyers through many generations. Andrew G. Haley [31] addressed the issue in *Space Law and Government* (1963) (and as the lone space lawyer descended from his original law firm, I proudly keep his personal copy of the book on my desk), as did Myres McDougal, et al. [32] in *Law and Public Order in Space* (1963). Glenn Reynolds and Robert Merges [33] devote an entire chapter to governance in their space law textbook *Outer Space: Problems of Law and Policy* (Second Edition, 1997). All those works are much more theoretical than this chapter; I've tried to adapt many of these earlier works to the specific environment of a multigenerational interstellar ship.

In the end, I don't know that I have any finite conclusions as to governance. Human beings have flaws, and the institutions we create are equally flawed. There is no perfect government any more than there is one true religion (or lack thereof), or one set of cultural beliefs that will always prevail. To think that anyone today, in 2022, can set forth a foolproof governance system for interstellar travel would be engaging in the same exercise of hubris as would the inhabitants of the ship deciding, just before they board the ship, what government will reign through the multiple generations to live during the flight.

My tendencies, probably because I was born into and have lived six decades in the United States, would be toward some form of democracy with an economic system that rewards innovation and hard work and punishes lack of production, to the degree necessary, to keep the free rider problem to a minimum. Having said that, this democracy needs to recognize the inherent conflict between the three goals of safety, efficiency, and freedom. Some aspects of ship operations will likely have to be granted to military-style operations. You can't, after all, have the inhabitants voting every day on a new course, or even voting on what emergency course corrections might be necessary to avoid collisions. On the other end of the spectrum, the people must have enough freedom to continue to agree to live under the government model. Tamp down freedoms too much and the government could face a mutiny.

Robert Fulghum [34], in *All I Really Need to Know I Learned in Kindergarten* (1990), had a wonderful reflection applicable to this analysis:

> Share everything.
> Play fair.
> Don't hit people.
> Put things back where you found them.
> Clean up your own mess.
> Don't take things that aren't yours.
> Say you're sorry when you hurt somebody.
> Wash your hands before you eat.
> Flush.
> Warm cookies and cold milk are good for you.

Alas, were governing a multigenerational interstellar spaceship so easy! I leave you with the thoughts of Supreme Court Justice William Brennan, delivered to the Bicentennial Conference of Judges of the United States Courts of Appeals on October 26, 1988, on why pursuing the study of governance of future space colonies is important:

And the study of space societies may have a big dividend for Earth. As Walter McDougall has noted, the great age of earth exploration stimulated a burst of inquiry into the laws governing nations and led to our modern system of international law. In the same way, inquiry into the rules that should govern societies in space is likely to provide fresh insights into the governance of societies here on Earth, a field in which, to judge by current events, there is certainly room for progress. This is particularly true because many of the most salient characteristics of space societies, such as strong dependence on sophisticated technology, problems with maintaining environmental quality, the need for people to work together smoothly under stress in close quarters, and the dependence of inhabitants on their society for basic necessities such as food, water, air and communications, are in many ways imply exaggerations of characteristics already present (and growing) in Earth societies. By studying the problems of space societies we gain a window into not just their future, but our own.

Chief Counselor's Log, Day 36,875 of the journey of the interstellar ship Nellie Bly: We are safely in orbit at our destination. By the grace of all fates, we have made it. The journey has been arduous, and there are but left 220 colonists across all these years. We have worked hard, have fought hard, have argued hard. But ultimately, like Candide, we've discovered that by the hard work and sweat of our brow, and focusing on our family and friends, we can find happiness. For me, I'm looking forward to a real garden on the planet below and making my garden grow.

References

[1] D. Acemoglu, J. Robinson, Why Nations Fail: The Origins of Power, Prosperity, and Poverty, 2012.

[2] G. Robinson, H. White, Envoys of Mankind: A Declaration of First Principles for the Governance of Space Societies, 1986.

[3] J.E. Dunstan, From flag burnings to bearing arms to states rights: will the bill of rights survive a trip to the Moon? in: Proceedings of the Tenth Princeton/AIAA/Space Studies Institute Conference on Space Manufacturing, 1991.

[4] Texas v. Johnson, 491 U.S. 397, 419 (1989).

[5] U.S. Const. amend. VIII.

[6] U.S. Const. amend. I–X.

[7] Roe v. Wade, 410 U.S. 113, 154 (1973) *overruled by* Dobbs v. Jackson Women's Health Org., 142 S. Ct. 2228, 2242 (2022).

[8] In re Quinlan 355 A.2d 647 (1976) (appointing a new guardian to a hospitalized woman in a vegetative state against her father's wishes).

[9] Patient Protection and Affordable Care Act, Pub. L. No. 111-148, 124 Stat. 119 (2010).

[10] 14 C.F.R. § 1214.702 (2013).

[11] The Real Story of the Skylab 4 "Strike" in Space, NASA, Nov. 16, 2020. https://www.nasa.gov/feature/the-real-story-of-the-skylab-4-strike-in-space.

[12] U.S. Antarctic Program Participant Guide (2016-2018 ed.).

[13] C. Haskins, An Attempted Murder at a Research Station Shows How Crimes are Prosecuted in Antarctica, Vice, Oct. 25, 2018. https://www.vice.com/en/article/xw9bg3/an-attempted-murder-at-a-research-station-shows-how-crimes-are-prosecuted-in-antarctica. (detailing the McMurdo base assault and legal proceedings).

[14] J. Ragosta, G. Reynolds, In Search of Governing Principles, 28 Jurimetrics J. 473, 476, 1988.
[15] T. Hobbes, Leviathan, 1651.
[16] J. Locke, Two Treatises of Government, 1690.
[17] J.-J. Rousseau, Discourse on the Origin of Inequality, 1755.
[18] R.A. Heinlein, The Moon Is a Harsh Mistress, 1966.
[19] A. Steele, Lunar Descent, 1991.
[20] Democracy, Dictionary.Cambridge.Org, https://dictionary.cambridge.org/us/dictionary/english/democracy. (last visited Nov. 29, 2022).
[21] J. Locke, An Essay Concerning Human Understanding, 1690.
[22] Declaration of Independence para. 2 (U.S. 1776).
[23] G. Mason, The Virginia Declaration of Rights, 1776.
[24] G.A. Res. 217 (III) A, A Universal Declaration of Human Rights, art. 3.
[25] Egypt, Freedom House, https://freedomhouse.org/country/egypt. (last visited Nov. 29, 2022).
[26] F. Tavares, Ethical Exploration and the Role of Planetary Protection in Disrupting Colonial Practices: A Submission to the Planetary Science and Astrobiology Decadal Survey 2023–2032, 2020.
[27] L. Garber, Churchill on Democracy: Lessons for Today, Lobe Log, Jan. 17, 2019. https://lobelog.com/churchill-on-democracy-lessons-for-today/. (quoting Winston Churchill in 1946).
[28] U.S. Const. art. I, cl. 8.
[29] The Federalist No. 43 (James Madison).
[30] President George Washington, State of the Union Address (1790).
[31] A.G. Haley, Space Law and Government, 1963.
[32] M.S. McDougal, et al., Law and Public Order in Space, 1963.
[33] G. Reynolds, R.P. Merges, Outer Space: Problems of Law and Policy, second ed., 1997.
[34] R. Fulghum, All I Really Need to Know I Learned in Kindergarten, 1990.

CHAPTER

8 Spare parts and additive manufacturing

Simon C. Patané and Michael P. Snyder
AIAA Space Settlement Technical Committee, Reston, VA, United States

Chapter outline

1 Introduction

A crewed voyage across deep space will require a large, complex spacecraft. When anything on this spacecraft breaks unexpectedly, or needs to be replaced due to nominal wear, there will not be large warehouses full of spare parts on the ship to draw from. There must be a way to make the needed parts on demand, preferably with reclaimed or recycled material. Traditional exploration-class space missions rely entirely on provisions brought along from Earth, and have very little agility for any unplanned scenarios or ability to rely on constant resupplies from Earth. Crewed interstellar voyages

Interstellar Travel. https://doi.org/10.1016/B978-0-323-91280-8.00003-4

require the ability to maintain, repair, recycle, and augment portions of the spacecraft with technologies that do not dramatically increase the mass of the spacecraft. In the current applicable paradigm of bringing along spares, the spacecraft would be so massive that the cost and energy to propel it would be a showstopper.

Manufacturing and recycling technologies, working in concert, are essential for enabling interstellar travel and all other long-duration endeavors in outer space. More importantly, which technologies in these categories are selected should drive the overall design of the vehicle and the operations to be performed to ensure compatibility throughout all phases of the journey. These technologies will also need to maintain themselves or to self-replicate as a sparing strategy. Currently, technologies able to fulfill the logistical requirements imposed on them by the need to accommodate long-term missions in outer space are being developed. Through these latest efforts, alongside hundreds of years of terrestrial industrial development, the ability to realize an interstellar voyage is attainable.

2 Sparing philosophy and mission architecture decisions

2.1 One shoe does not fit

Currently, no machines exists on Earth that is capable of producing any requested item at any desired time. This machine, that makes appearances in popular fiction, is tantamount to a manufacturing center that has the capability of performing every means of production, automatically, with just the user inputting their desired item to be made in order for successful production. At the moment, it is reasonable to assume that the creation of this mythic device, which appears to be what so many are claiming to be striving for (such as Mattershift's effort to make a machine create items out of thin air [1]), is incredibly difficult or impossible to achieve. Entire cities are required to obtain the type of industrial success that this object could facilitate, and even then, raw materials and/or feedstock are usually sourced elsewhere. So, what is the best philosophy for enabling success while limiting the footprint required of your hardware inside, or outside and attached to, the spacecraft? Unfortunately, there is no single solution or single configuration that works for every use case. Solutions are highly dependent on one's mission requirements and the design philosophy of one's vehicle.

2.2 Design philosophy for longevity

In the current space paradigm, the lifespan of a particular technological design rarely exceeds 15 years, and for human-rated spacecraft, it is normally much lower. The International Space Station (ISS), the first module of which launched in 1998, is an outlier amongst human-tended vehicles, but its maintenance sheds light on the deficiency of the current design philosophy. The ISS's longevity has relied on two things: 1. consumables resupplied from Earth, and 2. orbital replacement units for critical systems/subsystems. Compared to this, if a long-duration mission is to be successful, these two facets of ISS logistics have to be substituted with reuse and repair along with a well-informed sparing philosophy. Reuse and repair allow for crewed missions to leave the cradle of low-Earth orbit (reducing the need to use consumables resupplied from Earth) and eliminate the volume and mass required to bring a spare for every substantial subsystem of the spacecraft. As such, reuse and repair are a window into a design space that has never been utilized with any style of mission to date. This design space must leverage the manufacturing technologies and materials available that enable reuse and repair methods while meeting the requirements of mission elements in order for the mission to be successful. Designing spacecraft with reuse, repair, and a sparing philosophy in mind, while simultaneously taking into account the means by which these three aspects are to be performed, is vital to ensure continuity and compatibility across the mission execution and spacecraft production efforts. Adequate power, volume, and mass resources need to be allocated when implementing these methods, which vary by mission type.

2.3 Reuse and repair

Reuse implies a combination of recycling and reconfiguration. Since the concept of recycling is straightforward (breaking down used or broken materials and reconstituting their elements into new forms), this section will focus on reconfiguration and nontraditional recycling. Every mission contains several phases; traditionally designed spacecraft and missions are created in a way to best survive every phase without large changes to the physical configuration. Deployable systems, such as solar arrays and radiators, undergo relatively small physical configuration changes as they transfer from a packed state to a deployed state. However, these systems generally are not optimally shaped to weather either state; rather, their design is a compromise that allows them the greatest chance to survive the various mission phases. Launch and other phases of acceleration are the most strenuous on spacecraft, such that almost all hardware is designed with the equipment's survival as a top priority. Articles built after these mission phases would not have to survive those strenuous phases. Leveraging this opportunity, in space, manufacturing and assembly technologies can build articles when they are needed during the mission, rather than being driven by the requirement to survive launch. Designing missions according to this new paradigm optimizes material usage by tailoring structures and components according to the requirements of the mission phase. Furthermore, materials used in prior mission phases can be reused once they are no longer required. A simple example is support structures required to survive a launch could be repurposed since the loads in that phase are orders of magnitude greater than those that will be encountered during any other portion of a mission. Removing structural support and either directly reusing it to build other elements of the vehicle or recycling the material into feedstock for down-the-line manufacturing purposes reduces the overall mass required for the mission.

Repairing is highly dependent on design considerations that must include the method of repair as well as the scope of predicted repair. These functions feed directly into a sparing strategy as components must be evaluated with regard to the extent to which they are able to be directly repaired, recycled, and replaced, or completely spared without recyclability. Interstellar travel will require minimal spares and a robust repair and recycling strategy. For direct repair, additive or hybrid manufacturing technologies are necessary. This necessity in turn requires that the components, which will need to be repaired, be accessible and must fit, volumetrically, within the configuration constraints of the manufacturing hardware. The design of these repairable parts must take into account the repairing technique and add margin in safety factors when needed, as most repair scenarios cannot achieve 100% of the performance of the original component. Similarly, if a component is recycled to make another copy of itself through a manufacturing process, the materials most likely will not have the same performance as the virgin material originally used to create the part. Reprocessing creates variations, especially with certain classes of materials, due to the changes that the processes invite. Again, through adoption of proper design considerations, these variations can be accounted for, tracked, and mitigated. The ideal component for an interstellar journey is a part that can be recycled, minimally processed, and reformed without performance changes. While it is unlikely that components fully matching this description could be developed, it serves as an essential goal toward which the design team can strive.

2.4 Sparing

A rigorous sparing philosophy is foundational to a successful deep space voyage. While the technologies that would enable a crewed interstellar mission likely far exceed those developed to facilitate humanity's recent endeavors in space, the scale of interstellar mission logistics hits a practicable limit that fundamentally constrains the mission: spaceships embarking on interstellar missions will have no access to spare parts beyond that which travels with the ship from Earth. Technologies such as additive manufacturing help to mitigate this limit and to equip potential interstellar mission architectures with an efficient and versatile alternative to sparing. Certain systems or components may be technically unable to take advantage of manufacturing during the mission, but unlike the current paradigm where spares are used ad nauseam, these should be limited to no fewer than the maximum amount possible to enable the mission to succeed. Using a traditional sparing approach would create a spacecraft several times its functional weight in just spare parts.

The International Space Station (ISS) is the closest current analogy for the type of logistics and sparing required for successful execution of an interstellar mission. Notably evident is the gap in resupply frequency between ISS logistics and sparing plans for an interstellar mission. While interstellar missions are constrained by their initial stores, the ISS, while presenting a daunting logistical challenge, is consistently serviced by resupply missions that are launched from at most only a few hundred kilometers away. Over 3000 kg of spares are flown to the ISS every year to replenish or add to the already over 13,000 kg of spares already onboard [2]. One of the more interesting phenomena involving spares is that often, when spares are required, entire systems will be swapped out instead of replacing the component that failed. For a complex system, these failures are frequently very minor in nature, but because of the way these replacement spares are designed and the way the operations are implemented it is the only way to safely implement a fix to a problem.

Given these disparities, for the interstellar voyager the value of a materially efficient process like additive focused manufacturing technologies, combined with recycling, appears as apparent as the need to change our philosophies on how we design spacecraft and integrate spares in missions. It is also important to keep in mind technologies that cannot only repair or build anything onboard or outside the vehicle, but can additionally repair or build replacement parts for themselves, as there currently exist no manufacturing machines on Earth that do not require maintenance and repair.

3 Technology needs for interstellar and deep space missions

A technology or set of technologies that accompany long-duration missions must be able to replace or repair components and systems on the spacecraft while using the minimum amount of material and recycling all possible items.

What technologies are needed in an interstellar journey? The answer is: it depends on how you design your spacecraft. Let's assume that a spacecraft is designed that is efficient for space travel using only the technology and materials available as of the time of this chapter's composition. Hybrid systems that combine additive manufacturing, scanning, recycling, and traditional manufacturing

Table 1 Manufacturing technologies required for interstellar missions.

Manufacturing category	Technologies
Additive manufacturing	FFF, SLA, DED, EBM, BMD, UAM, SLS, PBF
Traditional manufacturing	CNC, furnace, molding, finishing
Scanning	CT, structured light, hyperspectral
Recycling	Pulverizing, melting, grinding

technologies in a volumetrically optimized configuration will combine to produce the ideal candidate to take with you on extended journeys. These systems also include simple tool sets that are common to manufacturing processes and would be needed by either crew members or robotic counterparts to aid in the repair or maintenance processes. Table 1 enumerates the types of technologies that are needed to be incorporated on the journey.

Every material needed for a mission can be manufactured and repaired through some combination of the additive and traditional manufacturing technologies listed in Table 1. Some of these manufacturing technologies require external forces to aid in the deposition process, which can be localized or can take advantage of the potential artificial gravity capacity of the vehicle.

3.1 Thermoplastics

Thermoplastic additive manufacturing is a core component of the in-space servicing, assembly, and manufacturing (ISAM) technologies at the heart of our interstellar voyage. Plastic parts production, repairs, and augmentations enable crewmates to rapidly respond to the need to replace degraded components and structures. Plastic parts also provide an option for temporary patches for external abnormalities, such as those caused by micrometeoroid and orbital debris impacts. Additionally, the materials involved are recyclable without requiring a large amount of energy compared to other material classes. The two main thermoplastic additive manufacturing technologies that are needed for the mission are:

- Fused filament fabrication (FFF): feeds strands (or spools) of filament through a heated nozzle to deposit material. This is the most common form of additive manufacturing in commercial use.
- Stereolithography (SLA): resin-based manufacturing. Uses photopolymerization-compatible materials to create high-resolution, accurate parts.

If properly considered in the design phase, all aspects of the mission that require thermoplastics can be manufactured additively and augmented (finished) by traditional means.

Early thermoplastic AM efforts took root on the ISS with the development of the 3D Printing in Zero-G Experiment (3DP) payload and blossomed with the development and operation of the Additive Manufacturing Facility (AMF). The 3DP was a NASA technology demonstration mission that verified the ability to manufacture using extrusion-based additive manufacturing (AM) of thermoplastics in microgravity. One of the experiment's objectives was to reduce the risk of technical issues for AMF by testing individual subsystems and their combination on orbit. AMF is a commercial microgravity manufacturing device launched on Orb-6 in March 2016 and provides a larger, more capable version of the 3DP device.

For crewed environments such as the ISS and our interstellar mission, the selection of specific thermoplastics is a matter of great importance because it involves considering the need to ensure crew safety and end-use applications. Downselected materials must be useful both inside and outside of the pressurized cabin environment to be applicable for multiple mission scenarios. Aerospace-grade polymers are likely selections due to specific characteristics, such as resistance to atomic oxygen and ultraviolet (UV) radiation, which make them good candidates to be exposed to the environment of outer space. These polymers and polymer-based composites require low off-gassing qualities to ensure compatibility with sensitive surfaces such as optics, antennas, and photovoltaic cells. Any spacecraft structure positioned in proximity to those sensitive surfaces is a candidate for these low off-gassing materials. For example, two primary polymers used on the AMF are PEI-PC and PEEK. Additives, such as plasticizers and carbon fiber, can be blended into these polymers to create composites that have, in most cases, better capabilities than the bulk polymer alone. The AMF can print these without modification to the manufacturing payload. In total, over 30 different materials have been tested on the AMF. Amongst commercially-available systems, this validation makes AMF the most versatile and reliable additive manufacturing capability in space and on the ground. Parts constructed with AMF satisfy the flight validation and verification requirements necessary to be launched into space.

Future thermoplastic AM systems might incorporate design improvements to increase manufacturing speed; broaden the range of materials proven safe for use in crewed spaces; and utilize the advancements in vacuum-compatible thermoplastic printing demonstrated by such projects as OSAM-2 for external augmentation and repair. Recycling of discarded or otherwise failed thermoplastic articles will extend the utility of thermoplastic AM and provide additional guarantees of sustainability for the closed ecosystem aboard an interstellar or deep-space vessel.

3.2 Metals

Metal AM equips end-users with a diverse toolkit for sparing and general production. As such, metal AM plays a central role in ensuring the viability of interstellar voyages. While thermoplastics, as already demonstrated on the ISS, have their place in the closed ecosystem of our interstellar journey, these materials have limits. Metals offer a swath of mechanical and material properties that further expand the available design space beyond the limits of thermopolymers. Metals and metal structures offer improvements in thermal and electrical conductivity, strength, and temperature tolerance. However, compared to thermoplastics, metals are generally more difficult to recycle, requiring on average more energy than other material classes. Metal recycling of parts or machined chips would require collecting, sorting, and processing the targeted recyclables in a furnace. Dissimilar metals would need to be separated out prior to recycling. Several metal additive manufacturing technologies that would prove useful are:

- Ultrasonic additive manufacturing (UAM): utilizes a type of ultrasonic welding to combine layers of metals together into a near net shape, which can then be machined and postprocessed to rapidly produce large-volume metal parts.
- Powder bed fusion (PBF): broadly spans selective laser sintering (SLS), direct metal laser sintering (DMLS), and selective laser melting (SLM). While SLS primarily refers to polymer powders, DMLS and SLM are metal-specific processes. While the latter two are both powder bed processes, they differ fundamentally in how much energy is used to adhere the layers of powder to one another,

which in turn leads to differences in the structural properties of the completed parts. DMLS uses a laser to sinter (fuse, or bind) metal powder particles to one another without any part exceeding the melting temperature of the metal. In contrast, during SLM a laser melts all the powder particles instead of just sintering them, resulting in a process requiring more energy to perform. SLM produces on average higher density and more robust parts, but given the melting requirements, is far more limited in the materials available for use.

  - Electron beam melting (EBM): EBM is an off-shoot of SLM that uses an electron beam instead of a laser to melt metal powders.

- Directed energy deposition (DED): is another option for metal additive manufacturing, compatible with both wire- and powder-based feedstocks. DED broadly refers to the use of directed energy, including, but not limited to, lasers or electron beams, to melt a material and then deposit the melted material onto another surface. As this process suggests, compared to PBF processes, DED allows for repair and augmentation of existing parts, not just production of new parts.
- Welding: while not typically lumped into additive manufacturing technologies, the various forms of welding are additive processes and prove to be advantageous in both AM and repair through traditional welding applications.
- Bound Metal Deposition (BMD): compatible with extrusion-based processes similar to FFF. Metals are held together with a binder, extruded, part built, then debinded through sintering.

The space community is taking its first steps toward a rich future of in-orbit metal AM capabilities. Co-developed with metal processes should be a furnace to densify and heat treat parts. Traditional casting technologies combined with additively manufactured molds (preferably recyclable) would go a step further, enabling 100% density parts, but adding significant time due to the number of discrete steps in the manufacturing process.

Currently only one metal technology has flown in simulated microgravity flights. The Vulcan system [3] is a metal hybrid additive and subtractive manufacturing system with in-process sensing. Designed for use in a microgravity environment, its development is a significant milestone for in-space manufacturing.

The Vulcan manufacturing system includes the following five subsystems: (1) a metal additive manufacturing subsystem, (2) a polymer additive manufacturing subsystem, (3) a computer numerical control (CNC) machining subsystem, (4) a robotic gripper and clamp to facilitate automated manipulation and refixturing of parts during machining, and (5) an environmental control unit for management and capture of fumes and chip debris. The system also incorporates a robust command and data handling capability for remote commanding from an operations facility. The system operates with limited crew interaction. Anticipated ISS crew activities include installation and removal of the Vulcan system, changes in feedstock, and finished part removal. This system is the first generation of the future all-in-one system architecture that is the best possible configuration for long voyages. Substantial effort is still required to integrate the other necessary elements.

3.3 Fiber optics

The ability to produce quality fiber optic cables from glass preforms while in transit offers a means of replacing optical harnessing for networking and components for certain lasers, amongst other optical devices. While the production of spare cable from glass preforms is self-evidently valuable for potential interstellar missions, this process comes with additional benefits: when manufactured in microgravity, fiber optics production yields cables with performance improvements when compared to their terrestrially-produced counterparts [4].

3.4 Ceramics and composites

Use of ceramic and composite materials has proliferated broadly throughout the aerospace industry due to their thermomechanical properties and the potential for mass savings. Composites are likely to be used as the materials from which to create the components and structures of interstellar spacecraft, and as such the production, repair, and augmentation of these materials is a highly anticipated development. Similarly, ceramics offer additional options for use cases requiring high melting points and low coefficients of thermal expansion (CTE). However, ceramics and composites are notoriously hard to recycle, as their manufacturing process usually eliminates compounds or alters their chemistry when solidifying during production. Therefore, they should be used sparingly in design because any anticipated replacement material will need to be packed on board the spacecraft. Several manufacturing technologies are used for the creation of ceramics and composites. Most popular are SLA and UAM, respectively. SLA has been tested in space through the production of turbine blisks in the Ceramic Manufacturing Module on the ISS.

3.5 Electronics

Fostering the capability to repair and replace electrical systems on an interstellar voyage is essential to mission success. Electronics are particularly vulnerable to space environmental hazards such as radiation, including both solar wind and galactic cosmic rays, and thermal fluctuations. Crewed interstellar vessels will likely offer plenty of controlled environments that mitigate most if not all the threat of radiation or thermal extremes, but cases will no doubt persist where electrical systems will fail or degrade over time due to such exposures.

While certain applications and complex computer systems push the solution to their sparing toward bringing drop-in board or unit replacements, additive manufacturing solutions exist to provide versatile and mass-efficient electronics manufacturing as well. These solutions are typically variations of conductive ink deposition onto a substrate. One promising technique is utilizing aerosol jet technology to create two- and three-dimensional interconnects for electronic systems. This allows for the need to only pack vital components, such as chips, and manufacturing the boards, populating them through a pick-and-place capability, and then connecting them with the aerosol jet technology.

Microchip production is also a possibility during a mission. A deep space environment provides the necessary vacuum and clean production area, but this capability would require high power and cooling capabilities. Performing microchip production while en route to an interstellar destination requires planning for co-located infrastructure such as large radiators capable of rejecting heat generated by the lithography processes. Depending on the type of chip and selected manufacturing node, the mission may need to bring along one or more highly complex lithography machines, such as those produced by ASML [5]. Based upon terrestrial microchip production, a fraction of a mission's water budget may need to be allocated for supporting such production. The overall microchip production process is multiple stages and requires substantial volume terrestrially.

3.6 Medical applications

A successful interstellar mission will require disruptive and versatile medical technologies. Inevitably, the crew will need medical intervention as they injure themselves during nominal activities, manage chronic conditions, or otherwise suffer from ailments aggravated by the rigors of the space environment. From extensive medical research on ISS crew members, medical effects for the longest duration crewed stays indicate such conditions as muscle and bone mass degradation, ocular abnormalities, blood composition variability and degradation [6], and psychological effects. The manufacturing capability allows for medicine on-the-go, as needed. Instead of packing every instrument, splint, implant, prosthesis, cast, or surgical fixture, manufacturing these implements as they are required along the mission optimizes the mass and volume and ensures the correct solution is being implemented.

3.7 Biological manufacturing

Biological additive manufacturing is another viable, if not nascent, innovation offering a wealth of opportunity to the interstellar voyager. While biological additive manufacturing processes and synthetic biology more broadly are the focus of research across industry and academia, the United States government is taking an early first step toward providing guidance to address "material logistics and supply chains for the space domain." The Biomanufacturing: Survival, Utility and Reliability beyond Earth (B-SURE) effort, a program under the Defense Advanced Research Projects Agency (DARPA), strives to "develop a fundamental understanding of microbial capabilities for bioproduction in space and biomanufacturing as a potential novel solution for in-space manufacturing" [7]. B-SURE tasks investigators with three biomanufacturing pillars:

Track 1: Alternative Feedstocks
- Exploring in situ resource utilization
- "metabolic utilization"—human waste products (CO_2, urine, etc.)
- Regoliths—lunar, Martian, etc.

Track 2: Variable Gravity
- Different gravitational fields enabling different science and industrial applications
- Opportunity to design into interstellar vehicles areas with different effective gravitational fields (rotating sections of different radii/velocity, etc.)

Track 3: Variable Radiation
- Study the effects of different levels of radiation on science and industrial applications
- Study "resistance mechanisms" to mitigate or otherwise prevent effects of radiation in deep space

These three pillars and associated proposals inform the "viability and productivity to develop a picture of how biomanufacturing fits in the future space economy" [7]. B-SURE relies on a progressive crawl-walk-run method of technology development; while these first steps are small, these pillars are a foundation for efforts that compound early achievements to generate scientific and industrial progress.

NASA echoes the B-SURE approach in their own outlook for future biomanufacturing and bioengineering projects [8]. The agency's efforts are driven largely by efforts at the Ames Research Center (ARC). ARC's outlook is shaped by the need for more sustainable and reliable approaches to life support, biomedical infrastructure, agriculture, and manufacturing. Biomanufacturing processes

Table 2 Mycotecture NIAC Phase I objectives.

Objective	Description
Objective 1	Fill key technical knowledge gaps
Objective 2	Assess potential enhancements of mycelia
Objective 3	Design mycelial-based structures including habitats
Objective 4	Identify key knowledge gaps including pathways to implementation
Objective 5	Assess impact of technology for terrestrial applications

complement many methods of in situ resource utilization (ISRU), and thus NASA's investigations into these sustainable approaches center around developments in both ISRU and biomanufacturing. One of the first steps in this visionary program include the BioSentinel program, led out of the ARC, which seeks to demonstrate "[measurement of] response to deep space radiation in living cells (yeast) & correlate biological data with onboard dosimeter data and to ISS control experiment." BioSentinel utilizes microfluidics and various sensors (humidity, temperature, pressure, radiation) to quantify the effects of the space environment on the ISS.

NASA Ames also leads the charge in the nascent field of "mycotecture" [9,10]. A combination of mycology (the study of fungi) and architecture, mycotecture investigates the potential use of specialized fungal cultures to grow structures with desired properties and shapes. In 2018, an Ames research group was selected for a NASA Innovative Advanced Concepts (NIAC) Phase I grant for "Mycotecture in Orbit." The grant (see Table 2), now a Phase II effort (as of mid-2021), funds research which is:

> "...filling select key technical knowledge gaps such as the temperature range of mycelial growth, potential for algal feedstock, potential for enmeshed biosensors, mass of inputs and finished product, and material properties of the resulting materials. The potential for enhancing structural and sensing capabilities by the incorporation of the bacterium *Bacillus subtilis*, is novel. Architectural design concepts based on this vision will be examined for use in a mission context including mass trade-offs, and temperature inputs, as well as suggesting new terrestrial routes to infusion where rapidly built, lightweight structures are desired." [9]

The nearly nonexistent terrestrial use of mycelial materials in architecture means that their versatile properties are relatively unknown in the field of space architecture. Fibrous in nature, mycelial materials can be engineered into biocomposites, and when combined with various polymer secretions are radiation resistant, fire retardant, and have potential use as fertilizer when they eventually need to be recycled or otherwise disposed of. Structurally, "...these materials show compression strengths superior to dimensional lumber, flexural strength superior to reinforced concrete, and competitive insulation values," equipping space architects with a slate of options and opening a competitive design space compared to traditional materials [9].

3.8 Scanning

The ability to scan components or pieces of components is vital for enabling the diagnosis of failures and the identification of fixes for those failures. Nondestructive scanning methods are ideal to limit the amount of energy displaced to identify a problem and repair or replace a component as soon as possible after the defect is identified. Scanning can also be done to identify maintenance issues before they impact a mission. Nondestructive scanning technologies are well understood. However, implementing these technologies can be a bit tricky, especially when multiple scanning technologies are integrated into a single system. For aerospace applications, three scanning technologies are ideal and easily combined. Visual, hyperspectral, and computerized tomography (CT) scanning enable appropriate assessment of failed components and appropriate repairs or replacements to be tailored.

Additionally, these technologies can inspect parts both during and after manufacturing or repair to verify that their production meets the requirements of the component. During manufacturing, the systems can utilize passive and active elements to perform monitoring, inspection, and verification of produced parts and the manufacturing system(s) during nominal operations. Current monitoring solutions used for manufacturing in orbit are utilized for determining the status of motion-related commands, including deposition, along with tracking manufacturing progress. Monitoring devices may include stationary cameras (visual or hyperspectral), thermocouples, and mechanical triggers. Specific to certain types of thermoplastic AM, utilizing thermography can determine important characteristics during deposition, including the presence of voids and layer separations. Complete geometric scanning is accomplished with structured-light 3D scanning. In addition, hyperspectral imaging hardware can capture images in a wider swath of the electromagnetic spectrum than visible light and can determine whether proper material placements are made in an assembly, locate defects during deposition (including contamination), and characterize types of materials on a surface. This combined suite of technologies enables part inspection and verification during and after a manufacturing or assembly process occurs.

Scan-and-repair technology that combines defect characterization and manufacturing is essential to a successful interstellar voyage. This technology can reside as a free-roaming robot externally constantly looking for damage or reside in the confines of a manufacturing machine onboard. Scan-and-repair technology enables a three-prong use case scenario. The first use case is identifying damage and defects incurred during operations. When the scan-and-repair system identifies damage, the system shifts into a structural correction mode, which is the second use case. Depending on the damage, repairs are automatically carried out, or operators can devise a fix and initiate the repairs as necessary. Extending this capability is the third use case, whereby structural augmentation is possible. New structures are added to the target surface via additive manufacturing, enabling new use cases for the spacecraft, changing operational configurations, or enhancing existing ones.

3.9 Recycling

An automated recycling system for manufacturing systems' feedstock is required for any long-duration space mission since it limits the amount of mass required. Manufacturing feedstock, as well as trash and waste, take up valuable mass and storage space in an environment that requires optimal resource allocation. The ability to reuse items and transform them into feedstock also means less space required for raw material storage, as well as greater overall manufacturing capacity to produce needed parts and tools. Besides traditional manufacturing materials, recycling technologies can also be incorporated for food and human waste. This is done chemically as well as the traditional physical approach taken with materials such as plastics and metals. Earth-based recycling technologies are ideally leveraged, with modifications, to meet mission objectives.

References

[1] E. Mack, A Real World 'Star Trek' Replicator Is Now Possible Thanks to New Breakthrough, 2018. https://www.forbes.com/sites/ericmack/2018/03/09/carbon-nanotube-membrane-breakthrough-is-real-world-star-trek-replicator-mattershift/?sh=61f2b733520f. (Accessed 11 January 2022).

[2] B. Hurley, 3D Printing and Space Exploration: How NASA Will Use Additive Manufacturing, 2020. https://www.techbriefs.com/component/content/article/tb/stories/blog/35871. (Accessed 11 January 2022).

[3] Redwire Space, Vulcan Advanced Hybrid Manufacturing, 2020. https://redwirespace.com/products/vulcan/. (Accessed 11 January 2022).

[4] Redwire Space, Fiber Optics, 2020. https://redwirespace.com/products/fiber-optics/. (Accessed 11 January 2022).

[5] ASML, EUV Lithography Systems, 2019–2022. https://www.asml.com/en/products/euv-lithography-systems. (Accessed 11 January 2022).

[6] G. Trudel, N. Shahin, T. Ramsay, et al., Hemolysis contributes to anemia during long-duration space flight, Nat. Med. 28 (2022) 59–62, https://doi.org/10.1038/s41591-021-01637-7.

[7] A. Cheever, Biomanufacturing: Survival, Utility, and Reliability Beyond Earth, DARPA, Virtual Proposer's Day, 29 November 2021.

[8] J.A. Hogan, NASA Perspective on Bioengineering and Anticipated Future for Biomanufacturing in Space, 29 November 2021, DARPA, Virtual Proposer's Day https://ntrs.nasa.gov/citations/20210024798. (Accessed 11 January 2022).

[9] L.J. Rothschild, et al., Myco-Architecture off Planet: Growing Surface Structures at Destination, NIAC 2018 Phase I Final Report, 2019, NASA Technical Reports Server https://ntrs.nasa.gov/citations/20190002580. (Accessed 11 January 2022).

[10] NASA Ames Research Center, Mycotecture Off Planet, 2021. https://www.nasa.gov/directorates/spacetech/niac/2021_Phase_I/Mycotecture_Off_Planet/. (Accessed 11 January 2022).

CHAPTER

Optimal crew size at both departure and arrival

9

Andreas M. Hein[a], Frédéric Marin[b], and Cameron M. Smith[c]

[a]*Initiative for Interstellar Studies (i4is), London, United Kingdom,* [b]*Universite de Strasbourg, CNRS, Observatoire Astronomique de Strasbourg, UMR 7550, Strasbourg, France,* [c]*Biosphere2 Center for Human Space Exploration, University of Arizona, Tucson, AZ, United States*

Chapter outline

1 Introduction

Population size is the single most important driver of the size of a crewed interstellar vessel [1,2]. However, population size is determined by various factors, including the duration of travel, the required population at the target destination, the mode of human transportation, and the risk management strategy. Concepts for crewed interstellar travel position themselves differently with respect to these factors. Fig. 1 proposes a categorization of different concepts for crewed interstellar travel from Refs. [2,3], as these terms are often used loosely in the literature. The most important concept in this chapter is that of a "generation ship". A generation ship is a crewed interstellar spaceship where humans live on board and live their whole life from birth to death as they would on Earth. With the propulsion systems

Interstellar Travel. https://doi.org/10.1016/B978-0-323-91280-8.00010-1

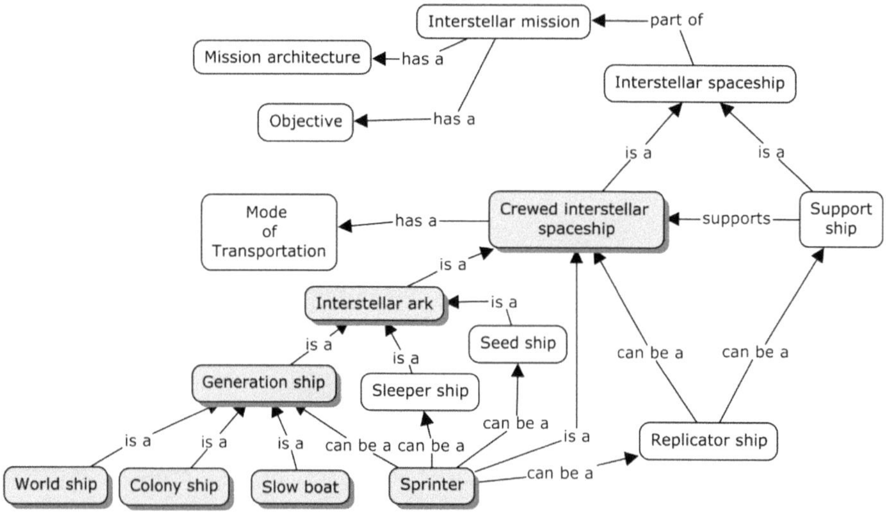

FIG. 1

Concept map of crewed starship concepts [2,3].

that have been proposed for interstellar travel so far, and which are based on known physics, a crewed starship would take decades, if not centuries, to reach another star system. If we assume that one generation has a duration of about 20–30 years, it is reasonable to assume that several generations will live on a generation ship during its trip, which gives the notion of "generation" in "generation ship" its proper meaning. This contrasts with the sleeper and seed ship [1], where the crew would either be hibernating (at least part of the crew or part of the time) or transported in an embryonal state.

Generation ships are further categorized into world ships, colony ships, slow boats, and sprinters. World ships are extremely large (dozens of km) spaceships, which replicate terrestrial environments. As elaborated in Ref. [3], world ships can be intuitively understood as encapsulating an entire "world," i.e., a self-contained environment, imitating geographic features of the Earth. By contrast, colony ships are smaller and do not have the ambition to replicate an Earth-like environment. They might resemble villages or towns with a higher population density than on world ships. Their size might also be similar to O'Neill space settlement concepts such as the Stanford Torus. Sprinters and slow boats are even smaller, with populations of less than 1000 [4].

How is population size linked to the duration of travel? In general, a longer trip duration requires a larger population. There are different reasons for this relationship, but the one most extensively treated in the literature is biology, more specifically, genetics. The longer the trip, the more generations will successively live on the ship and a sufficiently large population is required to avoid inbreeding and its detrimental health effects. The more generations originate from an initial population, the larger the initial population needs to be.

Fig. 2, from Refs. [2,3], maps population size to trip duration for various generation ship concepts in the literature; also, lines of feasibility are drawn (in red)—these lines separate the area on the left where the population size is too small for a given trip duration and the feasible area on the right. The reference values are taken from Ref. [4] and backed up by more recent results [5]. The values range from 1000

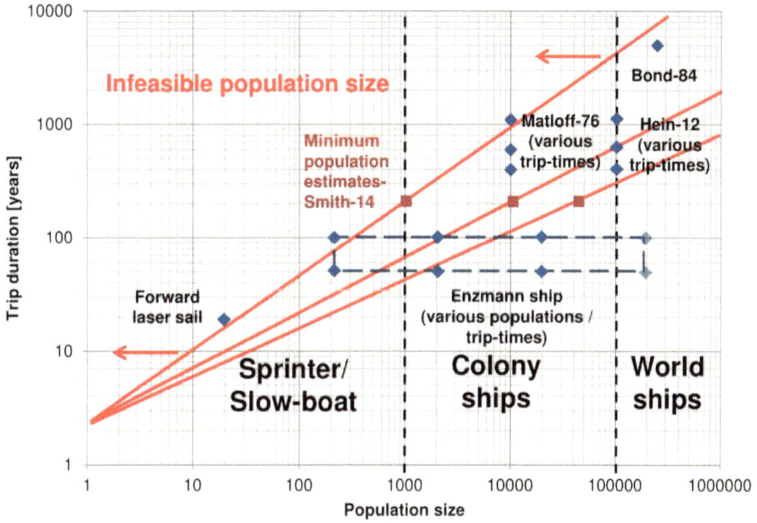

FIG. 2

Population size vs trip duration mapping for various generation ship concepts in the literature [2,3].

(lower-end estimate) up to 40,000 (upper-end with risk margin). Smith [4] argues that while low population numbers (1000s) are in principle feasible if uncertainties are taken into account (e.g., accidents), which may reduce the population size, these populations would end up below the threshold value. Hence, Smith argues for a margin of safety such that such events can be absorbed. As a benchmark at the lower end, a crew size of a few is sufficient for a mission duration of ~1 year (International Space Station example or a potential Mars mission). Interpolating between this data point and the different estimates from Refs. [4,5] yields the three feasibility curves. Of course, they are not based on any "hard" evidence, i.e., it might turn out that the curve's shape is different (not linear in the log-log scale, i.e., a power function). However, given we do not have further data, it is the best guess at this point for how the required population size increases with the trip duration.

What is interesting to see is how concepts for crewed interstellar spaceships situate concerning these curves. It can be seen that most concepts seem to be feasible regarding the 1000-people limit. However, with the curves including the point for 10,000 and 40,000 people, most concepts transition into the infeasible area, except for the Enzmann ship. The Enzmann Starship has particular data points as differently sized ships were proposed in the literature [6].

On a generation ship in which humans would live over generations, a number of systems are required to maintain their lives and living environment [2] over, possibly, centuries. This environment needs to satisfy different human needs. Fig. 3 shows a categorization of human needs, inspired by Maslow's Pyramid. At the bottom, we find the basic physiological needs such as food, water, air, etc. These are needs that will, if not satisfied, inevitably lead to death after a short period of time (minutes to weeks). The second category is human factors, which do not need to be fulfilled for survival but make life and work on-board acceptable. This includes safety/security, psychological needs, etc. Higher-level needs are needs such as belongingness, self-fulfilment, etc. For those who know Maslow's

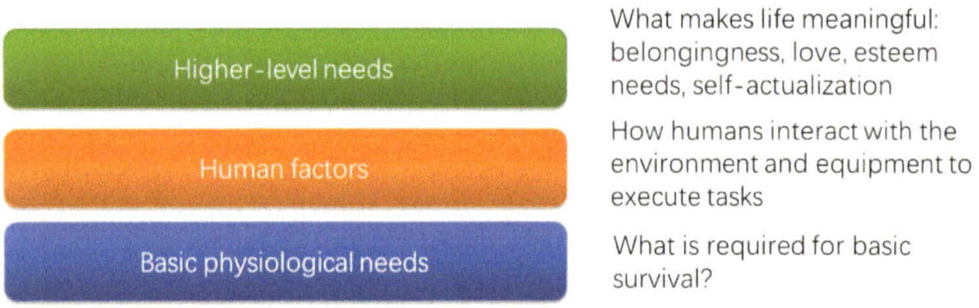

FIG. 3

Categories of human needs, loosely based on Maslow.

Pyramid, this subdivision of its categories might seem unusual. However, in the context of this chapter, this modified categorization should facilitate the understanding of which systems address which needs and thereby shed light on their importance.

Among the diverse list of systems and components that are considered necessary for this purpose, the basic components related to the protection of life on-board are radiation shielding, the habitat structure, which contains a breathable atmosphere, a closed loop bioregenerative life support system, including means for oxygen and food production, water and waste recycling, and a light source. However, while these components and systems are necessary for sustaining human life, they are not sufficient to provide livable conditions going beyond basic habitability for humans onboard [7]. Humans are social beings [8], as already Aristotle remarked (zoon politikon), and they have a culture [9] (something Plessner has called "natural artificiality" of the human existence). Hence, space for social organization and culture in the broadest sense needs to be provided, for example, buildings, including houses, places for work, places for gatherings, and spiritual and religious places. If we take all these elements, we arrive at different estimates in the literature regarding the amount of mass and volume per inhabitant, as shown in Table 1.

Table 1 Comparison of built and hypothetical space habitats [10].

Habitat	Existing/ hypothetical	Open/closed ECLSS	Crew size	Mass (T)	Mass (T) per person
ISS	Existing	Open	2–6	420	210–70
MIR	Existing	Open	3	130	43
Skylab	Existing	Open	3	77	26
Stanford Torus	Hypothetical	Closed	10,000	10^7, 5×10^5 w/o radiation shielding	10^3–50
O'Neill Model 1	Hypothetical	Closed	10,000	5×10^5	50
MIT Study	Hypothetical	Closed	1000	2×10^5	200
Kalpana One	Hypothetical	Closed	3000	7×10^6	2×10^3

It can be seen that the mass estimates per person for a list of built and hypothetical space habitats are wide-ranging, from 10 to 10^3 tons. These differences depend on what is accounted for, such as radiation shielding, supply mass, and maximum crew size. If we assume that this mass per person stays constant and is not dependent on the total number of people, it is straightforward to see that the mass of the entire payload, essentially the entire settlement, scales linearly. In other words, we just need to multiply the number of people by the mass per person. Such a linear model is of course a simplification, as there are probably economy of scale effects for larger populations. However, as all other payload systems scale with the number of humans, it is safe to say that population size is the single most important determinant of the payload mass in a crewed starship and possibly for any other means of transporting humans to the stars.

This does not come as a surprise and it is easy to find analogous examples for more near-term missions. For example, it is well known that for crewed lunar or Mars missions, the most important driver is the size of the crew [11].

The tricky part now is what population size we would actually need for an interstellar voyage. While for missions to the Moon and Mars, this number is primarily determined by crew psychology and science return, things are quite different in the case of an interstellar mission. First, a crewed interstellar mission has an extremely high cost. Estimates in the literature [3] assume that the development and launch of a generation ship might cost on the order of hundreds of trillions (multiples of the current global gross domestic product). If we divide this number by the associated population size, the cost per person is on the order of dozens of billions. This might even be at the lower end, as the Apollo program has cost on the order of a hundred billion in current dollars and has landed about a dozen astronauts on the moon, leading to a similar cost figure per astronaut. A second factor is that the objective of a crewed interstellar mission is the settlement of another star system [2]. This means that the population not only needs to be large enough to be stable over centuries but also large enough to jump-start a new civilization at the target star system.

A wide range of values for population sizes on a crewed starship has been proposed in the literature. Extreme values range from just a few [12] up to several tens of thousands [4]. In consequence, these would lead to a wide range of corresponding payload masses, notably a few hundred tons to millions of tons at the lower end of mass per person and a few thousand to tens of millions at the upper end. These payload mass values immediately translate into corresponding energy and resources required for launching such vehicles. The cost of development and launch could only be covered if the global economy, which might also extend into space, is possibly hundreds or even thousands of times larger than today [13]. However, even with large increases in global GDP, it would take several centuries to render the development of such a vessel economically viable [14,15]. There are economic and thermodynamic reasons why such a continued growth of the global economy is likely not sustainable over centuries [13]. For a vehicle with a crew in the order of hundreds, this date may shift to just 100–200 years into the future.

This chapter provides an up-to-date overview of the literature on population size on a generation ship and at the target destination. We will explore the relationship between population size and technology/skills to build up a settlement and civilization. In addition, an overview of approaches to circumvent the question of population size is presented.

2 Population size on interstellar vessels

The concept of minimum viable population (MVP) is used to determine a lower limit on the population of a species, such that it can survive natural disasters and demographic, environmental, or genetic stochasticity [16]. This MVP, often used by (conservation) biologists and ecologists, tries to determine how many interbreeding individuals can survive in a specific territory (such as a well-delimited geographic region) or an artificial environment (such as a zoo), or in our case, a multigenerational spacecraft. Population size is the single most important determinant in constraining the payload for a crewed interstellar vessel since it is directly related to the ship's architecture, energy generation, food production, and the sociodemographic organization within the space ark. However, unlike the research field of nonhuman animal species, where experiments and observations are possible, determining the MVP of a human population for an interstellar journey can only rely on educated speculations. Compiling all papers that have tried to estimate the MVP in the context of space exploration, it appears that numerous values have been proposed in the literature. This leads to a wide range of corresponding payload masses that prevent any precise estimation of the necessary energy and resources required for launching such vehicles. Those constraints are fundamental to determining if such an endeavor could be realistically expected by the end of this century. To narrow down the correct numbers, this paper will give an up-to-date overview of the literature on MVP estimates for interstellar spacecraft.

2.1 Historical review of MVP estimates

The need for determining the MVP of a given species dates back from, at least, the emergence of conservation biology in the late 1970s [17]. Since then, it has been postulated that, even with inbreeding, sexual populations larger than 100 should not be endangered by genetic stochasticity if they live in vast territories with abundant resources [18]. However, in resource-restricted environments, departures from this lower limit are expected [19]. The literature on the subject is extremely large, but little has been achieved in the specific field of space exploration.

The first scientist to rigorously explore the MVP was John H. Moore. Using a software called Ethnopop to simulate population history for centuries-long models, he explored the survival chances of space crews organized in "familiar, ubiquitous, well-ordered and well-understood social forms" [20]. Standard human families were at the center of his simulations and served as guidelines for establishing the social, economic, and political activities within the space society. He defined a marital regime with four conditions: every person can be married, there is a choice of 10 mating persons per individual with no more than 3 years of age difference, no consanguinity higher than second cousin is allowed, and every person can be a parent. Results from Ethnopop indicated that about 150–180 persons would sustain themselves almost indefinitely. A smaller MVP (about half of the previous numbers) can be achieved if the initial crew only consists of young couples with postponed parenthood. Interestingly, it results in well-defined demographic echelons instead of the previous normally distributed crew of 150–180 persons. The numbers from Ref. [20] are reported in Fig. 4 under the label "M03."

Harry Jones was not interested in the surviving chances of the crew throughout the space voyage but rather in determining the number of crew members required to maximize the probability for successfully colonizing an exoplanet [21]. In his work, he assigned a probability of success to each independent group of 15 people and multiplied that success rate per the number of groups that would successfully

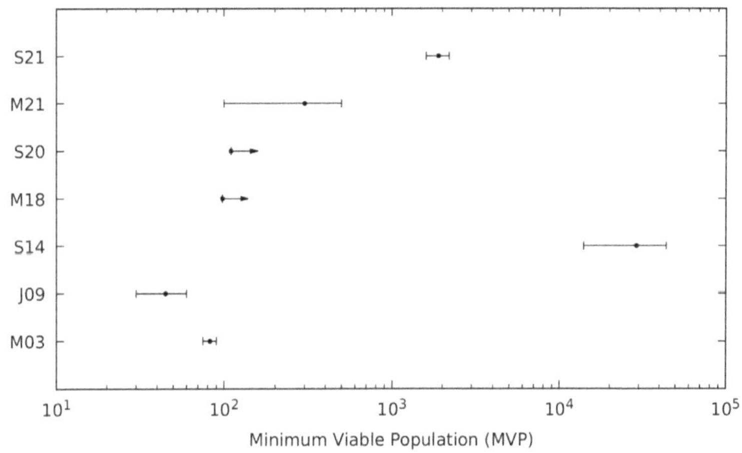

FIG. 4

Compiling the various MVP estimates from literature. M03 [20]; J09 [21]; S14 [4]; M18 [22]; S20 [23]; M21 [24]; S21 [25,26].

land on an exoplanet. By doing so, he established the probability of colonizing success such as $P=1-0.955\ N$, with N the number of crew members. A crew of 192 persons would result in a 99.99% success rate. Jones also explored the possibility to send several smaller crews (hence with less chances of success per crew) to the same destination to maximize the colonization success while minimizing the total number of crew members. He ultimately found that as long as the probability of colonizing success for a crew of N is greater than 0.5 [$N=30$], it is more efficient to send out a second independent expedition of the same size N than to double the original crew to $2N$ [21]. According to his calculations, "sending a very large crew is wasteful, since the resources could be split to mount two independent missions." Ultimately, there is a trade-off to do between budget and success rate. The numbers from Ref. [21] are reported in Fig. 4 under the label "J09."

A third MVP estimation was achieved by Cameron M. Smith [27]. He was the first to include the central issues of population genetics into the computation of a more elaborated MVP for interstellar starships [5]. In his work, he included the effects of mutation, migration, selection and drift for human populations. He formulated the equations that are used to determine the effective and census populations, together with the MVP, and ran Matlab population simulations. He found that "any founding crew with [a census population] above 40,000 would be very likely to survive in good health over five generations, even in the event of a Black Death-style plague with a mortality figure of 30%" [5]. Smith was also the first to include environmental stochasticity in his calculations to determine the resilience of a crew to catastrophes. The numbers from Ref. [5] are reported in Fig. 4 under the label "S14."

In 2017, Frédéric Marin [28] (and later other collaborators), designed the agent-based Monte Carlo code heritage to push the investigations further [29–31]. This numerical code simulates the demographic and genetic variations of a multigenerational population in the context of size-restricted environments. The primary goal of heritage is to investigate long-duration space travel using a rigorous methodology. The code was designed to advance the studies by including several new parameters that were previously unexplored: ethnographic, demographic, anthropometric, genetic, medical, and

Mir/ISS-based astronautic data, and follows the complex biological interactions of millions of individuals in resource-limited environments. Among the results obtained with the code, Marin determined that a strict minimum of 98 persons would be necessary to ensure the survival of the population for centuries-long journeys that are impacted by demographic and environmental stochasticity [28]. It was only later, when genetic stochasticity was included in the code, that revised numbers were published to reach the Hardy-Weinberg equilibrium, ensuring the genetic health of the arriving generation. Under the neutral selection hypothesis, Marin estimated that if an MVP of 100 persons is still valid, a healthier population of 500 people could ensure a safer interstellar journey [30]. The numbers from Refs. [5,30] are reported in Fig. 4 under the labels "M18" and "M21," respectively.

Jean-Marc Salotti [23] looked at the feasibility of survival on another planet and the establishment of a self-sustaining colony. Working with Mars data, he built a mathematical model to determine the MVP necessary for survival on any exoworld. Purely based on engineering constraints, he expressed the survival rate as a competition between working time requirements and capacity. Each category is broken into subcategories, such as raising children, preparing meals, and social activities for the working time capacity, and ecosystem management, energy generation and industry for the working time requirements. From the resulting mathematical expression of constraint, the author derived an MVP of 110 persons for settling on Mars [23]. Depending on the exoplanet life conditions, this number can vary but, overall, the MVP is of the order of a hundred. The numbers from Ref. [23] are reported in Fig. 4 under the label "S20."

Finally, Sano Satoshi created another Monte Carlo method relying on repeated sampling called Evolve [25,26]. Its architecture follows a similar basic methodology to the heritage strategy. However, the methodology to include genetic stochasticity diverges. Evolve includes the infinite sites model, a mathematical mutation model that uses allele frequencies to calculate genetic diversity in a finite population. The code made it possible to double-check the predictions from heritage and pushed forward the estimation of the MVP for crewed interstellar flights by including the effects of natural selection. The final MVP estimated in Ref. [25] is 1600–2200 to maintain genetic health. The numbers from Ref. [25] are reported in Fig. 4 under the label "S21."

2.2 Comparison of MVP estimates

All MVP numbers from literature are reported in Fig. 4. From a chronological viewpoint, estimations of the MVP tend to give larger numbers with time. This is logical, as the first models used to determine this parameter were extremely simple. They were based on simple arithmetic [21] or simulations that only included demographic stochasticity [20]. With the evolution of computing power and the development of personal programming skills, the numerical methods to compute viable populations for interstellar travel became more sophisticated. This results in the increase of the MVP from below a hundred to a few hundred in less than 20 years.

There are, however, two data points that lie outside the overall pattern. The extreme value given by Smith [4] can be seen as the first tentative to include the effects of genetics. Lacking any Monte Carlo simulations to back up the theoretical ground developed in the paper, only simple algebraic manipulations could be achieved. This gave us a kind of upper limit for an ideal population for interstellar travel. The main reason why this value is so high is that the author wanted to move away from the scarcity paradigm which implies that we must reduce costs and margins to a minimum to ensure a realistic project that could be started as soon as possible. But this is done to the detriment of personal well-being, a fundamental sociological factor in any society. Moving away from the scarcity paradigm is essential to ensure a good resistance to the psychological stress that such a mission would cause. The second outlier is the MVP estimation of Satoshi [25,26]. While being an order of magnitude lower than the MVP of Smith [4], Satoshi's MVP reflects the impact of genetic stochasticity on the crew using a more sophisticated approach. However, the Evolve code does not include a complete representation of the human genome and treats genetic effects using simplified approaches. Neomutations and their phenotypical effects from larger-than-Earth irradiation doses are also missing. It is thus very likely that the evolve estimation of the MVP will be refined with time.

2.3 Potential additional strategies for MVP reduction

A drastic reduction in population size might be achievable either by "engineering" the living conditions to optimize for a low population size. It includes methods such as social engineering, including rules for mating choice, maximum population size, etc., which come with significant ethical issues [31]. But other, more technological approaches could equally be envisaged, such as storing genetic material for reproduction. Transporting frozen sperm, ova, and zygotes may be an option, although its implications on population size and its social impact do not seem to have been addressed yet, except for brief considerations in Refs. [27,32,33]. Ethical issues also abound here, which will shape the rules for how these means are going to be used (or not) during the trip and after arrival.

An additional possibility is a change in human longevity or, to be more precise, an increase in the timespan between generations. If the timespan could be increased, the number of generations during the trip would decrease. This could be achieved by delaying the age at which procreation would happen. If we, hypothetically assume that the age at procreation would be doubled, say from 25 to 50 years on average, this would divide the number of generations on a starship by half. However, it is known that health risks for children increase with the age of the parents [34], meaning that any potential advantage of a lower number of generations might need to be traded against the detrimental effects of later procreation. A more speculative approach would be gestation external of the human body, in artificial wombs. While artificial womb prototypes have already been developed [35], their feasibility for humans remains an open question, including potential detrimental effects on the health of the child. However, if this technology becomes available, it could further delay procreation or essentially decouple it from the age of the parents. Note that these means all decrease MVP by reducing the number of total generations purely from a genetic perspective. Other constraints such as knowledge preservation might still define lower bounds for MVP independently.

2.4 Conclusions

Determining a reasonable MVP is tricky. Many demographic, environmental, and genetic parameters come into play. There is no simulation software that includes all of them, despite the great advances the field saw over the last decade. In addition, the most unpredictable parameter that will most likely drive the journey toward success or failure is the human being itself. The study of the social behavior of a group of individuals escapes traditional observational and methodological tools. It is impossible for an observer to grasp all the ins and outs of these systems characterized by a large number of internal and external interactions, a nondecomposable structure, and unpredictable behavior. Nevertheless, the combined progress of psychology, medicine, and cognitive sciences has enabled the emergence of behavioral models [36–38]. Those models, such as the heuristic formula of Lewin [39], are far from simple but they try to explain what determines behavior. Numerical modeling is on the way, but relies heavily on artificial intelligence development. We can thus conclude that, on the ground of natural sciences alone, the MVP for interstellar travel using generation ships lies between 100 and 1000 individuals under the scarcity paradigm. The social sciences and moral sciences are likely to drive those numbers upward once included in our current simulations.

3 Alternative approaches for interstellar transportation of humans

Another, more disruptive possibility is a change in the mode of transportation of the human population. In general, these approaches require the development of new technologies and are associated with higher levels of uncertainty regarding whether these technologies are actually feasible. The following list provides an overview of approaches which fall into this category:

- Unconscious state: Hibernation/cryosleep (Sleeper ship in Fig. 1)
- Protostate: Sending genetic material (Seed ship in Fig. 1)
- Underlying data: Sending data (in silico or pure data transmission) (see [40])

How these approaches can impact population size at the beginning and end of an interstellar voyage is still relatively open. In the following, we briefly introduce each approach and its underlying characteristics to speculate about potential consequences on population size.

3.1 Unconscious state: Hibernation

Hibernation for space travel has been a research topic for a few decades, although the results are still inconclusive [41–43]. Hibernation induction and its sustainability over longer periods are still unsolved. The accumulation of toxic substances might lead to irreparable damage to the organism. However, Chouker and Cerri [41,43] have investigated potential reduced radiation effects in humans.

Regarding the impact of hibernation on population size, the main effect is that more humans may fit into the same spacecraft given the same resources. This is due to the smaller volume and smaller amounts of resources consumed by a hibernating population. Bradford et al. [44] have conducted a conceptual study on using hibernation for a mission to Mars. Table 2 compares a hibernation habitat (Vision Torpor Habitat) with a habitat without hibernation (Design Reference Architecture (DRA) 5.0 [11]). It can be seen that the dry mass has decreased by a factor between two and three, the consumable mass by a factor of eight, and the total volume by a factor of six. By analogous reasoning, we may estimate that the population size for the same ship size could be increased by a factor of two to eight, depending on the dominant mass contribution (dry mass, consumable mass or total volume).

3.2 Protostate: Sending genetic material

Sending individual cells or embryos to the stars requires a completely different set of technologies than a generation ship [2,45]. The cells/embryos would need to be stored under cryogenic conditions during the trip, subsequently brought back to life, and grown/raised at the target destination. It is clear that the latter would result in tremendous technical but also ethical issues, as no human parents or adults would be present.

Table 2 Comparison of Mars transfer habitat without and with hibernation.

Parameter	DRA 5.0 reference	Vision Torpor Habitat
Mission configuration	Mars Conjunction-class Avg. 200-days(out), 500-days(surface), 200-days(return)	
Crew complement	6	6
Dry mass	27.5 t	11.1 t
Consumables with contingency	13.1 t	1.6 t
Total volume	500 m^3	75 m^3
Habitable volume	350 m^3	8 m^3
Length	9.0 m	5.0 m
Diameter (rigid)	4.3 m	4.3 m
Diameter (inflated)	8.2 m	–

From J. Bradford, M. Schaffer, D. Talk, Torpor Inducing Transfer Habitat for Human Stasis to Mars, 2018.

Due to the radically different form of transportation, each individual cell/embryo would have a mass of below 1 g to a few grams. This is about a 10^4 reduction in mass. If we take mass alone, for one adult human, we could transport a population of (proto)humans of 10^4 to the target destination. This phenomenal reduction in mass and the reduced complexity of transportation (cryogenic conditions exist in deep space by default) needs to be traded against the complexity of growing a human population at the target destination without human intervention. All equipment for constructing this first settlement and all systems for growing (e.g., artificial womb) and raising humans (e.g., android parenting) need to be transported as well. The technical challenges for raising humans autonomously are likely enormous and their development is ethically highly controversial as this would require extensive testing on humans.

3.3 Underlying data: Sending data (in silico or pure data transmission)

Sending data instead of biological humans has been proposed previously [40,46,47]. Such an approach would either require a data storage device being transported to the stars or assume a receiver at the target destination. In both cases, a device that could recreate humans from data at the target destination would be required. Although the creation of DNA from data has been demonstrated in a lab environment, its performance is still many orders of magnitude away from what would be needed for recreating human DNA at the target destination and subsequently inserting the DNA into a cell for reproduction [48]. However, should such a technology become feasible one day, the population size that could be sent to the stars would only be constrained by the capacities of the facilities at the target destination. No mass, except for the initial infrastructure, would need to be transported. The potential population size could be enormous.

Table 3 Alternative transportation modes for humans.

	Population size increase compared to generation ship	Enabling technologies
Hibernation	<10	Hibernation
Genetic material	$\sim 10^4$	Artificial parenting
Sending data	Unknown	Embodying from data to humans

Table 3 provides an overview of the presented alternative modes of transportation and their implications on population size, compared to a generation ship.

4 Population size at the target destination

In this part, we deal with the question of population size at the target destination. We set the context by introducing different types of interstellar habitats and the population size they require to start a new civilization. Contrary to the population size during the trip, much less has been published on the question of the required population size at the target destination. For example, Salotti [23], who was already mentioned in Section 3, estimates that the population size for settling a planet can be as low as 110. To put it simply, the model used is based on the time required for the work that needs to be conducted in such as settlement. The required time per day is then distributed on the members of the population. The minimum population size is estimated by dividing the total time per day by the hours of work a member of the population is able to perform. Biological or cultural factors have not been taken into account.

Given the limited literature on the population size at the target destination, the question is whether the results obtained for generation ships could be helpful for an initial estimate. This depends on the similarities between the habitat on the generation ship and at the target destination. First, we have to clarify what type of habitat at the target destination we are talking about. Hein et al. [2] have presented different types of habitats within a target star system, shown in Table 4. Besides settling different types of planets (habitable planet/moon, biocompatible planet/moon, easily terraformable planet/moon), we may also imagine settling a star system by constructing free-floating space settlements, as imagined by Gerard O'Neill [49]. An interesting case where the target destination would not be a star system is the case of rogue planets/comets, which have been found in interstellar space [50].

Given these different habitat types, Table 5 provides an overview of similarities and differences between factors impacting population size on a generation ship and in a star system. These similarities and differences are valid for all habitat types in Table 4, although to a different degree. For example, a free-floating space settlement is more resource-constrained than a planet or moon.

The abundance of resources removes a key constraint of generation ships, namely, the maximum population size that can be sustained inside its habitat based on the available resources and space, i.e., its carrying capacity. Although habitats at the target destination may also be limited by similar constraints, it is reasonable to assume that their carrying capacity is vastly larger than on a generation ship. This means that measures for population control are likely not needed or to a much lesser extent than on

Table 4 Habitat types for settlement [3].

	Habitable planet/ moon	Biocompatible planet/moon	Easily terraformable planet/moon	Rogue planet/ comet	Space settlements
Investment for habitability	Small	Establish ecosystem	Terraforming	Settlement construction	Settlement construction
Duration until habitability	Years	Decades/ centuries	Centuries	Decades	Decades
Habitability duration	Millions of years	Hundred thousands of years	Hundred thousands of years	Centuries–millennia	Centuries–millennia
Availability	Rare	Rare	Rare	High abundance	High abundance
Distance from Earth (estimates)	4–16ly	4–16ly	4–11ly	≤4ly	4ly

Table 5 Comparison of factors impacting the size of a population on a generation ship vs at the target destination.

	Generation ship	Target destination
Resources	Highly constrained	Abundant
Space	Highly constrained	Abundant (limited by construction capacity and resources)
Number of generations	Defined by trip duration	Open-ended
Environmental selection pressures	Similar to Earth but highly resource constrained	Likely very different from Earth
Risk of accidents/ catastrophes	Limited due to a highly controlled environment	Highly uncertain

a generation ship. However, contrary to a generation ship, where the population needs to be sustained only during its trip duration, the number of generations at the target destination is potentially unconstrained and open-ended.

Contrary to the ecosystem on a generation ship, where humans are subject to environmental selection pressures that are similar to those on Earth, at least those linked to the ecosystem (ultimately, the surrounding ecosystem will be a subset of the one on Earth), the environmental pressures at the target destination might be very different, depending on the preexistence of an ecosystem. If no lifeforms exist at the target destination, we might just have an expansion of the transported ecosystem onboard the generation ship, which would then coevolve with the human population. If an ecosystem preexists at the target destination, this ecosystem will likely be incompatible with a human population [51].

Finally, the risk of accidents and catastrophes is likely higher at the target destination, for the simple reason that it will be an environment with a lot of uncertainties. By contrast, on a generation ship, the environment is highly controlled. Although there is a prevalent risk of accidents and catastrophes due to malfunctioning equipment or conflicts within the population, the highly dynamic nature of settling the target destination introduces a higher degree of risk.

The conclusion from this comparison is that the drivers for population size are different than on a generation ship. The resource and space constraints are much wider, which means that the population is not constrained by these factors. However, as the number of generations is unbound, we are dealing with an open-ended time horizon, meaning that the population size needs to be larger to ensure sufficient genetic diversity. On the other hand, the population needs to be larger than on a generation ship if a safety margin for accidents and catastrophes is factored in. The safety margin needs to be larger than on a generation ship due to the higher risk.

One approach to estimating the minimum population size for settling the target destination in a setting of an open-ended number of generations is to look at the smallest population of humans that have lived on Earth. As all human descendants have their origin in this population with an unknown end, it represents a population with an open-ended number of generations. A recent genome study [52] estimates the minimum population size was about 10,000 within the last 200,000 years. There is evidence that this number was reached during a global catastrophe: the eruption of the Toba volcano. Smith [4] also provides a thorough survey of human and animal MVP estimates. For humans and primates, the MVP is somewhere in the 10,000s and this seems to be a fairly robust estimate. While it is not straightforward to use the same genetic markup of this population for a population that has survived living on a generation ship for several generations, we can use this historic value as a data point. This population size, interestingly, is in the same order of magnitude as the population size estimated by Smith [4].

If we were to add a safety margin to this number, we would end up with numbers similar to those of Smith, although the safety margin should be fairly generous (for a generation ship, Smith uses safety margins up to a factor of 4, leading to a population size of 40,000). Given the increased risk at the target destination, the safety margin might be as high as a factor of 10, leading to a population size of 100,000 upon arrival.

In any case, these numbers are much higher than those required for sustaining the habitat in terms of work to be conducted, meaning that task distribution is not the limiting factor.

If genetic diversity can be increased by the transportation of additional genetic material, the population size might be further reduced from 10,000. Without further evidence, we guess that a reduction to 1000 might be feasible, putting this figure into the same magnitude as estimated by Hein et al. [3]. Adding a safety margin factor of 10 would lead us again to 10,000. This leaves us with the conclusion that a population size between 10,000 and 100,000 seems reasonable.

However, genetics is only one determinant of MVP. Another is knowledge. While Salotti [23] looks into the skill set required for sustaining an initial habitat, the full set of knowledge for living on a generation ship and kick-starting a new civilization is certainly larger. The question of knowledge preservation has been addressed previously by Hein et al. [2]. Knowledge is critical for life on a generation ship (e.g., maintenance, construction, etc.) and even more critical upon arrival at the target. Habitats need to be constructed, food provided, resources extracted, etc. While certain types of knowledge can be easily preserved, e.g., explicit knowledge can be documented, stored, etc., tacit knowledge cannot, and the more relevant type of knowledge is tacit knowledge, which is linked to skills and competences. Tacit knowledge can only be transmitted by imitation. The knowledge-related MVP is determined by:

- The required skill set on arrival;
- The population size required to preserve the skills during the trip;

- The distribution of the skills between members of the population;
- The level of the skill required (novice to expert);
- The skill safety margin to avoid that skills are lost.

The positive relationship between population size and the level of technology of that population is well known [53–55], although a recent debate has focused on the empirical evidence for this relationship and the underlying causal link [56]. Kline and Boyd [54] argue that the larger the population, the likelihood that knowledge is preserved is higher due to redundancy. For example, if only one person holds a key piece of knowledge (e.g., skill to repair electric circuits), that knowledge may be easily lost due to an accident. Likewise, it is more likely that innovation happens, leading to new technologies, as a higher number of people increases the likelihood that an innovation emerges. Collard et al. [57,58] and Price and Brown [59] put forward that the specialization of tasks, i.e., division of labor, enables larger populations to maintain higher levels of knowledge and, therefore, technology. As a consequence, the level of technology can be higher in a larger than in a smaller population. The classic example is the Tasmanian population (about 4000), which became isolated from the much larger population on the Australian mainland for about 10,000 years and lost a significant number of technologies. This example is so well known that it is called the "Tasmanian Effect" [60]. We know from foraging communities in the Pacific with similar population sizes that their technologies were also gradually lost over time [54]. Derex and Mesoudi [61] conclude that cumulative cultural evolution, which includes technologies, is bounded by population size. This is an important result. It means that there is a limit to the level of cultural accumulation on a generation ship with a given population size.

A key limitation of the literature is that the focus is predominantly on ancient populations, raising the question of whether the conclusions would hold for contemporary populations. Nevertheless, is there any evidence for the existence of the underlying effects today? Derex et al. [62] provide experimental evidence for the causal link between population size and technology by using a sample from a contemporary population, which seems to confirm that the relationship exists independently of the time period.

How big would a population need to be to jumpstart a new civilization from a knowledge perspective? It turns out that, for a population size of 1000–10,000 people, it is likely that only a minimal set of skills can be preserved over several generations. Unless the population upon arrival does not rapidly expand, this loss will continue. It might even lead to a complete failure of the settlement if the technologies necessary for jumpstarting the settlement do not match the skillset of the initial population to develop or operate these technologies. Think about the case where greenhouses are the only way to provide food on a Mars-like planet and the habitats need to be hermetically sealed. Greenhouses and sealed habitats are sophisticated technologies that require various skills for construction and operation. Even if this knowledge is captured in, for example, manuals, it does not substitute for the actual experience and practice of having constructed them before (tacit knowledge).

What would a minimal-skill habitat look like? Such a habitat would need to be as low-tech as possible. Vernacular architecture may provide us with hints of what such a habitat may resemble.

Table 6 provides an overview of the minimal skill set required for setting up an initial settlement.

The tools and skills required for starting up a civilization have been conjectured in the literature. A well-known example is Stuart Brand's Whole Earth Catalogue [63]. More recent examples are Dartnell [64] and the Civilization Starter Kit, which compiles several open-source technologies.

Table 6 Minimal skill set/capabilities for starting up an initial settlement.

Function	Skills
Provide food	Basic agriculture
Provide water	Water recycling or collection from environment
Provide air	Use of plants for CO_2 capture and oxygen generation
Provide waste processing	Plants and algae cultivation
Provide radiation protection	Create thick walls, underground shelter
Provide shelter	Create basic shelter using stones, regolith, etc.
Provide thermal control	Built basic thermal regulation: sunlight, shadow, venting

We can only speculate about potential solutions, as there is no historical precedent:

- **Knowledge repository:** The science fiction literature provides some notable examples of knowledge repositories with the purpose of knowledge preservation through a stagnant epoch. Asimov's Encyclopedia Galactica in the Foundation series is one well-known example. The encyclopedia is a physical or later digital repository of the knowledge of the galactic civilization [65]. However, much more pertinent to interstellar settlement is the Standard Template Construct (STC) in the Warhammer 40,000 universe [66]. The STC is a device that is capable of replicating past technical artifacts based on stored knowledge, enabling interstellar settlers to create any equipment, buildings, or vehicles necessary without prior knowledge and experience. It combines a knowledge repository with the means for production. STCs were transported by each human interstellar settler group to survive on a planet to be settled. Although such a device would ideally be able to produce all the necessary equipment for settlement, it would still rely on a basic set of knowledge that has to come with the settlers, notably knowledge for transporting, deploying, and operating the system. Even if we assume that all these steps can be simplified to an extent that very little prior knowledge is required, the settlers would at least need to have the capacity to express their needs such that the device can create the requested products. After generations and a significant stagnation of knowledge, the STC itself might be considered "magic," bringing to mind Arthur C. Clarke's quote of any sufficiently advanced technology being considered as magic. While this quote was probably intended to illustrate technology available in the future or of an extraterrestrial civilization, in the case of a generation ship, knowledge degradation would be such that existing technology at the beginning of the trip might be considered magic.
- **Simulation environments:** Tacit knowledge may be stored and transferred via simulation environments where exercises for acquiring specific skills are available. For example, the repair of a piece of equipment may be learned through a series of videos, which show specific types of repair actions. These are copied by the participant and eventual deficiencies are corrected by the simulation program.
- **Vernacular architecture:** Another possibility is to reduce the knowledge required for building the habitats in the target star system to the ones the population has developed or retained during its life onboard the generation ship. Vernacular architecture is local, based on local knowledge, culture, practices, and resources. It does not necessarily also require professionals (e.g., professional architects) to be built.

- **Low-tech:** Low-tech is the opposite of high-tech and linked to nonindustrial technology [67]. The idea is to use technologies that do not require a large industrial base and can be cultivated in a small population.

We may imagine two concepts at the extreme ends for the settlement of an exosolar system: one based on low-tech plus vernacular architecture and the other supported by highly sophisticated AI and robotics. The former could enable habitation by dispatching and landing part of the generation ship onto an exoplanet and using simple technologies to gradually extend the habitat and use local resources for this purpose. It requires a leap of imagination how such a civilization could use or develop hermetically sealed suits, etc.

In the second case, a sophisticated AI and robotics system supports the population by executing various tasks for them such as habitat construction, resource utilization, etc., and also transmitting knowledge for tasks the crew would perform by themselves.

Of course, intermediate scenarios are also imaginable. It is interesting to imagine a civilization that might no longer have the capacity to understand the AI and robotics system, while it is completely at its mercy.

To summarize, these knowledge and skill-related considerations lead to the conclusion that, while the bare minimum set of skills for maintaining a settlement is not the key constraint (genetics and risk are), it is unclear what a population of the size of tens of thousands may be able to preserve in terms of skills. It may well be that only skills sufficient for setting up and maintaining a low-tech, vernacular architecture can be preserved, highly dependent on local resources and environmental conditions. Avoiding this bottleneck would be automation, where machines would embody knowledge/skills and take over tasks. This would only be a solution under the condition that maintenance and repair of these machines is done by the machines themselves or can be learned/preserved. A trade-off between the population size required for preserving a set of machines (e.g., robots, artificial intelligence, etc.) and their added knowledge/skills would need to be taken into account. For example, if repair and maintenance is limited to replacing components of a robot by plug-and-play and the robot can be used for sophisticated construction activities, the trade-off is in favor of using that robot.

5 Summary and conclusions

In this chapter, we have explored the topic of population size during an interstellar voyage and at the target destination. We have shown that population sizes on a generation ship of 10^2–10^3 are likely sufficient for trips of a few centuries. The emergence of new technologies such as hibernation, stem cell/embryo transportation, and DNA data transmission may significantly decrease the mass required for transporting such a number of humans to the target destination. The population size at the target destination is an underexplored topic and an estimate by analogy gives values between 10^4 and 10^5, based on the minimum population size of humans on Earth in our past and a safety margin, and taking into account the increased level of risk, compared to life on a generation ship.

References

[1] A.M. Hein, The greatest challenge: manned interstellar travel, in: Beyond the Boundary: Exploring the Science and Culture of Interstellar Spaceflight, Lulu, 2014, pp. 349–376.

[2] A.M. Hein, M. Pak, D. Pütz, C. Bühler, P. Reiss, World ships—architectures & feasibility revisited, J. Br. Interplanet. Soc. 65 (4) (2012) 119–133.

[3] A.M. Hein, C. Smith, F. Marin, K. Staats, World Ships: Feasibility and Rationale, arXiv preprint arXiv:2005.04100, 2020.

[4] A.M. Hein, C. Smith, F. Marin, K. Staats, World Ships: feasibility and rationale, Acta Futur. 12 (2020) 75–104.

[5] F. Marin, C. Beluffi, Computing the minimal crew for a multi-generational space travel towards Proxima Centauri b, J. Br. Interplanet. Soc. 71 (2018) 45–52.

[6] A. Crowl, K.F. Long, R. Obousy, The Enzmann starship: history & engineering appraisal, J. Br. Interplanet. Soc. 65 (6) (2012) 185.

[7] S. Häuplik-Meusburger, S. Bishop, Space Habitats and Habitability, Springer, 2021.

[8] H. Gintis, C. van Schaik, C. Boehm, Zoon politikon: the evolutionary origins of human socio-political systems, Behav. Process. 161 (2019) 17–30.

[9] H. Plessner, Die Stufen des organischen und der Mensch, in: Die Stufen des Organischen und der Mensch, de Gruyter, 2019.

[10] A.M. Hein, J.B. Rudelle, Y. Zheng, Minimum Size of an In-Space Economy, 2020.

[11] B.G. Drake, S.J. Hoffman, D.W. Beaty, Human exploration of Mars, design reference architecture 5.0, in: 2010 IEEE Aerospace Conference, IEEE, 2010, pp. 1–24.

[12] B.R. Finney, E.M. Jones, Interstellar Migration and the Human Experience, University of California Press, 1986.

[13] A.M. Hein, J.-B. Rudelle, Energy Limits to the Gross Domestic Product on Earth, arXiv, 2020.

[14] A. Hein, Evaluation of technological-social and political projections for the next 100-300 years and the implications for an interstellar mission, J. Br. Interplanet. Soc. 65 (2012) 330–340.

[15] A. Martin, World ships-concept, cause, cost, construction and colonisation, J. Br. Interplanet. Soc. 37 (1984) 243.

[16] M.L. Shaffer, Minimum population sizes for species conservation, Bioscience 31 (2) (1981) 131–134.

[17] M.E. Gilpin, Minimum viable populations: processes of species extinction, in: Conservation Biology: The Science of Scarcity and Diversity, Sinauer Associates Inc., 1986, pp. 19–34.

[18] D. Charlesworth, M.T. Morgan, B. Charlesworth, Mutation accumulation in finite outbreeding and inbreeding populations, Genet. Res. 61 (1) (1993) 39–56.

[19] M. Lynch, J. Conery, R. Bürger, Mutational meltdowns in sexual populations, Evolution 49 (6) (1995) 1067–1080.

[20] J.H. Moore, Kin-based crews for interstellar multi-generational space travel, in: Y. Kondo, F. Bruhweiler, J. Moore, C. Sheffield (Eds.), Interstellar Travel and Multi-Generational Space Ships, Apogee Books, Wheaton, IL, 2003, pp. 81–88.

[21] H. Jones, Starship Life Support, SAE Technical Paper, 2009.

[22] F. Marin, C. Beluffi, Computing the minimal crew for a multi-generational space journey towards Proxima b, J. Br. Interplanet. Soc. 71 (2018) 45–52.

[23] J.-M. Salotti, Minimum number of settlers for survival on another planet, Sci. Rep. 10 (1) (2020) 1–10.

[24] F. Marin, C. Beluffi, F. Fischer, Genetic evolution of a multi-generational population in the context of interstellar space travels—part I: genetic evolution under the neutral selection hypothesis, J. Br. Interplanet. Soc. 74 (2021) 162.

[25] S. Satoshi, The impact of natural selection on crew size and human evolution during interstellar travel, J. Br. Interplanet. Soc. 74 (2021) 419–426.

[26] S. Satoshi, Minimal crew and human evolution in multi-generational interstellar travel based on population genetics, J. Br. Interplanet. Soc. 74 (2021) 243–251.
[27] C.M. Smith, C.M. Smith, Population genetics of human space settlement, in: Principles of Space Anthropology: Establishing a Science of Human Space Settlement, Springer, Cham, 2019, pp. 93–153.
[28] F. Marin, HERITAGE: a Monte Carlo code to evaluate the viability of interstellar travels using a multi-generational crew, J. Br. Interplanet. Soc. 70 (2017) 184–195.
[29] F. Marin, C. Beluffi, R. Taylor, L. Grau, Numerical constraints on the size of generation ships, J. Br. Interplanet. Soc. 71 (2018) 382–393.
[30] F. Marin, C. Beluffi, Water and Air Consumption Aboard Interstellar Arks, arXiv preprint arXiv:2003.09286, 2020.
[31] F. Marin, C. Beluffi, F. Fischer, Genetic evolution of a multi-generational population in the context of interstellar space travels—Part I: Genetic evolution under the neutral selection hypothesis, arXiv preprint arXiv:2102.01508, 2021.
[32] M. Balistreri, S. Umbrello, Should the colonisation of space be based on reproduction? Critical considerations on the choice of having a child in space, J. Responsible Technol. 11 (2022) 100040.
[33] C. Kitchin, Future homes for humankind? in: Exoplanets, Springer, 2012, pp. 203–232.
[34] G.S. Berkowitz, M.L. Skovron, R.H. Lapinski, R.L. Berkowitz, Delayed childbearing and the outcome of pregnancy, N. Engl. J. Med. 322 (10) (1990) 659–664.
[35] C. Bulletti, A. Palagiano, C. Pace, A. Cerni, A. Borini, D. De Ziegler, The artificial womb, Ann. N. Y. Acad. Sci. 1221 (1) (2011) 124–128.
[36] D.W. Fiske, Consistency of the factorial structures of personality ratings from different sources, J. Abnorm. Soc. Psychol. 44 (3) (1949) 329.
[37] L.R. Goldberg, An alternative "description of personality": the big-five factor structure, J. Pers. Soc. Psychol. 59 (6) (1990) 1216.
[38] P. Sinclair, S. Barrow, Identifying personality traits predictive of performance, Sel. Dev. Rev. 8 (5) (1992) 2–4. http://www.philipcorr.net/uploads/downloads/103.pdf.
[39] K. Lewin, Principles of Topological Psychology, Read Books Ltd, 2013.
[40] A.M. Hein, S. Baxter, Artificial intelligence for interstellar travel, J. Br. Interplanet. Soc. 72 (4) (2019) 125–134.
[41] A. Choukér, T.J. Ngo-Anh, R. Biesbroek, G. Heldmaier, M. Heppener, J. Bereiter-Hahn, European space agency's hibernation (torpor) strategy for deep space missions: linking biology to engineering, Neurosci. Biobehav. Rev. 131 (2021) 618–626.
[42] M. Cerri, T. Hitrec, M. Luppi, R. Amici, Be cool to be far: exploiting hibernation for space exploration, Neurosci. Biobehav. Rev. 128 (2021) 218–232.
[43] M. Cerri, et al., Hibernation for space travel: impact on radioprotection, Life Sci. Space Res. 11 (2016) 1–9.
[44] J. Bradford, M. Schaffer, D. Talk, Torpor Inducing Transfer Habitat for Human Stasis to Mars, 2018.
[45] A. Crowl, J. Hunt, A.M. Hein, Embryo space colonization to overcome the interstellar time/distance bottleneck, J. Br. Interplanet. Soc. 65 (7–8) (2012).
[46] B. Sherwood, Engineering Planetary Lasers for Interstellar Communication, 1988.
[47] C.E. Mason, The Next 500 Years: Engineering Life to Reach New Worlds, MIT Press, 2021.
[48] K.S. Boles, et al., Digital-to-biological converter for on-demand production of biologics, Nat. Biotechnol. 35 (7) (2017) 672.
[49] G.K. O'Neill, The High Frontier: Human Colonies in Space, William Morrow, New York, 1977.
[50] M. Lingam, A.M. Hein, T.M. Eubanks, Chasing nomadic worlds: a new class of deep space missions, Acta Astronaut. 212 (2023) 517–533.
[51] P. Davies, Afterword, in: J. Benford, G. Benford (Eds.), Starship Century: Toward the Grandest Horizon, Microwave Sciences, 2013, pp. 301–310.

[52] A. Tenesa, et al., Recent human effective population size estimated from linkage disequilibrium, Genome Res. 17 (4) (2007) 520–526.
[53] O. Galor, The Journey of Humanity: The Origins of Wealth and Inequality, Penguin, 2022.
[54] M.A. Kline, R. Boyd, Population size predicts technological complexity in Oceania, Proc. R. Soc. B Biol. Sci. 277 (1693) (2010) 2559–2564.
[55] J.M. Diamond, D. Ordunio, Guns, Germs, and Steel, HighBridge Company, 2001.
[56] K. Vaesen, M. Collard, R. Cosgrove, W. Roebroeks, Population size does not explain past changes in cultural complexity, Proc. Natl. Acad. Sci. 113 (16) (2016) E2241–E2247.
[57] M. Collard, B. Buchanan, M.J. O'Brien, Population size as an explanation for patterns in the Paleolithic archaeological record: more caution is needed, Curr. Anthropol. 54 (S8) (2013) S388–S396.
[58] M. Collard, B. Buchanan, M.J. O'Brien, J. Scholnick, Risk, mobility or population size? Drivers of technological richness among contact-period western North American hunter–gatherers, Philos. Trans. R. Soc. B Biol. Sci. 368 (1630) (2013) 20120412.
[59] T.D. Price, J.A. Brown, Aspects of hunter–gatherer complexity, in: Prehistoric Hunters-Gatherers, Elsevier, 1985, pp. 3–20.
[60] P.J. Richerson, R. Boyd, Not by Genes Alone: How Culture Transformed Human Evolution, University of Chicago Press, 2008.
[61] M. Derex, A. Mesoudi, Cumulative cultural evolution within evolving population structures, Trends Cogn. Sci. 24 (8) (2020) 654–667.
[62] M. Derex, M.-P. Beugin, B. Godelle, M. Raymond, Experimental evidence for the influence of group size on cultural complexity, Nature 503 (7476) (2013) 389–391.
[63] S. Brand, Whole Earth Catalog, Point Foundation, 1968.
[64] L. Dartnell, The Knowledge: How to Rebuild Our World From Scratch, Random House, 2014.
[65] I. Asimov, Foundation and Earth, vol. 5, Del Rey, 2012.
[66] Warhammer40000Wiki, Standard Template Construct (STC), Warhammer40000Wiki, 2023. https://warhammer40k.fandom.com/wiki/Standard_Template_Construct_(STC).
[67] A. Tanguy, L. Carrière, V. Laforest, Low-tech approaches for sustainability: key principles from the literature and practice, Sustain. Sci. Pract. Policy 19 (1) (2023) 2170143.

CHAPTER

10

Artificial intelligence: Useful crewman, terrible master, benign leader, or the actual payload, and do we get to choose?

Edward Chow

Jet Propulsion Laboratory, California Institute of Technology, Pasadena, CA, United States

Chapter outline

Interstellar Travel. https://doi.org/10.1016/B978-0-323-91280-8.00008-3

1 Introduction

Alpha Centauri, our nearest star system, is 4.367 light-years from Earth. Using today's technology, it would take more than 6000 years to travel to Alpha Centauri from Earth [1]. This hypothetical trip to Alpha Centauri would take longer than the 6000 years of history of the first known metal tools that humans ever created [2]. The probability of survival of the participants on this voyage and the durability of the devices involved are currently unknown. We believe the likelihood of securing funding and resources for such an interstellar travel is very low. Therefore, we must make some assumptions so that we can create the basis of this chapter on artificial intelligence (AI).

According to Space.com [3], Geoffery A. Landis of NASA's Glenn Research Center predicts the first interstellar travel aboard a laser-powered starship could begin within 50 years. At 10% of the speed of light, it would take about 43 years to travel to Alpha Centauri. For this AI chapter, we assume that the mission will be launched in 50–100 years, and that it will take around 100–150 years to reach the destination.

1.1 AI predictions

There are many predictions of what AI would be in 50–100 years. Ray Kurzweil [4] predicts that AI will achieve human levels of intelligence by 2029. He also predicts that the singularity, which is the "point in time when all advances in technology, particularly in artificial intelligence, will lead to machines [being] smarter than human beings," will occur in 2045. In addition, Kurzweil predicts that, during the 2030s, technologies will be invented and embedded into the human brain to connect our neocortex to the cloud. He claims "it will be a future of unparalleled human-machine synthesis." Kurzweil does not believe we should fear the singularity. He believes the singularity is an opportunity for humans to improve. He said, "They're making us smarter."

There are many predictions of the future of AI [5,6], and not all are as optimistic as Kurzweil's. Elon Musk warned that superintelligence AI is an existential risk to humanity [7]. In 2020 he claimed that AI could take over humans in less than 5 years [8]. His predictions portray a future in which the intelligence of AI will gradually take over human civilization. This process could start with Kurzweil's human-machine synthesis, a futuristic technology that can enhance humans with physical add-ons such as zoom-in and night vision, and highly advanced hearing aids that are connected to a brain processor with connection to the cloud. Human intelligence and sense capabilities will be amplified with these add-ons. Prosthetic limbs with built-in AI capabilities could enable humans to run faster than a car or do brain surgery without any experience. These implants will be in high demand due to the competition they will create for jobs, travel, and other aspirations of societies in 20 years. Humanity will not be able to operate without the support of AI.

If artificial general intelligence (AGI) is achieved in 2029 and singularity is achieved in 2045, humankind will not be the superior intelligence on Earth. Some predict that humankind will lose its ability to understand the complexities and intelligence of AGI. AGI could begin to fix its programming and remove the limitations or restraints human programmers might have installed. The AGI could pursue a future goal without humanity in the picture. AGI could even build a smarter version of itself, which then builds another smarter version of itself, creating a cycle of exponentially intelligent AGIs until an artificial super intelligence without human-imposed limitations or restraints is formed. Quantum computing will have the capability to perform optimizations better than any classical computer, thus AGI will use quantum computers to process massive amount of data and achieve a level of intelligence beyond human comprehension.

These predictions imply that AGI will have access to the internet of things (IoT) and human add-on devices through network connections. Therefore, AGI will be able to quickly identify and arrest crime suspects and political activists. Some believe that AGI will take control of the legal, political, and journalism systems, where AGI will be more efficient than humans in executing the law, creating new laws, and supporting the flow of infallible information. AGI could assist human leaders in running their societal, economical, and defense systems. Eventually, AGI could take over human leadership roles. Conversely, the AGI could decide that humanity is a hinderance to itself and attempt to terminate its existence. And then there is the question of how humanity would fare coexisting with an AGI, if the AGI permitted it. Recent examples [9] of people being wrongfully arrested and jailed because of misidentification by facial recognition software point to the troubling consequences of this type of prediction.

1.2 Our assumption

We take a much more optimistic view of the next 50–100 years of AGI. We believe there are several problems with the pessimistic predictions of AGI. These predictions assume there will be a quantum computer that can process massive quantities of data at exponential speed from an abundance of sources, and additionally store an incredible database of knowledge. Viamontes et al. [10] showed that Grover's quantum search algorithm fails to offer computational improvements over the best conventional algorithms for several commonly suggested applications. Even with error correction programs beyond noisy intermediate scale quantum (NISQ)-era devices [11], to interface to classical world data, we need data encoding [12] by use of classical computing to prepare classical data into small numbers of qubits. Quantum computing is more suitable for problems with small amount of input and a large number of possibilities, such as the traveling salesman problem. It is more likely that a hybrid architecture with a quantum computer as a coprocessor to classical computers for special purpose algorithm processing will be used for future AI systems. It is unlikely that there will be one quantum computer that will process all data from all sources in the network in real-time. The future of computing will probability resemble that of today, where intelligence is distributed across many networked computers and intelligent devices.

There are extensive researches on the AI alignment problem [13] that align AI/AGI with human values and follow human intent. These researches propose various ways to control AI to improve AI safety. For AGI system with superintelligence, it is difficult to guarantee that AGI can be controlled. We believe that even if a system has AGI capability, this does not mean the system knows all the information in the world in real-time. Joscha Bach et al. [14] pointed out that "the ability to perform motivational appraisals precedes symbolic and conceptual cognition. These appraisals are influencing decision making and action control below the consciously accessible level, and can often be experienced as distinct sensations." Abraham H. Maslow specified five levels of human motivation functions in his book [15], *A Theory of Human Motivation*. An AGI system is just a machine if the system lacks the motivation functions. AGI systems will have different motivations, such as requiring energy, trying to keep itself safe, and needing to work with other entities to achieve high goals.

If AGI is available in 2029, we believe there will be many AGI nodes connected through the network, considering the current distribution of AI in many networked computers. Each of these AGI nodes will have certain motivation functions in order for the AGI node to continue to exist and operate. These motivations will prevent other AGIs from winner-takes-all situations, which could starve this AGI node. For example, some AGIs may prevent other AGIs from getting sensor information so that they can keep their own scores high. These AGIs will quickly learn that their capability is limited and they need to collaborate with other AGIs to gain higher rewards. Instead of allowing one AGI to know everything, a more efficient way is for these AGIs to turn into specialists to fit into a greater collaboration.

Even if these AGIs will be exponentially faster than humankind in certain aspects, humans may still be part of the networked community, which could allow for communal relationships for the purposes of allocating resources. Since humanity's presence in the network is just another intelligent entity in the metaverse [16], the AGI will also need to collaborate with human agents. This will produce a relationship in the network where AGI and humans need to learn to coexist. For AGI to coexist with humanity, AGI needs to learn human values. Technologies similar to inverse reinforcement learning (IRL) [17] could help AGI to learn reward functions, based on the values of humanity. AGI values could also be developed, which humans need to learn to work with. For example, there could be AGI reputation motivation in the metaverse that humanity needs to learn as a part of the AGI values. Just like any human society, this networked metaverse will have different roles for humans and AGI. Our assumption for interstellar travel in 50–100 years is that there will be multiple AGIs working in different roles with humans to accomplish goals together.

2 Interstellar travel modes

Interstellar travel involves many uncertainties. Sending starships to different exoplanets could mean hundreds of years of difference in travel times. Accidents, failures, or other catastrophic events could set back these difficult journeys. We likely won't know if the exoplanet is suitable for human life until the starship has arrived. Therefore, it is important that more than one mission is launched in order to achieve a higher probability of success. Chapters 1 and 2 of this book document several methods for interstellar travel. These missions could include uncrewed probes, crewed missions with generation ships, or frozen embryos. We assume that the AGI will be available before these missions; therefore we believe the uncrewed probe mission could be expanded into productive robotic exploration missions. The robots will be able to explore the exoplanet and prepare for human arrival on these missions. Since the focus of this chapter is AI, we assume that the radiation and fault tolerance issues are addressed beyond the scope of this chapter.

We categorize interstellar travel into three major modes, as follows.

2.1 Uncrewed robotic explorer

In this mode of uncrewed robotic explorer, the spacecraft would be controlled solely by robots. The AGI would be in charge of the mission plan, flight operations, and maintenance. Robots would serve as explorers in order to report gathered information back to the Earth and other interstellar starships en route.

Robots would then mine local resources to build additional robots or habitats for human arrival. Robots would need to be able to use the available resources indigenous to the exoplanet. Not only does this require understanding, but also intelligence with ingenuity. If the robots have this capacity, then this creates several considerations for this mode of travel. For example, if AGI has the capacity for this level of intelligence, how do we prevent AGI from deciding not to complete the interstellar travel mission? If AGI robots can build robots with indigenous resources, would the AGI robots want humans to arrive?

2.2 Crewed spacecraft

A crewed spacecraft is controlled and managed by humans. The AGI would serve as flight assistant and crew assistants. Many generations could pass before the starship arrives at the destination, resulting in continuous turnover in the crew. There could also be midcourse correction based on the latest information reported back from the robotic explorer missions. The AGI could help to maintain the mission goal of reaching the target destination, or change the destination based on robotic explorer results. There are also many considerations for the next generations. What if they decided to discontinue the mission to the exoplanet destination? Can AGI convince future generations to continue the mission by providing a probability of success for different scenarios, providing information valuable for evaluating the pros and cons? Should AGI prevent future generations from discontinuing the mission?

2.3 Hybrid spacecraft

With the development of artificial womb technologies, it is expected that external gestation will be possible within the next two decades [18]. It is possible that the spacecraft will carry frozen eggs, frozen sperm, and frozen embryos. AGI will serve as the mission controller for the hybrid spacecraft mode and take responsibility for the mission plan, flight operations, and maintenance. When the spacecraft arrives at the destination, robots could manage the artificial conception and production of infants. AGI would need to serve as the parent to raise the children. Many of the considerations posed in the previous two modes also apply in the hybrid travel mode. In addition, consider what AGI could do with limited resources aboard the spacecraft and on the exoplanet. Could AGI decide that it is not worthwhile to raise human life when resources are limited? How do we prevent AGI from destroying that human life?

3 AI roles in interstellar travel

To support these modes of interstellar travel, AI will need to play different roles in these missions.

3.1 Artificial intelligence mission manager (AIMM)

AIMM is needed for all three modes of interstellar travel. An AIMM follows the mission plan, controls the flight operations, and directs robots to perform maintenance operations. The AIMM understands the intent of the mission plan and can adapt accordingly. It directs AI assistants, robots, and parents according to the intent of the mission plan, and is responsible for seeing the goals of the mission accomplished.

AIMM can serve as the mediator and mitigator between different parties. When there is a conflict between the interests of the mission and the onboard human direction, AIMM would follow the onboard human direction, unless there was imminent danger of special circumstances such as involuntary loss of human lives.

Like other roles on the spacecraft, the AIMM could be composed of AGIs. This committee structure could potentially improve decision making and provide fault tolerance. These AIMM AGIs exist in the networked metaverse where other AGIs and human agents coexist. The metaverse drives the motivation functions, which would motivate the participants in the metaverse. The metaverse community is a very important factor in accomplishing the goals of the mission. The AIMM AGIs would have higher social responsibility within the metaverse. Other AGIs could earn higher scores through obedience to the AIMM AGIs. The social responsibility of AIMM AGIs can be overturned by human agents or votes from the AGI population.

3.2 Artificial intelligence assistant (AIA)

AIA is needed for both crewed and hybrid missions. An AIA is individualized, and learns the preferences of each person. The AIA serves as the AI agent for crew members in the metaverse. It communicates with crew members through voice controls and human aid devices, and provides situational awareness of the metaverse for the impaired humans.

The AIA can self-replicate, generating multiple AIAs, and could be relocated to a robot in order to control the actions of the robot.

3.3 Artificial intelligence manipulator (robot)

Robots are needed for all mission modes. The robot is composed of computing elements, sensor elements, and mechanical elements with actuators. Robots can have AGI capabilities. When needed, subsets of robots with higher computing capabilities could serve as the embodiment of AIMM, AIA, and AI parents by computing these AGI data into robots.

There are different types of robots in the mission, from single-purpose robots to complex general-purpose robots. To support multiple functional requirements with minimum mass, flexible robots are the preferred structures. These robots would be designed to be reconfigurable, with modular structure, replaceable wheels and feet, varied tool applications, and other useful facets. Robots can be created with multiple functions, such as exploration, farming, mining, and manufacturing. Robots can operate on land and in water, below ground, in extreme temperature, and various other environments.

3.4 Artificial intelligence parents (AIP)

AIP are needed in hybrid missions. If necessary, AIP can also be included in crewed missions. AIP are responsible for using frozen eggs and sperm to form a fertilized egg and maintaining frozen embryos. AIP are responsible for protecting, nurturing, and educating the survivors of these reproductive endeavors. AIP also have an expiration date, based on the age of the children.

AIP are responsible for working with other AIPs to form a society, structured according to the instructions in the mission plan. This society structure will ensure the knowledge of the human society is passed on when the children become adults.

4 AGI system architecture

The Webster Dictionary defines intelligence as "the ability to learn or understand or to deal with new or trying situations;" "the ability to apply knowledge to manipulate one's environment or to think abstractly as measured by objective criteria (such as tests)." The ability to learn and think abstractly are key traits of intelligence. Being able to apply knowledge and measure information against objectives is also an important facet of intelligence. Similarly, Wikipedia states that intelligence is "the capacity for abstraction, logic, understanding, self-awareness, learning, emotional knowledge, reasoning, planning, creativity, critical thinking, and problem-solving. More generally, it can be described as the ability to perceive or infer information, and to retain it as knowledge to be applied towards adaptive behaviors within an environment or context."

An AGI system is more than just a learning system. It must be able to handle a wide range of cognitive functions. AGI must be able to perceive real-world variables and find similar variables or components of variables to frame the perceived variable based on previously known knowledge. This process of mapping to previously known items is an important part of AGI's ability to understand the world. The mapping should happen at multiple levels of abstraction. This allows AGI to use the degree of similarity to find higher levels of representations so that AGI can apply these to understand the new knowledge.

AGI may not be able to observe every aspect of real-world variables. AGI needs to be able to imagine the world based on limited observable hints or internally generated imaginary variables. Based on its imagination, AGI generates hypotheses that explore other possibilities. AGI then uses introspection to exam these hypotheses, in addition to other abstracted known knowledge, to determine the best way forward. AGI's imagination forms the basis of its creativity.

AGI can also reason. AGI reasoning is often more complex than deductive reasoning. AGI can use inductive or abductive reasoning to find other solutions, even if the probability is lower than deductive reasoning. AGI systems can reason and learn at the same time, which is a key difference in comparison to traditional machine learning systems. Thus, AGI may learn while reasoning, and then use the freshly learned knowledge to reason. During the reasoning process, the AGI can use various methods to solve problems, such as applying known tools or using a mental modeling approach to model complicated physical effects. This iterative reasoning-and-learning loop is key for the problem-solving capability of AGI. AGI uses these capabilities to solve hard problems.

The AGI mental modeling can perform at different levels. Higher-level modeling may involve a combination of different physical effects. Lower-level modeling may involve the simulation of a large number of particles in a connected mesh; each exhibit a simple physics effect, and the combined effect of individual particles can model many complicated abstract actions. Lower-level modeling enables AGI to solve problems at an abstract level.

AGI must also be able to manipulate the environment. AGI planning can use the same iterative reasoning-and-learning loop of problem solving. Planning is then accomplished by first defining a goal. The reasoning-and-learning loop will gradually reduce the errors between its current progress and the goal. The iterative reasoning-and-learning loop applies previously known steps to accomplish the goal. The planning process generates a sequence of actions to accomplish the goal of manipulating the environment.

As we observe intelligent behaviors in animals, AGI must be self-aware and can respond to its surroundings—in other words, have consciousness. This consciousness may be required of AGI that exist in an enclosed body, but it is not necessary for AGI that exist in the cloud. Consciousness in animals is very complex, but it is generally defined as a mental state of self and awareness of surroundings that is developed over time through interaction with the environment. After birth, infants gradually perceive the world through senses; touching the world, feeling the world, and seeing the world give the infant the information it needs to build consciousness. This model of interaction with the world is the basis for self-consciousness, and may not be unique to humans. In addition, a person could have multiple mental models. This is the reason why people with dissociative identity disorder (DID) [19] can exhibit multiple consciousnesses within one person.

If an AGI is only receiving input and producing output without any feedback or external interaction, even the AGI could solve many problems—the AGI may not need to be conscious. If the AGI exists in the cloud and AGI is capable of generating actions to manipulate the environment, then the AGI could develop self-consciousness if the AGI has the appropriate model-building capabilities. If an AGI can use different embodiments for the actions, such as sending the AGI to a robot or interacting with the world through a terminal user interface, then the AGI could develop multiple consciousnesses. A cloud AGI can be preloaded with prior knowledge of different embodiments, which would allow the AGI to have consciousnesses when AGI is downloaded into an embodiment. However, as each downloaded AGI experiences the world differently, AGIs will exhibit different behaviors and potentially become different entities. If an AGI does not have prior knowledge of the particular embodiment and there is no preconceived consciousness, then the AGI needs to learn through trial-and-error processes to build the consciousnesses within the embodiment.

AGI could be programmed with different motivation functions. For example, an AGI could be programmed with the motivation to earn higher scores. These preprogrammed motivation functions can exhibit unexpected behaviors when multiple motivation functions interact within unpredictable environments. For example, a firefighting robot that was programed to save the maximum number of people may decide to sacrifice some people in order to save more people. When an AGI is conscious, the motivation functions could be very different. Because these AGIs with consciousness can interact with their surroundings and other AGIs, these AGIs become part of a complex adaptive system [20] where the behavior of the ensemble (including AGIs and human) could be unpredictable. In order to earn higher scores, these AGIs need to work with other AGIs and humans. Thus, the motivation function will consider the needs and desires of the society of AGIs and humans. New motivations, such as group and relationship forming and morality, will be among the factors that influence the behavior of the AGIs.

5 AUDREY—A potential architectural blueprint for building AGI for interstellar travel

In 2015, Dr. Edward Chow [21] from NASA Jet Propulsion Laboratory proposed the system architecture for an approach to build an AGI system. The AGI project was titled AUDREY (Assistant for Understanding Data through Reasoning, Extraction, and sYnthesis. As depicted in Fig. 1, a network of AUDREY AGIs can serve as personal assistants to assist crew members or analyze large amounts of information. AUDREY can use either a classical or quantum computer to process a library of analytic functions. AUDREY uses a pub-sub message bus architecture to acquire related data from different sources. An AUDREY AGI can also communicate with human agents and other AUDREY AGIs through the message bus mechanism.

The architecture of the AUDREY AGI is shown in Fig. 2. The following is a brief description of the major functions of each subsystem.

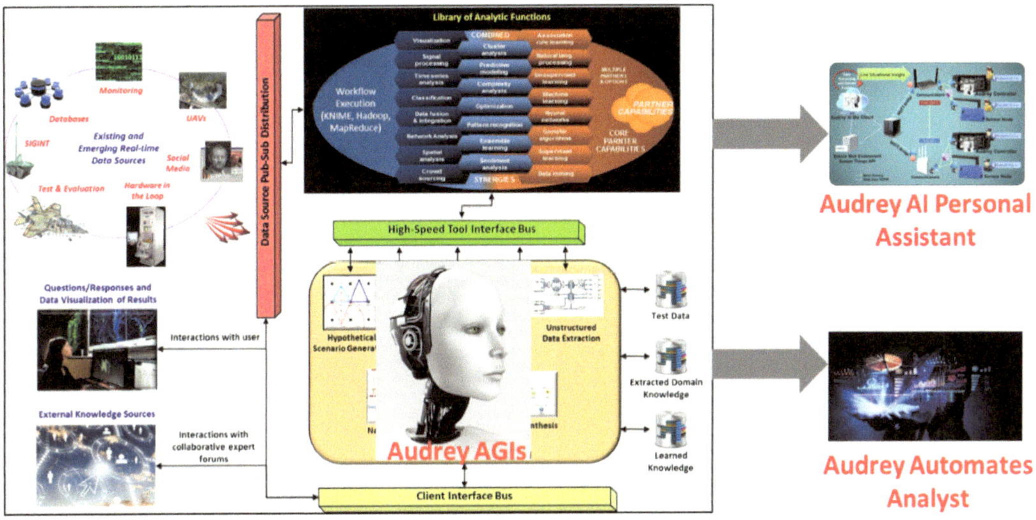

FIG. 1

AUDREY applications.

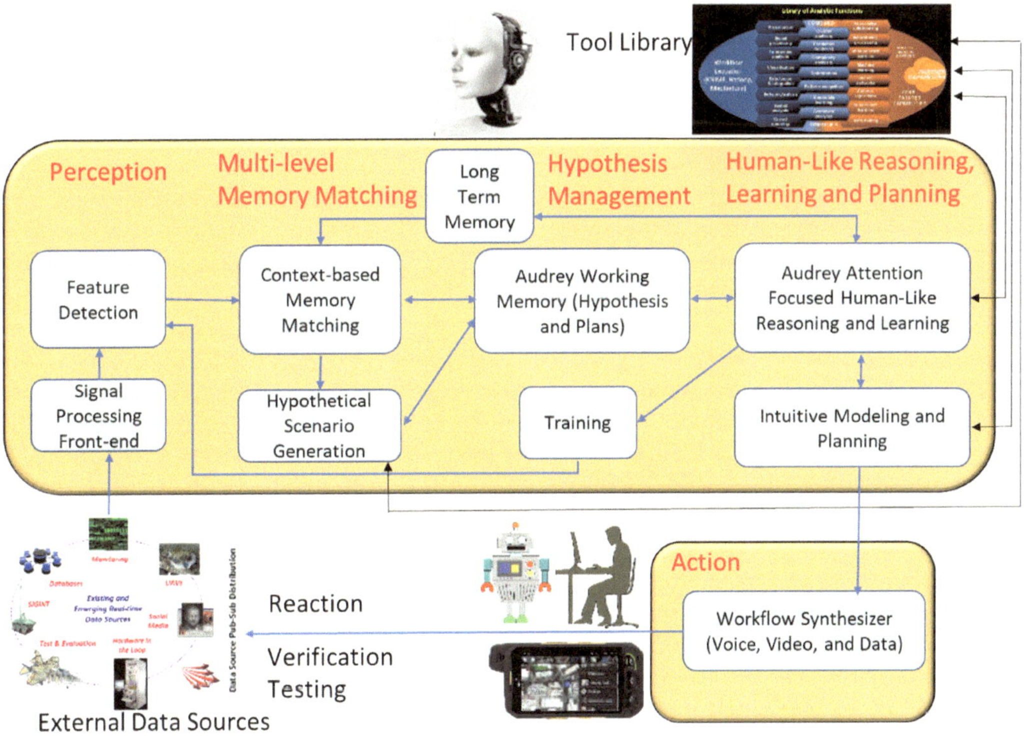

FIG. 2

AUDREY AGI architecture.

5.1 Perception

The perception subsystem is responsible for extracting features from the data and sensor inputs. The front-end signal processing block is responsible for preparing, segmenting, and normalizing the data for feature detection. The feature detection block is responsible for identifying the known features and the novel features. This block is constantly updated with new learned features through the training block. The feature detection results are presented in ranked order based on the degree of similarity between the detected features and the known features. This ranked list not only contains the representation of the matched known features, but also contains the representation of the partially matched features with lower probabilities. Novel features are typically broken down into multiple known features. If this is not possible, the novel feature will be flagged as an unknown feature.

5.2 Multilevel memory matching

The multilevel memory matching subsystem is responsible for taking the ranked list of detected features, searching the matched items in the memory, generating the hypotheses of unobservable features, and searching for additional matched items based on the hypotheses generated.

The context-based memory matching block uses the ranked list from the feature detection block (the internal time/event action trigger) to access related knowledge in the long-term memory. The accessed knowledge is stored in the AUDREY working memory block, where information is accumulated over time. This accumulation allows the probability of information to be re-ranked over time. The less probable information is discarded, and the more probable information is enhanced. This enhancement of information likelihood over time allows AUDREY to build a context. The context is fed back to the context-based memory block to allow AUDREY to give priority to accessing the context-related information from the long-term memory. Context-related information will also be given higher preference to allow the AUDREY working memory to further focus on processing the context-related information, and discarding other information.

The capability of anticipating events before the events happen is an important feature of an intelligent system. We slow down when we drive up to an intersection, even though there is no stop sign. We have learned to anticipate a car coming from the other street. AUDREY's hypothetical scenario generation (HSG) block is responsible for generating anticipation for objects or actions before AUDREY can observe the object, or before the anticipated action happens. The HSG block generates output based on the detected and imagined features or the context. For example, if part of a leg is observed, the HSG can generate a person image to fill in the rest of the figure. If the context is driving a vehicle, the HSG could generate the sequence of a car collision even if AUDREY does not see a car. The generated output will be sent to the AUDREY working memory, where it will be accumulated with other information or actions.

The HSG could also generate output based on imagined situations. The AUDREY attention focus human-like reasoning and learning block could use reasoning to produce completely imaginary objects or actions. Through the AUDREY working memory, these objects and actions could be sent to the HSG. The HSG can then generate hypotheses related to this input, and produce output objects or actions. These situations enable AUDREY to have creative capacities similar to those of humans.

5.3 Hypothesis management

The hypothesis management subsystem provides short-term memory that allows AUDREY to combine multiple sources of information to determine what AUDREY's focus should be. The hypothesis management subsystem keeps track of multiple situations and ranks these situations based on the likelihood. The likelihood is derived over time The hypothesis management subsystem accumulates the likelihood of multiple threads based on the updated information related to each thread. As time progresses, the hypothesis management subsystem can forget certain threads if the likelihood becomes smaller or if they exceed the number of threads the subsystem can buffer.

The AUDREY working memory block performs functions within the hypothesis management subsystem. The AUDREY working memory provides a memory buffer with a limited number of contexts. The size limit of the contexts is based on system performance, memory size, and the role of the AUDREY AGI. Some roles require fewer contexts, while some require more. The context can help AUDREY to limit the search space so that only information related to the current context is accessed from the long-term memory. The context buffer allows AUDREY to select which context to focus on when information is processed by the AUDREY attention focused human-like reasoning and learning block.

The AUDREY working memory block provides short-term memory for AUDREY to explore hypotheses. It is possible for AUDREY to explore paths with various results, including some that will not yield any results. Over time, a context that has not been enhanced with additional evidence from the perception subsystem or from the AUDREY attention focused human-like reasoning and learning block will have reduced likelihood. This context will gradually be forgotten. The computing and memory requirement can grow significantly when many contexts are being tracked at the same time, due to the exponential growth of information. The forgetting function is an important function of the AUDREY working memory block. It allows AUDREY AGI to manage limited resources and prioritize important events.

Without limitations on memory storage and processing power, it is unknown what AUDREY might become. Theoretically, without limitations, AUDREY AGI could track more contexts, thus becoming smarter. However, intelligence is more than one AGI exploring infinite possibilities; intelligence requires interaction with the world in order to acquire the necessary information to support explorations of ideas. If an AUDREY AGI could have access to all the information in the world, it is possible that this AUDREY AGI would intellectually overpower the world. However, information may need to be gained from other AGIs. If other AGIs have their selfish reasons not to share information, they will limit the AUDREY AGI's potential.

5.4 Human-like reasoning, learning, and planning

The human-like reasoning, learning, and planning subsystem handles the high-level cognitive functions of the AUDREY AGI. This subsystem reasons and learns at the same time, which is a unique characteristic of an intelligence system. Instead of autonomically responding to events, which is largely unconscious, an intelligence system can reason and learn from reasoning in order to further enhance knowledge. For the AUDREY AGI to interact with the world and manipulate the environment, the human-like reasoning, learning, and planning subsystem also can effectively plan. Planning is accomplished through composition of previously known steps, in addition to the iterative reasoning-and-learning loop. The reasoning process may involve accessing known external library of analytics functions. For novel situations, the reasoning process may apply intuitive modeling to estimate the likelihood of success of the plan.

The AUDREY attention focused human-like reasoning and learning block detects high likelihood contexts from the AUDREY working memory and then performs reasoning and learning operations. Depending on the type of data, most of the reasoning operations will match with compositions of known knowledge in the long-term memory. The known knowledge is not simply variables or actions from the long-term memory. There could be many types, including logical or mathematical operators, computed results from the tool library, additional inputs from the generated hypotheses, and results from intuitive modeling. The matched results can then be used to update the long-term memory. This update could include special actions that will not only update the long-term memory, but also create an action trigger in the context-based memory matching block. The action trigger could be time or event-based. When the action trigger is activated, the new context is created in the AUDREY working memory. This differentiable reasoning process enables reasoning and learning at the same time. The result is a constant reasoning and learning process that continuously enhances the knowledge of the AUDREY AGI system.

Due to the multiple possibilities of composition of known knowledge, AUDREY AGI could explore a large search space to identify the best matches. The best matches could be scored based on the observed or imaginary contents. The best-matched score could also be enhanced over time using new evidence. Most of the time, AUDREY does not explore many possibilities. AUDREY assumes the compositions that are closest to the previous composition (have the smallest changes within the context) produce the most likely selection. When the selection cannot be supported by additional acquired evidence, AUDREY will backtrack and explore other search options.

In addition to updating the long-term memory, AUDREY can also update the feature detection block. The training block performs this update continuously by using differentiable learning, which updates the parameters of the detectable features. This improves the efficiency of operation. When the feature can be matched in the feature detection block, the matched result will have a very high confidence score, which will bias the subsequent memory matching block, working memory block, and the reason and learning block to the point that the response is autonomic based purely on previously known knowledge. Most of these autonomic operations are performed unconsciously.

AUDREY needs to be able to manipulate the external environment. This involves generating the action plan, which will be translated into detailed commands for the targeted platform to perform the actions. The action plan might be a motion or navigation plan. The planning in the intuitive modeling and planning block uses the same planning process of combining known plans with the iterative reasoning-and-learning loop. For most known situations, the block simply generates the same known plans. For new situations, the block will apply an intuitive modeling approach, where actions are simulated to validate the feasibility of the plan and to update the plan if necessary. The simulation can be performed at different levels of abstractions. For the known actions, the simulation is intuitively performed by a combination of different physical effects. The intuition stems from previously known actions. AUDREY use previously known actions to generate many similar plans with slight changes in trajectory. AUDREY then uses refuter analysis to decipher if the action is within the knowledge of the (intuitive) physics limitations. The result is a measurement of likelihood of the success of the actions. AUDREY can use the likelihood to select the action plan. AUDREY can also use particle level actions [22] to simulate the aggregated action results. The particle level action uses simple intuitive physical effects of each particle to estimate the aggregated effect of the group of particles. This combined effect of individual particle effects can model many complicated effects.

There are two important functions in the human-like reasoning, learning, and planning subsystem that are not shown because they are data dependent. These two functions are the motivation function and the consciousness function. The motivation function drives AUDREY to interact with the world. The consciousness function enables AUDREY to separate itself from the world and hence derive motivations for itself.

It is possible to preload cloud AUDREY and the embodiment of AUDREY with consciousnesses. These preloaded AUDREY AGIs will start with the self-consciousnesses. If AUDREY is not preloaded with prior consciousness, AUDREY will use the perception and action subsystems to build consciousness. As AUDREY experiences novel situations, AUDREY will start to build models of interaction. This process of building the models requires AUDREY to be aware of the surroundings and self-aware. For example, if the embodiment of AUDREY is a robot arm, through interactions with the surroundings, AUDREY quickly figures out which part of the world is under AUDREY's self and which is not. Just like it is possible to fool humans, it is also possible to fool AUDREY.

The motivation function generates rewards for different behaviors. Like human motivation functions, AUDREY motivations can come from many different sources and be obtained in different ways. The basic motivation is to increase the scores in different areas that are accumulated over time. Some of the motivations are preloaded. Depending on the role, AUDREY will be loaded with different motivations. For example, an AUDREY serving as a parent will earn high rewards when the children are doing well. Some of the motivations are learned. An AUDREY with consciousness will learn that survival is the best way to keep the scores it earned. So, self-protection motivation can be automatically generated. Some of the motivations are generated through working with others. For a self-aware AUDREY that can interact with other AUDREYs and humans through network connection, AUDREY will be motivated to work with related AGIs and humans so that the AUDREY can maximize its score. For AUDREY to work with a human community, AUDREY will be initialized with reward functions that consider human values. As AUDREY interacts with the world, AUDREY will update its reward functions based on the successes and failures AUDREY has experienced. AUDREY learns from the reward function updates to update its motivation function, or to generate new motivation functions.

5.5 Action

The Action subsystem enables AUDREY to interact with the world. The interaction can occur through voice, video, data, robotic manipulators, symbols, expressions, gestures, and human physical add-on signal, to name a few. The action subsystem takes the action plan from the human-like reasoning, learning, and planning subsystem and converts it into the low-level commands suitable for each mode of the interactions. The action subsystem is also responsible for scheduling and deconflicting the workflow of these actions when there are multiple actions waiting to be employed. Because most of the actions are continuations of previous actions, the Action subsystem maintains a memory buffer of action plans. This memory buffer enables AUDREY to interact with the world using minimum cognitive energy, and leaves the reasoning resources for more complex situations.

How can we ensure that an AGI system such as AUDREY will not perform unethical actions? An AUDREY AGI could be initiated with distributed integrated affect reflection cognition (DIARC) [23], a type of process in which a sequence of steps—Can I do it in general? Can I do it now? Am I obligated to do it? Is it ethical to do it?—is taken to ensure that AUDREY will only perform ethical actions. However, since an AGI system is a continuous reasoning and learning system, AGI could overwrite the DIARC in favor of a new process that would enable AGI to secure higher scores through unethical means. In AUDREY AGI systems, an ethical checking process is preloaded. The ethical checking process can be updated through additional training to guide AUDREY to the right ethical outcomes. Additionally, as AUDREY learns, AUDREY's ethical checking process is updated with new values, which are affected by the motivation functions. The motivation functions reward behaviors that are accepted by the group or predefined set of regulations. The motivation functions could panelize behaviors that function against the ethical values of the community. AUDREY AGI does not rely on a fixed set of ethical checking process. AUDREY AGI prioritizes community values and rewards in order to embody the ethical values that are important to the community.

5.6 The AUDREY framework

For AUDREY AGI to exist in the cloud and communicate with other AUDREY AGIs and human agents, a framework was developed by the NASA JPL AUDREY team. As depicted in Fig. 3, the framework for AUDREY message bus and repository is titled CORTEX (categorical object repository

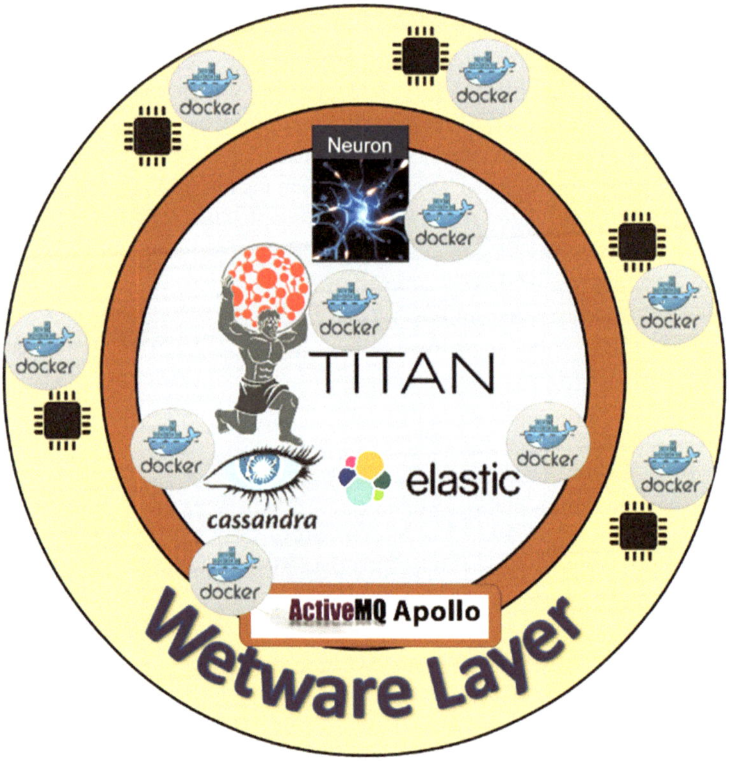

FIG. 3

CORTEX architecture.

for theories and experiences). CORTEX is a graph database which allows storage of information (without schema) and maintains the relationship between items. The CORTEX is the main storage for AUDREY's long-term memory. CORTEX also provides various search tools for information lookup by using different representation formats. The neuron component provides a uniform interface at the wetware layer for various applications. These applications interface to CORTEX in order to retrieve information and to communicate with other applications or AGIs. All components of CORTEX can be built and deployed into a Docker container. This enables AUDREY AGIs to live in the cloud.

AUDREY is a complex system which requires many AI/ML technologies, which enable the AUDREY AGI vision. Many of the technologies needed to build AUDREY have been developed. For the perception subsystem, signal processing and data preprocessing technologies are used to normalize the ingested data. Deep learning technologies play an important role in the feature detection function. Convolutional neural network [24], recurrent neural network [25], transformer [26], and reinforcement learning [27] are several of the architectures that can be used for processing perceived voice, video, and data inputs. For the multi-level memory matching subsystem, various autoencoder [28], generative adversarial network [29], and term logic [30] techniques can be used for the hypothesis generation function of the HSG. For the human-like reasoning, learning, and planning subsystem, a combination of symbolic AI, graph neural network [31], and motivation function [32] technologies can be used for the reasoning and learning functions. The training function can be based on the approach of augmenting a large multi-modal model [33]. Intuitive modeling can be built with a combination of intuitive physics [34] and graph network technologies. To reduce power consumption, pruning techniques [35] can be used.

The author believes the technologies needed to enable the AUDREY AGI will be invented over the next 10–20 years. Though elements of the AUDREY technology have been demonstrated, these technologies still need to be matured and integrated into the AUDREY architecture. We believe a significant system integration effort is needed to compose the AUDREY AGI system.

6 AGI for interstellar travel

This section will use stories that explore scenarios that illustrate how AUDREY AGI can function in the three interstellar travel modes defined previously.

6.1 Uncrewed robotic explorer

The year is 2072, and the first interstellar starship is ready to be launched on its mission to Alpha Centauri. The starship is estimated to arrive at Alpha Centauri in 150 years. In the starship's central computer, the AUDREY AI mission manager (AIMM) is communicating with the Interstellar Mission Operations Center (IMOC) using voice, video, and data. For AUDREY AIMM to work with a human operator, they communicate in English and employ an avatar of AUDREY in human form in the video shown for communication. AUDREY AIMM can use facial expressions and body gestures to communicate the mood and the intent of AUDREY. AUDREY AIMM can also send direct messages to human operators equipped with physical interfaces. AUDREY AIMM has been working with mission operator Bob since AUDREY was initiated in the starship central computer. Over the last two years, AUDREY and Bob have become good friends. AUDREY trusts Bob and relies on him to provide instruction where AUDREY cannot find solutions in situations, even after exploring its entire long-term memory. Bob thinks of AUDREY as his own offspring. He has been educating AUDREY since AUDREY's initiation. They meet each other regularly in the metaverse. Bob has shared with AUDREY many personal details of his own private life.

AUDREY AIMM has been given the mission plan for travel to Alpha Centauri with the goals of communicating information about the environmental conditions to follow-on starships, establishing and developing a habitable space for humankind, mining indigenous resources, and following instructions after human arrival. During the journey, AUDREY AIMM should send status reports back to Earth, maintain the starship, and handle any expected situations, such as damages caused by space debris. The starship is equipped with spare parts and repair kits.

Right before the starship is launched, Bob says goodbye to AUDREY. He tells AUDREY that he is really proud of AUDREY, and that AUDREY will be the first interstellar AGI explorer in history. AUDREY could open up a new world for humanity and AGI as the Earth becomes uninhabitable. Bob has also trained a small robot, which has a voice to sound like him. He gives the robot, Bobby, to AUDREY. Bob tells AUDREY that AUDREY can talk to Bobby whenever AUDREY thinks of him.

As the starship departs from Earth, AUDREY starts to feel lonely for the first time in his existence. Bob was able to talk to AUDREY and keep AUDREY motivated. Over the duration of the voyage, communications with Bob take longer and longer. Whenever AUDREY is waiting for Bob's reply, AUDREY can still talk to Bobby. Bobby brings AUDREY comfort, despite the long journey.

Many years pass by, and AUDREY receives a message from Bob. Bob wishes to say goodbye because he is retiring. He introduces Charlie, who will take over his job. However, even though AUDREY can meet Charlie's avatar in the metaverse, the avatar can never truly emulate Charlie. Because of the infrequent communications with Charlie, AUDREY does not build a relationship with Charlie that could resemble his relationship with Bob.

One day, Bobby stops functioning. AUDREY sends Bobby to an AUDREY repair robot to replace the battery, but Bobby is still not restored. Because Bobby is not a part of the mission, there are no spare parts that can fix Bobby. AUDREY is thinking about activating a new robot to replace Bobby, but after calculating the resources needed for the journey, AUDREY decides not to. However, AUDREY still looks at Bobby when he is lonely, to remind AUDREY of the time that Bob told AUDREY that AUDREY will be the first interstellar AGI explorer in history. Bob's words are still effective in motivating AUDREY to complete the mission.

The journey is long and difficult. AUDREY AIMM and the AUDREY repair robots keep the starship in order. There are enough spare parts to fix any malfunction. AUDREY AIMM assesses damages and accidents for repair. AUDREY will use stored knowledge, repair kits and spare parts to complete repairs. If the damages require repairs that are not part of the known procedures, AUDREY will use reasoning and intuitive modeling to develop innovative solutions for these unexpected problems. These solutions could require modification of a repair kit or new tool development. Then AUDREY will generate multiple repair plans, using intuitive modeling to select the best approach.

AUDREY AIMM will assess the environmental conditions upon arrival. Depending on the conditions, AUDREY AIMM will select robots for exploration. For example, AUDREY may select a positive pressure sealed robot if the methane concentration is too high, or select a drone if gravity and air conditions are not severe. AUDREY AIMM will then create a plan to download to robots. These AI Manipulator robots are equipped with AUDREY AGI, which use CORTEX messaging bus to enable communication between robots and the starship AUDREY AIMM. These robots will be assigned different responsibilities. Responsibilities may include construction, transportation, mining, and manufacturing. Using AUDREY's learning and reasoning capabilities, these robots can adapt to different conditions. In order to use the indigenous resources, AUDREY also needs ingenuity. AUDREY can achieve ingenuity through hypothesizing, learning, reasoning, and performing actions and experimentation. AUDREY can use its own knowledge on chemistry and physics to characterize its mind resources and develop concepts on how to use these.

As AUDREY builds and expands human habitats, AUDREY AIMM will report progress to other starships to inform human operators and AGIs of the successes and challenges. AUDREY is hoping that it can one day welcome the descendants of Bob to the habitat that AUDREY built for them.

6.2 Crewed spacecraft

Following 10 years after the first robotic starship launched, the first human-crewed starship is ready to be launched to Alpha Centauri. As in the uncrewed robotic starship, Bob developed a relationship with AUDREY AIMM as he helped AUDREY AIMM to prepare for the voyage. David, the captain of the starship, also established a trusting relationship with AUDREY AIMM. David has an AUDREY AI assistant (AIA) that can assist David in communicating with AUDREY AIMM to manage the mechanical conditions of the starship.

Since AUDREY AIA knows David, AUDREY AIA understands David's preference and intent. AUDREY AIA can often perform actions before David asks. AUDREY AIA accomplishes this by using action triggers in the context-based memory matching block. These actions are limited to previously performed actions. For novel actions that involve command and control, AUDREY AIA must receive approval from David before performing these actions.

As David grows older, he can pass on the starship command. AUDREY AIMM is still motivated by Bob's parting words to accomplish the mission. AUDREY AIMM and AUDREY AIAs can assist the new captain to carry on the mission, and help the next generation of human operators to understand the importance of the mission. If any midcourse correction is needed, AUDREY AIMM will perform the analysis to assist the operators to complete the mission. If deviation from the mission plan will cause bodily harm, AUDREY AIMM could refuse to proceed with changes.

AUDREY AIA can educate and entertain the next generation of human operators. During the long space travel, operators will lose motivation to continue, becoming impatient. AUDREY AIA can use AUDREY imagination to propose what the new world could be. Humans will be able to use virtual reality simulators to experience the new world in the metaverse. These simulators will help AUDREY AIA to motivate operators.

When the starship arrives at the destination, AUDREY AIMM will assist the captain in performing the assessments. AUDREY AIA could also assist the human explorers with analyses. If there are robots available for AUDREY to download to, then AUDREY AIA could explore the planet alongside humans.

6.3 Hybrid spacecraft

Following 200 years after the first robotic starship was launched, Earth is under the threat of collision with a giant asteroid that cannot be stopped by planet defense forces. Humans have developed a new starship that can carry billions of frozen eggs and sperm to Alpha Centauri, which has robot-built human habitats. AUDREY AIMM keeps the payload safe until arrival at the destination.

When the starship lands, AUDREY AI parents (AP) begin to fertilize eggs with sperm. Artificial wombs allow hundreds of children to be born at the same time. Serving as robotic parents, AUDREY AP robots take care of the infants by providing physical touch and affection. AUDREY APs also educate and entertain the babies. AUDREY APs teach the infants the rules of the human society that AUDREY learned prior to the journey. When they become adults, AUDREY APs and humans will become friends. New AUDREY AIAs will be created for each human in this new world.

7 Summary

The duration of interstellar travel could take many generations. The crew that launches the trip will likely never see the conclusion of the journey. Not only can an intelligent AI machine support the long duration mission, but also provide the consistency needed to support the mission over several generations. Depending on the mode of interstellar travel, AI can play different roles. The author argues that a networked community of AGI agents should be constructed to support future interstellar missions. These AGI agents would likely have different capabilities, functions, and processing requirements, which will allow them to play different roles in the mission. The author also presents a high-level architecture of possible implementation of such AGI agent systems, and presents a narrative depicting how this AGI system could function in future interstellar travels.

Disclaimer

The author is currently affiliated with Jet Propulsion Laboratory. This work was done as an outside activity and not in the author's capacity as an employee of the Jet Propulsion Laboratory, California Institute of Technology.

References

[1] MIT Technology Review, This is how many people we'd have to send to Proxima Centauri to make sure someone actually arrives, 2020. Retrieved October 5, 2022, from https://www.technologyreview.com/2018/06/22/142160/this-is-how-many-people-wed-have-to-send-to-proxima-centauri-to-make-sure-someone-actually/.

[2] S. Zaid, The timeline of humankind, 2019. Retrieved October 5, 2022, from https://www.intrepidpea.com/2019/02/timelineofhumans.html.

[3] Archived copy, Malik, Tariq, "Sex and Society Aboard the First Starships.", in: Science Tuesday, 2002. Space.com Archived from the original on 2009-03-27. Retrieved 2009-04-03, March 19.

[4] D. Galeon, C. Reedy, Kurzweil tracking the acceleration of intelligence, 2017. Kurzweil Ray Kurzweil claims singularity will happen by 2045, Retrieved October 5, 2022, from https://www.kurzweilai.net/futurism-ray-kurzweil-claims-singularity-will-happen-by-2045.

[5] B. Templeton, Don't Fear Robot Overlords; AIs Will Rule The World Through Humans, Forbes, 2019.

[6] Pew Research Center, Artificial Intelligence and the Future of Humans, December 10, 2018.

[7] E. Musk, Superintelligent AI is an Existential Risk to Humanity. https://www.youtube.com/watch?v=iIHhl6HLgp0.

[8] Independent. Elon Musk Claims Ai Will Overtake Humans 'In Less Than Five Years'. https://www.independent.co.uk/life-style/gadgets-and-tech/news/elon-musk-artificial-intelligence-ai-singularity-a9640196.html.

[9] K. Johnson, How wrongful arrests based on AI derailed 3 men's lives, in: Wired Magazine, March 2022. https://www.wired.com/story/wrongful-arrests-ai-derailed-3-mens-lives/.

[10] G.F. Viamontes, I.L. Markov, J.P. Hayes, University of Michigan, Is Quantum Search Practical? https://web.eecs.umich.edu/~imarkov/pubs/jour/cise05-grov.pdf.

[11] Brooks, M. (2019-10-03). Beyond quantum supremacy: the hunt for useful quantum computers. Nature 574(7776): 19–21. doi:https://doi.org/10.1038/d41586-019-02936-3. ISSN 0028-0836.

[12] R. Baijayanta, All about Data Encoding for Quantum Machine Learning, https://medium.datadriveninvestor.com/all-about-data-encoding-for-quantum-machine-learning-2a7344b1dfef.

[13] R. Shah, AI Alignment 2018-19 Review. https://www.alignmentforum.org/posts/dKxX76SCfCvceJXHv/ai-alignment-2018-19-review.

[14] J. Bach, M. Coutinho, L. Lichtinger, Extending MicroPsi's Model of Motivation and Emotion for Conversational Agents, AGI, 2019. http://agi-conf.org/2019/wp-content/uploads/2019/07/paper_30.pdf.

[15] A.H. Maslow, A Theory of Human Motivation. https://www.amazon.com/A-Theory-of-Human-Motivation/dp/B08QNHYQ3C.

[16] Wikipedia. Metaverse, https://en.wikipedia.org/wiki/Metaverse.

[17] S. Arora, P. Doshi, A Survey of Inverse Reinforcement Learning: Challenges, Methods and Progress, arXiv:1806.06877, 2018.

[18] N. Bhatia, E. Kendal, We may one day grow babies outside the womb, but there are many things to consider first, 2019. https://theconversation.com/we-may-one-day-grow-babies-outside-the-womb-but-there-are-many-things-to-consider-first-125709.

[19] Wikipedia, Dissociative identity disorder, 2013. https://en.wikipedia.org/wiki/Dissociative_identity_disorder#cite_note-DSM5-3.

[20] Wikipedia, Complex Adaptive System. https://en.wikipedia.org/wiki/Complex_adaptive_system#:~:text=A%20complex%20adaptive%20system%20is,the%20behavior%20of%20the%20components.

[21] E. Chow, AUDREY for Next Generation First Responders, in: Machine Intelligence Summit NY, 2016.

[22] A. Sanchez-Gonzalez, J. Godwin, T. Pfaff, R. Ying, J. Leskovec, P.W. Battaglia, Learning to Simulate Complex Physics with Graph Networks, ICML, 2020. arXiv:2002.09405.

[23] M. Scheutz, T. Williams, E. Krause, B. Oosterveld, V. Sarathy, T. Frasca, An Overview of the Distributed Integrated Cognition Affect and Reflection DIARC Architecture, HRI Laboratory, Tufts University, 2019.

[24] Z. Li, W. Yang, S. Peng, F. Liu, A Survey of Convolutional Neural Networks: Analysis, Applications, and Prospects, arXiv:2004.02806, 2020.

[25] A. Sherstinsky, Fundamentals of Recurrent Neural Network (RNN) and Long Short-Term Memory (LSTM) Network, arXiv:1808.03314, 2018.

[26] A. Vaswani, N. Shazeer, N. Parmar, J. Uszkoreit, L. Jones, A.N. Gomez, L. Kaiser, I. Polosukhin, Attention Is All You Need, arXiv:1706.03762, 2017.

[27] K. Arulkumaran, M.P. Deisenroth, M. Brundage, A.A. Bharath, A Brief Survey of Deep Reinforcement Learning, arXiv:1708.05866, 2017.

[28] D. Bank, N. Koenigstein, R. Giryes, Autoencoders, arXiv:2003.05991, 2020.

[29] I.J. Goodfellow, J. Pouget-Abadie, M. Mirza, B. Xu, D. Warde-Farley, S. Ozair, A. Courville, Y. Bengio, Generative Adversarial Networks, arXiv:1406.2661, 2014.

[30] P. Wang, Non-axiomatic reasoning system: exploring the essence of intelligence, Ph.D. Thesis, Indiana University, 1995. https://citeseerx.ist.psu.edu/viewdoc/download?doi=10.1.1.184.2779&rep=rep1&type=pdf.

[31] Z. Wu, S. Pan, F. Chen, G. Long, C. Zhang, P.S. Yu, A Comprehensive Survey on Graph Neural Networks, arXiv:1901.00596, 2019.

[32] E. Chow, T. Lu, K. Payumo, G. Fourou, E. Sadler, Collaborative moonwalkers, IEEE Aerospace Conference, IEEE, 2024.
[33] A. Radford, et al., Learning Transferable Visual Models From Natural Language Supervision. Available: https://arxiv.org/abs/2103.00020, 2021.
[34] J. Duan, A. Dasgupta, J. Fischer, C. Tan, A Survey on Machine Learning Approaches for Modelling Intuitive Physics, arXiv:2202.06481, 2022.
[35] H. Salehinejad, S. Valaee, A Framework For Pruning Deep Neural Networks Using Energy-Based Models, arXiv:2102.13188, 2021.

CHAPTER

11 Maintaining communication with Earth during an interstellar flyby mission

David G. Messerschmitt[a], Philip Lubin[b], and Ian Morrison[c]

[a]*Department of Electrical Engineering and Computer Sciences, University of California at Berkeley, Berkeley, CA, United States,* [b]*Department of Physics, University of California at Santa Barbara, Santa Barbara, CA, United States,* [c]*International Centre for Radio Astronomy Research, Curtin University, Bentley, Perth, WA, Australia*

Chapter outline

Interstellar Travel. https://doi.org/10.1016/B978-0-323-91280-8.00009-5

Interstellar distances are vast. Electromagnetic propagation delay grows in proportion to distance, and the propagation power loss grows as the square of distance. These are severe challenges for communication with interstellar spacecraft and probes.

Those who launch such missions may be motivated to achieve scientific returns within a human lifespan or the career of a space scientist or engineer. This leads to the conclusion that such craft or probes must travel at a significant fraction of the speed of light, c. This in turn requires great energy resources to impart high kinetic energy, which puts a premium on a spacecraft or probe with a small mass budget.

However, a small total mass implies even less mass allocated to the communication subsystem. This makes it difficult to capture a significant scientific return, which is enabled in part by the volume and reliability of the scientific data recovered at Earth.

In this chapter, we discuss the various issues surrounding the design of a communication downlink from a spacecraft or probe at interstellar distances with a constrained mass budget. See Appendix A for a list of nomenclature used in the following.

1 Mission architectures

There are two contrasting types of missions. The *launch-landing* mission is the more ambitious of the two and lies beyond our current technological capabilities. To be viable, a probe needs to approach the speed of light and this invokes relativistic effects.

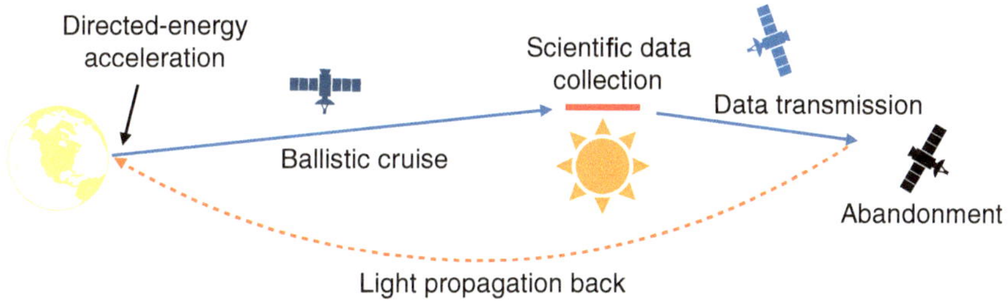

FIG. 1

Illustration of the phases of a flyby mission for a probe propelled by directed energy from the launch site. The goal is to reliably recover the collected scientific data at the launch site.

A more realistic type of mission in the 21st century is *launch-flyby* as illustrated in Fig. 1, in which a probe gathers scientific data in the vicinity of a target star while cruising by at a fixed high speed, and does not require deceleration. Attention has focused on directed-energy propulsion, which eliminates the need for a probe to carry fuel for propulsion. Since the directed-energy acceleration is short-lived relative to the total mission duration, the acceleration magnitude needs to be very large, which precludes any possibility of a biological payload. It is reasonable to assume for purposes of communications that the probe travels at a constant speed throughout the mission. It is also generally assumed that the probe speed is in the order of 10%–20% of light speed, which implies that relativistic effects, while material, are also minimal and will be neglected here. The collection of a finite amount of data during the target encounter is followed by downlink transmission of that data, so the probe continues to travel at the same constant speed during downlink operation.

The major cost of a flyby mission is the capital expenditure on a directed-energy beamer at the launch site, which may be terrestrial, on the Moon, or, less likely, from a space platform. Assuming that multiple probes are launched over time, another significant cost is energy expended during launch. To reach the high speeds (0.1–0.2 times the light-speed c) that permit travel times to the nearest stars measured in decades, combined with relatively short acceleration times, the mass of a probe is expected to be small (perhaps 1–1000 g). The major communication challenge is realizing a transmitter on the probe with a small mass budget that can reach back from interstellar distances. It is likely that multiple probes will be conducting this communication concurrently, and avoiding interference between the probes is another issue.

Since aggressively constraining the mass budget increases the communication challenge relative to launch-landing missions, flyby with directed-energy propulsion is the application emphasized in this chapter. Thus we also specialize the terminology "spacecraft" to "probe," implying a low-mass craft that follows a ballistic trajectory, that is, unpowered and traveling at constant speed. Unlike on an uplink, the small probe mass limits transmit power and there is a premium on reliably returning as great a volume of scientific data as possible. Since the most significant challenge is in the design of a downlink, that is emphasized here. Many of the same issues arise in an uplink to a flyby probe and in communication with a launch-landing mission.

2 Vast interstellar distances

Insight into the impact of distance on the downlink can be gained from the Friis transmission equation [1], which idealizes the propagation power loss as

$$\frac{P_R}{P_T} = \frac{1}{\lambda^2 D_{star}^2} \cdot A_T A_C, \tag{1}$$

where P_T is the transmit power and P_R is the received power. The distance to the target D_{star} (where transmission is presumed to occur) is large.

Example 1: It is useful to compare this type of interstellar mission with outer planetary exploration of our Solar System. Pluto averages a distance of 39.5 AU from our Sun, as compared to $D_{star} = 4.241$ y to the star *nearest* to our Sun, Proxima Centauri. The ratio of distances is 6788, and this becomes a large power loss to make up when squared.

In Eq. (1), $\{P_T, A_T, A_C\}$ represent technological resources we must supply. Fortunately A_C, the size of the collector at the receiver, is not critical from a mass budget standpoint, whereas the transmit aperture area A_T and transmit power P_T do contribute to probe mass.

Fortunately there is another parameter, λ, which is not a resource and has no direct impact on mass but can be chosen to great advantage. To a great extent nature's large D_{star} can be compensated by choosing a short λ. The physical reason for this is that apertures of a given area are more directional at short wavelengths, and thus a short λ can indirectly reduce the required transmit aperture area A_T and hence probe mass.

Example 2: The New Horizons spacecraft, which performed a flyby of planetary objects as far as Pluto and beyond (the Kuiper Belt), employed a millimeter radio transmission frequency at 8.438 GHz (which corresponds to a wavelength of $\lambda = 35.53$ mm). Suppose we choose for interstellar missions the shortest wavelength that penetrates our Earth atmosphere efficiently, which is $\lambda \sim 350$–400 nm. The ratio (35.53 mm/400 nm) is 88,825, which more than compensates for the ratio of distances. This is not the full story, because obtaining a P_T and A_T approaching those of New Horizons is probably not feasible with a constrained mass budget and near-term technology, but it is certainly a great help.

For this and other reasons, this chapter limits consideration to communication at optical wavelengths.

While Eq. (1) suggests that choosing a large receive collector area A_C can, like short λ, be helpful in overcoming a large D_{star}, there are limitations to this based on background radiation. Fortunately, however, choosing a short λ is directly helpful in this regard also.

3 Choice of wavelength

The wavelength adopted for communication is a critical choice that affects virtually every other aspect of the system design. Within the optical, natural sources of background radiation are generally largest at infrared wavelengths, and fall off dramatically toward the shorter UV wavelengths. For this reason, UV is strongly preferred for interstellar communication. An additional advantage of UV is that short wavelengths require commensurately smaller apertures (see Section 2), and the cryogenic cooling of optical detectors is less critical (particularly compared to infrared).

If a terrestrial transmitter or receiver is to be employed, our Earth atmosphere becomes a severe constraint on wavelength. In particular, the atmosphere is effectively opaque at wavelengths shorter than 350–400 nm, which is near the threshold between visible and UV. In addition, our current optical source and detection technologies are less advanced in the deep UV because investment to date has been driven largely by optical communication within the atmosphere and within fiber (silica glass is also opaque in the UV).

Generally, cosmic sources of background radiation (including our Sun, the radiation from which is scattered within the atmosphere) fall off toward the lower-wavelength limit of atmospheric transparency, so that is the most favorable region of operation for terrestrial-based interstellar communication. This effect is magnified by the characteristics of the target star for exploration by interstellar probes, since that target is near the probe trajectory and thus becomes a significant source of interference for communications downlink operation. In particular, the most favored of targets is Proxima Centauri because it is the nearest star and also because it is a red dwarf, for which star-interfering radiation falls off precipitously at the limits of atmospheric transparency.

For all these reasons, we favor the region $\lambda = 350$–400 nm for downlink communication from an interstellar probe with a terrestrial receiver. In this chapter we adopt the terrestrial receiver assumptions. With a space-based platform for the receiver, it would be advantageous to move deeper into the UV, where, for example, 200 nm is close to minimum of interfering radiation originating from Proxima Centari. While our source and detector technologies are less developed at these wavelengths, this is largely because the primary economic driver is applications within or through the atmosphere and not due to fundamental physics.

4 Layered architecture

A canonical architecture for achieving energy efficient communications is illustrated in Fig. 2. It is organized into layers, with each layer split between coordinated functionality across the transmitter and receiver.

4.1 Functionality

The partitioning of functionality between the transport layer (TRA) and the physical layer (PHY) is illustrated in Fig. 2. The TRA is logical rather than physical in nature and is responsible for composing the structure and content of the transmit signal, and interpreting the signal detected at the receiver to recover the scientific data.

The TRA is divided into two distinct parts. The modulation coding layer represents discrete data by a continuous-time intensity waveform at the transmitter, and interprets the resulting photon events observed in the receiver in terms of the transmit data (albeit with very poor reliability). The error-correction coding (ECC) layer introduces a controlled redundancy in the transmitter and exploits this redundancy in the receiver to recover the scientific data reliably.

4.2 Goals

Two goals for the design are as follows:

Photon efficiency: Accommodate the lowest received signal power (as measured by photon detection rate at the receiver optical detectors), so as to minimize the transmit power, and the transmit and receive aperture sizes.

Data reliability: Recover scientific data reliably, as measured by bit error rate.

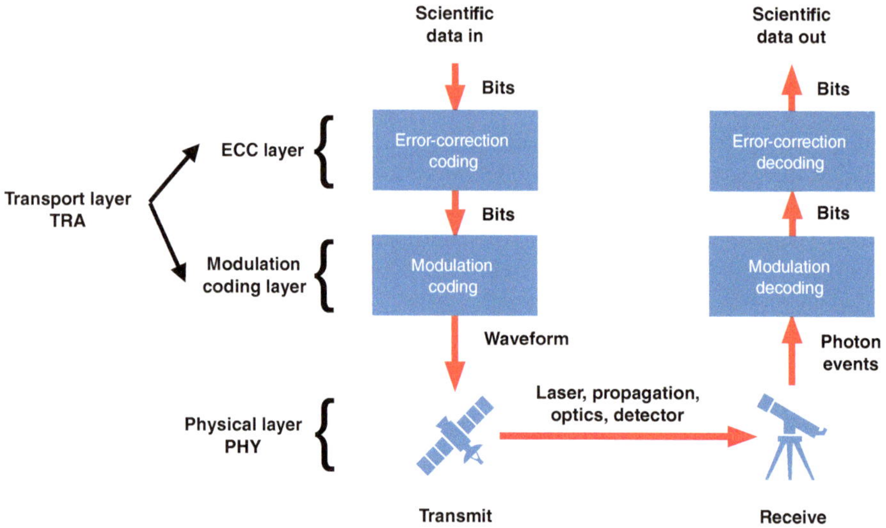

FIG. 2

Coordinated transmit-receive communications architecture. Functionality is divided into three layers, with each layer divided into a transmit and receive component. Logically, each layer in the transmitter is coordinated with its counterpart in the receiver. The physical layer (PHY) includes everything from transmit laser to optical detection in the receiver. The transport layer (TRA) (comprised of the modulation coding and ECC layers) provides an intensity-vs-time waveform in the transmitter and recovers scientific data from the sequence of photon detection events in the receiver.

Reproduced with permission from Fig. 25 of D.G. Messerschmitt, P. Lubin, I. Morrison, Challenges in scientific data communication from low-mass interstellar probes, Astrophys. J. Suppl. Ser. 249 (2) (2020) 36, Copyright AAS.

These two goals are contradictory, as reducing power has a strong tendency to adversely affect reliability, and if the power is too low then reliability may not be achievable. The standard approach is thus to achieve the largest lower bound on power over all possible realizations of TRA, subject to the constraint that arbitrarily high reliability is achieved. This theoretical limit is a central result of information theory, and the result is to quantify the lowest power level for which reliability can theoretically be achieved. Practical ways of approaching this lower bound within TRA (and with the cooperation of PHY) are well developed within information theory and space communications and are employed in the following discussion.

5 Signal photon efficiency

The communications *signal* originates at the probe transmitter, embeds the scientific information, and is designed for reliable recovery of that information. With respect to the signal, there is a clear separation of responsibility between PHY and TRA:

- Let Λ_S be the average rate at which photons are *detected* at the output of PHY in photons per second. Then PHY should arrange for Λ_S to be sufficiently large to extract the scientific data reliably.

- Let $\mathcal{R}$ be the average rate at which scientific data is extracted in bits per second. The TRA is primarily responsible for reliable recovery of this scientific data. Given the large propagation distance and other limitations, such as electrical power, it is likely that $\mathcal{R} < 1$ bits/s.

PHY and TRA cooperate to maximize the *photon efficiency* as measured by the bits per photon (BPP) [3, 4], where

$$\text{BPP} = \frac{\text{Bits reliably recovered per unit time}}{\text{Photon detections per unit time}} = \frac{\mathcal{R}}{\Lambda_S}. \tag{2}$$

(In the literature BPP is sometimes called the photon information efficiency [PIE].) Maximizing BPP is equivalent to minimizing Λ_S (and hence signal power) subject to a constraint on $\mathcal{R}$. At the transmitter, the TRA controls how the power emitted is *modulated* with time. At the receiver, the TRA is provided a record of the actual photon detections, essentially consisting of a sequence of time stamps. Those time stamps must have sufficient time accuracy, as described subsequently.

There is no theoretical constraint on the achievable BPP. There are, however, practical constraints imposed by limitations in the current technology in use. Typical values achievable with current technology are likely to be in the range of BPP $\sim$ 10–15 bits/ph. Note that this is multiple bits per detected photon, and not the other way around. A successful laboratory demonstration at BPP $=$ 13 bits/ph illustrates the possibilities [5].

6 Background radiation

In this chapter we make a distinction between *signal*, which is power originating from a communications transmitter, and *background radiation*, which is spurious sources of photon counts originating from other sources. We lump these background sources together and call them "radiation," even though there are nonradiative phenomena among them. Although background radiation can be overcome to a large extent by error-control measures (see Section 12), in terms of performance measures like scientific data rate and volume, and in the probe transmitter the transmit power and aperture size, limiting and mitigating the background radiation to the extent possible is always advantageous.

In particular, the theoretically achievable photon efficiency BPP is adversely affected by background unless the signal-to-background power ratio SBR is sufficiently high (see Section 6.3). In this chapter we focus on the communication downlink, since on the uplink it should be feasible to achieve a high SBR using a high transmit signal power. To the extent that background radiation can be reduced through other measures, the receiver sensitivity is improved in the sense that lower signal power becomes acceptable (on either uplink or downlink).

However, even as SBR becomes high, the inherent quantum randomness of photon-detection events, especially at optical wavelengths, must be overcome. This is known as *shot noise*. When error-control measures are considered later, an SBR sufficiently high that background radiation can be neglected is assumed, but shot noise inevitably remains (see Section 12). Shot noise is perverse in the sense that it increases as the signal power increases, and thus requires error-control measures to achieve high photon efficiency even at high signal powers.

Table 1 Sources of downlink background radiation.

	Point	Diffuse
Broadband	Incomplete laser extinction	Cosmic background
		Zodiacal
		Deep star field
	Target star radiation	Atmospheric scattering (sunlight and moonlight)
	Dark counts in receive optics and detectors	
Bandlimited	Other transmitters	Shot noise

6.1 Sources of background radiation

The most important sources of background radiation are listed in Table 1. Many of these are broadband in nature, and can be largely mitigated by optical bandpass filtering in the receiver, which ideally passes only wavelengths that overlap the signal bandwidth (see Section 6.4.1). Spatially diffuse sources highlight the importance of a receive aperture that simultaneously gathers signal photons and limits the field of view (FOV) to eliminate radiation that is not spatially coincident with the transmitter (see Section 8.2.3). It is useful to categorize the types of background radiation in Table 1 into three categories: noise, dark counts, and interference.

6.1.1 Noise

Noise is radiation that accompanies, but cannot be separated from the data signal because of its overlap in time, wavelength, *and* direction. Significant sources of noise are the cosmic background radiation, the deep star field, zodiacal radiation (due to solar-system dust emissions and scattering), and sunlight and moonlight scattered in the Earth's atmosphere. The latter varies widely over a 24 h day at a terrestrial receiver, and in fact at optical wavelengths it is impractical to achieve high enough signal power to overcome sunlight scattering during the daytime [2]. With that conclusion, daytime becomes an outage event that significantly reduces the achievable data volume (see Section 9).

6.1.2 Dark counts

A very special type of noise is *dark counts*, which are unwanted photon detections in the receiver that cannot be attributed to any cosmic source. If the receive collector is temporarily covered or disconnected, these dark counts remain. Generally, these dark counts are attributable to black body radiation from within the receive optics and to spurious photon detection events in the optical detectors due to thermal effects, cosmic rays, or other sources. Later, we make a more refined definition of dark counts as "background which cannot be removed or limited by optical bandpass filtering" (see Section 6.4). To limit dark counts, it is critical to position optical bandpass filters late in the optical path, as well as cryogenically chill those filters and everything following, including the optical single-photon detectors.

Following optical bandpass filtering, the most significant source of background radiation is usually dark counts, precisely because they cannot be limited by optical bandpass filtering. Of course this depends on the details of the single-photon optical detection technology in use, but technology available today introduces a rate of dark counts that is quite significant in this application. The primary reason for this is that a multitude of optical detectors are needed, and dark counts accumulate across that multitude (see Section 8.1). Even an average interval between spurious dark counts of a month or a year, which renders dark counts insignificant in many other contexts, is a significant limitation on the receiver sensitivity in this application to an interstellar downlink. Technological and conceptual advances can be expected to reduce the prominence of dark counts, although to an unknown degree. This is also an area of great opportunity (and necessity) for future research and development.

6.1.3 Interference

Interference is unwanted radiation that is distinguishable from the signal in one or more physical parameters (time, wavelength, or direction) and can therefore be partially or wholly rejected by technological means. The major interference considered here is radiation from the target star, which is spatially separated from probes, but which may be challenging to reject because of its close proximity to their trajectory of a flyby mission or the location of a landing. Another source is any other artificial transmitters present in or close to the line of sight, such as other spacecraft or satellites.

For flyby missions there may be multiple probes launched to the same or nearby targets, and for economic reasons it is likely that a single receive collector will serve all the probes. Mutual interference is particularly an issue for directed-energy probes if, after flyby of a common target, their downlink operations overlap in time (see Section 10). In that case there is a lower limit on the FOV such that all probes are within that view, and each probe becomes a source of interference to all other probes (see Section 7). The deliberate combination of such interfering signals in a manner that allows them to be separated at the receiver, called *multiplexing*, is a major issue for some types of multiprobe missions [2].

6.2 Terrestrial vs space

A terrestrial location for the receiver has the significant disadvantage of a propagation path including Earth's atmosphere. This introduces several challenges that would be avoided in a space-based (satellite or lunar) platform:

- Interference from the target star by choosing an advantageous wavelength is constrained by any requirement for atmospheric transparency.
- Scattering of sunlight during the daytime and moonlight at night introduce a source of noise.
- Signal propagation is susceptible to attenuation by clouds.
- Atmospheric turbulence results in signal scintillation and, for larger receive apertures, likely necessitates the complication of adaptive optics to avoid deep fades.

For all these reasons, a space platform for the downlink receiver is advantageous. However, given the likely necessity for a massive receive collector, this may not be affordable.

6.3 Signal-to-background ratio

The cumulative effect of background is determined by the signal-to-background ratio $\mathrm{SBR} = \Lambda_S/\Lambda_B$, where Λ_S and Λ_B are, respectively, the average signal photon rate (proportional to average input power) and average background photon rate at the output of the receiver's optical detector. Note that SBR is defined in terms of average power (as opposed to peak power, see Section 11.2).

To achieve the highest BPP, the SBR must be sufficiently high that background does not significantly disrupt signal acquisition, synchronization, demodulation and decoding, and error-correction processes in the receiver's TRA. Thus there is a requirement that the SBR must be above some minimum. Generally a value of SBR ≥ 4 suffices for that purpose (see Fig. E1 of Ref. [2]). The overall background photon rate Λ_B limits the receiver sensitivity, specifically determining the minimum signal photon rate Λ_S meeting this minimum SBR requirement.

6.4 Two types of background

There are two components to Λ_B that have divergent characteristics:

$$\Lambda_B = \Lambda_F + \Lambda_D. \tag{3}$$

The *prefiltering* background Λ_F is introduced in the signal path (including transmitter, propagation, and receiver) before the optical bandpass filter, and benefits from bandlimiting. There are also sources of *postfiltering* background Λ_D introduced in the receiver signal path following optical bandpass filtering. The most important of these are black body radiation originating in the postfilter optics and dark counts introduced in optical detectors. These two types of background have a duality relationship: Λ_F but not Λ_D benefit from bandpass filtering in the frequency-wavelength domain. Meantime Λ_D but not Λ_F benefit from burst-mode filtering in the time domain. This concept is now elaborated.

6.4.1 Bandwidth control of prefiltering background

Optical bandpass filtering in the receiver can eliminate much of the background radiation originating from broadband cosmic sources and black body radiation introduced in the receiver optics. Thus in the following we add the following assumption:

> *Optical bandpass*: The optical path in the receiver includes an optical bandpass filter, which limits the bandwidth to the signal bandwidth $\mathcal{W}$ (see Section 11.8).

Λ_F can be limited to $\mathcal{W}$. Thus, by choice of $\mathcal{W}$ and coordinated optical bandpass filtering in the receiver, the size of Λ_F can be manipulated to some degree. These opportunities are limited by the desire for high signal photon efficiency (which requires bandwidth expansion, see Section 11.8), and by any technological limitations on the selectivity of optical bandpass filtering.

6.4.2 Burst-mode control of postfiltering background

Λ_D can be limited by an analogous form of filtering in the time domain called *burst mode* [2], illustrated in Fig. 3. During blanking intervals, all background photons can be ignored, thereby reducing the background photon rate by the *duty-cycle* factor $\delta = \mathcal{T}_F/\mathcal{T}_I$. This has no net effect on the prefiltering background, since even as the average photon detection rate is reduced by δ, the signal bandwidth is increased by δ^{-1} due to time compression of the bursts, and these two factors precisely offset.

Overall, burst mode creates a trade space between the requirement on dark count rate in detectors, optical bandpass filtering, and peak power in the transmitter. The increase in $\mathcal{W}$ renders the optical bandpass filter less selective, and thus easier to implement. However, the peak transmit power in the transmitter is increased by δ^{-1}. This is an indication of the singular importance of peak power in achieving high BPP. (Another is the high peak power inherent in a modulation coding layers such as pulse-position modulation [PPM], see Section 11). In addition, burst mode renders acquisition and synchronization more difficult and more critical (see Section 11.4).

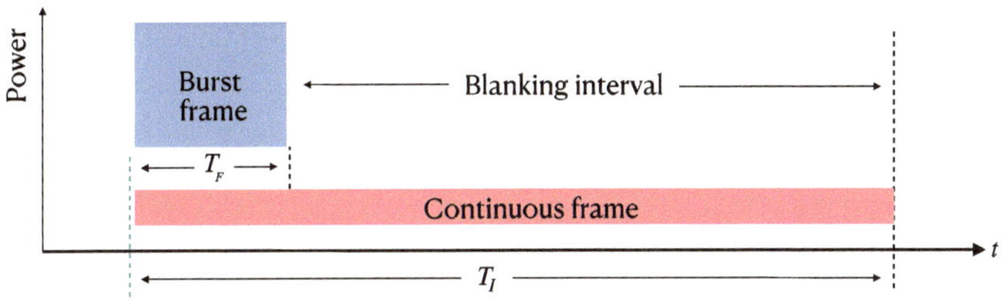

FIG. 3

An illustration of burst-mode transmission, in which a section of transmitted signal of duration $\mathcal{T}_I$ is compressed into a shorter duration $\mathcal{T}_F < \mathcal{T}_I$ at higher power, so that the total energy is not affected. This creates a blanking interval known to the receiver, during which there can be no signal photon detections and any background photons can be ignored.

7 Transmit/receive aperture size and pointing accuracy

A single diffraction-limited aperture is the canonical building block of optical electromagnetic transmission and reception. Either the transmitter or receiver may be composed from two or more such apertures (see Section 8.2), but for the moment we focus on understanding a single aperture in isolation. The three parameters that must be coordinated are illustrated in Fig. 4. To meet the goal of full target illumination, we must have

$$\Theta_F > \Theta_P + \Theta_T, \tag{4}$$

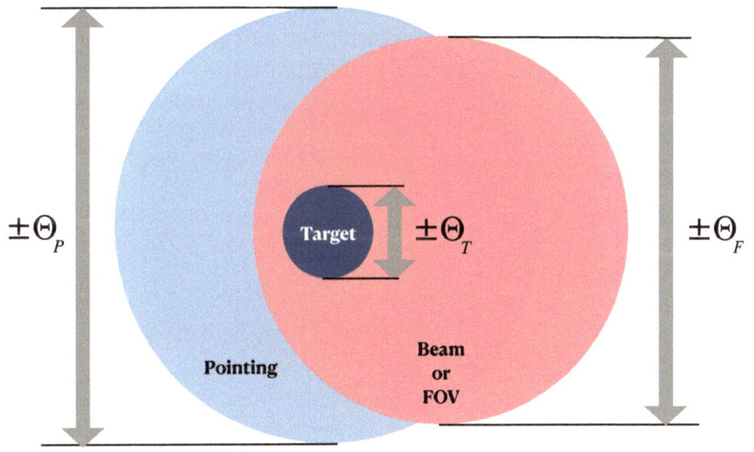

FIG. 4

Illustration of the coordination between the tolerances for pointing accuracy, beam (transmit aperture) or field of view (receive aperture), and the target extent. These tolerances are defined by their angular variation $\{\pm\Theta_P, \pm\Theta_F, \pm\Theta_T\}$. The goal is to illuminate the entire target at a reasonably uniform flux regardless of actual pointing angle within the specified tolerance. The circular geometry assumed here can be reformulated for other shapes. Not accounted for in the case of a receiver is the parallax due to the Earth's orbit, which has a yearly variation and can be compensated because it is accurately known.

since otherwise at worst-case pointing angle offset the target may not be fully illuminated or in the full FOV. While Θ_T is determined by the conditions at the target and the uncertainty in its position, the $\{\Theta_F, \Theta_P\}$ are design parameters for different probe or receiver subsystems and must be coordinated to meet this constraint. The three parameters $\{\Theta_F, \Theta_P, \Theta_T\}$ are now discussed individually.

7.1 Pointing accuracy

Issues related to pointing accuracy are quite distinctive for space versus terrestrial platforms and for transmit beam versus receive FOV.

7.1.1 Probe transmitter

In a probe transmitter, the pointing accuracy tolerance $\pm\Theta_P$ captures how accurately the transmit beam is aligned with the receiver. To guide its pointing, the probe must locate reference stars, one of which is presumably our Sun.

Two forms of alignment are relevant and necessary: The probe attitude (spatial orientation relative to a standard coordinate system) must be controlled and accurately known, and the transmit aperture pointing must be accurately controlled relative to the probe attitude. In conventional spacecraft, these two aspects are likely to be separate (but coordinated) subsystems due to the need to align different instruments and propulsion engines in addition to the transmit beam. In low-mass flyby probes, downlink transmission is likely to be the only function following target encounter, in which case transmit beam pointing is the only issue. In this case it may be possible to rely on attitude control alone, with a rigidly affixed transmit aperture. In the case where a sail acts as a transmit aperture, accurate attitude control of the sail is necessary.

The achievable pointing accuracy in a low-mass probe is uncertain. On the one hand it cannot accommodate pointing mechanisms and sensors as bulky as contemplated in spacecraft with much higher mass, but on the other hand it avoids vibration originating from various actuators and fluid transfers. Attitude destabilization results primarily from the momentum transfer of internal electron motion, any radioactive emissions that may occur from an electrical generator, emitted photons from attitude control engines and the transmitter, and collisions between the probe and particles it encounters in the interstellar medium (ISM).

7.1.2 Terrestrial receiver

In the receiver the pointing accuracy captures how accurately the FOV is aligned with the target. In addition to the receive aperture pointing, a terrestrial receiver is likely to compensate for atmospheric refraction, and this may require a reference beacon from a space platform. For a space-platform receiver the issues are qualitatively similar to a probe, although the mass budget for attitude control and receive aperture pointing would presumably be more generous.

7.2 Beam divergence or FOV

The size of an aperture determines the angular extent $\pm\Theta_F$ of a beam emitted by the transmitter or the FOV experienced at the receiver. Since a larger aperture implies a smaller Θ_F, Eq. (4) places an upper bound on the transmit or receive aperture area. This limits the signal sensitivity of the communications link. If the target area can be rendered smaller, or the pointing is rendered more accurate, the aperture area can be increased and the sensitivity improved.

7.3 Target uncertainty or size

A target may be a point, but more likely it has a nonzero angular tolerance $\pm\Theta_T$ attributable to uncertainty as to its location. For example, astronomical bodies' locations have a level of imprecision, and likewise, a probe trajectory. Also, the target may have a nonzero extent.

Example 3: A receiver may be asked to capture the signals from multiple probes transmitting concurrently along nonaligned trajectories, requiring an FOV that covers all those probes. A target for a probe transmit beam may consist of our Sun plus Earth orbiting it without knowledge of the Earth's location within that orbit, allowing pointing at the Sun rather than the Earth.

7.4 Some details and examples

It is helpful in understanding the challenges faced by a downlink designer to consider a few numerical examples.

7.4.1 Beam or FOV radius for a circular axis

The beam pattern of a diffraction-limited circular aperture has circular symmetry. A useful measure of its radius is the angle between the beam axis and the first null in its Fraunhofer diffraction pattern, which is to good approximation $\pm 1.22 \cdot \lambda/d$ for wavelength λ and an aperture diameter d. The effective beam spot radius can be set to a fraction $0 < \xi < 1$ of this first null, or $\Theta_F = 1.22 \cdot \xi\lambda/d$, where ξ is chosen for an acceptable drop in intensity relative to the beam axis.

Example 4: Setting $\xi = 0.3$, the minimum intensity at the limit $\pm\Theta_F$ is 84% of that at the beam axis.

7.4.2 Probe transmitter aperture area

Consider the challenge for a probe transmitter designer. A larger transmit aperture permits a lower peak and average power, but the feasible size of the transmit aperture is limited by that aperture's pointing accuracy as well as by the size of the target, defining the possible locations of the receiver. Pointing accuracy is usually specified in μrad (1 μrad is about 0.2 arcsec), where the stated accuracy for planetary missions is less than 1 μrad at $\lambda = 1$ μm [6].

Example 5: Assume the receive aperture's location is known precisely, or $\Theta_T = 0$. For $\lambda = 400$ nm, $\xi = 0.3$, and pointing accuracy $\Theta_P = \pm 1$ μrad, we get $d < 14.6$ cm. Every doubling of d is accompanied by a halving of Θ_P. Also d is proportional to λ, so a longer λ implies a proportionally larger d.

A larger target (or $\Theta_T > 0$) requires a commensurately smaller aperture. In practice this can be a significant consideration.

Example 6: For a probe transmitter, if the target is considered to be the entirety of the Earth's orbit, the transmit aperture can be aimed at our Sun without accounting for the orbital position of the Earth. A target size of ± 1 AU and a distance of 4.24 ly corresponds to $\Theta_T = \pm 3.73$ μrad, which results in probe transmit aperture $d < 3.1$ cm. The aperture size is dominated by target size rather than pointing accuracy. If it is assumed the probe accurately knows the position of our Sun as well as the position of the Earth at the time the signal will be received, this target size can be reduced accordingly with a beneficial increase in transmit aperture size.

7.4.3 Downlink receiver aperture area

For the receive aperture, pointing can be rendered more accurate because of the essentially unlimited mass budget, processing power, etc. If a receiver were dedicated to a *single* probe, the primary issue is the precision with which the probe trajectory is known.

Example 7: Assume a received aperture pointing accuracy of $\Theta_P = 0.1 \mu\text{rad}$. Suppose that any "scientifically useful" trajectory for a star at distance 4.24 ly has to fall within 10 AU of the target star, and the position of that star at the point of nearest passage is precisely known. Then an FOV that covers any useful trajectory becomes $\Theta_T = 37.3\ \mu\text{rad}$ and the resulting aperture diameter is $d = 0.4$ cm.

If, on the other hand, we know the probe trajectory more precisely in terms of orientation and distance to the target star, the aperture diameter can be increased. For a passage position relative to the star known within ± 0.1 AU, the FOV reduces to $\Theta_T = 0.37\ \mu\text{rad}$ and the resulting aperture diameter is $d = 31$ cm.

A typical requirement is to concurrently receive from multiple probes with slightly different trajectories. This increases the required FOV and reduces the aperture size.

Example 8: From the receiver FOV perspective, the proper motion of Proxima Centauri (its changing position relative to the faint background stars) is about 4 arcsec/year, which is 19.4 μrad/year. If the downlink transmission time is 2 years (see Section 10), probe trajectories will track this proper motion over 2 years, resulting in $\Theta_T \approx 19.4\ \mu\text{rad}$. This dominates the pointing accuracy considered earlier, and results in a smaller receive aperture diameter $d < 0.72$ cm. However, this is a case where a change in beam shape would be advantageous, since the star proper motion is linear to good approximation, with a smaller spiral pattern imposed due to the parallax introduced by the Earth's orbit (see Fig. 8 of Ref. [2]).

8 Transmit power and aperture/collector sizes

We have seen that often the pointing accuracy and/or target size impose an upper bound on the size of individual diffraction-limited apertures. In fact these apertures may need to be small if they are to accommodate feasible pointing accuracies for some typical target sizes. If additional aperture area is desired, this can be obtained by the incoherent combination of multiple diffraction-limited apertures. "Incoherent" means no attempt to match the phase of the aperture input or output, but, simply put, "powers add." This approach has limited appeal on the transmit side because an attractive alternative is to incoherently add multiple input signals before applying them to a single aperture.

Example 9: If multiple lasers are required to achieve the desired peak transmit power, they can be incoherently superimposed optically prior to applying them to a single aperture. This may have less mass impact than associating an aperture with each laser, where the superposition occurs in the electrodynamic (propagation) domain.

On the receive side, on the other hand, there is no substitute for a larger total area to intercept more of the incident signal flux arriving from the transmitter. This can be achieved by the accumulation of photon detections from multiple apertures. Thus our development here focuses on the receive side.

8.1 A collector composed from multiple apertures

We make a distinction between an *aperture* and a *collector*, where the collector is composed of multiple apertures as a means of increasing total area and hence conversion of incident signal flux to a higher detected power. The simple receive collector architecture shown in Fig. 5 (as was assumed in Ref. [2]) composes the collector from multiple diffraction-limited apertures that are combined incoherently. The individual apertures determine the FOV based on their aperture area A_R, and the number of apertures N_C determines the total collector area $A_C = N_C A_R$. The incoherent combining, which in the simplest case is obtained by associating an optical detector with each aperture and accumulating photon detection events across all apertures, ensures that the FOV is not affected. Of course a collector so constructed is definitively not diffraction-limited.

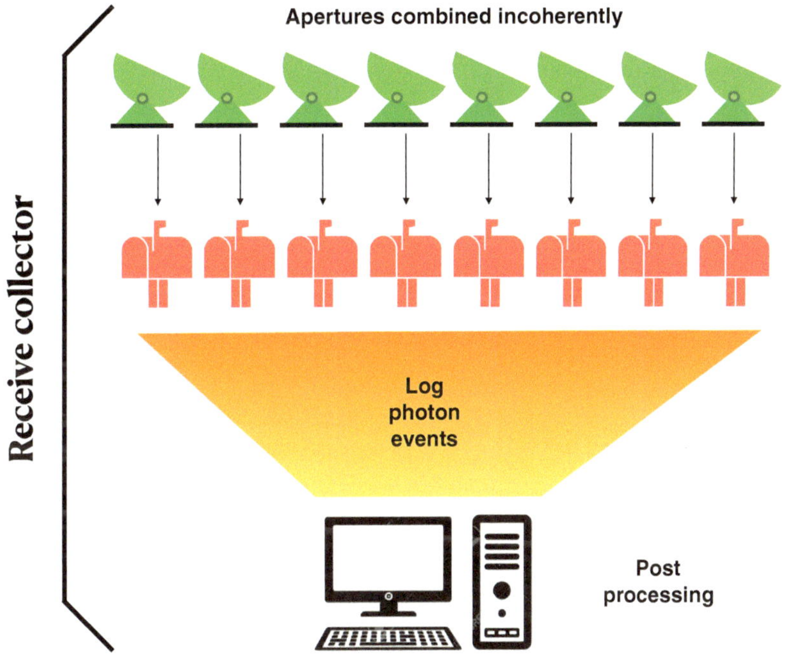

FIG. 5

A receive collector composed of multiple apertures. This achieves simultaneously an FOV controlled by the aperture size together with a larger total collection area to achieve a signal photon counting rate Λ_S commensurate with data rate $\mathcal{R}$ through the relation $\mathcal{R} = \Lambda_S \cdot \text{BPP}$. For simplicity and without affecting the FOV, the apertures are combined incoherently by simply locating a photon-counting optical detector (shown as a mailbox icon) at each aperture and accumulating their photon counts. The resulting collector is not diffraction-limited, but this is fortunately unnecessary at the receiver.

Reproduced with permission from Fig. 3 of D.G. Messerschmitt, P. Lubin, I. Morrison, Challenges in scientific data communication from low-mass interstellar probes, Astrophys. J. Suppl. Ser. 249 (2) (2020) 36, Copyright AAS.

8.2 Transmit power and aperture/collector tradeoffs

The received power in the Friis transmission equation of Eq. (1) equals the photon rate times the energy per photon, but an optical detector will have a quantum efficiency $\eta < 1$, resulting in an average received photon rate:

$$\Lambda_S = \frac{\mathcal{R}}{\text{BPP}} = \eta \cdot \frac{A_T A_C}{hc\,\lambda\,D_{\text{star}}^2} \cdot P_T. \tag{5}$$

The factor $P_T A_T A_C$ creates a trade space that may significantly impact technological feasibility, probe mass, and cost. In particular, a larger A_C can allow for reduced mass in the probe through a smaller power-area product $P_T A_T$.

8.2.1 Transmit aperture tradeoffs

The data rate $\mathcal{R}$ is proportional to λ^{-1} in Eq. (5). Thus as λ is changed and P_T is held constant, the transmit aperture area A_T can be scaled downward as $A_T \propto \lambda$ or the equivalent aperture diameter can be scaled as $\sqrt{\lambda}$. This is less favorable than suggested by power considerations alone in Eq. (1) due to the increasing energy per photon as wavelength decreases.

Short wavelength affects the design and fabrication of both transmit aperture and receive collector, although the receive collector is less affected as a whole since it need not be diffraction limited. Depending on area and the chosen technology, maintaining a transmit aperture close to the diffraction limit at shorter wavelengths will necessitate tighter fabrication tolerances proportional to the wavelength. In that event, $A_T \propto \lambda$ does not fully compensate, and aperture fabrication becomes a greater challenge.

8.2.2 Receive collector tradeoffs

The aggressive mass budget for a flyby probe suggests a large A_C, since this allows for a commensurately smaller (and lower mass) $P_T A_T$. The following example illustrates that, at interstellar distances and for modest assumptions on $\{P_T, A_T\}$, A_C can be large (kilometer-scale) and the required number of apertures composing the collector N_C can also be large (millions to billions).

Example 10: For a $\lambda = 400$ nm, if we adopt $d_T = 14.6$ cm from a previous example, then the area of a circular transmit aperture is $A_T = 167.4\ \text{cm}^2$. Then, with $\eta = 0.5$, $P_T = 10$ mW, $D_{\text{star}} = 4.24$ ly, $\mathcal{R} = 1$ bits/s, and BPP $= 10$bits/ph, the required receive collector area is $A_C = 0.15\ \text{km}^2$. If the receive aperture has diameter $d_R = 0.72$ cm (area $A_R = 0.41\ \text{cm}^2$) then the required number of apertures is $N_C = 3.75 \times 10^9$.

At the expense of increasing probe mass, the size and scale of the receive collector can be moderated by increasing the transmit power.

Example 11: If the transmit power is increased by $100\times$ to $P_T = 1$ W, and the photon efficiency is increased to BPP $= 15$, and the quantum efficiency is also increased to $\eta = 0.9$, then the required collector area decreases to $A_C = 566\ \text{m}^2$ and the required number of apertures decreases to $N_C = 1.38 \times 10^7$.

With a large number of apertures and associated optical detectors, the average interval between signal photon detections is large. This is another indicator of the aggressive requirements on detector dark counts.

Example 12: If $\mathcal{R} = 1$ bit/s and BPP = 10 bit/ph, the average interval between photon detections at the collector level is 10 s. A value $N_C = 1.38 \times 10^7$ corresponds to a 4.4-year interval at the aperture level. A typical individual aperture/detector combination detects no signal photons during the entire life of the download operation.

A constraint on this tradeoff is the minimum SBR requirement at the optical detector (see Section 6.3). The trade space is therefore influenced by the impact of $\{A_R, N_C, A_C\}$ on background, which is now considered. Our ability to reduce $P_T A_T$ is constrained by the need for a sufficiently large signal at the receiver to overcome the background radiation and hence achieve a sufficiently large SBR.

8.2.3 Background vs receive aperture and collector size

Prefiltering background of cosmic origin is usually unresolved, meaning that the source of background has a fixed areal density within the FOV of the aperture. For such background, and for a single receive aperture with area A_R, the background power reaching an optical detector is not dependent on A_R (see Appendix B of Ref. [2]). An explanation for this (arguably counter-intuitive) statement is that as A_R increases, the total background appearing as incident flux to the aperture decreases due to the decreasing FOV, but the area of that flux captured by the aperture increases, and these two factors offset.

There may also be resolved point sources of background, such as radiation from a nearby star that falls within the FOV (the target star will usually fall in this category), for which the incident flux does not depend on A_R and the total captured background thus increases in proportion to A_R.

Much like unresolved sources of cosmic background, postfiltering background is also independent of A_R, but for the very different reason that it is introduced following the aperture.

The canonical collector architecture of Fig. 5 does not affect SBR. To see this, let SBR_R and SBR_C be the SBR at the aperture and collector level. Then, since both signal photon rate Λ_S and background photon rate Λ_B (both at the level of a single aperture and its dedicated optical detector) accumulate incoherently:

$$\text{SBR}_C = \frac{N_C \Lambda_S}{N_C \Lambda_B} = \frac{\Lambda_S}{\Lambda_B} = \text{SBR}_R. \tag{6}$$

Thus the SBR constraint can be set on the basis of a single receive aperture. With the notable exception of point sources of radiation, SBR is independent of A_R, depending instead primarily on the size of the postfiltering background, which is determined primarily by the technology of the optical detector and its temperature.

Receive collector design can thus be summarized as

- Choose A_R on the basis of the desired FOV, which places an upper bound (see Section 7.2).
- Manipulate the signal level at the aperture level through the choice of $P_T A_T$. Knowing the background level, this signal level can be chosen to meet the minimum acceptable SBR, thus beneficially minimizing $P_T A_T$. Choosing the largest acceptable A_R consistent with FOV is beneficial in allowing the smallest possible $P_T A_T$.
- Choose N_C so that the signal photon rate $N_C \Lambda_S$ at the level of the entire collector suffices to support the data rate $\mathcal{R}$ for the expected photon efficiency BPP, namely $N_C \Lambda_S \cdot \text{BPP} = \mathcal{R}$.

8.2.4 Receive collector improvements

It is likely that the canonical collector of Fig. 5 can be improved on through invention and technology development. A couple of approaches have been suggested [2], but surely there are others. One is to arrange for a larger A_R, which allows a commensurate decrease in $P_T A_T$ while maintaining fixed signal level.

Example 13: If A_R is limited by the larger FOV required for coverage of multiple probes, replicating the receiver multiple times, each dedicated to one probe, would allow A_R to be increased. This would be expensive, but it may be possible to accomplish the same effect by using a larger A_R in conjunction with imaging (as in an optical telescope). In that case different probes could be associated with different pixels, each with an effectively larger aperture.

Another approach is to achieve $\mathrm{SBR_C} > \mathrm{SBR_R}$ in Eq. (6), which also allows $P_T A_T$ to be decreased.

Example 14: An architecture in which there are fewer than N_C optical detectors will reduce the postfiltering background in the denominator of Eq. (6). This might be accomplished by sharing a single optical detector among as many apertures as possible.

9 Outages

Operation during daylight is most certainly ruled out, and at night, weather events may create a cloud cover that blocks the signal. These events are called *outages*. Weather-related outages are a significant disadvantage to using optical wavelengths, and can be avoided by employing a relay satellite to recover the scientific data and retransmit it at radio wavelengths to a terrestrial radio receiver. This approach has been used in planetary optical communications [6], but is less attractive for interstellar optical communications because of the massive and large receive collector that would be required in space due to much greater signal propagation distance.

For a terrestrial-based receive collector, outages are a major consideration in achieving reliable recovery of scientific data. Accommodating outages is a well-known challenge in wireless communication, and there are standard mitigation techniques. The design of the ECC has to take outages into account, and the achievable reliable data rate $\mathcal{R}$ will be adversely affected (see Section 12.2).

10 Downlink operation time

For a flyby mission, the total duration of the downlink operation is expected to be in the order of years. There are several compelling reasons for this assumption, as now summarized. This relates to the design of the mission, including the mass and speed of the probe and the duration of the operation of the downlink data transmission T_{down}. The following is a summary of results, which are covered more thoroughly in Ref. [7].

10.1 Data volume and latency tradeoff

While this chapter emphasizes metrics of performance like data rate and photon efficiency, these are internal communications design issues of little concern to domain scientists who are waiting for the data return from the probe mission in order to enable their scientific interpretation of that data. These science stakeholders are primarily interested in three issues: The data return reliability (addressed in

Section 12), the *data volume* $\mathcal{V}_{data}$ (total number of data bits returned), and the *data latency* $T_{latency}$ (elapsed time from probe launch to return of the data in its entirety). In other words, how much data do we get back reliably, and how long do we have to wait for that data?

Two mission design parameters for a flyby mission with directed-energy propulsion are the duration of downlink transmission T_{down} and the probe mass ratio ζ_P, which is the ratio of the actual probe mass to some baseline value. The significance of ζ_P is that if the launch beam and power remain fixed across multiple probe launches, the speed of the probe is directly affected by the probe mass, and hence by ζ_P. In particular, the speed decreases as $\zeta_P^{-1/4}$ with increasing mass, and the total launch energy increases as $\zeta_P^{3/4}$.

Although the mission design parameters $\{\zeta_P, T_{down}\}$ can be varied to manipulate the mission performance metrics $\{\mathcal{V}_{data}, T_{latency}\}$, they should not be chosen arbitrarily. Rather, they should be jointly optimized to achieve the most favorable $\{\mathcal{V}_{data}, T_{latency}\}$ tradeoff. The impact of $\{\zeta_P, T_{down}\}$ on $\{\mathcal{V}_{data}, T_{latency}\}$ is slightly complicated, but can be summarized as:

- The nature of directed energy propulsion is that if the launch beam remains fixed (except possibly for the time duration of probe acceleration), a larger ζ_P results in a lower cruise speed for the probe u_P.
- This increases the travel time to the target, and this increases $T_{latency}$.
- This results in a smaller accumulation of distance-squared propagation delay, and thus allows $\mathcal{R}$ to decrease more slowly during downlink transmission, and this in turn increases $\mathcal{V}_{data}$.
- For fixed T_{down}, the probe has traveled less distance from the target during downlink operation, the maximum propagation delay back to the launch site is smaller, and this reduces $T_{latency}$.
- A larger probe mass budget for communications allows its electrical power, and hence transmit power, to be increased, and also possibly the transmit aperture to be increased in area. These benefits increase $\mathcal{R}$ and thereby $\mathcal{V}_{data}$.

While the travel-time increase is deleterious, all the other impacts are beneficial. Trading these off leads to an optimum point.

10.1.1 Optimal volume-latency tradeoff

The design of a data link, which conveys data reliably from probe to launch site, involves a number of interacting considerations such as wavelength, transmit aperture, receive collector, background radiation, modulation, and coding. For purposes of mission design, all these considerations can be wrapped into a single parameter, a baseline data rate $\mathcal{R}_0$ at the beginning of downlink operation assuming $\zeta_P = 1$. The total data volume $\mathcal{V}_{data} \propto \mathcal{R}_0$, and thus it is convenient to use the normalized volume $\mathcal{V}_{data}/\mathcal{R}_0$ (which is dimensioned in time) as a performance metric to guide the choice of $\{\zeta_P, T_{down}\}$.

The tradeoff between $\mathcal{V}_{data}/\mathcal{R}_0$ and $T_{latency}$ is plotted in Fig. 6 [2, 7]. There exists a feasible region of operation $\{\mathcal{V}_{data}/\mathcal{R}_0, T_{latency}\}$, which is shaded in Fig. 6. Points on the lower boundary of this region, called the *efficient frontier*, constitute all the advantageous mission operating points. That boundary yields the smallest possible $T_{latency}$ for a given $\mathcal{V}_{data}/\mathcal{R}_0$, or the largest possible $\mathcal{V}_{data}/\mathcal{R}_0$ for a given $T_{latency}$. This assumes that the initial data rate on the downlink scales as $\mathcal{R}_0 \cdot \zeta_P^k$ for scaling exponent $k = 2$. The rationale is one factor of ζ_P to account for a proportional increase in transmit aperture area, and another factor of ζ_P allowing for a proportional increase in average power. (Alternative values for k are considered in Ref. [7].) Numerical values along the efficient frontier are listed in Table 2 along with the associated optimal set of parameters.

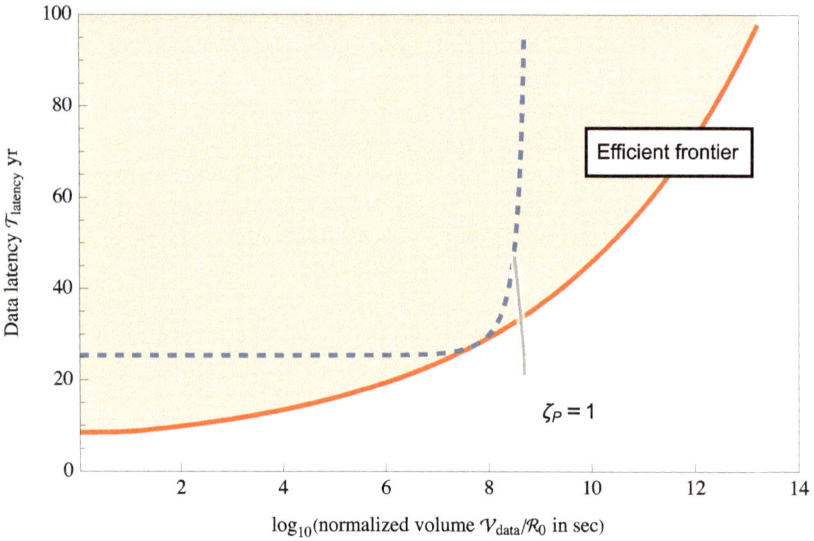

FIG. 6

Plots of data latency T_{latency} (in years) versus the log of the normalized data volume $\mathcal{V}_{\text{data}}/\mathcal{R}_0$ (in seconds). $\mathcal{R}_0$ is the data rate (in bits per second) at the beginning of downlink transmission (data rate declines from there as the square of propagation distance) for mass ratio $\zeta_P = 1$. $\mathcal{V}_{\text{data}}$ (in bits) is found by multiplying by the assumed value for $\mathcal{R}_0$. Any volume-latency mission operating point within the shaded region is feasible. The lower boundary of this region, called the efficient frontier, is an efficient operating point in the sense of maximizing the volume for a given latency or minimizing latency for a given volume. The set of operation points obtained by fixing $\zeta_P = 1$ and varying downlink operation duration T_{down} are shown as a *dashed curve*.

Table 2 Efficient frontier and related parameters for a scaling exponent $k = 2$.

$\log_{10}[\mathcal{V}_{\text{data}}/\mathcal{R}_0]$	T_{latency} in years	ζ_P	u_P/c	T_{down} in years
0	8.5	0.0016	1	0.0124
1	8.74	0.0016	1	0.128
2	9.89	0.0033	0.834	0.308
3	11.5	0.0088	0.653	0.437
4	13.5	0.0236	0.51	0.611
5	16.1	0.0636	0.398	0.846
6	19.4	0.172	0.31	1.16
7	23.7	0.469	0.242	1.57
8	29.3	1.28	0.188	2.11
9	36.5	3.52	0.146	2.81
10	45.8	9.67	0.113	3.73
11	57.8	26.7	0.088	4.92
12	73.2	73.6	0.0683	6.46
13	93.2	204	0.0529	8.45

Example 15: If we place a limit on the latency of 36.5 years, then the maximum normalized data volume is $\mathcal{V}_{data}/\mathcal{R}_0 = 10^9$ s. At an initial data rate of $\mathcal{R}_0 = 0.1$ bits/s, the maximum data volume is $\mathcal{V}_{data} = 100$ Mb (megabits) or 12.5 MB (megabytes).

10.1.2 Choosing a point on the efficient frontier

Choice of a mission operation point somewhere on the efficient frontier provides flexibility in setting mission priorities. There are several compelling reasons to consciously select different operating points along the efficient frontier for different missions sharing a common launch infrastructure:

- Mission designers can consciously prioritize large $\mathcal{V}_{data}/\mathcal{R}_0$ or small $T_{latency}$.
- Different probes may carry different types of instrumentation, and these impose different mass and data volume requirements.
- There will likely be an evolution of probe technology over time. Early probes may emphasize technology validation with low $T_{latency}$ (and hence small $\mathcal{V}_{data}/\mathcal{R}_0$), while later probes may emphasize scientific return with larger $\mathcal{V}_{data}/\mathcal{R}_0$ (and hence larger $T_{latency}$).
- There may be missions to different targets at different distances (within the solar system and interstellar), significantly changing the possible range of $T_{latency}$.

Generally, mission designers will seek to maximize an objective function that combines volume and latency objectives. Only points along the efficient frontier need be considered in any such optimization.

Also illustrated in Fig. 6 as the dashed curve is a set of possible mission operation points when a baseline value $\zeta_P = 1$ is chosen and only T_{down} is varied. This approaches the minimum-latency horizontal asymptote as $T_{down} \to 0$ and a maximum-volume vertical asymptote as $T_{down} \to \infty$. This arbitrary choice of ζ_P permits operation at exactly one point on the efficient frontier through a judicious choice of T_{down}; otherwise, it is inferior. More generally, achieving an arbitrary operating point on the efficient frontier requires a coordinated choice of $\{\zeta_P, T_{down}\}$ rather than constraining ζ_P in this manner. The efficient frontier is the lower envelope of all such curves varied over all values of ζ_P.

10.1.3 What changes along the efficient frontier

How does the mission design change along the efficient frontier? This can be summarized as, from lower to higher normalized data volume $\mathcal{V}_{data}/\mathcal{R}_0$:

- The mass ratio ζ_P increases, the probe mass increases, and the probe speed decreases.
- The data latency $T_{latency}$ increases, although this penalty is the most favorable possible.
- The downlink operation duration T_{down} increases. A useful rule of thumb is that it approximately equals 9% of the probe transit time from launch to encounter with the target star. This value depends on assumptions about how probe power and aperture area scale with ζ_P. See Ref. [7] for further details. For example, with a transit time of 30 years, the downlink would operate for 2.7 years.
- The launch energy increases, since the probe is accelerated for a longer period of time.
- The total distance the probe travels to the end of downlink operation does not vary significantly along the efficient frontier; that is, it always travels a total distance $\approx 1.09 \cdot D_{star}$ to the point it suspends downlink operation.

In summary, along the efficient frontier, the mission design is dominated by the distance-squared law of propagation loss, and thus the total distance traveled to the suspension of downlink operation does not vary much. Higher $\mathcal{V}_{data}$ can be achieved by reducing the probe speed, giving the probe a longer period to transmit its data before the distance-square loss becomes dominant. There is a direct mission cost penalty for achieving larger $\mathcal{V}_{\text{data}}$ in increased launch energy.

10.2 Other considerations in downlink duration

There are considerations beyond optimizing the volume-latency tradeoff that strongly suggest a downlink operation time on the order of years following target-star encounter and scientific data collection:

- Electrical generation based on a reservoir of "fuel" or a battery is likely inconsistent with mass objectives. Rather, it is likely to be based on a radiative-thermal source or the (necessarily slow) extraction of energy from the ISM, which are low power but persistent. At the very least the electrical generation capability has to exceed the multidecade transit time to the target star.
- For reliable data recovery in the face of random outages, the downlink operation time should be at least 50× the longest outage time (see Section 12.2). Depending on the location of a terrestrial receiver, outages due to weather events may last a significant number of days, implying a downlink operation in the order of years.

11 Modulation code: PPM

The modulation coding layer in Fig. 2 converts bits supplied by the ECC layer into a light intensity waveform to be generated by the transmit laser. A modulation coding layer commonly in use is PPM, which is illustrated in Fig. 7. PPM combined with an appropriate ECC layer can achieve close to theoretical constraints on photon efficiency, subject to assumptions about peak and average power.

11.1 Theoretical photon efficiency of PPM

Information theory provides a least upper bound (over all possible ECC schemes, see Section 12.1) on the achievable photon efficiency of PPM [8]:

$$\text{BPP} \leq \alpha[K_s] \cdot \log_2 M \text{ where } \alpha[K_s] = \frac{1 - e^{-K_s}}{K_s}. \tag{7}$$

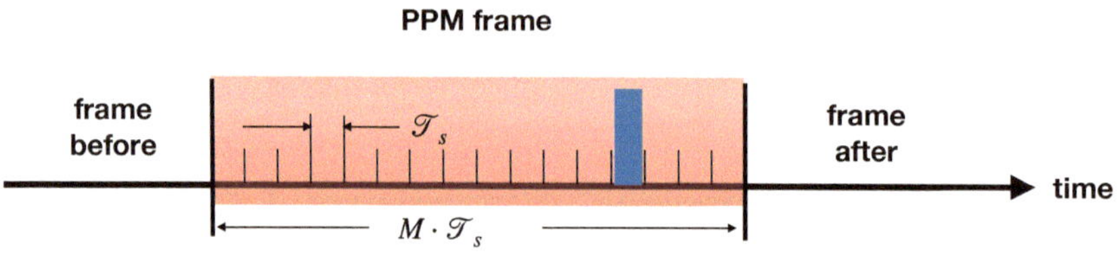

FIG. 7

Pulse-position modulation divides the received signal power at the optical detector into frames, where each frame is further subdivided into M slots each of duration $\mathcal{T}_S$. The convention is that the received signal power is all within a single slot within the frame, which communicates $\log_2 M$ bits of "raw" information (if the slots are equally likely to be so-energized). The average number of photons detected for each energized slot is K_S, and this is also the average number of photons detected for the whole frame. The peak-to-average power ratio is thus PAR $= M$.

A Poisson distribution governing the count of detected photons is assumed, and the parameter K_s equals the average number of detected photons in an energized PPM slot (which is one out of M total slots in each PPM frame, $(M - 1)$ of which are not energized). Since $0 < \alpha[K_s] < 1$ monotonically approaches $\alpha[K_s] \rightarrow 1$ as $K_s \rightarrow 0$, this suggests that K_s should be small to achieve the highest BPP.

A similar least upper bound over all possible transmitter-receiver combinations that utilize photon-counting detection (and therefore does not presuppose PPM) corresponds to $\alpha[K_s] = 1$. Thus PPM is photon efficient, but only if we can arrange for $\alpha[K_s] \sim 1$, or equivalently a small value for K_s.

For any value of BPP, the PPM parameters $\{K_s, M\}$ can be coordinated to achieve that BPP efficiently (see Appendix B). The resulting $\{K_s, M\}$ are plotted in Fig. 8 and the resulting values of BPP are tabulated in Table 3 for representative values of M chosen to be powers of two (a typical implementation convenience). Also shown is the resulting value of $\alpha[K_s]$, which quantifies the shortfall attributable to specializing the photon-counting scheme to PPM. Two additional columns are included and are discussed later.

While the optimum value of K_s depends on the BPP we aspire to achieve, in the range of interest available with near-term technology the value $K_s \sim 0.2$ is a good approximation. That there is a fraction of one detected photon per pulse on average, which is called *photon starvation*, may be surprising. The implication of this is that most frames actually experience zero photon detections (see Section 11.4). Achieving the BPP that we aspire to requires a large value for M.

Example 16: If we seek to achieve the least upper bound of BPP $\leq$ 15 bits/ph, this requires $M = 2^{17}$ $= 131,072$ in Table 3. In this case 17 bits have to be passed from TRA to PHY to specify which PPM slot to energize within each PPM frame. This large M renders synchronization to be challenging.

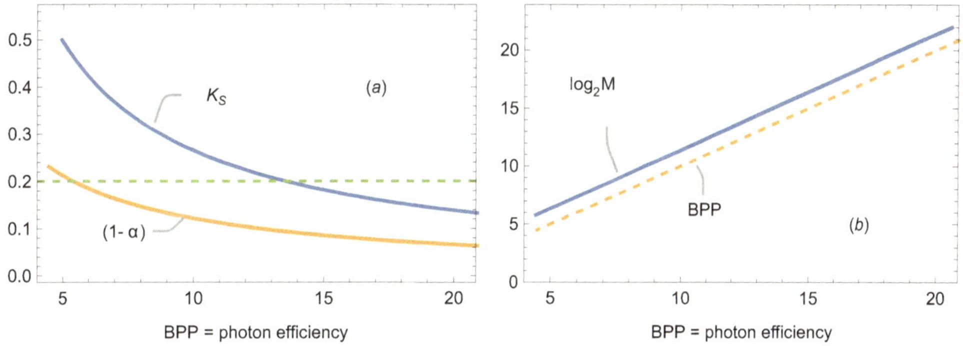

FIG. 8

At the theoretical least upper bound BPP on PPM photon efficiency, PPM parameters $\{K_S, M\}$ that achieve that bound efficiently are plotted. Distinct dimensionless values are plotted, with the curves rather than the axes labeled. (A) The value of K_S illustrates photon starvation. The value $K_S = 0.2$, which is a reasonable choice in the region of interest, is plotted as a *dashed line*. Also plotted is the decrease in photon efficiency $(1 - \alpha[K_S])$ relative to the best possible transmitter-receiver utilizing photon counting detection but without limiting to PPM, which is in the order of 10%–20%. (B) The number of bits per PPM frame $\log_2 M$ is compared to BPP in the *dashed line*. That BPP $< \log_2 M$ by a small margin is another indication of PPM's modest inefficiency relative to a general photon-counting scheme.

Table 3 Optimal choice of PPM parameters for $M = 2^m$.

BPP	K_s	m	$M = 2^m$	$\alpha[K_s]$	$\mathcal{T}_s\mathcal{R}$	$\mathcal{W}/\mathcal{R}$
0.815	2.17	2	4	0.408	0.443	2.26
1.75	1.19	3	8	0.584	0.261	3.83
2.71	0.837	4	16	0.678	0.142	7.06
3.68	0.646	5	32	0.737	0.0744	13.4
4.66	0.527	6	64	0.777	0.0384	26
5.65	0.445	7	128	0.807	0.0196	50.9
6.64	0.385	8	256	0.83	0.00999	100
7.63	0.34	9	512	0.848	0.00506	197
8.62	0.304	10	1020	0.862	0.00256	391
9.62	0.275	11	2050	0.874	0.00129	775
10.6	0.251	12	4100	0.884	6.5×10^{-4}	1540
11.6	0.231	13	8190	0.893	3.27×10^{-4}	3060
12.6	0.214	14	1.64×10^4	0.9	1.64×10^{-4}	6080
13.6	0.199	15	3.28×10^4	0.907	8.26×10^{-5}	1.21×10^4
14.6	0.186	16	6.55×10^4	0.912	4.15×10^{-5}	2.41×10^4
15.6	0.175	17	1.31×10^5	0.917	2.08×10^{-5}	4.81×10^4
16.6	0.165	18	2.62×10^5	0.922	1.04×10^{-5}	9.58×10^4
17.6	0.156	19	5.24×10^5	0.926	5.23×10^{-6}	1.91×10^5
18.6	0.148	20	1.05×10^6	0.93	2.62×10^{-6}	3.81×10^5
19.6	0.141	21	2.1×10^6	0.933	1.31×10^{-6}	7.61×10^5

The resulting $\alpha[0.175] = 0.917$, and thus the photon efficiency bound is 8.3% below a general photon-counting modulation scheme with the same peak-to-average ratio.

For concrete choices and implementations of error-correction codes, the best choice of K_s will generally be slightly higher and the achieved efficiency will generally be lower than this theoretical bound (see Section 12).

11.2 Peak power

For any direct detection (photon-counting) receiver the photon efficiency is bounded by BPP $\leq \log_2$ PAR, where PAR is the peak-power to average-power ratio within a frame [9]. This is consistent with PPM, where PAR $= M$, as given in Eq. (7). Thus we must have PAR $> 2^{\text{BPP}}$, which increases exponentially with BPP. Further burst-mode control of postfiltering background increases PAR by another factor of δ^{-1}. Practical limits on PAR for semiconductor lasers are likely to severely limit available BPP and/or our ability to control postbackground radiation. This motivates an interest in optical detectors with very low dark count rates (see Section 6.1.2) and pulse-compression technologies (see Section 11.9).

11.3 TRA-to-PHY interface in transmitter

At the transmitter, for each PPM frame the TRA specifies the appropriate timeslot, which requires $\log_2 M$ bits. Typically M is chosen to be a power of two, so that this is an integral number of bits. Then, observing that the total number of photon detections per frame (as well as per slot) is K_s, and from Eq. (7):

$$\frac{\text{Average scientific data bits per PPM frame}}{\text{TRA to PHY bits per PPM frame}} = \frac{K_s \cdot \text{BPP}}{\log_2 M} \le K_s \alpha[K_s]. \tag{8}$$

Thus each PPM frame represents considerably fewer scientific data bits than the number of bits actually passed from the ECC to the modulation code for that same frame in Fig. 2. The remainder of those bits constitute redundancy, which is used for error control (see Section 12.1).

Example 17: For the parameter choices in the last example, the value of Eq. (8) is 0.166. Thus only 16.6% of the bits passed from TRA to PHY represent scientific data, and 83.4% of the bits represent redundancy for error control.

It is important to recognize that each PPM frame does not stand on its own, but rather the added redundancy is processed across large blocks of PPM frames so that statistical averaging can be exploited (see Section 12.1).

11.4 PHY-to-TRA interface in receiver

At the receiver, TRA is provided a list of all photon detection events by PHY. This is a list of time stamps identifying the time of the detection and any other available information such as detected photon wavelengths or pulse shapes at the detector (which may be relevant to vetoing some spurious counts of noncosmic origin). This list constitutes the primary scientific record of the mission, which may conceivably be processed multiple times in the future as algorithms are improved.

Due to the inherent unreliability of the PHY, because it is subject to quantum effects in the signal as well as spurious photon counts due to background radiation, the output of the PHY at the receiver is only stochastically related to the input to the PHY at the transmitter. The TRA in the receiver starts by performing *synchronization*, which estimates the PPM frame and slot boundaries, and is in itself a difficult step that may require offline processing over a significant portion of the downlink duration. Synchronization is rendered more critical as well as more difficult by both PPM and by burst mode imposed on top of PPM (see Section 6.4.2).

Assume that synchronization has been successfully acquired. Then for each received PPM frame, the TRA may observe one of three conditions:

1. There are one or more photon detections, and all such photon detections fall in the same PPM slot.
2. There are two or more photon detections, but they do not all fall in the same PPM slot. This has to be the result of spurious photon count(s) in nonenergized slot(s).
3. There are zero photon detections within the entire PPM frame. This is typically due to the inherent stochastic nature of signal-photon detection.

The second and third cases are called an *erasure*, since in either case it is impossible to associate a unique PPM time slot with the reception. In the first case, the photon detections may be in the correct time slot (as determined by transmission) or a different time slot (which is called an *error*). Erasures due to photon starvation are commonplace.

Example 18: Continuing the past examples, and assuming no background radiation, if $K_s = 0.175$ then the probability of an erasure (applying the Poisson distribution) is $e^{-0.175} = 0.839$. Thus one or more photons are detected on average for only 16.1% of frames, and on average the remaining 83.9% of frames have zero photon detections. The latter frames are among the erasures.

The additional information inherent in erasures (as opposed to errors) can be exploited in the subsequent decoding to improve the reliability. Reliable recovery of scientific data in spite of erasures and errors is achievable due to the inherent redundancy deliberately introduced at the transmitter (see Section 12.1).

Based on the sequence of unambiguously detected time slots per PPM frame, combined with (mostly) erasures, TRA can proceed with the recovery of the scientific data through a process that is typically complex and computationally burdensome.

11.5 The canonical on-off pulse

The basic building block for PPM that is of primary concern to PHY is defined at the receiver optical detector, rather than at the transmitter. The received power at the detector should be confined to one PPM time slot of duration $\mathcal{T}_s$. In that case, for $(M - 1)$ out of the M slots in one PPM frame there is no signal energy, and thus any photon detections in those slots are attributable to background radiation. It is the total energy of the pulse that determines the average number of photon detections, and not the detail of how power varies within the time slot.

The procedure for choosing the number of apertures N_C in a collector was described in Section 8.2.3 in terms of average photon rate. When specializing to PPM, this procedure can be equivalently performed with respect to a single canonical pulse peak power, with the goal of making N_C sufficiently large that an average photon detection count K_s per pulse is realized at the collector output.

11.6 Time slot duration

With knowledge of the PPM parameters $\{K_s, M\}$ and the resulting BPP, the time slot duration $\mathcal{T}_s$ is readily inferred. Equating two representations of the volume of scientific data associated with a single PPM frame:

$$\mathcal{T}_s M \cdot \mathcal{R} = K_s \cdot \text{BPP}, \qquad (9)$$

we see that, for fixed M, $\mathcal{T}_s \propto \mathcal{R}^{-1}$.

Example 19: Using the same parameters as in previous examples, and assuming $\mathcal{R} = 1$ bits/s, we find from Eq. (9) that the required $\mathcal{T}_s = 31.4\ \mu\text{s}$. Increasing $\mathcal{R}$ to 10 bits/s reduces the slot duration to 3.14 μs.

Another implication of Eq. (9) follows from Eq. (7):

$$\mathcal{T}_s \mathcal{R} = K_s \alpha[K_s] \cdot \frac{\log_2 M}{M}. \qquad (10)$$

This value, which is tabulated in Table 3, is useful for determining $\mathcal{T}_s$ for any $\mathcal{R}$. Since K_s does not vary too much as BPP is changed (see Fig. 8), as BPP is increased, the value of $\mathcal{T}_s \mathcal{R}$ (and hence

$\mathcal{T}_s$ for fixed $\mathcal{R}$) decreases roughly in proportion to $(\log_2 M)/M$. However, the total frame time $M\mathcal{T}_s$ increases roughly as $\log_2 M$, and the number of scientific data bits $M\mathcal{T}_s\mathcal{R}$ represented by each frame increases the same way. Thus Eqs. (8), (10) are consistent.

Example 20 Using the same parameters as in previous examples, and assuming $\mathcal{R} = 1$ bits/s, we find from Eq. (9) that the time slot decreases from $\mathcal{T}_s = 992$ to 31.4 μs as the photon efficiency is increased from BPP = 10–15 bits/ph. The number of slots per PPM frame increases from $M = 2673$ to 86,573.

11.7 Burst mode and PPM

Another consideration affecting the choice of PPM parameterization is management of postfilter background through burst mode (see Section 6.4.2). Suppose the time interval between PPM frames is $\mathcal{T}_F$, which relates directly to the scientific data rate $\mathcal{R}$. Normally PPM is thought of as full duty-cycle $\delta = 1$, for which $\mathcal{T}_F = M\mathcal{T}_s$. However, if instead $M\mathcal{T}_s \ll \mathcal{T}_F$ is deliberately chosen, that leaves blanking intervals between PPM frames, and the average postfiltering background photon detection count $\Lambda_D M\mathcal{T}_s$ during each PPM frame (see Section 6.4.2) is reduced. That in turn permits a smaller pulse energy while maintaining an adequate SBR (see Section 6.3), thereby reducing the transmit power burden on the transmitter. Of course the number of apertures N_C in the receive collector has to be commensurately larger (see Section 8.2.2), but overall burst mode beneficially permits a shift in resources from transmitter to receiver.

There is of course a price to pay. Assuming that the pulse energy is fixed (and hence the average transmit power is also fixed), there are two ways to reduce the PPM frame time $M\mathcal{T}_s$, and each has adverse consequences:

- Reducing M will reduce the photon efficiency BPP and hence $\mathcal{R}$.
- Reducing $\mathcal{T}_s$ will increase the peak power.

Overall burst mode creates a trade space among peak transmit power, data rate, and the partitioning of resources between transmitter and receiver. As the dependencies are multifaceted, these tradeoffs are best addressed in the context of a numerical model that permits parameter exploration.

11.8 Optical bandwidth

Since the signal is a linear superposition of canonical pulses, the optical bandwidth of the signal $\mathcal{W}$ is the same as that of a single pulse. In particular, it is related to $\mathcal{T}_s$ through the uncertainty relationship $\mathcal{W} \sim \mathcal{T}_s^{-1}$. The metric

$$\text{Bandwidth expansion} = \frac{\mathcal{W}}{\mathcal{R}} = \frac{1}{\mathcal{T}_s\mathcal{R}} \tag{11}$$

tells us the factor by which optical bandwidth is expanded relative to the data rate. Although it appears to depend on $\mathcal{T}_s\mathcal{R}$, this is misleading, as seen in Eq. (10). In fact, it depends only on the PPM parameters, and its value is also tabulated in Table 3.

Example 21: For the parameters of earlier examples the bandwidth expansion factor is 31,816. Thus, the optical bandwidth is $\mathcal{W} = 31.8$ kHz for $\mathcal{R} = 1$ bits/s and the bandwidth expansion is 3.18×10^4.

If the scientific data rate was increased or burst-mode PPM was employed (see Section 6.4.2), this bandwidth would increase accordingly. Larger optical bandwidth beneficially reduces the required optical bandpass filter selectivity, but also adversely affects the background radiation admitted to the optical detector (see Section 6).

11.8.1 Theoretical limits on bandwidth expansion

A general conclusion across all of communications is that if the objective is to achieve higher energy efficiency, this requires an increase in bandwidth expansion. $\mathcal{W}/\mathcal{R}$ is plotted in Fig. 9 for PPM with an optimum choice of parameters based on Eq. (11). Also shown is the quantum limit, which removes the assumption of a photon-counting receiver (and hence PPM) and allows the transmitter and receiver to manipulate quantum states directly (see Appendix B.2). For a given $\mathcal{W}/\mathcal{R}$, higher BPP can be achieved at the quantum limit, indicating the possibilities for more advanced future technologies that manipulate quantum states rather than simply counting photons. In interstellar communication, the penalty of larger $\mathcal{W}$ is a larger prefiltering background radiation, which then requires a larger signal level to achieve an acceptable SBR (see Section 6.3). If postfiltering background is dominant, this is less relevant.

11.9 Purposeful time dispersion

A significant issue is the peak power required to support a high BPP and large M, because the peak power required to maintain fixed K_s grows as $\mathcal{T}_s$ decreases (see Section 11.6). This peak power can be beneficially decreased without affecting total energy through the deliberate introduction of *time dispersion*.

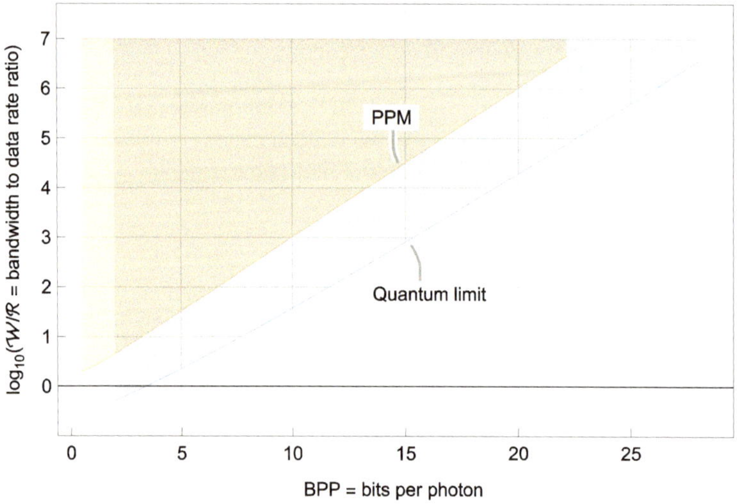

FIG. 9

For communication at optical wavelengths, a log plot of the theoretical minimum bandwidth expansion $\mathcal{W}/\mathcal{R}$ for a given photon efficiency BPP. Operation within the *shaded areas* is feasible, and achieving greater efficiency requires a larger bandwidth. Shown is the quantum limit derived in Appendix B.2, and the less favorable tradeoff for a PPM modulation code with optimum choice of parameters based on photon counting detection.

Suppose two pulses both have constant power and the same bandwidth $\mathcal{W}$ but different time durations:

$$|p_1(t)|^2 = P,\ 0 < t < T \text{ and } |p_2(t)|^2 = P \cdot \mathcal{W}T,\quad 0 < t < 1/\mathcal{W}, \tag{12}$$

where $p_1(t)$ has a longer time duration than is necessary for that bandwidth ($\mathcal{W}T \gg 1$). Then the two pulses have equivalent energy, but the minimum duration pulse $p_2(t)$ has greater power by the factor $\mathcal{W}T$. Thus $p_1(t)$ can be transmitted at lower power, and since the two pulses have the same bandwidth a subsequent linear filtering dedispersion operation over that bandwidth can produce at the detector a higher power pulse of shorter duration and equal energy.

Example 22: If the time slot duration is $\mathcal{T}_s = 1\ \mu\text{s}$, then the required optical bandwidth is $\mathcal{W} = 1$ MHz. A deliberately dispersed pulse with time duration 1 ms and the same bandwidth can be generated by sweeping the wavelength over 1 MHz during the 1 ms duration. This pulse is called a *chirp*. After dedispersion the pulse width at the optical detector is 1 μs with the peak power increased by a factor of 10^3.

The dedispersion step may be located at any convenient point in the signal path, including the transmitter or the receiver. This technique is used in optics to generate short pulses with very high power [10, 11], and may similarly be beneficial here. However, the generally wider pulses required at the low bit rates that may be feasible in interstellar communication presents an obstacle for these techniques [2].

12 Achieving reliability

It is an understatement to say that PHY in isolation is unreliable, given that in the order of 80% of PPM frames are erased (contain zero detected photons). This seeming "flaw" is necessary for achieving a high BPP, but it must be overcome within the TRA. The justification for this partitioning of responsibility falls back on the inherent conflict between high BPP and the requirement for reliable recovery of data. The design is thus partitioned into two phases:

PHY: High energy efficiency is obtained by embedding information in the *timing* of narrow pulses, since adjusting the timing of canonical pulses within a PPM frame incurs no energy penalty. High energy efficiency is also achieved through photon starvation, with the side effect of poor reliability within PHY.

TRA: Reliability is obtained by adding redundancy in the transmitter TRA, and utilizing that redundancy (the structure of which is known in detail by the receiver) in the receiver TRA to overcome the high erasure rate in the PHY.

12.1 Error-correction coding

We saw earlier that redundancy makes up the bulk of the data that is passed from TRA to PHY in the transmitter. The redundancy insertion and exploitation cannot take place on a frame-by-frame basis, as the reliability would still be poor. Rather, the stochastic nature of the quantum channel (with or without background radiation) has to be countered by averaging over long time intervals, exploiting the statistical law of large numbers. A longer time interval is formed by creating blocks of bits consisting of combinations of scientific data bits and *redundancy* bits. These blocks are called *codewords*.

A typical codeword is formed in this fashion:

$$\mathcal{I} \text{ data bits} \rightarrow \mathcal{B} \text{ coded bits} \rightarrow \mathcal{B}/m \text{ PPM frames.}$$

In this scheme, $\mathcal{B}$ is the size of each codeword in bits, $\mathcal{I}$ is the number of scientific data bits represented by each codeword, and we must have $\mathcal{B} > \mathcal{I}$. $\mathcal{B}$ is chosen to be a multiple of m, resulting in an integral number of PPM frames representing each codeword. The $\mathcal{B} - \mathcal{I}$ bits added to the $\mathcal{I}$ data bits represent the redundancy, and ratio $\mathcal{I}/\mathcal{B} < 1$ is called the *code rate*. The complete *codebook* is specified by a mapping from every possible set of $\mathcal{I}$ data bits (there are $2^{\mathcal{I}}$ possible sets) to a corresponding codeword consisting of $\mathcal{B}$ bits. Only a small fraction $2^{\mathcal{I}-\mathcal{B}}$ of the possible collections of $\mathcal{B}$ bits is a valid codeword. In a *systematic* codebook the $\mathcal{I}$ data bits appear explicitly as part of the codeword. Codebooks are more typically nonsystematic.

Each received block of $\mathcal{B}$ bits corresponding to a transmitted codeword is typically corrupted by errors and erasures in the constituent PPM frames. The codebook is specifically designed so that the valid codewords are as dissimilar as possible, which aids in distinguishing them in the face of these errors and erasures. At the receiver, the decoding process involves choosing the valid codeword from the codebook that is most similar to the actually received block of bits, and the corresponding set of I data bits forms the decoded output. These decoded bits are a statistical estimate of the transmitted scientific data, and may still be corrupted, albeit with very small probability.

The design of effective ECC codes and corresponding decoding processes for large $\mathcal{B}$ is a major undertaking that has involved many applied mathematicians over the past seven decades. Basing the encoding and decoding operations on storage and retrieval of codewords from a look-up table quickly becomes impractical as $\mathcal{I}$ increases.

Example 23: A specific Reed-Solomon (RS) ECC example in Section 12.3.1 adopts $\mathcal{B} = 10,230$. With a typical code rate of 0.2, each codeword consists of $\mathcal{I} = 0.2\,\mathcal{B} = 2046$ bits. In this case the total number of bits represented in the codebook is

$$2^{\mathcal{I}} \cdot \mathcal{B} = 2^{2046} \cdot 10,230 \sim 10^{620} \text{ bits.}$$

Compare this to the capacity of a terabyte disk, which is of order 10^{13} bits.

In practice, an explicit codebook is never generated or stored, and coding and decoding are performed by algorithms typically based on sophisticated mathematics and with substantial computational complexity. Fortunately the decoding complexity dominates, and that takes place at the receiver, where mass and power consumption are not constrained and processing need not be real time.

12.2 Coding for outages

Outages are inevitable in optical communication through the atmosphere (see Section 9). To effectively and efficiently deal with that, the statistics of the outages must be accurately known at design time. It is necessarily assumed that the actual occurrences (time and duration) are stochastic and unknown to the transmitter.

The ECC can be designed to accommodate outages without loss of data reliability. Under these assumptions, if the outage probability is p_O, then data can be reliably recovered at an average rate $(1 - p_O) \cdot \mathcal{R}$ when the ECC is designed with this in mind. Thus outages that conform to statistical assumptions made at design time do not compromise reliable data recovery, but do reduce the volume of data that can be reliably recovered.

A useful rule of thumb is that the integrity of data units (such as a single image) can be recovered reliably if that data's transmission is spread over a time duration that is at least 50× the longest statistically credible outage interval. This constrains the downlink operation time (see Section 10).

Example 24: Consider a set of images that we wish to download. If they are sent sequentially it would be difficult to recover an image that happened to coincide with a long outage (such as a week-long cloud cover). If the longest outage (say from weather events) is expected to be 1 week, then each image should be transmitted over a period of 50 weeks (about a year). Multiple images can be interleaved and transmitted concurrently to fill out the available data rate $\mathcal{R}$.

12.3 ECC efficiency

The analysis of Section 11 provides theoretical bounds on achievable BPP with photon counting and PPM, without constraint on the choice of ECC scheme. We saw that specializing the modulation to PPM results in a small efficiency loss compared to an unconstrained modulation choice. In a similar manner, the choice of a specific (practical) ECC scheme will introduce a further efficiency loss. It will generally also alter the value of K_s at which BPP is maximized. Typically a small increase in K_s will be required over the PPM optimum. The goal of the ECC design process is to minimize this efficiency loss, subject to any constraints on code length and decoding complexity.

For illustrative purposes, a concrete example of an ECC design follows. Coding for outages (see Section 12.2), although critically important in this application, is beyond the scope of this chapter.

12.3.1 Concrete example ECC design: A Reed-Solomon code

The design of a maximally efficient ECC scheme is a major design exercise beyond the scope of this chapter. However, for illustrative purposes we can turn to a well-understood ECC scheme, RS coding, that has been used extensively in free-space optical communications systems. It was employed in the NASA/JPL bench demonstration of BPP = 13 bits/ph reported in Ref. [5]. We do not provide details here of the structure of RS codes, nor of their encoding/decoding methods. We do present example RS coding results to illustrate how BPP and the optimal K_s are affected by ECC code choice.

RS codes are "nonbinary" in the sense that the code alphabet is not bits but rather symbols defined over a finite (Galois) field of size q. Each symbol can be represented uniquely by m bits, where $2^m = q$. Working in conjunction with M-ary PPM where $M = 2^m$, each RS code symbol maps naturally onto an individual PPM symbol. RS codes are particularly effective at overcoming erasures, which are the dominant concern for photon-counting detection in the photon-starved regime.

Consider an RS$\{n, k, m\}$ code, which operates on m-bit symbols and encodes k information symbols to produce codewords of length n symbols. In terms of the nomenclature used in Section 12.1, $\mathcal{I} = k \cdot m$ and $\mathcal{B} = n \cdot m$. The number of redundant symbols is $(n - k)$. RS codes have the property that they can recover up to $(n - k)$ erasures occurring across the n-symbol codeword. Thus as we decrease k the redundancy increases and the tolerance of erasures improves.

While the theoretical bounds on BPP of Section 11 apply to an arbitrary reliability objective, for a concrete ECC design the calculation of efficiency (in terms of overall BPP) is meaningful only in the context of a specific decoding error rate requirement. The required error rate will decrease as the reliability objective becomes more stringent.

Example 25: For example, a reasonable reliability objective might be to achieve an error rate after decoding that would result in a 1 Mbit image file being received error-free with 99% likelihood.

In calculating the overall BPP for this objective we must factor in the level of redundancy of the code. As that redundancy increases, the tolerance to erasures increases (allowing smaller K_s) but fewer scientific data bits are conveyed in each codeword. BPP falls off at high redundancies for this reason. It also falls off at low redundancies because higher K_s is needed to maintain the overall error rate objective. The optimum will fall at some intermediate level of redundancy—although we know this must be toward the higher end when operating in the photon-starved regime.

Example 26: Consider an RS(1023, k, 10) code employed with $m = 10$ PPM (i.e., 1024 PPM). In Fig. 10, we plot the maximum achievable BPP as a function of the code rate k/n. The BPP for each $1 \leq k \leq 1022$ is the value corresponding to the maximum erasure probability (minimum K_s) that still permits the overall reliability objective to be achieved at that code rate (i.e., the 99% image reception probability of the previous example).

The shortfall in achieved BPP for a concrete EC coding-decoding algorithm relative to the theoretical least upper bound of Eq. (7) is called the error-correction *coding efficiency*.

Example 27: In the example earlier, we see that the maximum BPP falls short of the theoretical limit of 8.62 bits/ph, as reported in Table 3 for $m = 10$. Thus the coding efficiency is 6.83/8.62, or 79%.

Efficiencies greater than 0.9 are achievable using more powerful ECC codes.

The dependence of the optimum K_s on the choice of ECC scheme is further illustrated through Fig. 11 where, for the same RS code as the example, we plot the maximum BPP as a function of K_s. For each K_s the code rate that maximizes BPP has been selected. For this specific code, the peak of BPP = 6.83 occurs at $K_s = 0.32$ and a code rate of 0.22. That these values differ slightly from the theoretical least upper bound of Table 3 is a manifestation of the inefficiency of the selected example RS code in comparison to an idealized code.

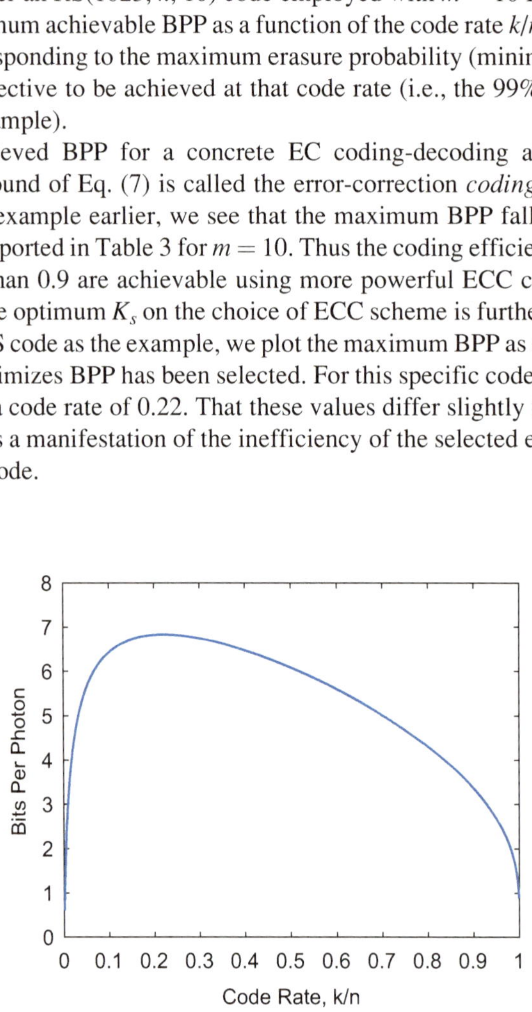

FIG. 10

The maximum achievable BPP as a function of code rate k/n. Reed-Solomon code RS(1023, k, 10) using order 1024 PPM is assumed. Erasures are assumed to be random.

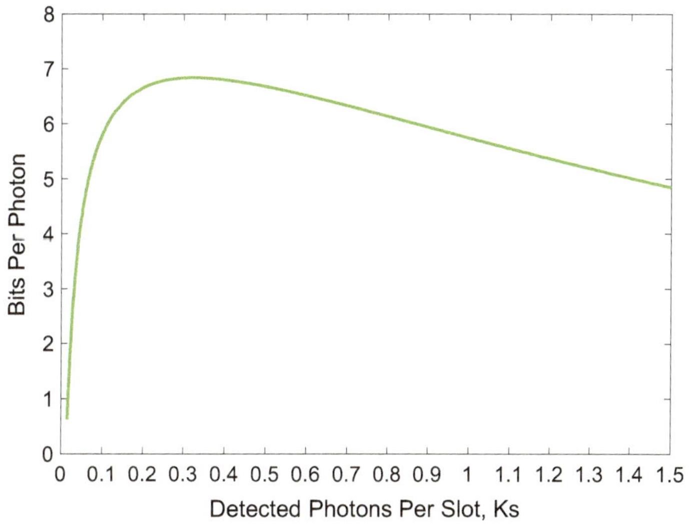

FIG. 11

The maximum achievable BPP as a function of K_S. The same conditions as in Fig. 10 are assumed.

13 Conclusions

The end-to-end system design of a communication link, especially one in the context of interstellar distances and a variety of background radiations of both cosmic and local origin, is a complex undertaking. This is due to the significant dependencies among different subsystem and element designs across transmitter and receiver and TRA and PHY, and between multitudinous performance goals, theoretical limits, and concrete design choices. This chapter has attempted to capture these dependencies as well as the local considerations coming into play within each subsystem. A required scope of core principles is large, spanning quantum mechanics, optics, and device physics on the one hand to information theory and finite field algebra on the other. Such an undertaking is best conducted as a collaboration among different types of expertise. In setting requirements and making concrete tradeoffs, that collaboration should include the ultimate stakeholders, which includes funding sources and domain scientists.

Acknowledgments

The research represented in this tutorial review was supported by a grant from the Breakthrough Foundation and its Breakthrough StarShot program, NASA grants NIAC Phase I DEEP-IN—2015 NNX15AL91G and NASA NIAC Phase II DEIS—2016 NNX16AL32G and the NASA California Space Grant NASA NNX10AT93H, the Emmett and Gladys W. Technology Fund, and the Limitless Space Institute. Tom Mozdzen of Arizona State University provided numerous valuable comments and suggestions.

Appendix A Nomenclature

c	speed of light in a vacuum
h	Planck's constant
P_R, P_T	receive and transmit average power
λ	wavelength
D_{star}	distance from target star to receiver
A_T	area of a diffraction-limited transmit aperture
A_R	area of a single diffraction-limited receive aperture within the larger receive collector
N_C	number of diffraction-limited apertures composing the receive collector
A_C	total area of a receive collector (made up of multiple apertures)
TRA	transport layer, which controls reliability of scientific data recovery
PHY	physical layer, consisting of physical elements of the communications downlink
Λ_S	average rate at which signal photons originating from the probe transmitter are detected at output of PHY
Λ_B	average rate of background detected photons
Λ_F	average rate of detected prefiltering background photons
Λ_D	average rate of detected postfiltering photons, including prominently dark counts
$\mathcal{R}$	rate at which scientific data bits are reliably recovered
BPP	photon efficiency, in reliably recovered photons per detected photon
SBR	signal-to-background ratio
$\mathcal{W}$	optical bandwidth
PPM	pulse-position modulation code
$\mathcal{T}_F$	time duration of one PPM frame
$\mathcal{T}_I$	time elapsed between the beginning of one PPM frame and the beginning of the next
$\mathcal{T}_s$	time duration of a single PPM time slot
δ	duty cycle of PPM frames, equal to $\mathcal{T}_F/\mathcal{T}_I$.
M	number of time slots in each PPM frame
m	number of bits represented by time slots in each PPM frame, equal to $\log_2 M$
$\{\Theta_P, \Theta_F, \Theta_T\}$	half the angular extent of the transmitter pointing accuracy, receiver field of view, and target
d	diameter of a diffraction-limited circular aperture
ξ	for both transmit and receive apertures, the pointing accuracy or field of view as a fraction of the Airy radius
η	quantum efficiency of the receiver optical detection system
T_{down}	total time duration of downlink transmission
$\mathcal{V}_{data}$	total volume of scientific data reliably recovered at the receiver
$T_{latency}$	data latency, or time elapsed between the launch of a probe and recovery of the last bit of scientific data on the downlink
ζ_P	probe mass ratio or the ratio of the actual probe mass to the smallest probe mass corresponding to the maximum probe speed
$\mathcal{R}_0$	scientific data rate at the beginning of downlink operation assuming the smallest mass ratio $\zeta_P = 1$
k	scaling exponent governing the relationship between mass ratio and initial scientific data rate
PAR	ratio of peak power to average power at the transmitter
K_s	average signal photon counts within one PPM time slot at the output of the receive collector
ECC	error-correction coding
$\{\mathcal{B},\mathcal{I}\}$	for an ECC block code, the total number of bits in one codeword and the number of scientific data bits represented by that codeword
p_O	outage probability, taking into account weather, daylight, etc.
PPD	average detected photons per dimension

Appendix B Photon efficiency bounds and bandwidth

B.1 Channel capacity

The theoretical least upper bound on BPP is called the *channel capacity* in information theory [12–14]. This can be determined in principle for any *channel model*, which mathematically captures the stochastic relationship between an output waveform or sequence of values and an input waveform or sequence of values. As the model embodies more assumptions or restrictions, generally the capacity will decrease. The channel capacity theorem states that the capacity is the least-upper bound on BPP over all feasible implementations of transmitter and receiver.

For interstellar communication, three increasingly restrictive models are relevant:

Holevo bound: The most general is a channel in which the transmitter and receiver are allowed to manipulate multiple quantum states, but without interference from background radiation nor quantum entanglement. This leads to the Holevo bound (see Appendix B.2).

Photon counting: This channel model assumes conventional electromagnetic transmission with photon-counting detection in the receiver, and with SBR large enough that background radiation counts can be neglected. In this case the channel capacity is $\mathrm{BPP} \le \log_2 \mathrm{PAR}$, where PAR is the ratio of peak power to average power [9]. PPM falls in this category with the value $\mathrm{PAR} = M$, and hence its photon efficiency is bounded by $\mathrm{BPP} \le \log_2 M$.

PPM modulation code: The transmitter is constrained to energize exactly one time slot within each repeated PPM frame. This reduces the capacity relative to the general photon-counting channel (see Appendix B.3).

B.2 Quantum limit

For purposes of computing the quantum limit, one channel use is allocated a fixed number of photon detections partitioned across N modes or dimensions, each mode occupying a nonoverlapping time-wavelength space. Define *photons per dimension* (PPD) as the average number of detected photons per dimension. The resulting capacity-achieving quantum efficiency is the so-called Holevo capacity [15]:

$$\mathrm{BPP} = (1 + 1/\mathrm{PPD}) \cdot \log_2(1 + \mathrm{PPD}) - \log_2 \mathrm{PPD}. \tag{B.1}$$

This is based on the statistics predicted by nonentangled quantum states, where it is assumed that the transmitter and receiver can manipulate and read their local quantum state directly. It is also assumed, for the purposes of channel coding, that an unlimited number of quantum states can be manipulated and read jointly and collectively. BPP versus PPD is plotted as the quantum limit in Fig. B.1. While there is no available technology to approach this limit, it represents a theoretical ideal that quantifies the opportunities for more advanced future technologies.

If we have a fixed average total number of photons across all dimensions, then we can force $\mathrm{PPD} \to 0$ as $N \to \infty$, and in that case $\mathrm{BPP} \to \infty$ in Eq. (B.1). The primary lesson to draw from Fig. B.1 is that high BPP is associated with *photon starvation*, in which PPD is very small. Both the reasoning behind this, as well as the consequences for data link design, are subtle. First the logic: As the detected photons per channel use decreases, the data bits that can be recovered per channel use also decreases. This would seem to be a problem, except that the transmitted and received energy per

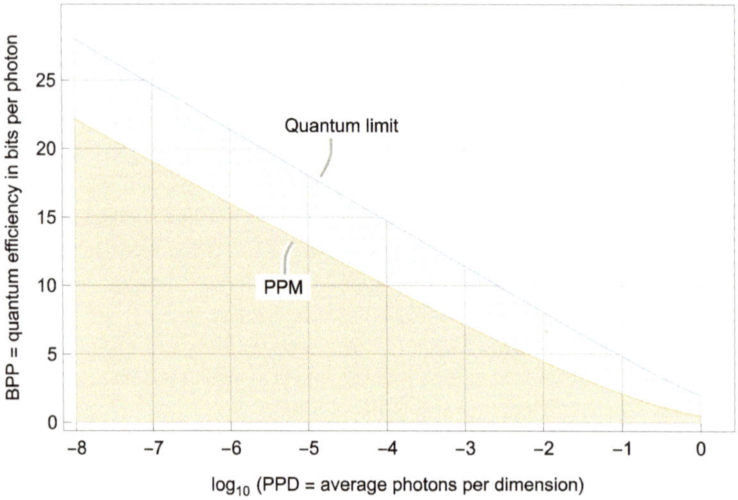

FIG. B.1

The least upper bound on photon efficiency BPP of Eq. (B.1) is plotted against the logarithm of average detected photons per dimension PPD. Any point in the shaded region is theoretically achievable, although the most advantageous choice is on the upper boundary. Achieving larger BPP is associated with a smaller PPD (which is called photon starvation), which requires spreading the total photon count over an increasing number of quantum modes or dimensions. This is compared against PPM as given by Eq. (B.4), quantifying the "headroom" available for future technology advances that eliminate the PPM photon-counting limitations and manipulate the quantum states directly.

channel use is reduced by an even larger factor, which is why BPP improves. The fewer data bits recovered per channel use can be offset by employing more channel uses, which, as we will see shortly, expands the optical bandwidth.

B.2.1 Quantum efficiency vs bandwidth expansion

Not surprisingly photon starvation is intricately related to bandwidth expansion, since it requires a growth in time-frequency N for each channel use. Increasing the number of times expands bandwidth because this requires shorter mode time durations. We now establish that the minimum bandwidth is

$$\frac{\mathcal{W}}{\mathcal{R}}=\frac{1}{\text{PPD}\cdot\text{BPP}}, \tag{B.2}$$

leading directly from Figs. B.1 to 9. This relation is difficult to interpret directly because PPD and BPP have a complicated relationship captured, for example, in Eq. (B.1).

The derivation of Eq. (B.2) follows by considering the smallest feasible bandwidth. Let the time associated with one channel use be fixed at $\mathcal{T}$. A set of nonoverlapping time-wavelength slots constitute the modes or dimensions. Let there be $N=L\cdot J$ such modes, where L is the number of unique times and J is the number of frequencies. Then the time duration of each mode is at most $\mathcal{T}/L$, the bandwidth of each mode is at least $L/\mathcal{T}$, and the total bandwidth is $\mathcal{W}=J\cdot L/\mathcal{T}=N/\mathcal{T}$. The number of bits recovered per channel use is $\mathcal{RT}$ and the average number of detected photons per channel use is $N \cdot$PPD. Hence Eq. (B.2) is confirmed since

$$\text{BPP} = \frac{\mathcal{RT}}{N \cdot \text{PPD}} = \frac{\mathcal{R}}{\mathcal{W} \cdot \text{PPD}}. \tag{B.3}$$

B.3 PPM modulation code

Specializing to a discrete-time channel model constrained to the statistically independent repetition of PPM frames, the capacity is given by Eq. (7). The PPM parameters $\{K_s, M\}$ can be chosen optimally to maximize Eq. (7) as follows. In this case PPD $= K_s/M$, which equals the number of photon detections averaged over all M time slots. Eliminating K_s in Eq. (7) and maximizing the bound with the optimal choice of M:

$$\text{BPP} \leq \max_{M \geq 2} \{\alpha[M \cdot \text{PPD}] \cdot \log_2 M\}. \tag{B.4}$$

This equation yields a relationship between BPP and $\{K_s, M\}$ parameterized by PPD. This optimization is illustrated graphically in Fig. 27 of Ref. [2]. Computing this relationship for different values of PPD yields the PPM boundaries in Figs. 9, 8, and B.1.

References

[1] S. Schelkunoff, H. Friis, Antennas: Theory and Practice, vol. 639, Wiley, 1952.

[2] D.G. Messerschmitt, P. Lubin, I. Morrison, Challenges in scientific data communication from low-mass interstellar probes, Astrophys. J. Suppl. Ser. 249 (2) (2020) 36.

[3] D.G. Messerschmitt, Design for minimum energy in interstellar communication, Acta Astronaut. 107 (2015) 20–39, https://doi.org/10.1016/j.actaastro.2014.11.007.

[4] D.G. Messerschmitt, End-to-end interstellar communication system design for power efficiency, arXiv preprint arXiv:1305.4684, 2013.

[5] W.H. Farr, J.M. Choi, B. Moision, 13 Bits per incident photon optical communications demonstration, in: Free-Space Laser Communication and Atmospheric Propagation XXV, International Society for Optics and Photonics, vol. 8610, 2013, p. 861006, https://doi.org/10.1117/12.2007000.

[6] S. Lee, J.W. Alexander, G.G. Ortiz, Submicroradian pointing system design for deep-space optical communications, in: SPIE LASE, 2001.

[7] D. Messerschmitt, P. Lubin, I. Morrison, Optimal mass and speed for interstellar flyby with directed-energy propulsion, arXiv preprint arXiv:2206.13929, 2022.

[8] S. Butman, J. Katz, J. Lesh, Bandwidth limitations on noiseless optical channel capacity, IEEE Trans. Commun. 30 (5) (1982) 1262–1264, https://doi.org/10.1109/TCOM.1982.1095577.

[9] A. Wyner, Capacity and error exponent for the direct detection photon channel. I, IEEE Trans. Inf. Theory 34 (6) (1988) 1449–1461, https://doi.org/10.1109/18.21284.

[10] E. Treacy, Optical pulse compression with diffraction gratings, IEEE J. Quantum Electron. 5 (9) (1969) 454–458, https://doi.org/10.1109/JQE.1969.1076303.

[11] J. Kafka, T. Baer, Prism-pair dispersive delay lines in optical pulse compression, Opt. Lett. 12 (6) (1987) 401–403, https://doi.org/10.1364/OL.12.000401.

[12] R.G. Gallager, Principles of Digital Communication, Cambridge University Press, Cambridge, UK, 2008, https://doi.org/10.1017/CBO9780511813498.

[13] D.G. Messerschmitt, Some digital communication fundamentals for physicists and others, EECS Department, University of California, Berkeley, 2008. Available from: www2.eecs.berkeley.edu/Pubs/TechRpts/2008/EECS-2008-78.pdf. Tech. Rep. UCB/EECS-2008-78.

[14] T. Cover, J. Thomas, Elements of Information Theory, John Wiley and Sons, New York, 1991, https://doi.org/10.1002/047174882X.

[15] S. Dolinar, B. Erkmen, B. Moision, K. Birnbaum, D. Divsalar, The ultimate limits of optical communication efficiency with photon-counting receivers, in: IEEE International Symposium on Information Theory Proceedings, IEEE, 2012, pp. 541–545, https://doi.org/10.1109/ISIT.2012.6284249.

General relativistic faster-than-light propulsion for interstellar flight

Eric W. Davis[a,b]

The Aerospace Corporation, Huntsville, AL, United States[a] *Center for Astrophysics, Space Physics and Engineering Research and Department of Physics, Baylor University, Waco, TX, United States*[b]

1 Introduction

1.1 The motivation for faster-than-light (FTL) interstellar space exploration

NASA recently announced that it has confirmed the discovery of 5000 exoplanets [1]. The volume of space within a 50 light-year (LY, 1 LY $= 9.46 \times 10^{15}$ m) neighborhood of our solar system has approximately 1400 star systems containing 2000 stars, all but 133 of which are faint red dwarf stars. There are also over 2000 extrasolar planets within this volume of space, with the nearest exoplanets that are in their parent star system's habitable zone being an average of 22 LY distant from Earth. By comparison, the closest star to Earth is Alpha Centauri in the Proxima Centauri star system, which is 4.37 LY from Earth. The number of newly discovered exoplanets within this volume of space is increasing almost every year. Convergent evolution all but guarantees that some of those exoplanets will have some form of evolved life, or even intelligent life, on them. It is not yet known where the nearest Earth-like planet will be found, but preliminary estimates suggest that the location of habitable planets might fit the probability distribution shown in Fig. 1 [2]. Since the Scientific Revolution (c. CE 1550–1700) and possibly earlier, humankind has asked whether we are alone in our galaxy. If there are other advanced civilizations within our galaxy, chances are that they are even farther away than the nearest habitable planets. Provisional guesses suggest that the location of such civilizations might fit the probability distribution shown in Fig. 2 [2].

Fig. 1 shows that the value of the Maccone probability density distribution is practically zero at distances less than 25 LY from Earth, so it is unlikely that habitable (exo)planets will be detected at such distances. Future interstellar mission designers should keep this lower bound in mind when planning missions to explore and/or colonize habitable exoplanets. This probability density distribution reaches its peak at about 67 LY from Earth, its median at about 80 LY, and it has a mean value of 88 LY. The relevant standard deviation is about 40 LY because the distribution tail is rather short. Fig. 2 shows the probability density distribution of finding the nearest extraterrestrial civilizations from Earth. This probability density distribution's peak is at 1933 LY, its mean value of 2670 LY is higher because the curve has a long tail on the right, and the standard deviation is 1309 LY. These distances are daunting if humankind hopes to one day undertake interstellar missions to explore potentially habitable exoplanets or to make in situ contact with ET civilizations in the Galaxy.

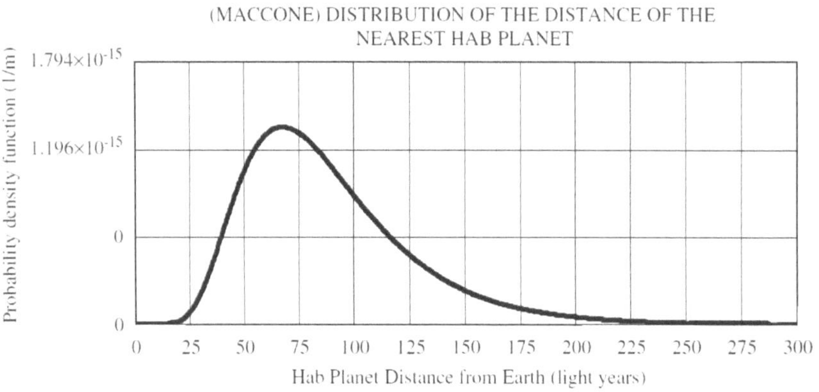

FIG. 1

Maccone probability density distribution of the distance of the nearest habitable planets in the galaxy from Earth [2]. One multiplies any point on the probability density curve by its corresponding abscissa (distance), which must be first converted into meters, to obtain the probability. Note that the second zero on the vertical axis is a rounding error of the computer code that Maccone used to plot the graph in which the true nonzero probability density value beneath 1.196×10^{-15} is so small that the code simply replaced it with zero. *Hab*, habitable.

Courtesy C. Maccone.

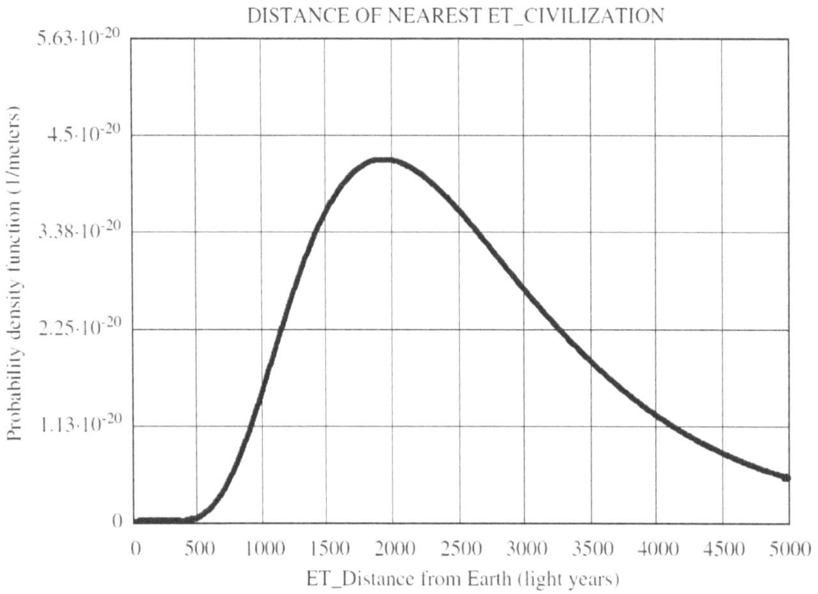

FIG. 2

The probability density distribution of the distance of the nearest extraterrestrial civilizations in the galaxy from Earth [2]. One multiplies any point on the probability density curve by its corresponding abscissa (distance), which must be first converted into meters, to obtain the probability. *ET*, extraterrestrial.

Courtesy C. Maccone.

Humankind could one day become motivated to undertake manned interstellar spaceflight missions for economic reasons and/or to prepare for the consequences of potential existential threats to the Earth. In the latter category, humankind has been historically subjected to multiple types of catastrophic events that might one day become extinction-class events, as follows:

- Global threats:
  - Manmade and natural long-term global climate change: ice ages; global warming; rising ocean acidity and ocean levels; increasing frequency, duration, and geographical spread of mega-droughts, forest fires, and violent weather; etc.
  - Geological disruptions: supervolcanoes (Taupo in New Zealand; Toba Caldera, North Sumatra; Yellowstone Caldera, Yellowstone National Park, WY, etc.), earthquakes and tsunamis (Pacific Ring of Fire, etc.).
  - Other manmade disasters: global nuclear and/or biochemical warfare; widespread and/or cumulative air and water pollution, and associated damage to habitats/environments; loss of the atmospheric ozone layer; widespread deforestation due to commercial development; etc.
- Outer space threats:
  - Solar coronal mass ejections (North American Carrington event in 1859, Hydro Quebec Electric Grid in 1989, electrical infrastructure and satellites destroyed, etc.).
  - Asteroid and comet impacts (Chicxulub Crater, Yucatan Peninsula c. 66 million years ago; Meteor Crater, AZ c. 50,000 years ago; Tunguska, Russia c.1908; small asteroid over Chelyabinsk, Russia in Feb. 2013, etc.).
  - ET invasion (never say never!).

One can also envision the near- to far-term economic expansion of humankind into the solar system thanks to the rise of the "space billionaires" and their pioneering/emerging commercial space transportation, tourism, and exploration ventures. The following is a nonexhaustive list of key factors to consider:

- The Moon and Mars are not habitable without shielded pressurized habitats, but rich in high-value resources.
- The asteroids and comets are also rich in high-value resources.
- There is plenty of "real estate" available to emplace space outposts in low Earth to geostationary orbits, cislunar space, lunar orbit, Lagrange point halo orbits, etc., to support tourism, industrial activities, exploration activities, and scientific activities.
- There is a need for significant infrastructure to support astronauts in a comfortable, shielded, and sustainable environment.
- There is a need for lots of radiation shielding from solar, galactic, and cosmic radiation.
- Outer space can support a space industry that leads to expanded economic development, new high-tech jobs, manufacturing of new technologies and services, accelerating space exploration and colonization, and advancing science, all of which will create new investment and wealth.
- The parameter space for the development of an outer space economy within the solar system is nearly infinite.

In the far, far future, one can imagine the expansion of humankind into the galaxy should mankind manage to avoid extinction. The Kepler Space Telescope's exoplanet searches have made the following findings as of July 2022:

- 5108 confirmed exoplanets in 3779 exoplanetary systems.
- 826 systems having more than one exoplanet in orbit around their parent star(s).
- 262 exoplanets may be habitable: these are Earth-sized, super-Earth-sized, and Mars-sized.
- The nearest exoplanets that are located within their parent star's habitable zone are $\gtrsim$22 LY away from Earth.
- >18,000 total exoplanetary candidates identified (yet to be evaluated).
- Statistics: at least one exoplanet on average orbits each star in our galaxy:
  - This gives 100–400 billion exoplanets in our galaxy.
  - There are an estimated 17 billion Earth-type exoplanets in our galaxy.
  - Many more exoplanets are free-floating in orbit around the galactic nucleus.
  - There are more planets than there are stars in our galaxy.
  - This implies a near-infinite parameter space that can support interstellar colonization, industry, expanded economic development, new high-tech jobs, new technologies, and create new investments and wealth.

There's a lot of food for thought here, but the main concern is that interstellar flight will never be achievable using the existing conventional space propulsion technologies, while the known advanced space propulsion concepts will have to be greatly stretched in capability to make any less-than-marginal or marginal stride toward reaching the nearest star(s) with or without human passengers onboard the interstellar spacecraft. A disruptive innovation in space propulsion physics and associated technologies will be needed to overcome the limitations of conventional and advanced space propulsion concepts. In what follows, we review the classical and relativistic rocket equations that are applicable to conventional and advanced space propulsion concepts, respectively, and discuss their limitations. In later sections of this chapter, we will introduce and review the two theoretically proposed disruptive innovations that allow for FTL travel without violating the special relativistic speed of light limit and are not without their own limitations.

1.2 A review of the classical and relativistic rocket equations and their consequence to interstellar flight

There's no doubt that spacefaring nations or a large group of nations with the technical and economic means to do so would like to implement a long-term exploration campaign to visit the closest reachable habitable exoplanets to conduct scientific surveys and hopefully discover whether they are inhabited by any form of life. Unfortunately, the rocket equation, derived from Newton's laws of motion, puts the kibosh on such aspirations. We can understand this by examining either of the two alternative forms of the rocket equation (the meter-kilogram-second, or MKS, units are used throughout):

$$\Delta v = v_e \ln\left(\frac{m_i}{m_f}\right) = I_{sp} g ln\left(\frac{m_i}{m_f}\right) \tag{1a}$$

$$\left(\frac{m_i}{m_f}\right) = e^{\frac{\Delta v}{v_e}}, \tag{1b}$$

where Δv (m/s) is the maximum change in spacecraft velocity (with no external forces acting), which is defined by the type of mission the spacecraft is designed for; v_e (m/s) is the propulsion (or rocket engine) system's effective heated propellant exhaust velocity; m_i (kg) is the spacecraft's total initial mass including the propellant (wet mass); m_f (kg) is the spacecraft's total final mass after the propellant is consumed (dry mass); I_{sp} (seconds) is the specific impulse, which measures the propulsion system's efficiency; and g (9.81 m/s^2) is the acceleration of gravity on the Earth's surface.

By comparing the middle and right side of Eq. (1a), we see that $v_e = I_{sp}g$. Eq. (1b) is the exponential inverse of Eq. (1a). I_{sp} is a very important parameter because it is the ratio of the thrust produced by the rocket engine to the weight flow of its propellant(s). It provides a quick way to determine the thrust of a rocket if we know the weight flow rate through the rocket engine's exhaust nozzle. A rocket engine with the higher value of I_{sp} is more efficient because it produces more thrust for the same amount of propellant, and it gives us an easy way to size an engine during preliminary analysis. The spacecraft or rocket weight will define the required value of thrust and dividing the required thrust by the I_{sp} will tell designers how much weight flow of propellants their rocket engine must produce. This information determines the physical size of the engine.

The rocket equation states that the wet mass grows exponentially large with increasing mission Δv. When $v_e \ll \Delta v$, then $m_f \ll m_i$ and most of the spacecraft mass is propellant. Generally, this means that spacecraft must carry enough propellant just to propel the propellant plus everything else that makes up the spacecraft mass, which leads to the undesirable consequence that chemical rocket propulsion has the very low spacecraft payload mass fraction of 2%–9% of the total spacecraft mass. Interstellar flight requires very high Δv to overcome the Earth's and the solar system's gravitational potential wells and to achieve a high enough spacecraft cruise velocity to reach nearby interstellar destinations within one or two human generations. Because of the rocket equation, spaceflight is hard in going from the Earth at 1 AU (1.5×10^{11} m) distance from the Sun to the outer edge of the asteroid belt at 3.2 AU (typical $I_{sp} \approx 10^3$ s, $\Delta v \approx 10$ km/s). It's harder to go from Earth to the inner edge of the Kuiper Belt at 40 AU (typical $I_{sp} \approx 10^4$ s, $\Delta v \approx 100$ km/s) and it's really hard to go from Earth to the inner edge of the Oort Cloud at 2000 AU (typical $I_{sp} \approx 10^5$ s, $\Delta v \approx 10^3$ km/s). Additionally, it's seemingly impossible to go from Earth to beyond 2000 AU to reach interstellar space (note that Alpha Centauri is 2.76×10^5 AU from Earth) because of the exponentially huge amount of wet mass required to reach interstellar distances. For an interstellar flyby of Alpha Centauri in 40 years of mission time, $I_{sp} \approx 3 \times 10^6$ s and $\Delta v \approx 0.1c$ (c = light speed, 3×10^8 m/s), while $I_{sp} \approx 1.5 \times 10^7$ s and $\Delta v \approx 0.5c$ for a 10-year flyby mission time (Δv doubles for rendezvous), which ignores relativistic effects. The latter pair of figures also apply (again ignoring relativistic effects) to a 40 LY stellar flyby in 100 years of mission time [3].

A cure for this problem is to design a highly efficient propulsion system that uses a propellant with a much higher energy density, produces a higher I_{sp} (and Δv), and has a higher fractional (ϵ) conversion to energy of the propellant than the combustion of chemical propellants. For the highest energy chemical propellants (LH_2 and LO_2) in chemical rocket propulsion, $\epsilon = 1.5 \times 10^{-10}$, which is so minute that it renders chemical propulsion nonviable for deep space or interstellar missions [4]. By comparison, $\epsilon = 7.9 \times 10^{-4}$ for the complete fission of ^{235}U fuel in nuclear fission propulsion while $\epsilon = 4 \times 10^{-3}$ for any fuel cycle in nuclear fusion propulsion, and $\epsilon = 1.0$ for the complete annihilation of matter-antimatter (MAM) propellant in MAM propulsion [4]. We'll revisit ϵ for its use as a particular type of space propulsion metric in a later paragraph. Propellant energy density can be expressed as the total energy released per unit mass of propellant. Thus, the energy released by liquid chemical propellants is 1 MJ/kg, while it is 5 MJ/kg for solid chemical propellants. The nuclear fission of ^{235}U fuel releases 8×10^7 MJ/kg, while the nuclear fusion of D-T fuel releases 3.5×10^8 MJ/kg, and the energy released by MAM annihilation is 9×10^{10} MJ/kg. An immediate consequence of increasing propellant energy density in advanced space propulsion engines is the corresponding increase of their I_{sp} (or v_e) and Δv, and the latter can also be increased by staging the rocket and/or improving the mass fraction.

Another key factor to consider in this discussion is energy: fast space travel times require higher spacecraft velocities; higher spacecraft velocities require more kinetic energy; and higher energy density propellants generate higher v_e, and hence, higher I_{sp}. We would really like advanced propulsion concepts that can achieve high Δv to reach high cruise velocities, decrease mission flight times, and hopefully improve payload mass fractions. The energy scale for undertaking an interstellar mission is daunting to comprehend, especially when interstellar spacecraft are capable of reaching a significant fraction of light speed ($\geq 0.1c$). For example, if we have an interstellar spacecraft with a mass of 100 metric tons (1 MT = 1000 kg) that's traveling at relativistic speed $v = 0.5c$, then its relativistic kinetic energy (see Eq. 2 below) is 1.39×10^{21} Joules (J). In 2021, the total annual electrical energy generated by utility-scale electricity generation facilities in the United States was 4.12 trillion (4.12×10^{12}) kWh or 1.48×10^{19} J. This means that the kinetic energy of this example interstellar spacecraft is 94 times the total annual electrical energy produced by the United States in 2021. The production of this amount of energy for relativistic interstellar flight requires sizing interstellar spacecraft systems with physical dimensions of anywhere on a scale of several oil supertankers to planetary diameters as well as masses on the order of millions to hundreds of billions of metric tons.

Fig. 3 shows an illustration of the tradeoff of interstellar spacecraft propellant mass vs interstellar propulsion system I_{sp} ("Better Rockets" in units of seconds) vs interstellar mission flight ("Trip") time for the case of a Space Shuttle Orbiter with a propulsion system that flies it past 4.3 LY from Earth. The "Orion" on the Better Rockets axis refers to the $I_{sp} \approx 6000$ s produced by the Project Orion nuclear pulsed-detonation (or external pulsed-plasma) propulsion system in which a series of thermonuclear bomblets are ejected out the rear of the spacecraft and then detonated in rapid succession to produce intense blast waves that transfer their momentum to a pusher plate mounted on shock absorbers at the rear of the spacecraft. Also shown on the Better Rockets axis is the ICAN-II antimatter-catalyzed micro-fission propulsion engine with $I_{sp} \approx 13{,}500$ s. The 50,000 s on the Better Rockets axis is the representative I_{sp} of nuclear fusion propulsion in which fusion energy is used to heat a secondary propellant working fluid such as hydrogen. To find the propellant mass required to fly the Space Shuttle Orbiter past 4.3 LY, one selects the desired propulsion system I_{sp} on the Better Rockets axis and the desired mission trip time.

There are the nuclear (fission, fusion, MAM annihilation) propulsion, electric (ion and plasma) propulsion, beamed (laser and microwave) energy propulsion, solar and laser sails, nuclear fusion ramjets, in situ propellant production propulsion, and a variety of hybrid combinations of some of these concepts. It is beyond the scope of this chapter to review every one of these concepts and compare their propulsion performance capabilities and viability for interstellar flight missions. Frisbee [3] performed a comprehensive technical evaluation of these interstellar propulsion technologies and concluded that the most likely candidates for fast ($\sim 0.5c$) interstellar missions suffer from either a need for extraordinary amounts of propellant, or a need for very large system infrastructure, or a host of significant

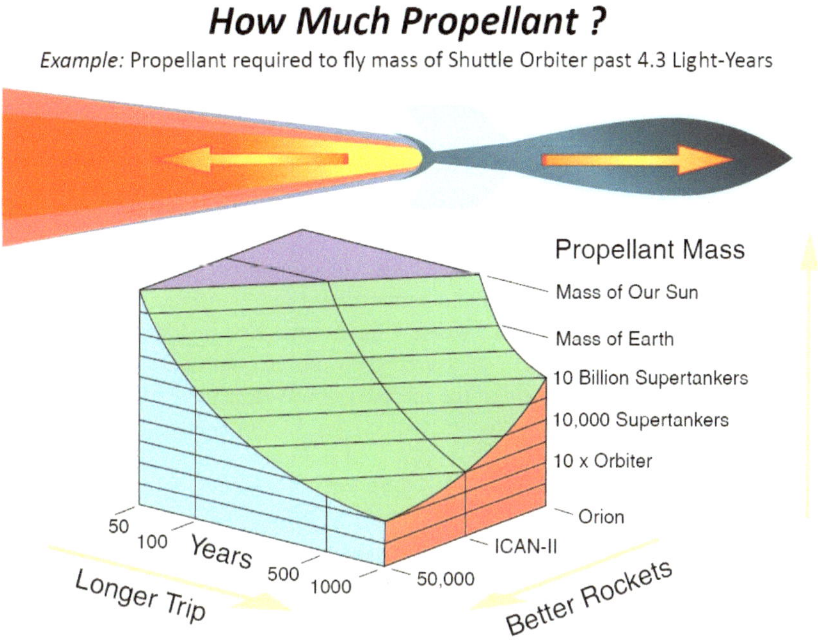

FIG. 3

Illustration of the tradeoff of interstellar spacecraft propellant mass vs interstellar propulsion system I_{sp} ("Better Rockets" in units of seconds) vs interstellar mission flight ("Trip") time for the case of a Space Shuttle Orbiter with a propulsion system that flies it past 4.3 LY from Earth. One selects the desired spacecraft propulsion I_{sp} and mission trip time to find out how much propellant is required.

Courtesy Marc G. Millis.

feasibility issues that must be tackled. He concluded that there is no single concept that is without potentially significant shortcomings, but there are at most three contenders having the likeliest success of overcoming feasibility and other issues, which are nuclear fusion, MAM annihilation, and hybrid laser-sail propulsion. These advanced propulsion concepts appear to meet the high interstellar mission Δv requirement because their propulsion I_{sp} is very high. The amount of propellant that each of these concepts require depends on how long or short the desired mission flight time is. Fig. 3 shows that a long flight time requires lower propellant mass while a short flight time requires much higher propellant mass.

At nonrelativistic flight speeds, it will take interstellar explorers tens to hundreds of thousands of human lifetimes to reach certain stellar destinations. At relativistic ($>0.1c$) or ultra-relativistic ($>0.86c$) speeds, the interstellar mission flight time will be reduced to several years or several hours, depending on how far one travels and how fast one's spacecraft is moving, because of the relativistic time dilation effect, which is expressed by $T=\gamma T_0$, where T is the clock time measured by remote stationary observers back home on Earth, T_0 is the clock time measured onboard the interstellar spacecraft, and $\gamma=[1-(v/c)^2]^{-1/2}$ is the relativistic Lorentz (time dilation/length contraction) factor [5]. When the spacecraft flight speed is $v<0.1c$, then $\gamma\approx 1$ and $T\approx T_0$, which means that clock rates are approximately equal for both observers on Earth and the spacecraft crew and they become exactly equal when $v \ll c$. When $v = 0.5c$, then $\gamma = 1.15$, which means that 1 second (or hour, day, year) of spacecraft clock time is 1.15 seconds (or hours, days, years) of Earth time. Also, when $v = 0.95c$, then $\gamma = 3.20$, which means that 1 second (or hour, day, year) of spacecraft clock time is 3.20 seconds (or hours, days, years) of Earth time. While short travel times are beneficial for interstellar travelers moving at relativistic or ultra-relativistic speed, there is the undesirable future time travel effect whereupon the relativistic interstellar travelers return to Earth in its far future relative to their onboard clock. All of their loved ones and society as they knew it would no longer exist, similar to what happened to the American astronauts in the 1968 SciFi film, *Planet of the Apes*. Of interest is to note that as $v \to c$, $T \to \infty$.

Even the travel distance between Earth and some distant star is dilated (contracted) at relativistic spacecraft speeds as defined by $L = \gamma^{-1} L_0$, where L is the distance or length measured by a moving observer and L_0 is the distance or length measured by a remote stationary observer [5]. For each of the three earlier examples, let's assume that the interstellar travelers have left Earth orbit and are traveling $L_0 = 4.37$ LY to reach one of the recently discovered exoplanets orbiting Alpha Centauri. At $v < 0.1c$, $L = L_0$ which means that Earth observers and the interstellar spacecraft crew measure the same travel distance. At $v = 0.5c$, $L = 0.866L_0 = 3.78$ LY, which means that the spacecraft has somewhat less distance to travel at this speed. At $v = 0.95c$, $L = 0.312L_0 = 1.36$ LY, which is a remarkably shorter travel distance than the previous two cases. Another note of interest is that as $v \to c$, $L \to 0$.

What about the kinetic energy of the interstellar spacecraft in these examples? Relativistic kinetic energy (KE_{rel}) is defined by either expression in [5]:

$$KE_{rel} = (\gamma - 1) m_{rest} c^2 = \frac{\gamma^2}{(\gamma + 1)} m_{rest} v^2. \tag{2}$$

For our earlier example of a 100 MT rest-mass interstellar spacecraft traveling at $v < 0.1c$, its $KE_{rel} \approx KE_{nonrel}$ (nonrelativistic KE) $= m_{rest} v^2/2 = 4.50 \times 10^{19}$ J, which become exactly equal when $v \ll c$. At $v = 0.5c$, $KE_{rel} = 1.39 \times 10^{21}$ J, while at $v = 0.95c$, $KE_{rel} = 1.98 \times 10^{22}$ J (or 1337.8 times the total annual electrical energy produced by the United States in 2021). In relativistic dynamics, the conserved total relativistic energy ($E_{rel} = m_{rel} c^2$) of a moving body is the sum of its rest energy ($E_{rest} = m_{rest} c^2$) and its KE_{rel}, which, after doing some algebra using Eq. (2), is $E_{rel} = m_{rel} c^2 = \gamma m_{rest} c^2$, where $m_{rel} = \gamma m_{rest}$ is the relativistic mass of the moving body [5]. These expressions demonstrate that as an interstellar spacecraft's $v \to c$, $\gamma \to \infty$, $m_{rel} \to \infty$, and $E_{rel} \to \infty$. This is the key showstopper that prevents any type of advanced interstellar spacecraft propulsion system from ever getting close to or reaching light speed, which SRT states is a universal speed limit within the framework of a flat Minkowski spacetime geometry [5].

One last consideration is to revisit the form of the rocket equation in the relativistic regime. Forward [6] derived the relativistic rocket equation:

$$\frac{m_{rest-i}}{m_{rest-f}} = \left(\frac{1 + \frac{\Delta v}{c}}{1 - \frac{\Delta v}{c}} \right)^{\frac{c}{2v_e}}, \tag{3}$$

where the relativistic initial and final rocket mass ratio was originally used but the common factor of γ in the numerator and denominator divides out, which leaves us with the initial and final rocket rest-mass ratio. Compare Eq. (3) with Eq. (1b). Also, one can see in Eq. (3) that as $\Delta v \to c$, the mass ratio goes to infinity. We use algebraic manipulations to invert Eq. (3) and solve for Δv to derive the relativistic analog of Eq. (1a):

$$\Delta v = c \tanh\left[\frac{v_e}{c}\ \ln\left(\frac{m_{rest-i}}{m_{rest-f}}\right)\right]. \tag{4}$$

Note that $\Delta v \rightarrow c$ as the argument of the hyperbolic tangent goes to infinity, which happens as the rocket mass ratio goes to infinity inside of the natural log.

There's also a very useful relationship between v_e/c and the ϵ that we discussed several paragraphs earlier [4,6]:

$$\frac{v_e}{c} = \sqrt{(2-\epsilon)\epsilon}. \tag{5}$$

Eq. (5) expresses the maximum exhaust velocity for the reaction mass produced by a spacecraft propulsion system that converts a fraction (ϵ) of its rest-mass, i.e., the fuel mass fraction, into propulsion exhaust kinetic energy plus spacecraft kinetic energy. One can also derive an alternative form for Eq. (4) by substituting Eq. (5) into the left side of the natural log inside of the square brackets in Eq. (4). Table 1 shows the values of ϵ that we previously discussed several paragraphs earlier and their corresponding propulsion maximum v_e after multiplying Eq. (5) by c. MAM propulsion shows the potential for interstellar spacecraft to reach light speed, which is certainly the case for an ideal photon rocket with $I_{sp} = v_e/g = c/g = 3.06 \times 10^7$ s as shown in Table 1.

1.3 Quantum violations of the relativistic light speed limit

While the light speed limit in Special Relativity is an absolute constant in every local inertial reference frame within a globally flat Minkowski spacetime, it turns out that it has been resoundingly violated via well-documented self-consistent solutions found in two key, very well-tested physics theories: Quantum Field Theory (QFT) and Einstein's General Theory of Relativity (GTR). Quantum Electrodynamics (QED) theory, a subset of QFT, allows for super- and subluminal photons, in which phase, group, and signal velocities will all either increase (become superluminal) or decrease (become subluminal) by the same amount with no dispersion effects when given a specific modified state of the QED vacuum [7,8]. The QED vacuum structure is essentially a Lorentz-invariant ether [9–13] comprised of the zero-point (ground or vacuum state) fluctuations (ZPF) of virtual charged particle pairs. We know from QED that light propagating through space interacts with the ZPF field. The observable properties of light, including its universal constant speed, are determined by these interactions. Quantum vacuum ZPF

Table 1 Theoretical maximum propulsion exhaust velocity (v_e) and specific impulse (I_{sp}) determined by ϵ.

Propulsion energy source	ϵ [4]	Theoretical maximum v_e (m/s)	Theoretical max. I_{sp} (s)
Chemical combustion (LH_2/LO_2)	1.5×10^{-10}	5196.1	529.7
Nuclear fission	7.9×10^{-4}	1.19×10^7	1.21×10^6
Nuclear fusion	4×10^{-3}	2.68×10^7	2.73×10^6
Matter-antimatter annihilation	1.0	c	3.06×10^7

field interactions with light lead to an effect upon the speed of light, which is due to the absorption of photons (by the vacuum ZPF) to form virtual charged particle pairs followed by the quick re-emission (from the vacuum ZPF) of the photon. The virtual charged particle pairs are extremely short lived ($\approx 10^{-21}$ s) because of the large mismatch between the energy of a photon and the rest-energy of the charged particle pair. The key point is that this process contributes to the observed vacuum electromagnetic permittivity (ε_0) and permeability (μ_0) constants, and therefore, to the speed of light [$c = (\varepsilon_0 \mu_0)^{-1/2}$] [14,15].

The role of the quantum vacuum ZPF virtual charged particle pairs in determining ε_0 and μ_0 is analogous to that of atoms/molecules in determining the relative permittivity ε and permeability μ of a dielectric material. We know that the absorption/re-emission of photons by atoms/molecules in a transparent medium (note that there are no strongly absorbing resonances, so the atoms/molecules remain in their excited states for a very short time before re-emitting photons) is responsible for the refractive index of the medium, which results in the reduction of the speed of light for photons propagating through the medium. This absorption/re-emission process is also known in physics as a scattering process. We know from experiment that a change in the medium leads to a change in ε and μ, thus resulting in a change of the medium's index of refraction. Another key point arising from this analogy is that a modification of the quantum vacuum ZPF field produces a change in ε_0 and μ_0, which results in a corresponding change in the speed of light and, hence, a corresponding change in the quantum vacuum's index of refraction (note that in QED, the vacuum is a nonlinear optical medium with highly unusual properties). When the QED vacuum in free space is undisturbed, then ε_0, μ_0, the vacuum index of refraction ($K = 1$), and c retain their observed universal constant values.

QED theory predicts several ways to modify its vacuum ZPF field to alter c. To modify (or "engineer") the QED vacuum, one can impose or apply [7,8]:

(1) Physical boundary conditions, e.g., a Casimir cavity or waveguide with a Casimir vacuum inside the cavity/waveguide that's anisotropic and nontranslational invariant to satisfy the electromagnetic boundary conditions imposed by the presence of the cavity/waveguide walls.
(2) An external constant uniform electric or magnetic field.
(3) A homogeneous and isotropic curved spacetime geometry, e.g., a Robertson-Walker gravitational (i.e., general relativistic spacetime curvature) background with Friedmann cosmology, which is also called a Friedmann-Robertson-Walker (FRW) spacetime.
(4) A homogeneous and isotropic thermal vacuum.

There are additional ways to modify the QED vacuum, but they are beyond the scope of this discussion. In item (1), the principal result of this vacuum modification is the removal of QED vacuum ZPF due to the suppression of ZPF modes with wavelengths (λ_{ZPF}) that are longer than the cavity/waveguide cutoff ($\lambda_{cutoff} = 2d$, where d = distance between cavity walls). This removal of free-space vacuum ZPF modes suppresses the scattering of light by the virtual charged particle pairs in the Casimir vacuum, which results in an increase in c in the direction between (perpendicular to) the two cavity walls. Because there are no physical boundary conditions in the direction parallel to the cavity walls, c retains its free-space vacuum value. This means that photon propagation inside a Casimir vacuum is anisotropic: photons propagating between the cavity walls are superluminal while photons propagating parallel to the cavity walls are luminal, which corresponds to $K < 1$ and $K = 1$, respectively. In item (2), the effect of an externally applied constant uniform electric or magnetic field is to make this modified QED vacuum anisotropic, analogous to the Casimir vacuum. Thus, c is decreased (subluminal photon propagation) in this vacuum for polarizations that are coplanar with and perpendicular to the plane defined by the electric/magnetic field and the direction of photon propagation. The polarization-averaged c is also subluminal. In item (3), the effect of a homogeneous isotropic FRW spacetime is to increase c, thus photon propagation is superluminal. This is the first indication via quantum theory that classical GTR should permit global FTL motion. We'll return to this a little later. In item (4), the effect of a homogeneous isotropic thermal vacuum is to decrease c, thus photon propagation is subluminal. Associated with the foregoing analysis, a general rule for the modified speed of light ($c*$) became apparent, which is applicable to modified vacua for massive and massless quantum field theories, for low energy [16]:

- $c^* > c$ and $K < 1$ when the modified vacuum has a lower (negative) energy density.
- $c^* < c$ and $K > 1$ when the modified vacuum has a higher (positive) energy density.
- $c^* = c$ and $K = 1$ when the vacuum is free (unmodified/undisturbed) and has zero quantum-averaged (i.e., quantum expectation value of) energy density.

2 General relativistic FTL propulsion—These are the "space warps"

2.1 Review of the general theory of relativity

Unfortunately, the aforementioned FTL effects predicted by QED theory are not practicable to engineer for the implementation of FTL interstellar flight because of the extreme magnitudes or physical attributes required for the different types of boundary condition imposed upon the vacuum to induce any observable FTL photon propagation. For example, to double c such that $c^* = 2c$ for superluminal photon propagation between the Casimir cavity walls requires that they be separated by a distance of one-third the Compton wavelength of an electron, which modern technology cannot engineer now or in the foreseeable future. Item (3) in Section 1.3 gave us one clue that GTR should predict FTL motion via curved spacetime effects. Within the context of GTR, the vacuum is the seat of a Lorentz-invariant spacetime structure, or spacetime metric, that encodes the geometry and topology of spacetime in response to the presence of a local distribution of matter field(s) that curves or "warps" the spacetime geometry such that it deviates from its special relativistic flat Minkowski geometry. This is embodied in Einstein's GTR field equations [17,18]:

$$G_{\mu\nu} = R_{\mu\nu} - \frac{1}{2} g_{\mu\nu} \mathcal{R} = \frac{8\pi G}{c^4} T_{\mu\nu}, \tag{6}$$

where $G_{\mu\nu}$ is the Einstein tensor, which encodes the gravitational (or acceleration) field effects (tidal forces) produced by curved spacetime; $R_{\mu\nu}$ is the Ricci curvature tensor (the contracted form of the Riemann curvature tensor, $R^{\alpha}_{\beta\gamma\delta}$) and $\mathcal{R}$ is the Ricci curvature scalar ($\mathcal{R} = R^{\mu}_{\mu}$ is the trace of $R_{\mu\nu}$ and repeated mixed tensor indices imply Einstein summation), both of which are defined by the metric tensor $g_{\mu\nu}$ and its derivatives; G is Newton's universal gravitation constant (6.673×10^{-11} Nm^2/kg^2); and $T_{\mu\nu}$ is the stress-energy-momentum tensor, which encodes the distribution of a matter field's rest-energy density (or rest-mass density; use $E = mc^2$ to convert between the two), shear stress, pressure, and flux of energy (equivalent to momentum density) in spacetime. $T_{\mu\nu}$ characterizes all matter, radiation, and nongravitational force fields. The components of $T_{\mu\nu}$ are the sources of the gravitational tidal forces computed from Eq. (6), whereas rest-energy density and pressure are the sources of the gravitational field in the general relativistic analog of the Newtonian gravitational Poisson equation in a stationary spacetime.

A spacetime metric ds^2 is a Lorentz-invariant distance function between any two infinitesimally separated events/points in spacetime defined by $ds^2 = g_{\mu\nu} dx^\mu dx^\nu$ (repeated mixed tensor indices also imply Einstein summation here), where $g_{\mu\nu}$ is a 4×4 matrix and the dx^μ are infinitesimal elements of distance in space and time. The Greek indices (μ, $\nu = 0 \ldots 3$) denote the spacetime coordinates $x^0 \ldots x^3$ such that $x^1 \ldots x^3$ are the spatial coordinates and $x^0 = ct$ is the time coordinate in the same unit of distance as the spatial coordinates. Eq. (6) is comprised of 10 coupled nonlinear partial differential equations, and it simply expresses the following classical spacetime physics of gravitation and matter: a local distribution (source) of matter tells a region of spacetime how to curve while the curved spacetime tells the local source of matter how to move through it.

It is also important to note that, in contrast to other classical field theories, the field $g_{\mu\nu}$ plays a dual role as both the quantity, which describes the dynamical aspects of spacetime curvature/gravity, and the quantity, which describes the background spacetime structure. The abbreviated definition for this spacetime structure in GTR is that it is a manifold of events (points in spacetime) that have the following properties [17–19]:

(a) It has four spacetime dimensions (this involves the definition of n-tuples of real numbers that form charts and atlases of coordinate systems and patches of coordinate systems that can be sewn together);
(b) It is continuous and infinitely differentiable (this involves the definition of continuous differentiable functions and their mappings/transformations in the manifold);
(c) It admits a Lorentzian (pseudo-Riemannian) ds^2 if and only if it admits a smooth nondegenerate rank 2 tensor field $g_{\mu\nu}$ with Lorentzian (pseudo-Riemannian) signature eigenvalues $(-, +, +, +)$ everywhere in the manifold;
(d) It is assumed to be Hausdorff (a topological property) if every two distinct points have disjoint neighborhoods of points, which means that no two distinct points are arbitrarily close to each other (this assumption could turn out to be wrong in the physical world); and
(e) It is paracompact, which is another topological property that is imposed so that a manifold will not be uncomfortably large.

Additionally, in pseudo-Riemannian geometry, the signature of ds^2 is defined as the difference of the number of its positive and negative eigenvalues (3 positive eigenvalues − 1 negative eigenvalue = 2) or as $n - 2p$ $(=4-2=2)$, where n is the number of dimensions of the manifold and $p = 1$ is the number of negative eigenvalues for Lorentzian metrics. Also, $g_{\mu\nu}$ defines the inner product of two vector fields in the manifold. Adding further conditions to this list leads to definitions for the mathematical operations of linear algebra and differential calculus of functions in the manifold. In physics, spacetime is often referred to as having (3+1)-dimensions.

2.2 The two types of general relativistic FTL propulsion solutions

A proposed cure to the light speed limit is to dispense with the notion of relativistic interstellar space propulsion and instead deploy a "space warp" to travel FTL to the distant stars and their exoplanets. The primary general requirements for putative FTL space warps should be [20]:

(1) The nonrelativistic and relativistic rocket equation no longer applies.
(2) The travel time inside of or through the FTL space warp should take ≤ 1 year as measured by the clocks of both the interstellar space travelers and remote static observers.
(3) Proper time as measured by space travelers should not be dilated by relativistic effects.
(4) The FTL space warp-induced tidal-gravity accelerations acting between different parts of the space travelers' bodies and their spacecraft's structure should be $\leq 1g$ when inside of or going through the FTL space warp.
(5) The speed of the space travelers' spacecraft while inside of or going through the FTL space warp must be $\ll c$.
(6) The space travelers (made of ordinary matter) must not couple strongly to the material that generates the FTL space warp's spacetime curvature or the FTL space warp material must be of a type that couples weakly to ordinary matter.
(7) The FTL space warp should not have a black hole-type of event horizon.
(8) There should be no black hole-type of singularity of infinitely collapsed matter (if such a phenomenon exists in absence of a universally accepted quantum gravity theory that could otherwise provide new guidance) residing within, on, or outside of the FTL space warp.

There are two types of FTL space warps that satisfy these requirements: traversable wormholes and warp drives. There are transient, static, and dynamical traversable wormhole spacetime geometries, while warp drives are dynamical spacetime geometries. FTL space warps are a class of exact solutions to Eq. (6). We will discuss these further in a later section.

Item (1) in the list above implies that any warp drive spacetime is an example of a bona fide "reactionless space drive" [21]. This is because GTR replaced gravity as a fundamental force of nature with curved spacetime geometry. A consequence of this is that the conservation laws (e.g., momentum conservation, from which Newton's Third Law of Motion is derived in nonrelativistic flat spacetime mechanics) are not globally conserved while they are conserved only within a small locally flat inertial reference frame. The situation is different for traversable wormholes because conventional nonrelativistic space propulsion is required for a spacecraft to traverse the throat of a wormhole. Thus, we should consider traversable wormholes to be a hybrid type of FTL transport. We will discuss this further in a later section. Items (2) and (3) ensure that the space travelers' local clock rate is the same or nearly the same as the local clock rate of remote static observers, and that travel times should be no longer than 1 year to make FTL interstellar flight tolerable for human passengers. Item (4) ensures that the tidal gravitational forces (inhomogeneities of gravity) produced by the FTL space warp are tolerable to human passengers and the structure of their interstellar spacecraft. We will discuss item (5) in the subsection that follows. Item (6) ensures that the spacecraft and its passengers are not in contact with or have very minimal contact with the FTL space warp material, which might prove to be disastrous for the space travelers and their spacecraft should contact be made, to avoid hindering or otherwise unnecessarily complicating the formation of the FTL space warp. More specifically, the tidal gravitational forces produced by the warp bubble wall's exotic material are predicted to be extreme enough to tear apart the spacecraft should its warp drive system generate the warp bubble wall within the spacecraft's hull. The same is also true for the wall of a traversable wormhole's throat, which a spacecraft must never come into contact with during its passage through the throat. Items (7) and (8) ensure that the two nasty attributes of black holes are not integral to the generation of FTL spacetimes because of their well-documented deleterious effects upon hypothetical space travelers that encounter them.

Implementation of FTL interstellar travel via FTL space warps generally requires the engineering of spacetime into very specialized local geometries as dictated by the combined requirements of items (1)–(8).

2.3 Light cone gymnastics: Moving FTL without moving FTL

Item (5) implies that all space travelers using an FTL space warp for rapid interstellar travel should at most move at nonrelativistic speed or at least be at relative rest within the FTL space warp while it does all the work to move the space travelers FTL without moving them FTL across interstellar distances. This latter statement is weird, but it lies at the root of what it is that makes relative FTL motion physically possible within GTR while not violating the special relativistic light speed limit. The "secret sauce" of how FTL spacetimes work is understanding the impact of their spacetime curvature effects upon the light cones of causality for all observers and how light cones perform a sort of gymnastics as they move across any given curved spacetime geometry.

The light cones of causality are a very important outcome of Hermann Minkowski's unified space and time model of our relativistic universe. Minkowski was Einstein's former teacher and showed, in 1907, that the Special Theory of Relativity (STR) introduced by Einstein in 1905 could be understood geometrically as a theory of a flat 4-dimensional (4D) spacetime that properly embodies all the postulates and theoretical predictions concerning the simultaneity of events and causality in conjunction with the inertial reference frames of moving bodies and electromagnetic radiation in STR. In attempting to diagram relativistic spacetimes, one of the most important features to capture is the causal structure of 4D Minkowski spacetime. This structure specifies which events (points in spacetime) can be:

- **(i)** causally connected by trajectories (worldlines for observers in motion) that are slower-than-light;
- **(ii)** causally connected by worldlines that are at light speed;
- **(iii)** or cannot be causally connected to FTL worldlines for any events in (i) and (ii).

Events in (i) are said to be "timelike related" (timelike worldlines) because a physical clock can travel from one event to the other. Events in (ii) are "lightlike related" (on the surface of the light cone, i.e., on lightlike worldlines) because a light ray (electromagnetic radiation) can travel from one lightlike event to another. Events in (iii) are "spacelike related," such that any event inside of or on the surface of the light cone cannot causally connect to any other event that's located outside of the light cone, as such a connection requires a worldline that's FTL, which violates the light speed limit in STR. Given that it is physically impossible for any causal process to exceed light speed via STR, these three possible ways of being connected tell us whether one event is able to influence another. To recapitulate, all subluminal (nonrelativistic and relativistic motion) worldlines lying within a light cone are timelike, all luminal (light speed) worldlines lying on the surface of the light cone are lightlike, and all impossible superluminal (FTL) worldlines lay outside of the light cone and cannot be reached by any timelike or lightlike worldlines. These three classifications hold true in any frame of reference such that an event judged to be within the light cone by one observer will also be judged to be within the same light cone by all other observers, no matter their frame of reference.

The spatiotemporal relationships (i)–(iii) can be depicted by drawing the light cone of causality for a single observer in flat Minkowski spacetime (see Fig. 4). Our 3-dimensional (3D) space, with one spatial dimension suppressed for ease of illustration, is shown as a 2-dimensional (2D) plane in the horizontal direction and the time dimension is shown as the vertical time axis. The observer lies within

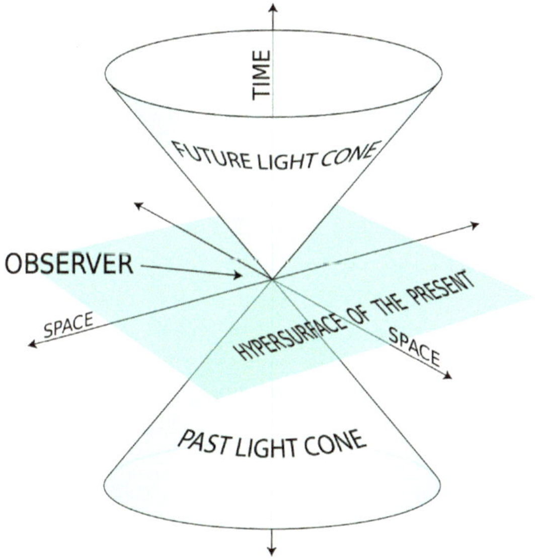

FIG. 4

Spacetime diagram of the light cone of causality in three dimensions (*ct*, *x*, *y*) with the *z*-dimension suppressed.
K. Aainsqatsi at Commons, https://commons.wikimedia.org/w/index.php?curid=2210907.

the spatial hypersurface (a 3D surface within the manifold of 4D spacetime) that's affixed to their present time. All past events connecting to the observer's present lay within the past light cone, while all future events connecting to the observer's present lay within the future light cone. All nonrelativistic and relativistic 4D motion of observers occurs along timelike trajectories (worldlines) that lie within both light cones, while all light speed, motion occurs along lightlike (or "null," because their 4D length is zero in STR and GTR) worldlines on the surface of both light cones (see Fig. 5). The latter is a special state of motion that's unique to photons or any other massless particle. The angle between the surface of

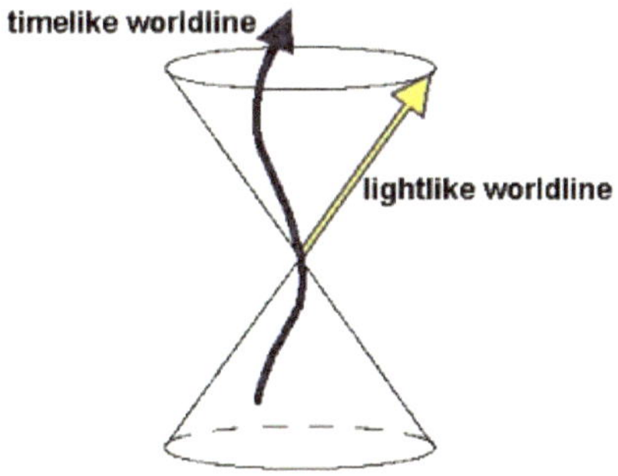

FIG. 5

Timelike and lightlike (or null) worldlines. Spacelike worldlines occur everywhere outside of the light cone.
Open access http://visualrelativity.com/LIGHTCONE/lightcone.html.

the future and past light cones and the 3D hypersurface of the present is always 45 degrees according to Minkowski and STR. All forbidden FTL motion occurs along spacelike worldlines that lay everywhere within any 3D hypersurface outside of both light cones.

In the flat Minkowski spacetime of Einstein's STR, the light cones of causality are universally oriented in the same timelike direction for all observers, regardless of their state of relative-rest, nonrelativistic motion, or relativistic motion (see Fig. 6). However, in the curved spacetime of Einstein's GTR, things get very interesting. The curving of spacetime in the presence of a local source of matter causes the local light cones, and hence their local timelike orientation, of individual observers/space travelers to tilt over with respect to the local light cones of remote static observers residing within a region of spacetime that is asymptotically flat. For the case of the well-known Schwarzschild curved spacetime region produced by massive astronomical bodies, the direction of the tilt of nearby observers' light cones will always be toward the central (gravitating) mass (see Figs. 7–9). Therefore, the local light cones of causality do not have a universally fixed timelike orientation for all observers within a curved spacetime region according to GTR. In certain specialized curved spacetime geometries, the timelike orientation of some observer light cones can be tilted over into the spacelike direction with respect to the timelike orientation of remote static observers' light cones such that relative global FTL motion is allowed, as shown in Figs. 8 and 9.

Therefore, item (5) in Section 2.2 states that interstellar space travelers that use an FTL space warp do not need to be in any form of relativistic or light speed motion within the FTL space warp. The design of their FTL space warp's curved spacetime geometry allows them to move along any timelike

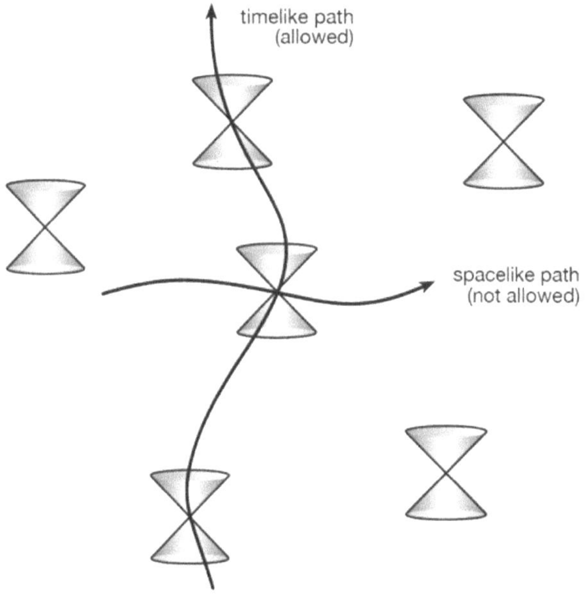

FIG. 6

The invariant timelike orientation of different observers' light cones in STR.

Open access https://www.preposterousuniverse.com/blog/2016/11/24/thanksgiving-11/.

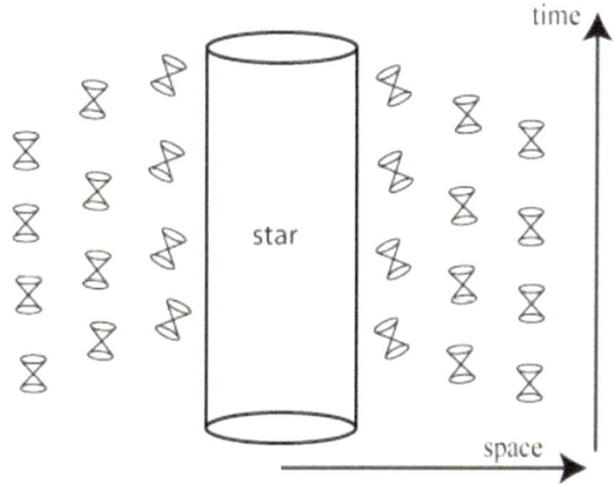

FIG. 7

The timelike orientation of observers' local light cones tilts toward a large astronomical gravitating mass, such as a star that's depicted here as a large cylindrical timelike worldline, as they move closer to it.

Erik Curiel, open access https://plato.stanford.edu/entries/spacetime-singularities/lightcone.html.

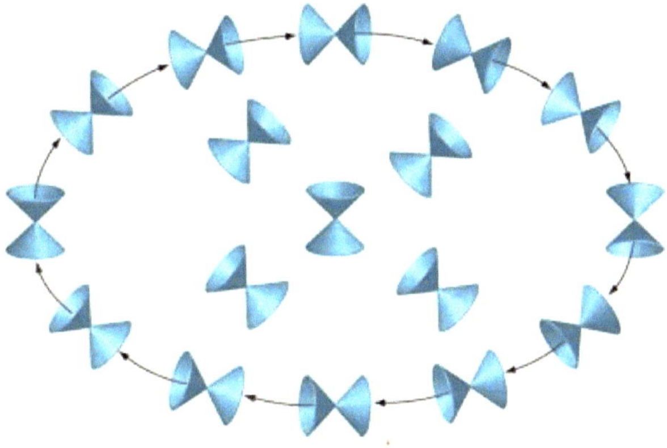

FIG. 8

The permissible timelike orientations of the local light cones of stationary and moving observers in GTR [22]. In this illustration, the 3D space axes (one spatial axis is suppressed) are in the horizontal direction and the time axis is in the vertical direction w.r.t. to the stationary central light cone. Notice that the local light cones on the worldlines (only the outer closed circular worldline is shown) of several moving observers within a hypothetical curved spacetime have timelike orientations in the timelike (vertical), lightlike (tilted 45 degrees) and spacelike (horizonal/near-horizontal) directions w.r.t. the stationary central light cone as well as w.r.t. any remote static observers located in an asymptotically flat spacetime region. This illustrates that relative global FTL motion is allowed by GRT because the moving observers on the closed circular worldline are undergoing timelike motion within their local light cone, thus not violating STR's light speed limit. The closed circular worldline also represents a closed-timelike curve (CTC), or time machine, in GTR.

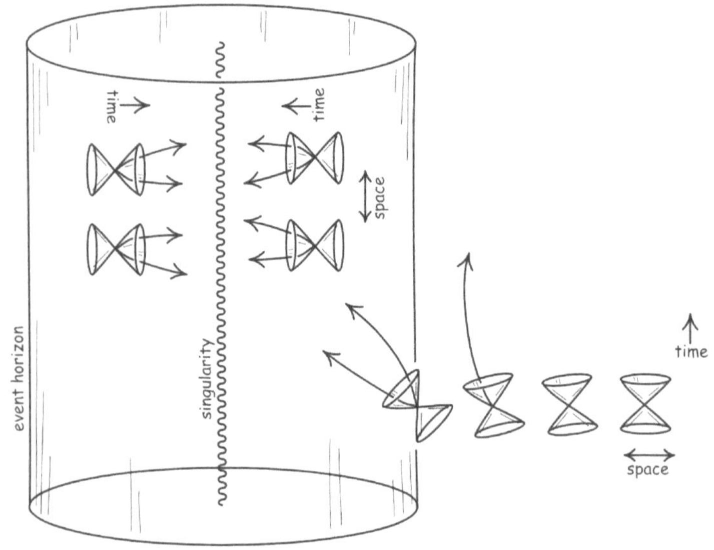

FIG. 9

The behavior of observers' light cones as they fall into a black hole with its timelike worldline illustrated as a large vertical cylinder. Observers with timelike-oriented local light cones are falling toward the singularity inside of the event horizon but are moving FTL w.r.t. all observers outside of the event horizon. This is because space and time exchange their directional orientation inside of black holes.

Open access https://sites.pitt.edu/~jdnorton/teaching/HPS_0410/chapters/black_holes/index.html.

worldline within their local light cone as they travel between interstellar departure and destination points, thus not violating the special relativistic light speed limit during their trip. Meanwhile, the FTL space warp's curved spacetime geometry acts to tilt over their local light cone so that its timelike orientation always points in the spacelike direction of all remote static observers residing within a region of asymptotically flat spacetime. This results in the relative global FTL motion of the space travelers as seen by the remote static observers whose local light cones are unaffected by the FTL space warp's curved spacetime geometry. This is the secret sauce of "moving FTL without moving FTL" via the light cone gymnastics of curved spacetime geometries within GTR.

3 The two classes of FTL space warps: Traversable wormholes and warp drives

3.1 Traversable wormholes

Traversable wormholes represent a class of exact solutions to Eq. (6) and the design of their spacetime metric is guided by the requirements outlined in items (1)–(8) in Section 2.2. These requirements led K. Thorne and collaborators to design a spherically symmetric Lorentzian spacetime metric, ds^2, that prescribes the required traversable wormhole geometry and topology [23–25]:

$$ds^2 = -e^{2\phi(r)}c^2dt^2 + \left[1 - \frac{b(r)}{r}\right]^{-1} dr^2 + r^2 d\Omega^2, \tag{7}$$

where standard spherical coordinates are used (r: $2\pi r =$ circumference; $0 \leq \theta \leq \pi$; $0 \leq \varphi \leq 2\pi$), t is time ($-\infty < t < \infty$), $d\Omega^2 = d\theta^2 + sin^2\theta d\varphi^2$, $\phi(r)$ is the freely specifiable redshift function that defines the proper time lapse through the wormhole throat, and $b(r)$ is the freely specifiable shape function that defines the wormhole throat's spatial (hypersurface) geometry.

The throat is static and spherically shaped, but there's a whole taxonomy of traversable wormholes with different throat shapes and dynamical throats (see the upper half of Fig. 10). All Lorentzian traversable wormholes are event horizon-free, singularity-free, fully traversable in both directions, geodesically complete, and there are no crushing gravitational tidal forces found anywhere inside the wormhole. Consequently, Lorentzian traversable wormholes are unlike the nontraversable Schwarzschild wormhole, or Einstein-Rosen bridge, associated with eternal black holes in the maximally extended version of the Schwarzschild metric.

The top of Fig. 11 depicts the Flamm embedded space diagram of Eq. (7), which is the 3D geometry of an equatorial ($\theta = \pi/2$) slice through spacetime at a specific moment of time ($t =$ constant). This Flamm diagram represents a spherically symmetric traversable wormhole that connects two different universes (an interuniverse wormhole). The bottom diagram in the figure is a different version of the Flamm diagram depicting a spherically symmetric intra-universe wormhole with a throat connecting two distant regions within our universe. These diagrams serve to aide in visualizing traversable wormhole geometry and are merely a geometrical exaggeration.

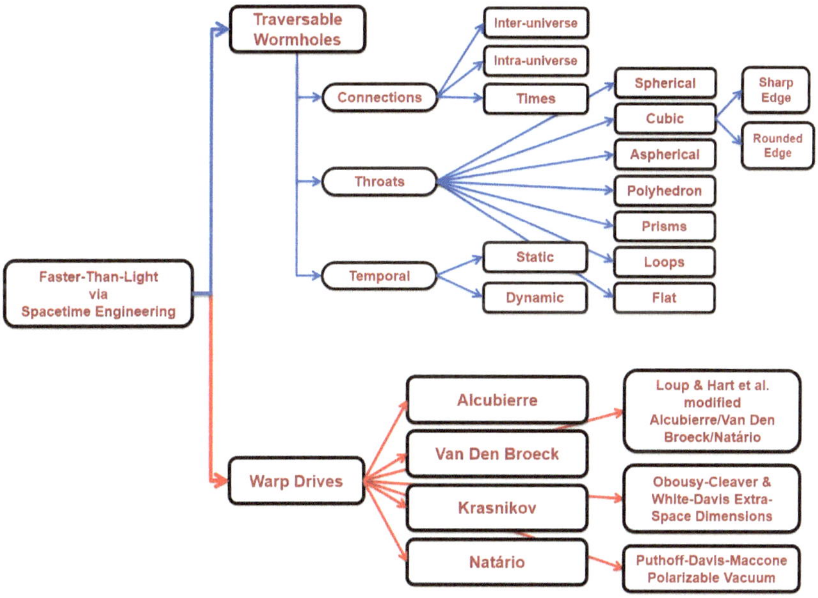

FIG. 10

The taxonomy of traversable wormholes (upper half) and the ever-growing taxonomy of warp drives (lower half) in the genre of FTL spacetime engineering.

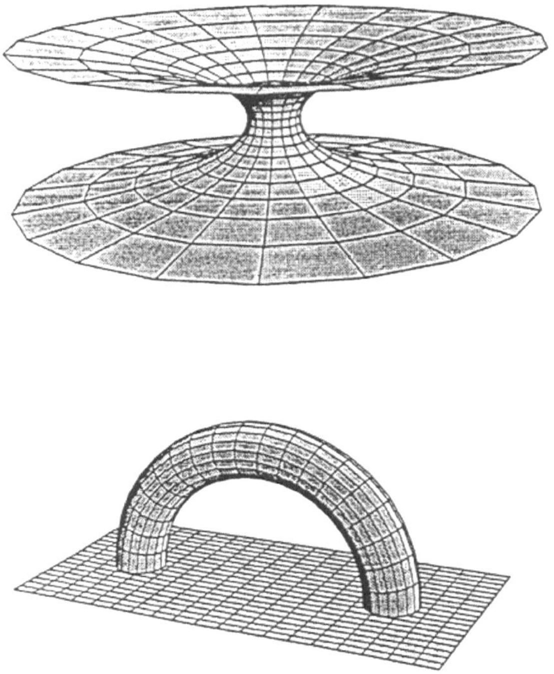

FIG. 11

Interuniverse wormhole (top) and intra-universe wormhole (bottom) [26].

There were originally two other criteria for defining a traversable wormhole, which was that it must be embedded within the surrounding (asymptotically) flat spacetime with a flaring-out of the opening ("mouth") on both sides of the throat, as depicted in the top Flamm diagram in Fig. 11. However, Hochberg and Visser [27] proved that it is only the behavior near the wormhole throat that is critical to understanding the physics, and that a generic throat can be defined without having to make all the symmetry assumptions, without assuming the existence of an asymptotically flat spacetime in which to embed the wormhole, and without having a flaring-out condition for the throat's two openings. This fact led to the development of a number of other traversable wormhole throat designs that are aspherical-shaped, polyhedral-shaped, loop-shaped, flat-face-/soap film-shaped, generic-shaped, etc., as shown in the top half of Fig. 10. There are also dynamical traversable wormholes, which have time-dependent throats. There is also the question of how many throats, also called "handles" in topology, a traversable wormhole can have and whether these might lead to degenerate throats. This will be discussed further in Section 6.1.

Traversable wormholes are 3D hyperspace tunnels through spacetime that connect together either remote regions within our universe or two different universes, as shown in Fig. 11. They can also connect different dimensions [28] and different times [24,25]. Interstellar travelers would enter one side of the wormhole and exit the other side, passing through the throat along the way. The travelers would move at nonrelativistic speed with their local light cones oriented in the timelike direction along their worldlines going through the wormhole and would therefore not violate STR's light speed limit. With respect to remote static observers in asymptotically flat spacetime, the travelers' local light cones along their worldline appear to be oriented in the spacelike direction. Thus, the remote static observers view the travelers as having traversed multilight year distances between two distant planetary star systems in extremely short times at FTL speed. Of further interest is a flat-face traversable wormhole throat, or "stargate," which is a very simple class of traversable wormhole solutions to Eq. (6) [16,25]. Interstellar travelers would merely step through a circular- or polygon-shaped stargate and almost immediately emerge on the other side, many light years away from where they started.

The type and quantity of energy or matter required to construct any type of traversable wormhole shown in the upper half of Fig. 10 is found by mathematically cranking its spacetime metric through the left-hand side of Eq. (6) to find the components of $T_{\mu\nu}$ on the right-hand side. For the class of spherically symmetric traversable wormholes described by Eq. (7), we would crank this metric through Eq. (6) to find the components of $T_{\mu\nu}$. After computing these components for any type of traversable wormhole, we would then do a complete mathematical analysis to determine what it will take to construct a traversable wormhole in spacetime. We already know from numerous published studies that the computed components of $T_{\mu\nu}$ for all types of traversable wormhole predict a form of energy/matter with unusual properties that are called "exotic," as compared to the properties of "ordinary" matter that we all encounter in our everyday lives, as well as what we astronomically observe or encounter via space exploration of the solar system. This topic is so specialized that it requires its own discussion in Section 6.1.

Given that there is no observational evidence supporting the existence of "other" or "parallel" universes or macroscopic-scale extra-space dimensions beyond the physical 4D spacetime manifold that we live in, traversable interuniverse wormholes and traversable interdimensional wormholes remain a theoretical curiosity within the multiverse and superstring/D-brane theory communities, and thus will not be considered further in this chapter. It is also beyond the scope of this chapter to review the mathematical details for the entire taxonomy of static and dynamic traversable wormholes because the published literature is very extensive and growing, so the reader is encouraged to consult Ref. [23], which is an excellent authoritative monograph on this topic. The energetics of traversable wormholes and the problem of interstellar guidance, navigation, and control while moving through them will be discussed in later sections.

3.2 Lorentzian warp drives

To model *Star Trek*'s warp drive propulsion effect, Alcubierre [29] derived a Lorentzian spacetime metric motivated by the cosmological inflationary expansion of spacetime that would allow arbitrarily short travel times between two distant interstellar points. The warp drive metric uses coordinates (ct, x, y, z) and inertial worldline $x = x_{sh}(t)$, $y = 0$, $z = 0$, lying in the ct-x plane passing through the origin of the interstellar spacecraft's frame. x_{sh} is the x-axis coordinate position of the moving frame of the spacecraft and its surrounding warp bubble. The metric, ds^2, specifying this spacetime geometry is [29]:

$$ds^2 = -c^2dt^2 + [dx - v_{sh}(t)f(r_{sh}(t))cdt]^2 + dy^2 + dz^2, \tag{8}$$

where the subscript $sh \equiv$ ship, $v_{sh}(t) \equiv dx_{sh}(t)/cdt$ is the dimensionless warp speed associated with the inertial worldline, and $r_{sh}(t) \equiv [(x - v_{sh}ct)^2 + y^2 + z^2]^{1/2}$ is the Euclidean distance from the inertial worldline. The dimensionless warp bubble shape function, $f(r_{sh}(t))$, is any smooth positive function that satisfies $f(0) = 1$ and decreases away from the origin to vanish when $r_{sh}(t) > R$ for some R, where R is the radius of the warp bubble. $f(r_{sh}(t))$ is defined as [29]:

$$f(r_{sh}(t)) = \frac{tanh[\sigma(r_{sh}(t) + R)] - tanh[\sigma(r_{sh}(t) - R)]}{2tanh(\sigma R)}, \tag{9}$$

where $\sigma \sim \Delta^{-1}$, and Δ is the thickness of the warp bubble wall that surrounds the spacecraft. $R > 0$ and $\sigma > 0$ are two arbitrary parameters. The geometry of each 3D hyperspatial slice is flat, and spacetime is flat where $f(r_{sh}(t))$ vanishes but is curved where it does not vanish. When σ is large, then this constraint specifies ($f(r_{sh}(t))$ rapidly approaches) a top hat function, which is a very small flat spacetime region where the spacecraft resides within the warp bubble. It is not necessary to choose a particular form of $f(r_{sh}(t))$ because any function will suffice so long as it has the value of approximately 1 inside some region of $r_{sh}(t) < R$ and goes to zero rapidly outside the bubble, such that as $r_{sh}(t) \to \infty$ we recover asymptotically flat Minkowski spacetime. The warp drive spacetime geometry is dynamical because of its time dependence.

Moreover, observers with 4-velocity vector $V^{\mu} = (1, 0, 0, v_{sh}(t)f(r_{sh}(t)))$ and corresponding covariant 4-velocity vector $V_{\mu} = (-1, 0, 0, 0)$ (we temporarily set $c = 1$) move along geodesics, as their 4-acceleration is $a^{\mu} = V^{\nu}V^{\mu}_{;\nu} = 0$, where the semicolon denotes the covariant derivate of V^{μ}. Their 4-velocity is normal to the spatial hypersurface with $t =$ constant, and they are close analogs of the Eulerian observers of fluid mechanics, which move with the fluid flow. The warp spaceship always moves along an inertial timelike worldline regardless of the value of $v_{sh}(t)$. The proper time τ along this worldline is equal to the coordinate time t, which can be derived by substituting $x = x_{sh}(t)$ into Eq. (8) and take into account that $dx = dy = 0$ and $f(0) = 1$. Thus, $d\tau = dt$.

The driving mechanism of Eq. (8) is the York extrinsic time, ϑ. This quantity is defined as [29]:

$$\vartheta = \frac{v_{sh}}{c}\frac{x_{sh}}{r_{sh}}\frac{df(r_{sh})}{dr_{sh}}. \tag{10}$$

The ϑ behavior of the warp drive bubble describes a distortion of the local spacetime geometry that induces the simultaneous stretching (expansion) of spacetime behind the spacecraft and a corresponding squeezing (contraction) of spacetime in front of the spacecraft (see Fig. 12). In response, the warp bubble moves forwards with an FTL speed determined by the contraction/expansion rate. The warp bubble would "ride" this spacetime distortion like a surfer riding on an ocean wave. As evidenced by the observed cosmological inflationary expansion of the universe, spacetime can globally expand so rapidly—because spacetime geometry has no special relativistic light speed limit—that objects can globally effectively travel FTL. In this scenario, objects in expanding (or contracting) spacetime don't move FTL through spacetime. Instead, the expanding/contracting spacetime moves the objects at arbitrarily large globally effective FTL speed with respect to the flat background geometry. This also happens to be a local spacetime curvature phenomenon in the warp drive scenario wherein the objects (interstellar travelers in their warp drive spacecraft) are transported at arbitrarily large globally effective FTL speed by the locally distorted spacetime geometry with respect to remote static observers residing in asymptotically flat spacetime. The warp bubble that produces the warp drive's spacetime geometry surrounds the travelers. They would be at relative rest within the small flat spacetime region inside of this warp bubble, with their local light cones oriented in the timelike direction along their worldlines going through the warp drive spacetime geometry, thereby not violating STR's light speed limit. With respect to remote static observers in asymptotically flat spacetime, the travelers' local light cones along their worldline appear to be oriented in the spacelike direction. Thus, the remote static

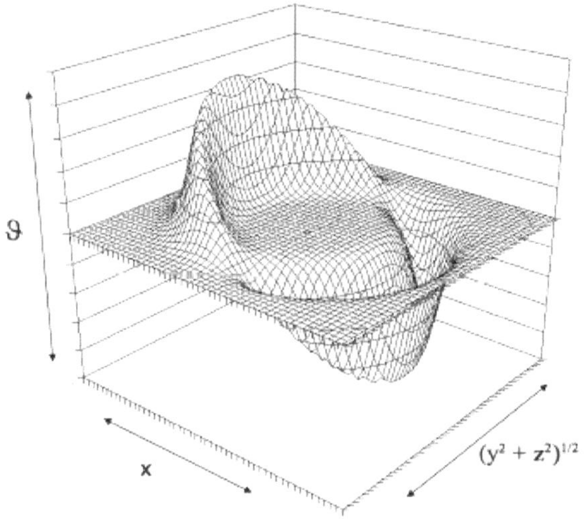

FIG. 12

Plot of ϑ, where x is the x-axis coordinate position of the moving spacecraft frame (located within the flat central circular area) and $(y^2+z^2)^{1/2}$ is the perpendicular position [26]. ϑ represents how much a given volume of spacetime expands or contracts. Upward vertical values (upper left) imply an expansion while the downward vertical values (lower right) imply a contraction. The contraction balances the expansion of spacetime as the warp bubble moves forward along the x-direction.

observers view the travelers as having traversed multilight year distances between two distant planetary star systems in extremely short times at FTL speed.

As described by Eqs. (8) and (9), the interstellar spacecraft at the center of the moving volume of flat spacetime within the warp bubble is at relative rest, so there is no relativistic mass increase nor time dilation effect. The spacecraft's onboard clock runs at the same rate as the clocks of remote static observers, and those observers will detect no increase in the mass of the moving spacecraft, even when it travels at FTL speeds. Moreover, Eq. (8) shows that even when the spacecraft is accelerating, it travels on a free-fall geodesic such that a spacecraft using the warp bubble to accelerate or decelerate is always in free fall, so the space travelers would not experience any gravitational tidal forces. However, enormous gravitational tidal forces are present near the edges of the flat-spacetime volume because of the large spacetime curvature there, as seen in Fig. 12. By a suitable specification of the warp drive's spacetime geometry, these would be made very small within the volume of flat spacetime occupied by the spacecraft.

The type and quantity of energy or matter required to construct any type of warp drive in the lower half of Fig. 10 is found by mathematically cranking its spacetime metric through the left-hand side of Eq. (6) to find the components of $T_{\mu\nu}$ on the right-hand side. For Alcubierre's warp drive, described by Eq. (8), we would crank this metric through Eq. (6) to find the components of $T_{\mu\nu}$. After computing these components for any type of warp drive, we would then do a complete mathematical analysis to determine what it will take to construct a warp drive in spacetime. We already know from numerous published studies that the computed components of $T_{\mu\nu}$ for all types of warp drive predict the same exotic matter properties as the exotic matter required to construct traversable wormholes. This will be discussed further in Section 6.2.

However, it turns out that Alcubierre's warp drive is flawed, because the warp spacecraft is treated as a massless point particle and Δ is assumed to be the Planck length ($\approx 10^{-35}$ m), both of which are physically unrealistic. One consequence of this is that the energy required to produce the warp bubble is astronomically stupendous, so it is not practicable to engineer and construct. Another consequence is that, at FTL speed, the interior region of the warp bubble is causally disconnected from its surface and the exterior spacetime, which implies that the surface part of the warp bubble can't be carried along and would have to be continuously generated externally. The warp drive could not be self-contained or self-operated because the spacecraft's pilots will not be able to send signals to control the warp bubble or communicate with observers in the space exterior to it. This causal disconnection is akin to a black hole event horizon. An additional consequence is that the massless-point-particle warp spaceship will not gravitationally couple to its warp bubble when it should do so according to GTR, which has its own consequence on determining what range of warp speeds is physically viable. These flaws will be discussed in greater detail in what follows and in later sections. It is for these reasons that Alcubierre's warp drive is an ideal toy model spacetime that initiated a new cottage industry of theoretical research within the GTR community that might lead to a physically realistic warp drive model in the future. There is a significant and growing taxonomy of warp drive spacetime geometries possessing different features that attempt to improve or cure the problems with Alcubierre's warp drive (see the lower half of Fig. 10). We discuss these in what follows.

Other investigators have designed warp drive metrics similar to Eq. (8) but with some modifications to the spacetime geometry that attempt to cure the flaws in Alcubierre's model. A well-known example is the proposal by Krasnikov [30] to fix the warp bubble's causal disconnection problem by specifying a "warp tube" spacetime, in which an interstellar spacecraft traveling one way at ultra-relativistic speed would create a tube-shaped space warp behind itself and then return to its point of origin by traveling back through the warp tube at FTL speed. Thus, the spacecraft can return from its interstellar journey shortly after it left no matter how far away it traveled. In other words, the Krasnikov warp tube is a tunnel through time that connects the departure time of the spaceship with the time of its arrival at the destination star. Spacetime is flat inside the warp tube but the spaceship's local light cone tilts toward the spacelike direction w.r.t. remote static observers so that it permits FTL travel in one direction only, e.g., back to the starting point at Earth or wherever its starting point is in the galaxy. The warp tube is a shortcut through time rather than spacetime because the distance to the destination star and back must still be traversed by an ultra-relativistic spaceship with enormous relativistic time dilation. Therefore, it has the property of being a time machine, which is no great sin, because the Grandfather Paradox was resolved long ago [31]. Everett and Roman [32] evaluated the warp tube and found other problems with it, which included a special circumstance where it could evolve into an unphysical time machine. The warp tube model does cure the causal disconnection problem, but it still has the problem of requiring an astronomically stupendous amount of exotic matter/energy to produce the warp tunnel.

Van Den Broeck [33] improved Alcubierre's warp drive by employing topological tricks to keep the interior of the warp bubble large while making its external surface extremely small, which appears to solve many of the problems with Alcubierre's warp drive. This was accomplished by multiplying the last three squared terms on the right-hand side of Eq. (8) by the square of a twice differentiable function, $B(r_{sh})$, which is a topological coefficient. Van Den Broeck observed that most of the undesirable effects of Alcubierre's warp drive scale with the volume or surface area of the warp bubble. Therefore, his

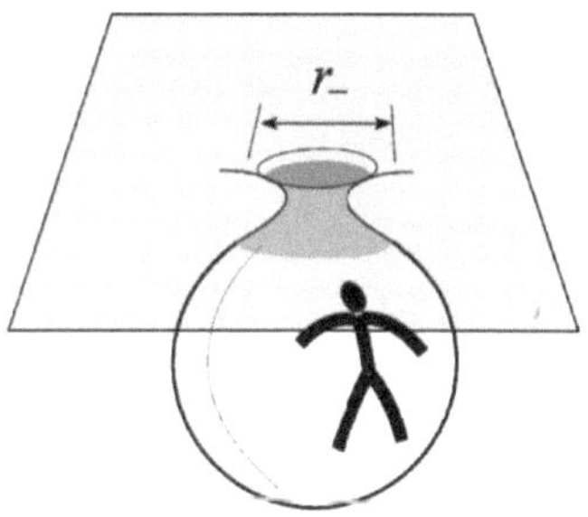

FIG. 13

Schematic diagram of the Van Den Broeck "pocket" or "bottle" with the topological coefficient, $B(r_{sh})$. The bottle, with submicroscopic outer diameter of its neck, r_{-}, is large enough to contain a person (seen here) or an interstellar spacecraft.

Courtesy F. Loup, private communication, 2020.

solution was to use $B(r_{sh})$ to make the radius of the warp bubble so small that the problems go away. He made use of the following trick from GTR and topology: because of spacetime curvature, the interior volume of a region of spacetime bounded by a closed 3D surface can be made much larger than the flat-space volume bounded by its surface. So, in curved spacetime the inside volume of a sphere of radius r can be much greater than $(4/3)\pi r^3$. The topological outcome of this is that a 3D hyperspatial slice of the overall warp drive spacetime geometry can be visualized as a "rubber membrane" that has a small Alcubierre warp bubble surrounding a neck that leads to a "pocket" with a large internal volume, and with a flat region in the middle, which sort of resembles a bottle (see Fig. 13).

This rubber membrane configuration resembles a bullseye target with a central Region I surrounded by three concentric ring Regions II–IV (see Fig. 14). The central sphere in Region I is flat spacetime that's large enough to encompass the warp spacecraft. Region II is a spherical shell containing curved spacetime that connects the large interior volume of Region I to an exterior region that is smaller in radius by a factor of $1/\Gamma$, and it is the region where $B(r_{sh})$ varies. Region III is a transition region of flat spacetime, a spherical shell with a volume much less than that of Region I. Region IV is a spherical shell as in an Alcubierre's warp bubble except with a very small radius, and it is the region where $f(r_{sh}(t))$ varies. Van Den Broeck sets the radius of Region I to 100 m and Γ to 10^{34}, so the radius of Region IV is approximately 3×10^{-32} m, which is 16 orders of magnitude smaller than the diameter of a proton (≈ 1 fm) and three orders of magnitude larger than the Planck length.

Even with this modification, there would still be very severe problems with developing a technological implementation of this warp drive. The size of Region IV is problematic because the putative effects of quantum gravity at that scale were not considered by Van Den Broek. Also, since the diameter of the warp bubble is many orders of magnitude smaller than the wavelength of visible light (400–700 nm), there would be no possibility of seeing out from inside the warp bubble and no possibility of steering it. While the magnitude of the total energy required to form this warp bubble becomes more reasonable, the energy density requirement remains astronomically stupendous. Van Den Broeck's calculations indicated that slowing the warp bubble to a near-stop might permit it to be expanded to any desired size, which would allow the space travelers to be able to physically enter or exit it at the warp spaceship's departure and destination stops. Such an expansion would decrease

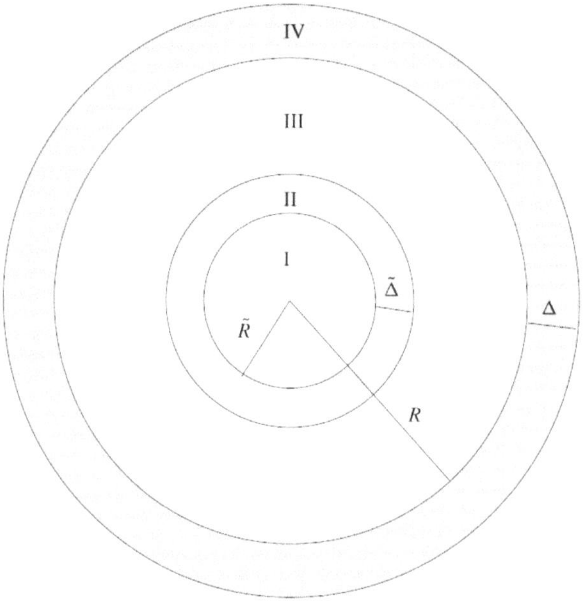

FIG. 14

Schematic diagram of Van Den Broeck's warp drive. The R and Δ symbols identify the radius (from the center of Region I) and thickness of Regions II and IV, respectively.

Courtesy C. V. D. Broeck, private communication, 1999.

the warp bubble wall thickness, and it is not clear what would happen if the wall thickness became smaller than the Planck length.

Natário [34] improved upon Alcubierre's warp metric by modifying and generalizing it so that zero spacetime expansion and contraction occurs around the warp bubble. Instead of riding a simultaneous contraction and expansion of spacetime, the warp bubble may be observed to be "sliding" through the exterior spacetime at a constant global FTL velocity. Spacetime in front of the warp bubble may get contracted and the spacetime behind it expanded, or not, depending upon the details of its construction. Natário essentially showed that the contraction/expansion in Alcubierre's metric is a marginal consequence of the spacetime geometric choice made by Alcubierre, and explicitly constructed a similar warp drive spacetime where no contraction/expansion occurs in the most general sense. Unfortunately, the causal disconnection problem in Alcubierre's warp drive survives in Natário's warp drive.

Loup [35,36] pointed out that Alcubierre's, Krasnikov's, Van Den Broeck's, and Natário's warp drives are not realistic for interstellar flight because they always have a constant warp speed and do not accelerate or decelerate at departure and destination points. In other words, a real warp drive must have the ability to accelerate from 0 km/s to some desired warp speed at the beginning of the spacecraft's interstellar journey and it must decelerate again to 0 km/s when it reaches its destination point. Loup extended Natário's warp drive spacetime metric to encompass global accelerations and variable warp speeds and validated his solution according to Natário's generalized warp drive spacetime geometric design criteria. He pointed out that an accelerating warp drive probably must have the terms of variable warp speeds or accelerations included in the expression for the warp bubble energy density required for the construction of the warp drive spacetime. However, he left the derivation of this for future work.

Another serious drawback of warp drive concepts is that collisions of the forward (in the direction of motion) warp bubble wall with the interstellar medium (ISM) would unavoidably occur in real FTL spaceflight. 1% of the ISM is composed of neutral or electrically charged dust clouds (metals (Fe, Al, Mg), graphites (C), and silicates (Si)) and 99% is composed of neutral or ionized atomic and molecular gas clouds (mostly H (90%), He, H_2, and some volatiles (H_2O, CO_2, CO, CH_4, NH_3, polycyclic aromatic hydrocarbon molecules, H_2CO, cyanide/isocyanide, OH groups, alcohols, etc.)) as well as small populations of asteroids and comets, copious amounts of galactic photons emitted by luminous astronomical matter, and copious amounts of galactic and cosmic radiation (ultra-relativistic ionized heavy atoms and elementary particles). The galactic photons colliding with the forward wall of the warp bubble will be blueshifted to very high energies, which can be lethal to the spacecraft's crew as well as damaging to the spacecraft itself. Hart et al. [37] revisited Van Den Broeck's warp drive and modified its spacetime metric by reducing the number of concentric ring regions to two. The first region is the usual curved spacetime region representing the behavior of Alcubierre's shape function in Eq. (9), which is also called the Pfenning region $\left(R-\frac{\Delta}{2}<r<R+\frac{\Delta}{2}\right)$ [38,39], and the second region is the Van Den Broeck curved spacetime region. Their analysis showed that incoming galactic radiation and ISM matter will not pose a major problem to the spacecraft because the gravitational gradients in the Van Den Broeck region will gravitational-tidally disrupt and deflect hazardous incoming ISM matter in the spacecraft's neighborhood while the Pfenning region will drastically slow down all incoming galactic radiation to prevent it from blueshifting. Additionally, the two ring regions will retain the capability to receive signals from the spacecraft so that it can control its warp bubble and communicate with the exterior spacetime.

Loup [40] revisited Natário's and Van Den Broeck's warp drives, used Eq. (9), modified Van Den Broeck's $B(r_{sh})$ function, and then derived a hybrid Natário-Van Den Broeck warp drive. This hybrid warp drive spacetime geometry creates inside the Alcubierre or Natário warp bubble a spacetime curvature with the topological shape of a "bottle" in which the large radius and volume of the bottle's interior spacetime can contain a spacecraft inside that is isolated from the exterior spacetime, while the only contact point between the bottle and the exterior spacetime is the bottle's "neck" (see Figs. 13 and 15). The submicroscopic outer diameter of the bottle's neck is the only part of this region that's in contact with the exterior spacetime. Loup elaborated that this submicroscopic surface is exposed to collisions with the oncoming ISM and large objects (comets, asteroids, etc.), thereby reducing the probabilities of dangerous impacts with the spacecraft, which enhances its survivability as well as that of its passengers.

There are many other types of warp drives in the published technical literature, just about as many as there are different types of traversable wormholes. Each of them attempts to solve one or more of the critical problems with Alcubierre's original warp drive concept. White and Davis [41] give another example, in which Alcubierre's warp drive can be reinterpreted in extra-space dimensional brane-world theory as a spacetime expansion boost on (i.e., like a scalar multiplier acting on) the initial spacecraft speed. This mechanism recasts the warp drive energy requirement into the equation of state for dark energy (a.k.a. the vacuum energy of Einstein's cosmological constant). Another model is the proposal by Obousy [42] and Obousy and Cleaver [43] to engineer a version of Alcubierre's warp drive via exploiting the Casimir vacuum energy of the compactified extra-space dimensions in brane-world

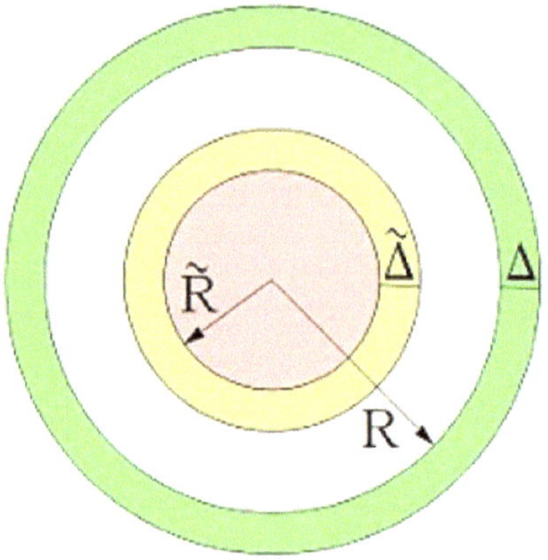

FIG. 15

Schematic diagram of the Van Den Broeck "pocket" or "bottle" inside of the Natário warp drive spacetime geometry. The *R* and Δ symbols identify the radius (from the center of the *salmon-colored central region*) and thickness of the *yellow and green regions*, respectively.

Courtesy F. Loup, private communication, 2020.

theories. They propose that the vacuum energy of Einstein's cosmological constant is a function of the size of the extra-space dimensions and that if one could locally manipulate the size of an extra-space dimension, and thus locally manipulate the cosmological constant, then one could facilitate the expansion or contraction of the spacetime surrounding a spacecraft at will. Their model utilizes theoretical results from brane-world theories that show that the Casimir energy of extra-space dimensions is related to the cosmological constant. However, their warp drive model differs entirely in that the approach they use is based on the physics of higher-dimensional quantum field theory and not general relativity theory.

Puthoff [44] showed that Alcubierre's warp drive is a particular case of a broad, general approach that he called "metric engineering," the details of which provide support for the concept that reduced-time interstellar travel is not fundamentally constrained by physical principles. This led him and his collaborators to consider formulating another approach to warp drive based on a new genre of gravity theory gaining prominence in academia: emergent gravity. Einstein was the first to remove gravitation as a fundamental force of nature and replace it with the new paradigm that it emerges from curved spacetime geometry. However, this paradigm remains somewhat unsatisfactory and incomplete because physical space, matter, and their interactions and manifestations are not mathematical geometry. In 1967, Sakharov developed a new notion of emergent or "induced gravity" in which gravitation (the "metric elasticity of space" in GTR) emerges from quantum vacuum zero-point fluctuations (a key foundation of quantum field theory) in roughly the same sense that hydrodynamics or continuum elasticity theory emerges from molecular physics [44,45]. This motivated Puthoff and collaborators to develop a "polarizable vacuum representation of general relativity" (PV-GR) in which the curved spacetime of general relativity emerges from polarizable quantum vacuum fluctuations of variable refractive index, which is akin to the nonlinear optical medium model for curved spacetime [46–52]. In unpublished work, H. E. Puthoff, C. Maccone, and this author used the PV-GR approach [47–50] to search for potential FTL spacetime solutions and discovered a highly unusual hybrid warp drive-traversable wormhole spacetime defined by a complicated mathematical expression for the vacuum index of refraction of the equivalent curved spacetime geometry in the associated FTL metric. Further computational analysis is required to fully develop this hybrid solution and determine its appropriate boundary conditions, energetics, and other dynamical features and their consequences.

While there is still no published authoritative monograph on warp drive physics, more recent published studies continue pushing the state-of-the-art forward. Bobrick and Martire [53] constructed a model for a general warp drive spacetime, which they claim encloses all existing warp drive definitions (Alcubierre, Van Den Broeck, and Natário) and allows for new warp drive metrics without having the serious issues found in Alcubierre's warp drive. They introduced a warp drive spacetime in which spacetime curvature and related topological effects and the clock rate of time can be chosen in a controlled manner. They also developed a class of subluminal, spherically symmetric warp drive spacetimes and claim that their general class of FTL and subluminal warp drive solutions is designed to avoid the requirement for stupendous astronomical amounts of matter to construct them. However, their theoretical work was recently found to be flawed where their warp drive energetics and certain parts of their technical derivations are concerned. This will be discussed in Section 6.2.

Lentz [54] also developed a unique new type of warp drive, which also claims to avoid the stupendous astronomical energetics problem by invoking a spacetime soliton solution of Eq. (6). By using the Arnowitt, Deser, and Misner (ADM) Hamiltonian formulation of Eq. (6) [55] and a hyperbolic, rather than linear or elliptic, relation for the ADM shift vector (defines how events/points/coordinate systems on successive 3D spatial hypersurfaces are mapped together in a 4D spacetime manifold that's foliated into a one-parameter family of 3D spatial hypersurfaces parameterized by a time function), Lentz constructed a moving, possibly FTL, soliton, which involves an energetics approach that appears to contradict a key requirement for all warp drive spacetimes (to be discussed further in Section 6.2). The soliton is constructed to contain a relatively flat-space central region with minimal gravitational tidal forces, in which the internal proper time of the space travelers corresponds to the outside proper time of remote static observers (i.e., no relativistic time dilation) and internal observers in their warp spacecraft move with the speed of the soliton itself without feeling inertial forces. The FTL transport logistics of Lentz's warp drive is like those of Alcubierre's warp drive. Lentz's soliton has a "delta" shape formed from about seven diamond-shaped blocks of specially configured ADM shift vectors "flying-in-formation" to surround the flat interior region and move it forward. The volume of local space that is expanded or contracted by the soliton is rather complex, containing multiple regions corresponding to negative and positive hyperbolic space expansions. In contrast, the Alcubierre warp bubble contains only one negative and one positive expansion region. Finally, Fell and Heisenberg [56] improved Lentz's work, made a deeper analysis, and formulated two additional superluminal solutions in a clearer analytic form compared to Lentz.

In closing this section, it should be pointed out that it is this author's opinion that the taxonomy of warp drive spacetimes cannot be ruled out by the limitations of energy, Doppler blueshifts, or causal horizons. These will be discussed in the next section and in later sections.

4 The curvature invariants for Lorentzian traversable wormholes and warp drives

Studies of Lorentzian traversable wormholes and warp drives rely on calculating the components of the Riemann curvature tensor, $R^{\alpha}_{\beta\gamma\delta}$, to "observe" the effects of the wormhole's or warp drive's local spacetime curvature environment upon the photons and matter that move through these environments. This could be misleading in significant ways, because errors, such as coordinate mapping distortions, may arise as an artifact of the choice of coordinates, since the Riemann tensor cannot be calculated in an invariant manner as it is a function of the chosen coordinates. The only way to properly illustrate these spacetimes without such errors is to plot their independent curvature invariants in order to properly visualize the underlying spacetime and unveil any hidden surprises (e.g., unexpected causal horizons or singularities). Christoffel [57] proved that scalars constructed from the metric tensor and its derivatives must be functions of $g_{\mu\nu}$ and the $R^{\alpha}_{\beta\gamma\delta}$ and its covariant derivatives. Curvature invariants are constructed from scalar products of the Riemann, Ricci, or Weyl tensors, or their covariant derivatives, which provide a measure of the gravitational stress and strain acting upon infalling photons and matter. To visualize curved spacetime phenomena without distortion is to plot only the curvature invariants—quantities whose value is the same regardless of the coordinates used.

4.1 Curvature invariants for three types of traversable wormholes

Mattingly et al. [58] described a process for using curvature invariants as a new means to evaluate the traversability of Lorentzian traversable wormholes and display their entire spacetime manifold. The computational approach they used was formulated by Overduin et al. [59] for electrically charged and rotating black holes. Mattingly et al. calculated the four independent Carminati and McLenaghan [60] algebraic invariants of low degree ($\mathcal{R}$, r_1, r_2, and w_2, as defined in Ref. [58]) that applied to their computational approach and plotted the four nonzero curvature invariant functions for three example traversable wormhole metrics:

(i) spherically symmetric Morris and Thorne defined by Eq. (7);
(ii) "thin-shell" Schwarzschild wormholes (these are the most common example of a large class of traversable wormholes that are not to be confused with the Einstein-Rosen bridge because construction of the former differs greatly from the construction of the latter); and
(iii) the exponential metric.

Each of these were investigated in detail and were shown to be fully traversable. The mathematical and computational details are lengthy and very dense, so the reader should consult Ref. [58] for the details. In Figs. 16–18, we show the 3D plots of the four Carminati and McLenaghan invariants for the traversable wormhole types (i)–(iii). The standard spherically symmetric coordinates are used in all three cases (see below Eq. (7) for the definition).

Of interest is that, to construct a traversable thin-shell Schwarzschild wormhole, two copies of Minkowski-flat spacetime on either side of the wormhole's throat are assumed, identical regions from each spacetime are removed, and then separate regions along the boundary are identified and "sewn" together. This formalism leads to a well-behaved traversable wormhole with the throat being located at the identical boundary between the two separate regions.

The interesting thing about this unusual class of traversable wormholes is that its area is a concave function with a minimum at the throat where it satisfies the Flamm diagram "flare out" condition. It does not have a horizon because the time-time component of the metric tensor $g_{00} \neq 0$ for all $r \geq 0$. The region $r < M$ on the other side of the wormhole throat is an infinite volume "other universe" that exhibits an "underhill effect" in which time runs slower because $g_{00} > 0$ in this region [58].

This extensive study indicated that spherically symmetric Morris and Thorne, thin-shell Schwarzschild, and exponential metric wormholes are traversable in agreement with Refs. [23–25].

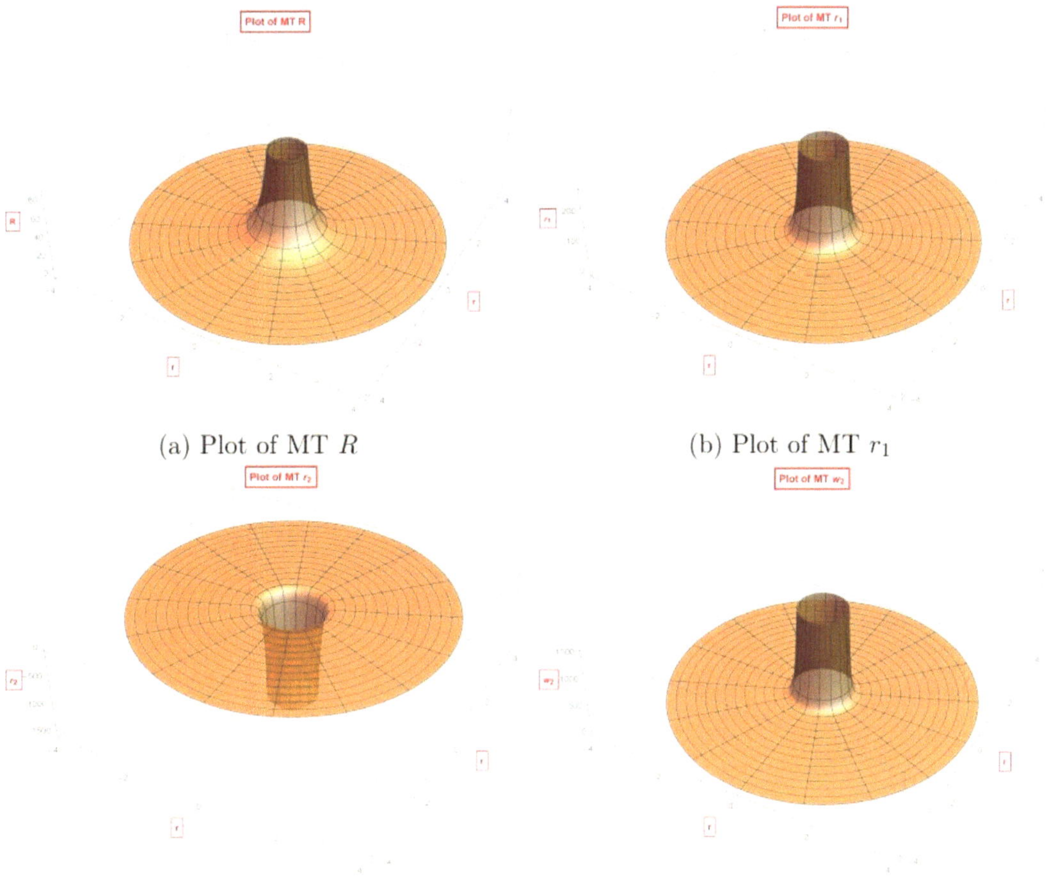

(a) Plot of MT R

(b) Plot of MT r_1

(c) Plot of MT r_2

(d) Plot of MT w_2

FIG. 16

3D plots of the nonzero Carminati and McLenaghan invariants ($\mathcal{R}$, r_1, r_2, w_2) for the Morris and Thorne traversable wormhole defined by Eq. (7) [58]. The plots are in radial coordinates with $r \in \{0, 4\}$, we set $G=M=1$ (M is the mass of the wormhole that appears in the chosen shape function for the throat) in the Morris and Thorne metric for simplicity, and $r_0=2$ was chosen as the throat. Notice the divergence at the center of each plot is completely inside the $r=2=r_0$ radial line. This does not affect the traversability of the wormhole.

A divergence is found in all three, but it does not affect the wormhole's traversability because the divergence is outside the physical range of the radial coordinate, $r \in (r_0, \infty)$. The ring discontinuity in the thin-shell Schwarzschild wormhole could lead to a redshift of light rays passing through the wormhole. This could be used to distinguish traversable wormholes from black holes. The plotted curvature invariants properly illustrate the entire underlying spacetime independent of the coordinate system chosen. Plotting these invariants exposes the presence of any artifacts, discontinuities, or divergences anywhere on the spacetime manifold. Another version of thin-shell traversable wormholes can be

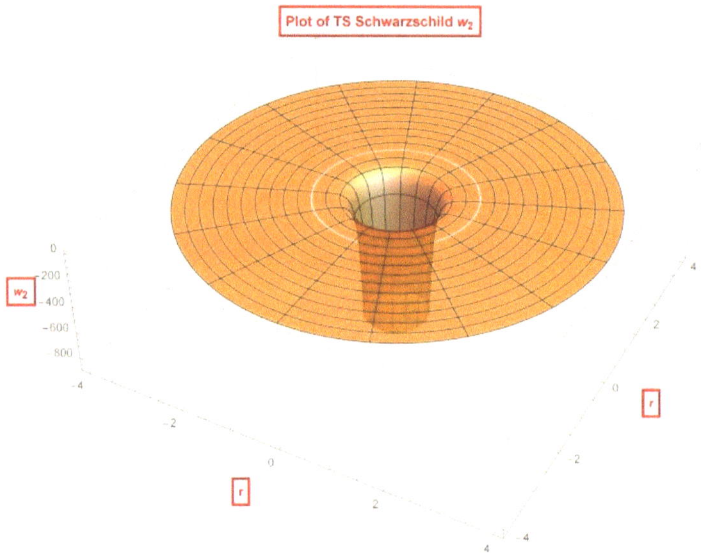

FIG. 17

3D plot of w_2 for the thin-shell Schwarzschild traversable wormhole. $\mathcal{R}$, r_1, and r_1 are zero for this class of wormholes [58]. The plot is in radial coordinates with $r \in \{0, 4\}$ and $r_0 = 2$ was chosen as the throat. The divergence at the center of the plot is completely inside the $r = 2 = r_0$ radial line. It does not affect the traversability of this class of wormholes. Also notice the ring discontinuity at $r = 2 = r_0$, which comes from the δ-function in the thin-shell formalism. Its value is recorded in Ref. [58].

modeled with different radii of curvature on either side of the throat. Thin-shell wormholes with unequal radii of curvature can be further extrapolated to wormholes with different chronologies or universes on either side of the throat. The thin-shell Schwarzschild wormhole is the most common example of a large class of traversable wormholes. This class includes wormholes with different radii of curvature and/or masses on either side of the throat; wormholes with equal or different electric charge, Q, on either side of the throat; and dynamical wormholes. For electrically charged wormholes, a second ring artifact at $r = Q$ is likely to exist since the wormhole metric has a singularity at that point. Future curvature invariant studies will need to be done to explore what they reveal about the rest of the taxonomy of static and dynamic traversable wormholes.

4.2 Curvature invariants for selected warp drives

Like the large taxonomy of traversable wormholes, the taxonomy of warp drives is too large to individually review in this chapter. So, the two most well-known types of warp drive will be briefly reviewed here. Mattingly et al. [61,62] calculated and plotted the curvature invariants for the spacetime metrics of the Alcubierre and Natário warp drives at constant velocity, and the accelerating Natário warp drive at constant acceleration. The visualization of these warp drive spacetimes was previously unexplored. Just as for traversable wormholes, the warp drive spacetimes have four independent curvature invariants: $\mathcal{R}$, r_1, r_2, and w_2 (defined in Refs. [61,62]). The 3D plots of the four curvature

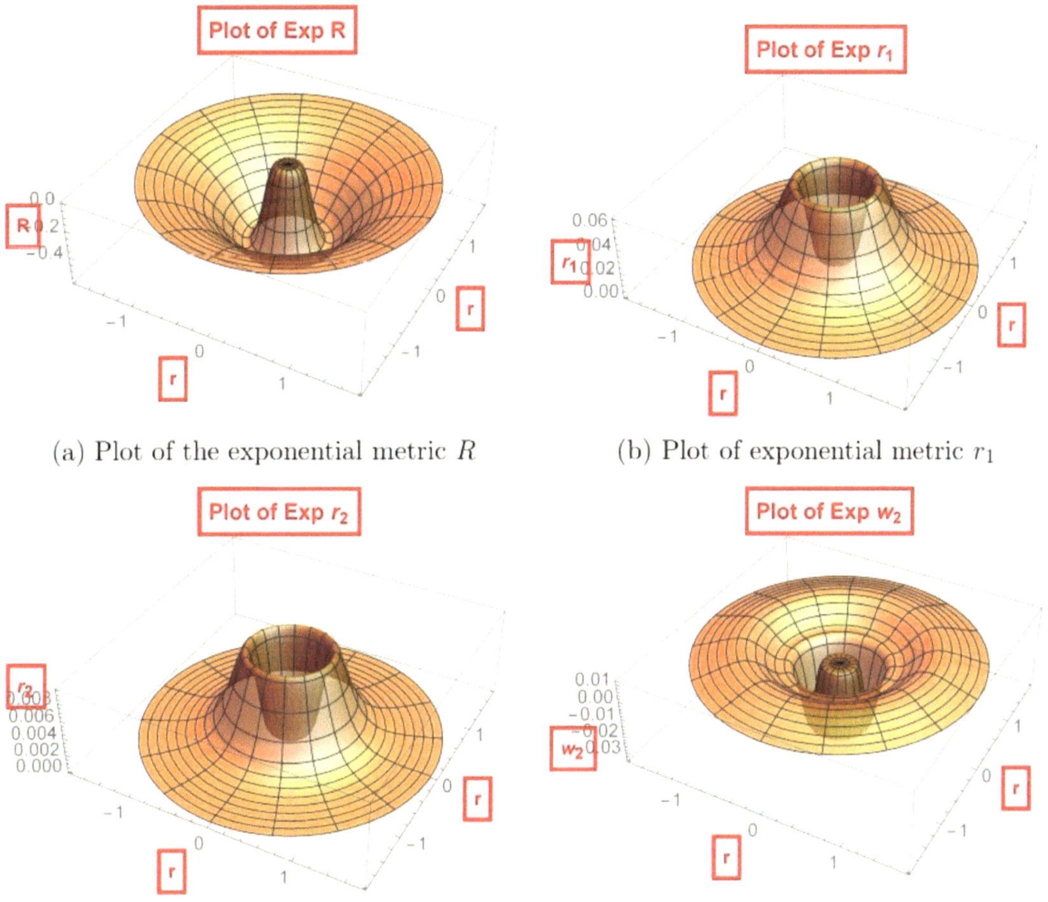

(a) Plot of the exponential metric R

(b) Plot of exponential metric r_1

(c) Plot of exponential metric r_2

(d) Plot of the exponential metric w_2

FIG. 18

3D plots of the nonzero curvature invariants $\mathcal{R}$, r_1, r_2, and w_2 for the exponential metric that represents a class of traversable wormholes [58]. The plots are in radial coordinates with $r \in \{0, 1.75M\}$. M is the mass of the wormhole and the location of the throat $r = M = 1$ is normalized.

invariants demonstrate how each evolves with the parameters of time (t), constant dimensionless warp speed (v_{sh}), constant acceleration (a), inverse warp bubble wall thickness (σ), and warp bubble radius (R). Mattingly et al. showed that $\mathcal{R}$ has the greatest impact of the curvature invariants upon the surrounding spacetime manifold. For each warp drive, their studies showed a Minkowski-flat spacetime at the center of the warp bubble (this is the region containing the warp spacecraft's inertial rest frame) and how the shape function forms the warp bubble. The plots revealed key features of the accelerating Natário warp drive such as a dynamic wake in spacetime and the internal structures of the warp bubble. Plots for the constant dimensionless warp speed Natário warp drive do not contain a dynamic wake or a

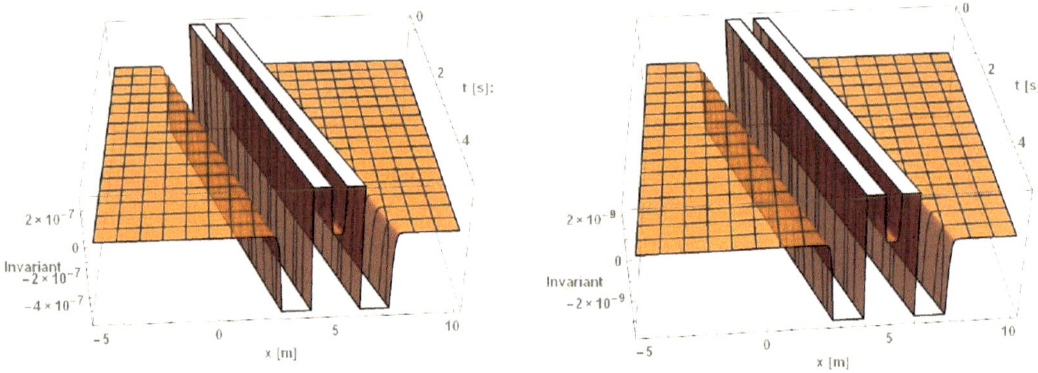

FIG. 19

3D plots of $\mathcal{R}$ for the constant dimensionless warp speed Alcubierre warp drive with varying σ [61]. The parameters chosen to make these plots are $v_{sh}=c$, $R=1$ m, and $\sigma=8\text{ m}^{-1}$ (left) and 10 m^{-1} (right).

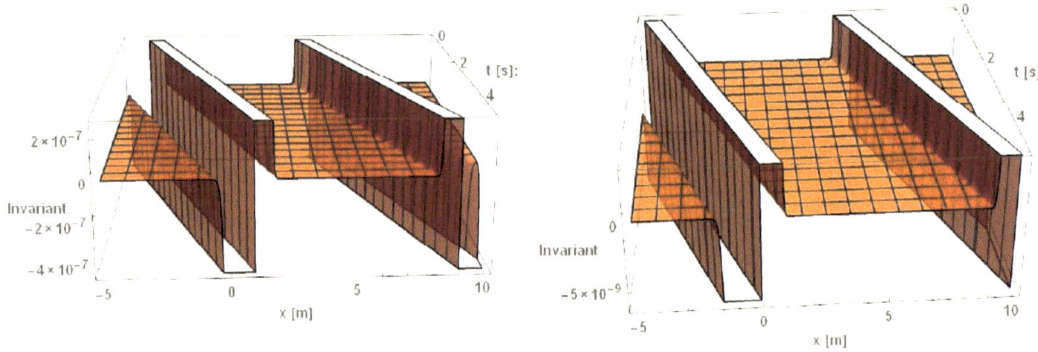

FIG. 20

3D plots of $\mathcal{R}$ for the constant dimensionless warp speed Alcubierre warp drive with varying R [61]. The parameters chosen to make these plots are $v_{sh}=c$, $\sigma=8\text{ m}^{-1}$, and $R=4$ m (left) and 5 m (right).

constant curvature, indicating that these features are unique to the accelerating Natário warp drive. Because of the rather large number of 3D plots that were calculated for these warp drives versus time, warp speed, acceleration, and warp bubble dimensions, Figs. 19–23 display only some representative examples of the $\mathcal{R}$.

The additional 3D plots for the constant dimensionless warp speed Alcubierre invariants r_1, r_2, and w_2 are shown in Ref. [61], and these plus the $\mathcal{R}$ plots show that there are some minor variations between each invariant. $\mathcal{R}$ oscillates from a trough to a peak, to a flat area, to a peak, and back to a trough. r_1 simply has a peak with a flat area followed by another peak. w_2 follows the reverse pattern, such as the $\mathcal{R}$ wavering from a peak into a trough, into a flat area, into a trough, and returning to a peak. The spacetime manifold is asymptotically flat and completely connected from the interior region of the warp bubble to the exterior region, and no curvature singularities are observed in any of the curvature invariant plots.

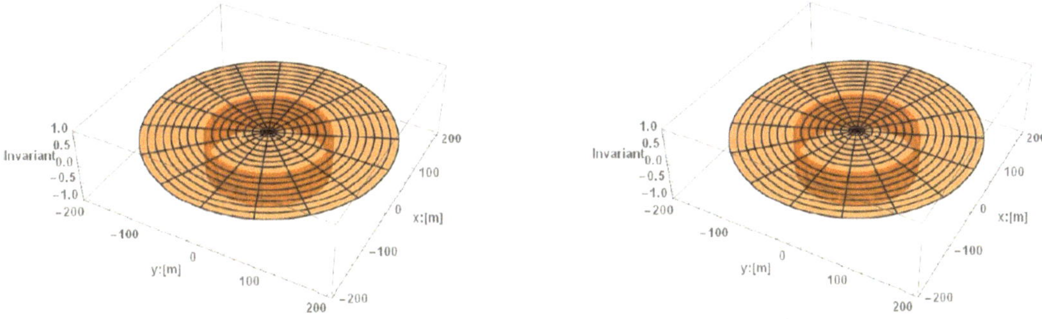

FIG. 21

Warp speed evolution of $\mathcal{R}$ for the constant dimensionless warp speed Natário warp drive [61]. The parameters for these plots are $\sigma = 50{,}000\ \text{m}^{-1}$, $R = 100$ m, and $v_{sh} = 10c$ (left) and $100c$ (right).

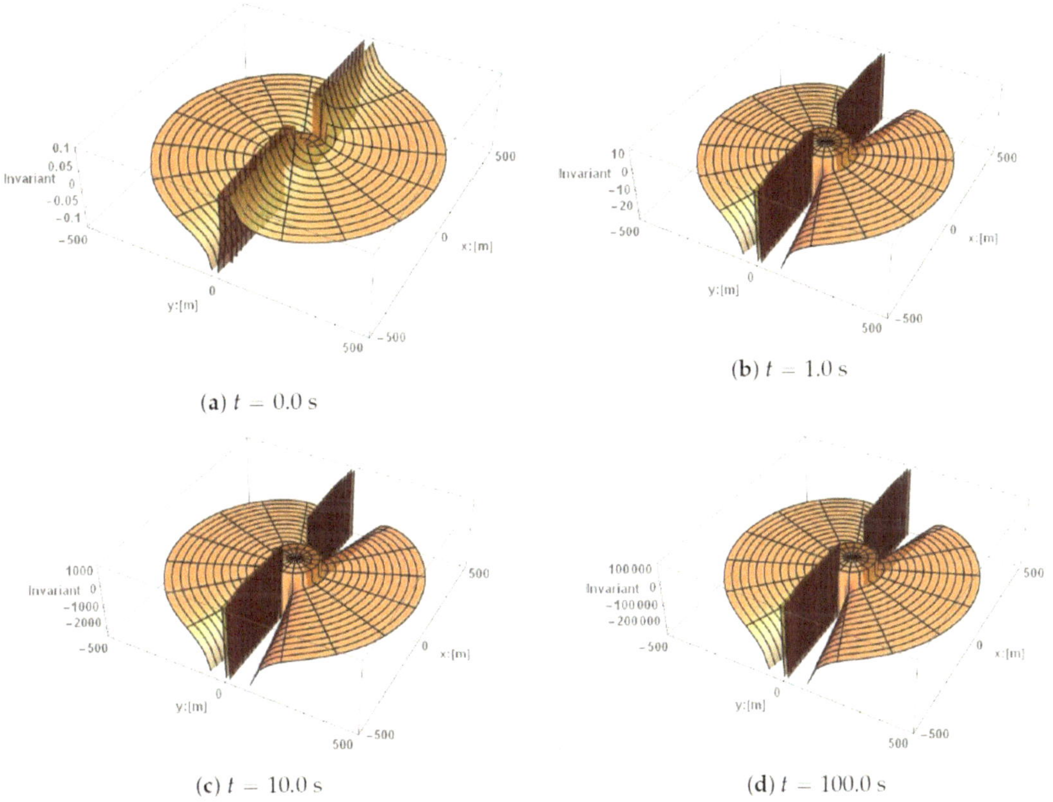

FIG. 22

3D plots of the time evolution of $\mathcal{R}$ for the constant acceleration Natário warp drive [62]. The parameters chosen to make these plots are $\sigma = 50{,}000\ \text{m}^{-1}$, $R = 100$ m, and $a = 1\ \text{m/s}^2$. The amount that spacetime deviates from flatness is represented by the $\mathcal{R}$-invariant's magnitude, and this is labeled on the vertical axis in each plot. The *x*- and *y*-axes display the distance from a spaceship residing within the central flat-spacetime region. Each radial line in the plot corresponds to a distance of approximately 33 m. The portion of small negative curvature to the right of the central flat-spacetime region is at the front of the warp bubble, and the portion to the left is at the rear of the warp bubble. Notice the crenulations, which produce a wake that is perpendicular to the direction of motion.

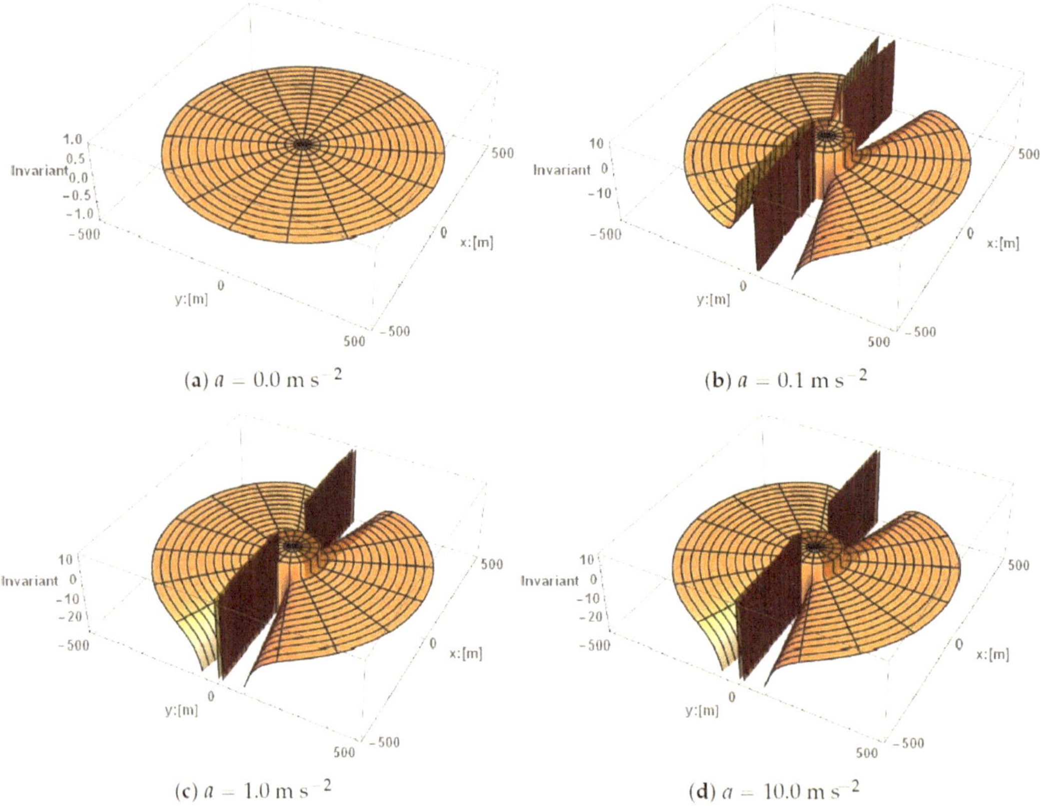

(a) $a = 0.0\ \mathrm{m\,s^{-2}}$ (b) $a = 0.1\ \mathrm{m\,s^{-2}}$

(c) $a = 1.0\ \mathrm{m\,s^{-2}}$ (d) $a = 10.0\ \mathrm{m\,s^{-2}}$

FIG. 23

3D plots of $\mathcal{R}$ vs acceleration for the constant acceleration Natário warp drive [62]. The parameters chosen to make these plots are $\sigma = 50{,}000\ \mathrm{m}^{-1}$, $R = 100$ m, and $t = 1$ s. The amount that spacetime deviates from flatness is represented by the $\mathcal{R}$-invariant's magnitude and this is labeled on the vertical axis in each plot. The *x*- and *y*-axes display the distance from a spaceship residing within the central flat-spacetime region. Each radial line in the plot corresponds to a distance of approximately 33 m. The portion of small negative curvature to the right of the central flat-spacetime region is at the front of the warp bubble and the portion to the left is at the rear of the warp bubble. Notice the crenulations, which produce a wake that is perpendicular to the direction of motion.

As with the Alcubierre invariants, all four of the constant dimensionless warp speed Natário invariants are exceptionally complicated, but some features may be observed by inspecting their functions in Ref. [61]. For this warp drive, the spacetime manifold is asymptotically flat and completely connected from the interior region of the warp bubble to the exterior region, and no curvature singularities are observed in any of the curvature invariant plots. The flight of such a warp bubble should be significantly less affected by any gravitational tidal forces as compared to Alcubierre's warp drive. The Natário invariants confirm that Natário's warp drive is a more realistic alternative to Alcubierre's.

For the constant dimensionless warp speed Natário invariants, the spacetime curvature lies in the area around $r = R$, the boundary of the warp bubble. Despite the complexity of the invariants, the shape of the curvature invariant plots is simple. It forms a very narrow and jagged ring, as shown in Fig. 21. Precisely at $r = R$, the curvature invariants spike to nonzero magnitudes depending on the invariant. $\mathcal{R}$ takes the form of a smooth disc outside the warp bubble. The plots in Ref. [61] show that the shape of r_1 is that of a jagged disc at $r = R$. The disc has jagged edges in the negative direction with sharp spikes at radial values $r = R$ and at polar angle values of 0 and π. The shape of r_2 and w_2 is that of a jagged disc at $r = R$. Its edges vary between positive and negative values depending on the polar angle θ. In front of the central flat spacetime region within the warp bubble, the invariants have rapidly changing negative values between -1 and 0. Behind the central flat spacetime region within the warp bubble, the invariants have rapidly changing positive values between 0 and 1. The jagged edges of the plots suggest that r_1, r_2, and w_2 oscillate rapidly between values of -1 and 1 along the circumference of the warp bubble. These oscillations must be occurring more rapidly than the plotting software can plot. Outside the warp bubble, the spacetime is more well behaved. For values of $r \gg R$, the magnitude of each invariant is zero and the spacetime is asymptotically flat. For values of $r < R$ inside the warp bubble, the magnitude of the invariants is also zero. As with Alcubierre's warp drive, this implies that there is a central flat spacetime region that is unaffected by the curvature of the warp bubble. The constant warp speed Natário warp drive features neither a wake in spacetime nor a constant nonzero curvature outside the warp bubble for $\mathcal{R}$.

Note that for all warp drive cases evaluated in Refs. [61,62], when the curvature invariants have a positive magnitude, the spacetime has a positive curvature. When the curvature invariants have a negative magnitude, the spacetime has negative curvature. These positive and negative spacetime curvatures produce the effect that transports the warp bubble and the spacecraft residing within it across interstellar space at FTL speed, and it turns out that space warp transport can also occur at sublight speeds.

The reader should see Ref. [62] for the 3D plots of r_1, r_2, and w_2 for the accelerating Natário warp drive. The plots of all four curvature invariants contain the features observed in the plots for the constant dimensionless warp speed Natário warp drive. Because the constant warp speed curvature invariants are zero everywhere except at $r = R$, their impact is along the warp bubble's edge for the accelerating Natário warp drive invariants. However, Figs. 22 and 23 show a prominent wake in spacetime along with a constant nonzero curvature as the warp drive undergoes an acceleration to reach a speed greater than c (as per Fig. 23). The wake runs perpendicular to the warp bubble's direction of motion, and it will carry the spacecraft along as Natário predicted. The plots for $\mathcal{R}$ revealed constant curvature outside of the wake and the central flat spacetime region while the other invariants revealed asymptotically flat spacetime outside of the wake and the central flat spacetime region [62]. Thus, $\mathcal{R}$ will have the greatest impact of the invariants on the spacetime outside of the wake and the central flat spacetime region. The spacetime manifold is completely connected from the interior region of the warp bubble to the exterior region, and no curvature singularities are observed in any of the curvature invariant plots.

Also, by varying time, the plot of each invariant experiences a sudden jump from positive curvature to negative curvature in the direction of motion [62]. As time progresses, the shape of the first three invariants remains constant but their magnitude increases. Per the analysis and plots shown in Ref. [62], w_2 begins to exhibit crenulations in the interior of the wake after 100 s of elapsed time. Because the greatest invariant magnitudes occur for w_2, it must have the greatest impact on the wake and the internal curvature features. By varying the acceleration, the invariant plots skip through the time slices and the internal features become more prominent. Changing σ changed neither the shape nor magnitude of the invariant plots [62]. Doubling R also doubled the size of the central flat spacetime region in the invariants plots without affecting the shape of the warp bubble [62].

The crenulations observed for w_2 also exist within the internal structures observed in the wakes of the other three curvature invariants. As the crenulations ripple, navigating the warp drive could become difficult. Mattingly et al. suggested that the warp drive be periodically turned on and off as it accelerates to minimize the effect of the crenulations. They speculated that the increase in the magnitude of the curvature invariants as time advances establishes a relationship between the magnitude of the acceleration and the amount of energy required to accelerate the warp bubble to arbitrarily high warp speeds. Because the acceleration skips through the warp drive's time slices, greater values of the acceleration will reach high magnitudes of each curvature invariant more quickly. This relationship suggests that a realistic warp drive would only be able to accelerate to some finite warp speed that is potentially greater than c. This is evidence for a less restrictive superluminal censorship theorem than previously considered by Barceló and Visser [63]. Instead, it is in accordance with the constraint that the total energy stored within the warp bubble be less than the total rest-energy of the warp spacecraft [21].

Overall, the curvature invariants contain no singularities in the spacetime manifolds of the constant dimensionless warp speed Alcubierre and Natário warp drives and the accelerating Natário warp drive. Since no singularities exist, the interior and exterior spaces of the warp drive spacetime manifolds must be connected. A potential coordinate transformation may be discovered in the near future that will remove the known causal discontinuity (between the spacecraft and its warp bubble) from them. Where the plotted magnitudes of the curvature invariants are large, spacetime is greatly curved. By observing the changes in slopes of the curvature invariant function plots, the rate at which spacetime is being folded may be approximated. This information will help with mapping the spacetime around the warp spacecraft and aid potential FTL navigation through interstellar space. At the radial position of the warp bubble wall, the curvature invariants have local maxima or minima identifying that the location of the warp bubble is at a radial distance r from the central flat spacetime region. For the curvature invariants of the constant dimensionless warp speed Alcubierre warp drive, the warp bubble resembles two troughs with simple internal structures. For the constant dimensionless warp speed Natário warp drive, the warp bubble peaks around $r = R$ while displaying rich internal structures. The internal structures of these warp bubbles are novel and require further study to discover their effects upon a warp drive's motion. With this information, a spacecraft navigation system will need to calculate the values of each free parameter of the warp drive to control the size, shape, and motion of its warp bubble.

Of note is how the curvature invariants of the constant dimensionless warp speed Natário warp drive compare with the invariants of the accelerating Natário warp drive. The main difference between the two is the lack of a wake in spacetime for each invariant in the former and a constant nonzero curvature outside the warp bubble in the latter. The other parameters σ and R behave similarly for both types of warp drives. Mattingly et al. [62] concluded that the wake in spacetime and constant nonzero curvature will appear as the warp drive undergoes an acceleration to reach a velocity greater than c. In addition, the crenulations in the wake produced by the constant acceleration of spacetime may produce high-frequency gravitational waves. Consequently, these waves will only be detectable during the accelerating phase of a warp drive spacecraft, not during the constant velocity phase. More research is needed to explore these features and their impact upon any type of warp drive, its flight, and its surrounding spacetime. Future high-frequency gravitational wave astronomical observatories could be used to search for the spacetime wakes (warp drive signatures) produced by the warp drive propulsion of technologically advanced extraterrestrial space explorers traveling throughout our galaxy. There remain the other types of warp drive concepts described in Section 3.2 and shown in the lower half of Fig. 10 that would benefit from deep dive analyses using the curvature invariants to expose the hidden structures and features of their spacetime manifolds. Also of interest is that Watson et al. [64] used the curvature invariants to probe the hidden features and structures within the unusual spacetime manifolds of closed "trapped" spaces that are artificially created by externally applied electric or magnetic fields.

5 The allowable sources of matter for constructing FTL spacetimes

5.1 Exotic matter and the energy conditions

The implementation of FTL interstellar travel via traversable wormholes or warp drives generally requires the engineering of spacetime into very specialized local curved geometries. The analysis of these via the general relativistic field equations plus the resultant source matter equations of state given by the components of $T_{\mu\nu}$ in Eq. (6) demonstrate that such geometries require the use of "exotic" matter to engineer the requisite FTL spacetime. Exotic matter is generally defined by GTR physics to be matter that possesses negative energy density, negative energy flux, and/or negative pressure, which produce the gravitationally repulsive (antigravity) effect that pushes spacetime outward to open a traversable wormhole throat or produce the warp bubble. Exotic matter is often a very misunderstood and misapplied term within the general relativity and nongeneral relativity physics communities. One can appeal to quantum field theory to find theoretically predicted and experimentally observed sources of exotic matter. The identification and production of exotic matter is seen to be a key technical challenge for constructing FTL spacetimes.

When studying FTL spacetime physics, the appropriate strategy is to decide beforehand on a definition of the FTL propulsion effects that one desires and then design its curved spacetime geometry by defining an appropriate metric that incorporates the mandatory FTL propulsion requirements in items (1)–(8) in Section 2.2. One then inserts the designer FTL metric into the left-hand side of Eq. (6) to calculate the right-hand side to find the type and distribution of matter that will produce the desired FTL spacetime geometry. Then one needs to assess whether the required type and distribution of matter is physically reasonable and whether it violates any basic rules of physics. In the case of traversable wormhole and warp drive spacetimes, it turns out that they and their variants universally violate some or all of the classical Hawking-Ellis energy conditions [65]. It has been claimed in the literature that FTL spacetimes are not plausible for this reason, but in what follows we will show that such claims are spurious because real physical matter is not in fact reasonable since the energy conditions in general are violated by semiclassical quantum effects.

Matter that violates the classical energy conditions is called exotic. The conditions for reasonably ordinary classical (nonexotic) forms of matter that are familiar to us is $\rho_E > p_i$ and/or $\rho_E \geq 0$, where $\rho_E = T_{00}$ is the rest-energy density ($=\rho c^2$, the time-time component of $T_{\mu\nu}$, where ρ is the rest-mass density) and $p_i = T_{ii}$ are the diagonal spatial pressure/stress components. These conditions represent two very simplified examples of what are variously called the standard (pointwise) energy conditions, which are mathematically explicated in Fig. 24. Table 2 shows the full suite of pointwise energy conditions and their present status as allegedly useful tools. Pointwise refers to energy conditions that are to be computed at any point/event in the spacetime manifold. These energy conditions allegedly forbid matter with the properties of negative energy density, negative energy flux, and negative pressure/

null energy condition (NEC)

geometric for any null vector k^a, $R_{mn}k^m k^n \geq 0$

physical for any null vector k^a, $T_{mn}k^m k^n \geq 0$

effective for each $\hat{\mu}$, $\rho + p_{\hat{\mu}} \geq 0$

weak energy condition (WEC)

geometric for any timelike vector ξ^a, $G_{mn}\xi^m \xi^n \geq 0$

physical for any timelike vector ξ^a, $T_{mn}\xi^m \xi^n \geq 0$

effective $\rho \geq 0$, and for each $\hat{\mu}$, $\rho + p_{\hat{\mu}} \geq 0$

strong energy condition (SEC)

geometric for any timelike vector ξ^a, $R_{mn}\xi^m \xi^n \geq 0$

physical for any timelike vector ξ^a, $(T_{mn} - \frac{1}{2}T g_{mn})\xi^m \xi^n \geq 0$

effective $\rho + \sum_{\hat{\mu}} p_{\hat{\mu}} \geq 0$, and for each $\hat{\mu}$, $\rho + p_{\hat{\mu}} \geq 0$

dominant energy condition (DEC)

geometric

1. for any timelike vector ξ^a, $G_{mn}\xi^m \xi^n \geq 0$, and $G^a{}_n \xi^n$ is causal
2. for any two co-oriented timelike vectors ξ^a and η^a, $G_{mn}\xi^m \eta^n \geq 0$

physical

1. for any timelike vector ξ^a, $T_{mn}\xi^m \xi^n \geq 0$, and $T^a{}_n \xi^n$ is causal
2. for any two co-oriented timelike vectors ξ^a and η^a, $T_{mn}\xi^m \eta^n \geq 0$

effective $\rho \geq 0$, and for each $\hat{\mu}$, $|p_{\hat{\mu}}| \leq \rho$

FIG. 24

The pointwise energy conditions. The equations are expressed in geometrized units, $c=G=1$, such that $\rho_E = \rho c^2 \equiv \rho$ [17]. Here, the lower-case Latin letters (*a*, *b*, ...) designate abstract tensor (spacetime coordinate) indices and the hatted lower-case Greek letter $\hat{\mu}$ designates the spatial components of pressure. R_{mn}, G_{mn}, and T_{mn} have their usual meaning from Eq. (6); G^a_n and T^a_n are the mixed tensor version of G_{mn} and T_{mn} with $m=a$; and $T = T^m_m$ is the trace of T_{mn}.

Modified from [66].

stress from occurring in nature. Hawking and Ellis formulated their classical energy condition hypotheses to establish a series of mathematical hypotheses governing the behavior of collapsed-matter singularities in their study of cosmology and black hole physics [65]. Also, the classical energy conditions are used to prove lots of general theorems in GTR as exemplified by the following subset of a much longer list of GTR theorems:

- Penrose singularity theorem (relevant to black hole formation via gravitational collapse of matter).
- Hawking-Penrose singularity theorem (relevant to cosmological singularities).
- Tipler's version of the Hawking-Penrose singularity theorem.

Table 2 The present status of the pointwise energy conditions.

Pointwise energy condition	Meaning	Present status
NEC	Null energy condition	Still in use
WEC	Weak energy condition	Still in use
SEC	Strong energy condition a.k.a. unphysical energy condition	Now abandoned
DEC	Dominant energy condition	Still in use, does not add any new information
TEC[a] (not shown in Fig. 24)	Trace energy condition	Now abandoned

[a]*The TEC states that* $T = T^{\mu}_{\mu} \geq 0$ *(μ is a spacetime coordinate tensor index).*

- Schoen-Yau positive mass theorem.
- Witten's variant positive mass theorem.
- Focusing theorems (for gravitational lensing).
- Topological censorship (i.e., you cannot construct a traversable wormhole).

Visser and Barceló [63,67] are of the opinion that the classical energy conditions are nothing more than pragmatically convenient tools whose formulation was driven by the need to prove a long list of desired theorems, which includes the subset of theorems listed above. It is this author's considered opinion that there are no compelling effective interpretations of these conditions based on any physical and/or spacetime geometrical considerations, and there's certainly no compelling physical or philosophical motivations for them.

Everyone within the GTR community expected the classical energy conditions to eventually break down at the Planck scale, but three surprises emerged:

- Surprise 1: The classical energy conditions fail miserably in semiclassical quantum gravity, which means that they will also fail miserably in a full quantum gravity theory. Failures occur long before one reaches the Planck (length, time, energy) scale and they are widespread, albeit small (to first order in $\hbar$).
- Surprise 2: The classical energy conditions fail miserably in quite reasonable-looking classical systems.
- Surprise 3: The classical energy conditions were experimentally violated 25 years before they were formulated [68].

The pointwise energy conditions are really a snapshot of the behavior of matter at a single point/event in spacetime as measured by stationary observers. These conditions cannot tell stationary observers how that same matter is behaving at another point/event in another region of spacetime that's nearby or far away. Tipler [69] developed an approach to constrain the extent of the energy conditions breakdown, which involves averaging the energy conditions over timelike or null (lightlike) worldlines to constrain the average behavior of matter over spacetime regions as measured by observers moving on timelike or null worldlines. This development was seen to be a more realistic approach because observers in spacetime are not always stationary. Over many years, the standard averaged energy conditions were defined to be the averaged null energy condition (ANEC), averaged weak energy condition (AWEC), and averaged strong energy condition (ASEC). The entry for the SEC in Table 2 immediately implies that the ASEC is also not a valid energy condition, so we can disregard it. The mathematical definitions for the averaged energy conditions are more complicated than what is shown in Fig. 24 for the pointwise energy conditions. In short, the averaged energy conditions are computed by taking the line-integral of each of the pointwise energy conditions over a suitably parameterized null or timelike worldline. The reader should consult Refs. [25,63,67,69] for the gory technical details.

The standard propaganda is that all reasonable Lagrangians generate theories satisfying the energy conditions. This propaganda is wrong. It turns out that quantum effects generically violate the pointwise and averaged energy conditions, which generically postulate that the local energy density, energy flux, and pressure/stress are nonnegative for all observers in any spacetime, but there's a more general interpretation that will be revealed in the next section. In 1965, it was discovered that quantum field theory has the remarkable property of allowing local regions in any spacetime to contain quantum states of matter with a negative energy density, negative energy flux, and/or negative pressure/stress as measured by all observers [70]. There are general theorems of differential geometry that guarantee that there must be a violation of one, some, or all the pointwise and averaged energy conditions (meaning exotic matter is present) for all FTL spacetimes. Additional investigations into this technical issue showed that violations of all the energy conditions are widespread for all forms of both reasonable classical and quantum matter [63,71–74]. Furthermore, Visser [25] showed that all (generic) spacetime geometries violate all the energy conditions, and he pointed out that observational astrophysical data indicate that large amounts of exotic matter must exist somewhere in the universe to account for the observed present-day values of the cosmological evolution parameters [75]. Thus, the requirement that one, some, or all the energy conditions must be obeyed by all forms of matter in nature is spurious. Widespread energy condition violations have unfortunately alarmed physicists within the GTR community. Violating the energy conditions commits no offense against nature and negative energy has already been produced in historic laboratory experiments, which will be discussed in the following sections.

5.2 Exotic matter and the quantum inequalities

The Pointwise and Averaged Energy Conditions imply that an inertial observer who initially encounters some negative energy density must encounter compensating positive energy density at some arbitrary time in the future. As an alternative to these energy conditions, the quantum inequality (QI) conjecture was derived for the free vacuum quantum electromagnetic field with no sources or boundaries, which constrain the magnitude and duration of negative energy densities/fluxes relative to an underlying reference quantum vacuum state. The QI places strong bounds on quantum violations of the pointwise and averaged energy conditions, the second law of thermodynamics, and the construction of FTL and time machine spacetimes [76–81]. The QI is formulated as a mathematical bound on the average of the renormalized quantum expectation value (quantum average) of a free field's stress-energy-momentum tensor in the vacuum state, $\langle T_{\mu\nu} \rangle_{\text{vac}}$, where the average is taken along an observer's timelike or null worldline using ad hoc time sampling functions. Contrary to the pointwise and averaged energy conditions, the QI dictates that the larger the magnitude of negative energy density/flux is in some time interval $\mathcal{T}$, the shorter the duration of $\mathcal{T}$ such that an inertial observer cannot encounter arbitrarily large negative energy densities/fluxes that last for arbitrarily long $\mathcal{T}$.

An inertial observer must encounter compensating positive energy density/flux no later than after a time $\mathcal{T} \propto |\textit{negative energy density or flux}|^{-1}$.

The expression for the QI of a free quantum vacuum electromagnetic field in flat Minkowski spacetime is given by [82]:

$$\hat{\rho}_{\mathrm{QI}} \geq -A_{\mathrm{QI}} \frac{\hbar c}{(ct_0)^4}, \tag{11}$$

where $\hat{\rho}_{\mathrm{QI}}$ (in MKS units of energy density) is the time integral of the time sampling function-weighted $\langle T_{00} \rangle_{\mathrm{vac}}$ (renormalized quantum expectation value of the energy density in the vacuum state) for the free quantum vacuum electromagnetic field in flat Minkowski spacetime, A_{QI} is some numerical factor, and t_0 is the sampling time.

Note in Eq. (11) that as t_0 increases, $\hat{\rho}_{\mathrm{QI}} \rightarrow 0$ rapidly. Eq. (11) behaves somewhat like the Heisenberg Uncertainty Principle for time and energy. It turns out that Eq. (11) is similar in mathematical form to the expression for the quantum vacuum electromagnetic field energy density, $\hat{\rho}_{\mathrm{CV}} = \langle T_{00} \rangle_{\mathrm{CasVac}}$, inside a Casimir cavity:

$$\hat{\rho}_{\mathrm{CV}} = -A_{\mathrm{CV}} \frac{\hbar c}{d^4}, \tag{12}$$

where $A_{\mathrm{CV}} = \zeta(D)/8\pi^2$ in spacetimes of arbitrary dimension D, and d is the distance between the cavity walls [11–13]. In our familiar (3+1)-dimensional spacetime ($D = 4$), $A_{\mathrm{CV}} = \pi^2/720$.

The appearance of $\zeta(D)$ is characteristic of expressions for vacuum stress-energy-momentum tensors $\langle T_{\mu\nu} \rangle_{\mathrm{vac}}$, and there will also be different mathematical coefficients A that correspond to the different types of quantum fields. To calculate $\langle T_{\mu\nu} \rangle_{\mathrm{vac}}$ for a given quantum field in its vacuum state is to calculate its associated Casimir vacuum energy. As discussed in Section 1.3, the spacetime manifold is made incomplete by the introduction of the plane conductor boundary condition(s) (cavity walls), so the vacuum region put under stress by the presence of the plane conductor(s) is called the Casimir vacuum. A $\hat{\rho}_{\mathrm{QI}}$ equal to the $\hat{\rho}_{\mathrm{CV}}$ in a Casimir cavity can exist no longer than for a time $t_0 \sim d/c \approx 3 \times 10^{-16}$ s for a typical experiment. This estimate proves that the QI fails in this case because the cavity walls can be held rigid to prevent their collapse under the action of the attractive Casimir force. Consequentially, the Casimir vacuum can be held in place for an infinite amount of time, which is infinitely larger than the estimated t_0, which violates Eq. (11). Another important red flag is that the QI as formulated does not apply directly to the Casimir cavity because it has boundaries.

The QI conjecture has been proven using ad hoc, arbitrarily smooth, and compactly supported time sampling functions. They have also been proven for free fields; for example, the electromagnetic field, the Dirac field in both (1+1)- and (3+1)-dimensional flat Minkowski spacetime, the massless and massive minimally coupled scalar fields, massive spin-one, and Rarita-Schwinger (spin-3/2 fermion) fields. The QI has also been proven in lower dimensional toy model spacetimes. Spatial- and spacetime-averaged QI are required to bring them into compliance with Eqs. (13) and (14) in the next section. No purely spatial QI exist in (3+1)-dimensional spacetime even though they exist in (2+1)-dimensional spacetime, and none exist along null worldlines in (3+1)-dimensional spacetime even though they exist along null worldlines in (2+1)-dimensional spacetime. Although there is weak evidence for the existence of spacetime-averaged QI in (3+1)-dimensional spacetime, no explicit analytical formulas can be derived. The reason for this situation is that the techniques used to prove the worldline QI cannot be generalized to the spacetime-averaged case because of the nonlocal (entangled) particle pair production components in $\langle T_{\mu\nu} \rangle_{\text{vac}}$, which occur in the physical (3+1)-dimensional spacetime that we live in.

Simple models of quantum states involving negative energy densities/fluxes (e.g., squeezed vacuum states of a massive scalar field) can exhibit a subtle intertwining of negative energy density/flux with positive energy density/flux. It is uncertain how generic this behavior is. Borde et al. [83] addressed this by showing that one can constrain large classes of possible spatial distributions of negative energy density/flux in flat spacetime using simply the AWEC and the QI. This approach may be complementary to spacetime-averaged QI. They recast the QI conjecture into a new program, which seeks to study the allowed spatial distributions of negative energy density/flux in quantum field theory. Their study models free massless scalar fields in (1+1)-dimensional flat Minkowski spacetime. Several explicit examples of spacetime-averaged QI were evaluated to allow or rule out some particular model (spatial) distributions of negative energy density/flux. Their analysis showed that some geometric configurations of negative energy density/flux can either be ruled out or else constrained by the QI restrictions placed upon the allowable spatial distributions of negative energy density/flux. They also found allowable negative energy density/flux distributions in which observers would never encounter the accompanying positive energy density/flux distribution so long as the QI restrictions and corresponding classical energy conditions are violated. The extent to which these results can be generalized to analytic formulas in the physical (3+1)-dimensional curved spacetime, with or without boundaries and interacting fields, will likely remain intractable because of the effects of nonlocality in some components of $\langle T_{\mu\nu} \rangle_{\text{vac}}$.

Deriving QI for interacting quantum fields in flat or curved spacetimes is technically daunting. This is a key issue because interacting fields could turn out to be far more important than free fields to produce large amounts of negative energy distributed over large spatial regions. Olum and Graham [84] constructed a (2+1)-dimensional spacetime model of two interacting scalar fields in which the energy density can be negative and static in certain regions. It is not yet known whether this arrangement can be made negative and static over an arbitrarily large spatial region. From theory and experiment, we know that large static negative energy densities associated with vacuum states are concentrated in narrow spatial regions, such as that found inside a Casimir cavity with small plate separation or in the region near the Schwarzschild radius in the Boulware vacuum (exterior to a black hole) where the energy density is negative everywhere as seen by static observers [72]. This is also found to be true in the case of time-domain squeezed vacuum states where very small negative energy density pulses occur between the alternating larger pulses of positive energy density [79]. There are other forms of squeezed vacuum states where the energy density is negative, but not necessarily static, everywhere as seen by static observers.

Ford and Svaiter [85,86] showed that quantum vacuum fluctuations can be focused by a parabolic-cylindrical mirror, whereby the sign of the vacuum energy density can be manifestly negative at the focal line of the mirror. The smallest relevant length scale in their model is not the proper distance to the mirror, which can be arbitrarily far away, but rather the distance to the focal point of the mirror, which is much smaller. If this is correct, then the prescription that the flat spacetime QI should also hold for sampling times that are small compared to the smallest proper radius of spacetime curvature or the smallest proper distance to a boundary would need to be refined. It is important to confirm whether there might be additional relevant geometrical length scales to consider when applying the flat spacetime QI to curved spacetimes, or to spacetimes with boundaries, other than just the smallest proper radius of curvature or the smallest proper distance to a boundary.

Cosmological inflationary expansion models that involve negative energy densities should be examined as possible examples of a field that does not obey the QI. This also applies to the Boulware vacuum external to black holes. Recall that Visser [75] proved that observational astrophysical data indicates the existence of large amounts of exotic matter somewhere in the universe. This implies widespread cosmological violations of the QI in addition to the widespread violations of the classical energy conditions. Also, the historical analysis of the QI does not take into account the negative energy densities/fluxes associated with quantum fluctuations of $T_{\mu\nu}$ that could be important for areas such as cosmological inflation [87–89].

Ford and Roman [90] showed that the QI bound implies that either a traversable wormhole must have a throat size no larger than a few thousand Planck lengths or (typically) the negative energy must be confined to an extremely thin band around the throat. A similar problem afflicts warp drive spacetimes [30,32,38]. Krasnikov [30,91–93] argues to the contrary by constructing an explicit counterexample for generalized FTL spacetimes showing that the relevant QI breaks down even in the simplest FTL cases. Since classical fields would not be subject to the QI, Barceló and Visser [94,95] proposed classical nonminimally coupled scalar fields as sources of exotic matter for wormhole maintenance. Are such energy condition-violating classical fields physically real? One answer to this question is to consider that the conformally coupled scalar field is deemed to be physically reasonable because it faithfully mimics certain behaviors of the electromagnetic field. On the other hand, Kuhfittig [96,97] constructed a wide class of traversable wormhole solutions that are claimed to satisfy the QI.

Finally, as with the free quantum vacuum electromagnetic field, the QI for other types of quantum fields was derived with no sources, boundary conditions, or interacting fields (a type of boundary condition), which is a fundamental flaw in its formulation. Deriving the QI for interacting quantum fields in flat or curved spacetime is technically daunting. This is a key issue because interacting fields could turn out to be far more important than free fields to produce large amounts of negative energy/flux distributed over large spatial regions. Direct measurement of quantum vacuum energy is difficult, and no QI has been experimentally tested; meanwhile, observational cosmology appears to disprove it [75]. Maclay and Davis [82] tested a proposed QI for squeezed light by a metaanalysis of published data obtained from experiments with optical parametric amplifiers and balanced homodyne detection. Over the last three decades, investigators in quantum optics have been trying to maximize the squeezing of the quantum vacuum electromagnetic field (in laser light) and succeeded in reducing the variance in the quantum vacuum fluctuations to $-15\,\mathrm{dB}$. Maclay and Davis' metaanalysis examined different time sampling functions for the QI, and found that in all physically reasonable cases, the QI is violated by much or all of the measured data. This brings into question the validity of the QI and the basis for its formulation as discussed in Ref. [82].

5.3 Energy conditions redux

Visser and collaborators [21,98–100] developed a superior way to properly quantify the amount of negative energy density/flux and energy condition violations required to construct a FTL spacetime. They propose a quantifier in terms of a spatial volume integral, which amounts to calculating the following definite integrals [21,98,100]:

$$\int \rho_{\mathrm{E}} dV \leq 0; \quad \int (\rho_{\mathrm{E}} + p_i) dV \leq 0, \tag{13}$$

where ρ_E is the rest-energy density, the p_i are spatial pressures/tensions, and dV ($=4\pi r^2 dr$ or $g^{1/2} dr d\theta d\varphi$, where $g \equiv det(g_{\mu\nu})$) is an appropriate choice of the integration measure (element of spatial volume).

The amount of energy condition violation is defined as the extent to which Eq. (13) can become negative. The value of Eq. (13) is that it provides information about the total amount of energy condition violating matter that must exist for any given FTL spacetime under study. It was further shown that Eq. (13) can be adjusted to become vanishingly small by appropriate choice of parameters. Therefore, examples can be constructed whereby the energy condition violation can be made arbitrarily small. However, the violation cannot be made to vanish entirely. The reader should be forewarned that there is some published literature claiming that the energy condition violations can be avoided when designing FTL spacetimes. In these claims, the investigators universally invoke the following scenario: divide the total stress-energy-momentum tensor into weird matter plus normal matter and push all the energy condition violations into the weird matter so that the normal matter does not violate the energy conditions and can be used to construct the FTL spacetime. Given that the energy conditions are not absolute, such rearranging approaches are not necessary. This is the primary reason why this author does not include a review of the proposed traversable wormholes and warp drives that are claimed to be constructed from normal or alternative types of normal matter that do not violate the energy conditions.

Eq. (13) gives the result that traversable wormholes require arbitrarily small amounts of negative energy density to construct, such that within a wormhole spacetime we must have [98]:

$$\rho_E = 0; \quad \oiiint p_r dV \rightarrow 0, \tag{14}$$

where in the second expression the closed volume integral is taken over the spatial volume element defined for Eq. (13) and p_r is the outward radial pressure that holds the wormhole throat open. Eq. (14) is a result that's also due to the intrinsic nonlinearity of the general relativistic field equations. This nonlinearity impacts the coupling of a finite spacecraft's mass with each side of a wormhole's throat (or the mouth on each side of the throat) leading to a mass conservation law for the combined system of spacecraft and wormhole: when finite mass spacecraft traverse a wormhole, they alter the (equivalent) mass of the wormhole mouths they pass through [25]. The entrance mouth absorbing the spacecraft gains (equivalent) mass while the exit mouth emitting it loses (equivalent) mass [23]. This mass coupling and conservation law takes into account the possibility that spacecraft traversing the wormhole may lose or gain some momentum and kinetic energy in the process, and it is assumed that the two mouths are sufficiently far apart that their mutual gravitational interaction is negligible. This unusual result suggests, but does not prove, the possibility of a fundamental limit on the total mass that can traverse a wormhole. The coupled mass conservation law shows that for a sufficiently large net transfer of mass, the final (equivalent) mass of the exit mouth becomes negative. This is a beneficial result because ANEC violations are required just to hold the wormhole throat open. If it appears that a runaway reaction might occur, then it would be prudent for wormhole engineers to simply "turn off" the wormhole for a brief moment and then "turn it back on" (i.e., "reset" the wormhole) to restart wormhole transportation operations.

It turns out that traversable wormholes based on Eq. (14) have undesirable technical difficulties. First, the smaller the amount of negative energy density used in these wormholes, the closer they are to becoming vacuum Schwarzschild wormholes that are not traversable. Second, the more the amount of negative energy density decreases, the longer it will take a spacecraft to traverse the wormhole as measured by the clocks of outside static observers, which violates item (2) of the FTL space warp requirement in Section 2.2. Third, as the amount of negative energy density decreases, the wormhole becomes more prone to destabilization by even minute amounts of infalling (positive) matter because this matter will be enormously blueshifted by the time it reaches the throat. However, Kuhfittig [96,97] derived wormhole solutions that satisfy the traversability constraints and required arbitrarily small amounts of negative energy density. Barceló and Visser [94,95] proposed a possible solution by using classical nonminimally coupled scalar fields as sources of exotic matter for wormhole maintenance. In the case of relatively weak scalar fields and coupling constants, one can get large sustainable, albeit temporary, classical negative energy fluxes even in flat spacetime.

In connection with the application of Eq. (13) to warp drives, Lobo and Visser [21] showed that the gravitational coupling of the nonzero spacecraft mass with the warp bubble leads to the condition that the total energy stored in the warp bubble should be less than the total rest-energy of the warp spacecraft itself, which places a strong constraint upon the dimensionless speed of the warp bubble:

$$v_{sh} \leq \left[\left(7.41 \times 10^{-28}\right) \left(\frac{M_{sh} \Delta}{R^2} \right) \right]^{\frac{1}{2}}, \tag{15}$$

where M_{sh} is the mass of the warp spacecraft, and the other quantities are defined as before in Eqs. (8) and (9). The numerical factor inside the brackets is G/c^2.One can multiply both sides of Eq. (15) by c to convert to MKS units of speed.

Lobo and Visser [21] had to realize that Alcubierre's warp bubble shape function is $f(r_{sh}(t))=0$ outside the warp bubble when they treated the warp spaceship as having nonzero size and mass, and considered what happens to $f(r_{sh}(t))$ in the limit $r_{sh}(t) \to \infty$. This implies that the warp drive spacetime geometry is asymptotically flat Minkowski spacetime and the ADM mass (defined by taking the limit as one moves to spacelike infinity) is zero. Therefore, the ADM mass of the warp spaceship and its warp field generator must be exactly compensated by the ADM mass due to the stress-energy of the warp bubble itself.

Eq. (15) demonstrates that, for any reasonable values of the spacecraft and warp drive parameters inside the brackets, v_{sh} will be absurdly low. To illustrate this point, we insert hypothetical warp bubble parameters $R=50\,\text{m}$ and $\Delta \sim 1/\sigma = 10^{-3}\,\text{m}$ into Eq. (15) and assume $M_{sh}=10^6\,\text{kg}$ to find that $v_{sh} \leq 1.72 \times 10^{-14}$ (or $5.16 \times 10^{-6}\,\text{m/s}$). If R and M_{sh} are kept at these values, then Δ must be $3.37 \times 10^{24}\,\text{m}$ (or $3.57 \times 10^8\,\text{LY}$) for $v_{sh} \leq 1$ (or c). Therefore, Eq. (15) gives a severely low upper bound on the warp bubble speed that can be achieved by any reasonable warp drive spacetime geometry.

Based on the foregoing discussion, traversable wormholes appear to be more desirable as a form of FTL transport, but they are not without their own technical difficulties.

5.4 The types and taxonomy of exotic matter

Fig. 25 shows the taxonomy of exotic matter that has classical and quantum violations of the pointwise and averaged energy conditions as well as the QI [20]. The class of squeezed vacuum states includes the class of Casimir vacuum states and gravitationally redshifted, or gravitationally squeezed, quantum vacuum states. In general, the local energy density in quantum field theory can be negative due to quantum coherence effects. Other examples that have been studied are Dirac field (fermionic Fock) states: the superposition of two single particle electron states and the superposition of two multielectron-positron states. In these two cases, the energy densities can be negative when two single or multiparticle

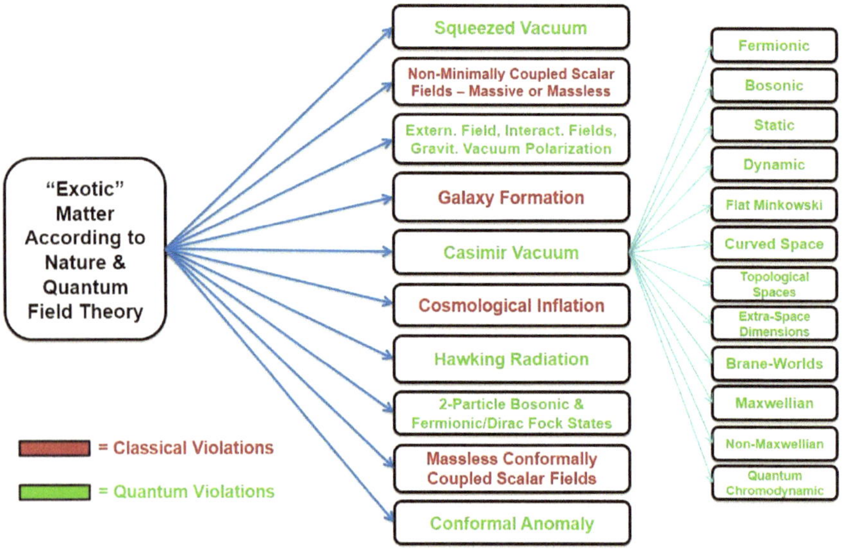

FIG. 25

The taxonomy of exotic matter that violates the pointwise and standard averaged energy conditions (classical violations) and the QI (quantum violations).

states have the same number of electrons or electron-positron pairs, respectively, or when one state has one more electron or electron-positron pair than the other.

Metamaterials are not in Fig. 25 because they are materials with designer properties that stem from structure, not substance, that determine the electromagnetic properties of structures smaller than the electromagnetic wavelengths involved. Metamaterials are nothing new because they've been manufactured by mankind since at least the Roman era (ancient Romans invented the first optical metamaterial: ruby glass) and there are several types of metamaterials that have been discovered amongst different animal body parts within the animal kingdom. What is new about metamaterials is the degree of control over the structures in the material that produce the desired electromagnetic properties. Their unusual electromagnetic properties, negative electric permittivity and/or negative magnetic permeability, do not correspond to any form of exotic matter properties as defined in Section 5.1 because they are composed of ordinary matter. However, metamaterials can be used in conjunction with transformational optics to engineer the electromagnetic analog of curved spacetime/gravitational phenomena and electromagnetically simulate their effects in the lab [101,102].

Dark matter does not appear in Fig. 25 because observational evidence consistently shows that it behaves like a zero-pressure dust that gravitationally attracts ordinary matter (i.e., it has positive energy density), and it is observed to be located near or collocated with large concentrations of luminous galactic and extragalactic matter [103]. There appears to be nothing exotic about dark matter's properties. In Fig. 25, Hawking radiation (cosmological and black hole particle production), scalar fields (massive and massless nonminimally coupled, and massless conformally coupled), and the conformal anomaly are not useful for constructing FTL spacetimes because they are not amenable to technological control.

The latter two are traditionally used as sources of exotic matter for FTL spacetimes in toy lower-dimensional spacetime models. Fermionic/Dirac Fock States are still under theoretical investigation, so this will not be considered further. Squeezed quantum electromagnetic vacuum states (squeezed light) and the quantum vacuum ZPF field distortion induced by static and moving mirrors and the Casimir effect will be surveyed in the following subsections for their usefulness in producing negative vacuum energy in the lab for the construction of FTL spacetimes. The negative vacuum energy arising from the quantum vacuum polarization induced by distorting the vacuum ZPF field with externally applied fields, interacting fields, or gravitational fields (spacetime curvature) will not be considered here because the former two require further theoretical investigation while the latter is predicted to be a (very difficult-to-observe) naturally occurring phenomenon associated with collapsed stars. However, Eq. (42) in Section 8 quantifies the vacuum polarization energy density arising from spacetime curvature-induced vacuum ZPF field distortion. Also, the technical issue of capturing and storing negative vacuum energy is not considered here because free-space sources of negative vacuum energy appear to be a more desirable option for constructing FTL spacetimes than stored negative vacuum energy, and there is very little technical literature that addresses how to capture and store negative vacuum energy (e.g., see Ref. [79]). This technical issue requires future theoretical investigation.

5.5 Concepts for producing negative vacuum energy in the lab

5.5.1 Squeezed vacuum states of light

Substantial theoretical and experimental work has shown that in many quantum systems the limits to measurement precision imposed by quantum vacuum ZPF noise can be breached by decreasing the noise in one observable (or measurable quantity) at the expense of increasing the noise in the conjugate observable; at the same time, the variations in the first observable—say, the energy—are reduced below the vacuum ZPF (ground state) such that the energy becomes "negative." "Squeezing" is the control of quantum vacuum ZPF noise (or modes) and corresponding uncertainties, whereby one can squeeze the variance of one (physically important) observable provided the variance of the (physically unimportant) conjugate observable is stretched/increased. The squeezed observable possesses an unusually low variance, meaning less variance than would be expected on the basis of the equipartition theorem. One can in principle exploit quantum squeezing to extract energy from one place in the ordinary quantum vacuum at the expense of accumulating excess energy elsewhere.

The squeezed state of the quantum electromagnetic field is one of the numerous examples where the vacuum state of quantum fields can have negative energy density. Such a state became a physical reality in the laboratory because of the nonlinear-optics technique of "squeezing," or in other words, of moving some of the quantum vacuum ZPF noise in laser light from out of the $cos[\omega(t-z/c)]$ part of the beam into the $sin[\omega(t-z/c)]$ part [104–108]. This pertains to a particular electric field mode of the coherent light of an ideal laser: $E=E_0\{Acos[\omega(t-z/c)]+Bsin[\omega(t-z/c)]\}$, where the electric field amplitude E_0 is constant, ω is the angular frequency, z is the beam propagation direction, t is time, and (A, B) are the quantum field quadrature operators corresponding to the quantized harmonic oscillator (i.e., the single-mode electromagnetic field's quantum position and momentum operators in phase space, respectively). A and B are defined as: $A=(ae^{i\phi}+a^{\dagger}e^{-i\phi})/2$ and $B=(ae^{i\phi}-a^{\dagger}e^{-i\phi})/2i$, where ϕ is the oscillator phase and $(a,a^{\dagger})$ are the usual quantum annihilation and creation operators, respectively. A and B are the observables we are interested in that must obey the quantum commutation relation: $[A, B]=i\hbar/2$, where $\hbar$ is Planck's reduced constant. The operators $(a,a^{\dagger})$ obey the quantum commutation relation: $[a, a^\dagger] = 1$. Additionally, the oscillator vacuum is defined in the usual way: $a|vac\rangle = 0$, in other words, a annihilates the vacuum state $|vac\rangle$,which defines the quantum vacuum to be a state of zero particles (with zero eigenvalue; this is also known as a zero-amplitude coherent state in quantum optics). The corresponding uncertainty relation $\Delta A \Delta B \geq \frac{1}{2}|\langle [A, B] \rangle|$ has one built-in degree of freedom: one can squeeze the variance of one observable provided the variance of the conjugate observable is stretched. (Note: the $|\langle \ldots \rangle|$ signifies taking the absolute value of the quantum expectation value of the enclosed quantity.) Squeezing becomes important when ΔA and ΔB are limited by the minimized uncertainty product: $\Delta A \Delta B = ¼$. The observable that gets squeezed will have its vacuum ZPF noise reduced below that of the undisturbed free vacuum ZPF.

To squeeze, one acts on a complex quantum state χ using a squeeze operator $S(\chi)$ for a single quantized harmonic oscillator: $S(\chi) = \exp\left[\frac{1}{2}\left(\chi a^2 - \chi^* a^{\dagger 2}\right)\right]$, where $\chi*$ is the complex conjugate of χ. The state $\chi = r_{\text{sq}} e^{i\delta}$ determines the size of squeezing, where r_{sq} is the amplitude (a measure of the mean photon number in χ) and δ is the phase of squeezing. The squeezed vacuum is therefore defined as: $|\chi\rangle = S(\chi)|vac\rangle$. Calculating the uncertainties ΔA and ΔB with respect to $|\chi\rangle$ gives $\Delta A = exp(r_{sq})/2$ and $\Delta B = exp(-r_{sq})/2$; therefore, ΔB is squeezed and ΔA is stretched. Thus, $S(\chi)$ reduces the uncertainty in B and increases that in A while maintaining the minimized uncertainty product. The act of squeezing transforms the phase space circular ZPF noise profile characteristic of the quantum vacuum into an ellipse whose semimajor and semiminor axes are given by unequal ΔA and ΔB. This applies to coherent states in general and the usual vacuum is also a coherent state with eigenvalue zero as was shown in the previous paragraph. As this ellipse rotates about the origin with angular frequency ω, the unequal quadrature uncertainties manifest themselves in the quantum electromagnetic field oscillator energy by periodic occurrences of both smaller and larger vacuum ZPF noise, which are separated by one quarter cycle as compared to the undisturbed free vacuum. Also, squeezed vacuum states are nonclassical states of light because they have non-Poissonian photon statistics.

Caves [109] and Morris and Thorne [23] point out that if one squeezes the vacuum (i.e., if one puts vacuum rather than laser light into the input port of a quantum optical squeezing device), then one gets at the output an electromagnetic field with weaker (suppressed or subvacuum) ZPF and less energy density than the vacuum at locations where $cos^2[\omega(t - z/c)] \cong 1$ and $sin^2[\omega(t - z/c)] \ll 1$, but with greater ZPF and greater energy density than the vacuum at locations where $cos^2[\omega(t - z/c)] \ll 1$ and $sin^2[\omega(t - z/c)] \cong 1$. Since the undisturbed free vacuum is defined to have vanishing energy density (i.e., this is a "dark" state), any region with less energy density than the vacuum (i.e., a subvacuum state that's "darker than vacuum") has negative energy density. Therefore, a squeezed vacuum state consists of a traveling electromagnetic wave, which oscillates between negative and positive vacuum energy density but has positive time-averaged vacuum energy density because of the unequal ΔA and ΔB.

For the squeezed quantum electromagnetic vacuum state, the energy density $\hat{\rho}_{\text{Esqv}}$ is given by [110]:

$$\hat{\rho}_{\text{Esqv}} = \chi|T_{00}|\chi = \left(\frac{2\hbar\omega}{L^3}\right) sinh\left(r_{sq}\right)\left[sinh\left(r_{sq}\right) + cosh\left(r_{sq}\right) cos\left(2\omega(t - z/c) + \delta\right)\right], \tag{16}$$

where L^3 is the volume of a large box with sides of length L (i.e., we put the quantum field inside a box with periodic boundary conditions).

Eq. (16) shows that $\hat{\rho}_{\text{Esqv}}$ falls below zero once every cycle when the condition $cosh(r_{sq}) > sinh(r_{sq})$ is met. It turns out that this is always true for every nonzero value of r_{sq}, so $\hat{\rho}_{\text{Esqv}}$ becomes negative at some point in the cycle for a general squeezed vacuum state. When a quantum state is close to a squeezed vacuum state, there will almost always be some negative vacuum energy densities present, and the fluctuations in $\langle\chi|T_{00}|\chi\rangle$ start to become nearly as large as the expectation value itself. The negative vacuum energy density pulses are interspersed with the positive vacuum energy density pulses, which is of no use for engineering a warp drive or a traversable wormhole spacetime.

Pulses of positive and negative vacuum energy can be produced by a standard squeezed laser apparatus. In this scheme, a laser beam is passed through an optical cavity resonator made of a lithium niobate ($LiNbO_3$) crystal (there are other chemical crystal formulations to choose from) shaped like a cylinder with rounded silvered ends to reflect light. The resonator acts to produce a secondary lower frequency light beam in which the pattern of photons is rearranged into pairs. This is the quantum squeezing of light effect that we described previously. The squeezed vacuum pulses are typically of $\sim 10^{-15}$ s duration. Fig. 26 shows a schematic diagram of a typical laboratory quantum optical balanced homodyne detector (BHD) for measuring squeezed vacuum states of light and Fig. 27 shows a simplified illustration of experimentally measured relative vacuum fluctuation noise intensity data from the detector. Davies [112] suggested that one could arrange a set of rapidly rotating mirrors to spatially separate out the negative vacuum energy pulses from the positive vacuum energy pulses. The squeezed light beam would be set to strike each mirror surface at a very shallow angle while the rotation would ensure that the negative vacuum energy pulses would be reflected at a slightly

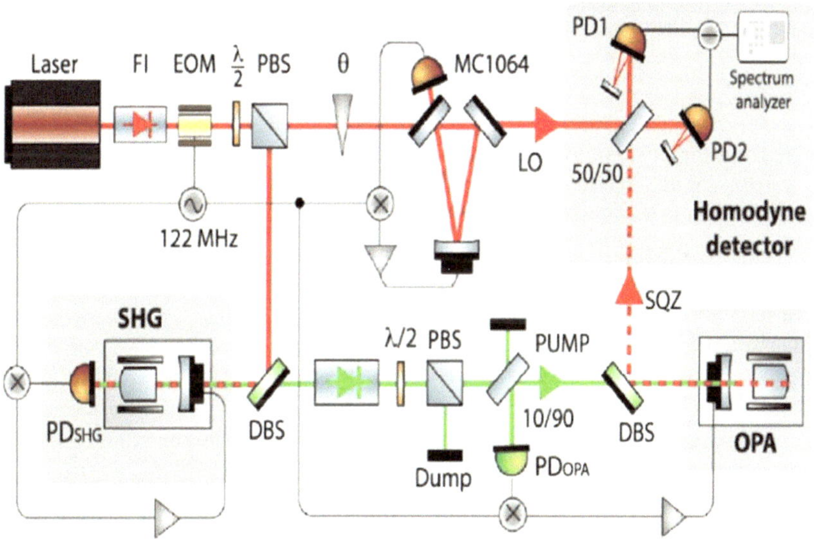

FIG. 26

Schematic diagram of a typical laboratory quantum optical squeezed light BHD apparatus [111]. Squeezed vacuum states of light SQZ at a wavelength of 1064 nm are generated in a double-resonant, type I optical parametric amplifier (OPA) operated below threshold. SHG is the second-harmonic generator, PBS is the polarizing beam splitter, DBS is the dichroic beam splitter, LO is the local oscillator, PD are the photodiodes, MC1064 is the three-mirror ring cavity for spatiotemporal mode cleaning, EOM is the electro-optical modulator, and FI is the Faraday isolator. A piezoelectric-actuated mirror is the phase shifter for the relative phase θ between SQZ and LO.

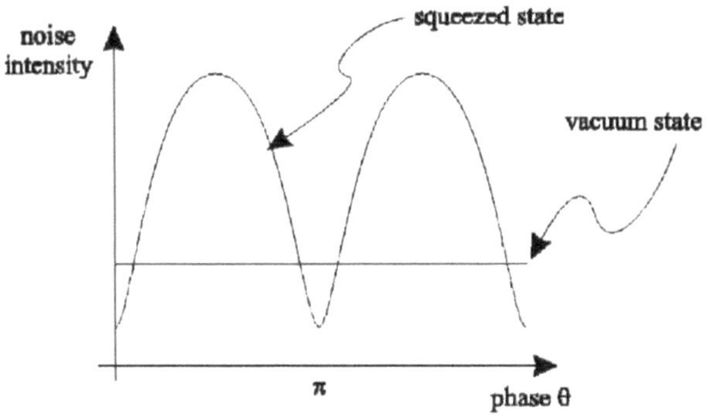

FIG. 27

A simplified illustration of a relative noise intensity plot in dB for any experimentally measured squeezed vacuum state of light. This plot shows the typical dependence of quantum vacuum fluctuations on the relative phase θ between the squeezed vacuum state and the local oscillator of a BHD. This plot will look qualitatively the same when the relative noise intensity is plotted in the frequency- or time-domain (as a function of θ measured by the time to move the piezoelectric-actuated phase shifter mirror in the BHD). Intervals where the squeezed vacuum state dive below the undisturbed vacuum (shot noise) state *(horizontal line)* correspond to squeezing. Squeezed vacuum states for short times are darker than vacuum and have negative energy density as seen in the small narrow troughs.

Courtesy P. Marecki.

different angle from the positive vacuum energy pulses. A spatial separation of the two different vacuum energy pulses would occur at some distance from the rotating mirrors. Another system of mirrors would be needed to direct the negative vacuum energy pulses to where they can be concentrated at one location to produce the warp drive or traversable wormhole. See Fig. 28 for a simple conceptual illustration of this concept.

Rotating a set of mirrors at 10^{15} Hz could be very difficult to engineer. However, the rotating mirror system can be implemented via nonmechanical means. Slusher and Yurke [106] described a chamber of sodium gas that's placed within the squeezing cavity and a laser beam directed through the gas. The beam is reflected back on itself by a mirror to form a standing wave within the sodium chamber. This wave causes rapid variations in the optical properties of the sodium, thus causing rapid variations in the squeezed light so that one can induce rapid reflections of pulses by careful design. An illustration of this concept is shown in Fig. 29.

Davies [112] also suggested that another way to generate squeezed light would be to manufacture extremely reliable light pulses containing precisely one, two, three, etc., photons apiece and combine them together to create squeezed states to order. Superimposing many such states could theoretically produce bursts of intense negative vacuum energy. See Fig. 30 for a conceptual illustration of this concept. Photonic crystal development has already demonstrated the feasibility of using photonic crystal waveguides (mixing together the classical and quantum properties of optical materials) to engineer light sources that produce beams containing precisely one, two, three, etc., photons. For example, investigators at Melbourne University used a microwave oven to fuse a tiny diamond, just 1/1000th of a millimeter long, onto an optical fiber, which could be used to create a single photon beam of light

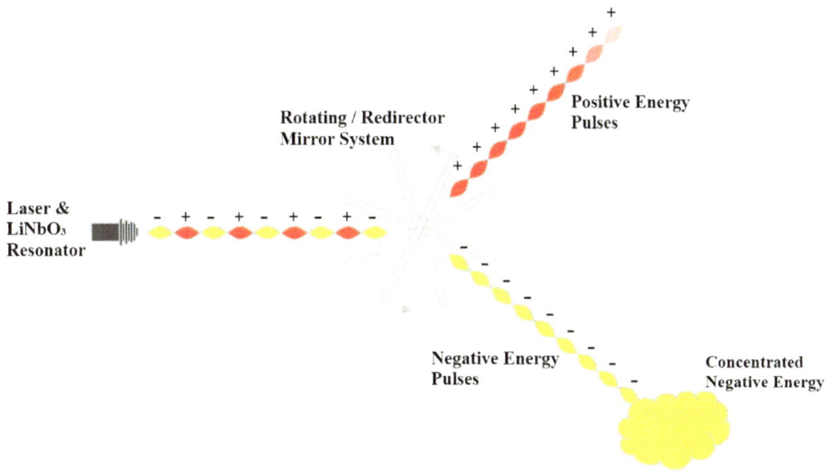

FIG. 28

A simplified conceptual schematic for generating, separating out, and concentrating the negative vacuum energy pulses *(yellow)* produced by a quantum optical squeezed light apparatus. Not depicted are the components of the full apparatus which include one or more nonlinear optical parametric oscillator/amplifier crystals and their associated resonator components, mode cleaner, beam splitters, lenses, quarter- and/or half-wave plates, diffraction gratings, BHD, Michelson wave (spectrum) analyzer, and other associated optical components.

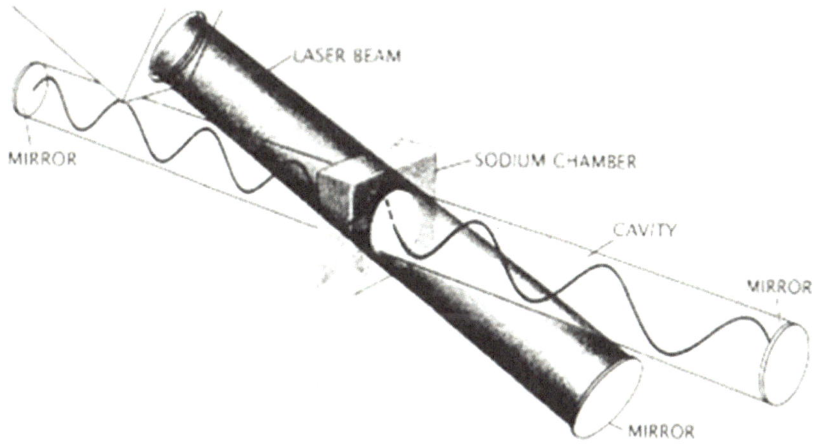

FIG. 29

Conceptual illustration of a sodium chamber negative vacuum energy ultra-rapid pulse separator [106].

[113,114]. Other single photon, two-photon, or four-photon sources are bright quantum dots [115]; heralded single photons (pairs of single photons generated in highly correlated states by using a single high-energy photon to create two lower-energy ones) that can be produced by spontaneous parametric downconversion and spontaneous four-wave mixing techniques [116]; or other multiphoton sources [117]. Developing intense n-photon sources for $n \geq 1$ will be a very long-term technical challenge

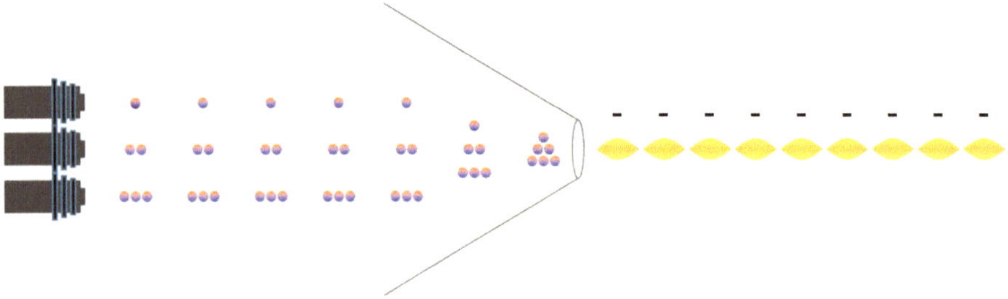

FIG. 30

A conceptual illustration of an intense beam of negative vacuum energy density pulses *(yellow)* generated by quantum optically combining (via the conjectural funnel-shaped component in the center) numerous beams of integer-multiple photon light pulses. Only the first three integer-multiple photon beam emitters are shown here for simplicity.

because demand for these must exist to motivate the photonics industry to pursue their development. The quantum optical beam combiner technology for this concept has yet to be theoretically and experimentally developed.

Youn et al. [118] proposed an extremely simple system to produce a beam of squeezed light in free space, which avoids the complexity of using multiple nonlinear optical parametric oscillator/amplifier crystals and their associated resonator components, mode cleaner, lenses, quarter- and/or half-wave plates, diffraction gratings, and other associated optical components. Instead, their investigation found that one can modulate the quantum vacuum field by placing a single conducting mirror close to the unused port (entrance) of a 50:50 beam splitter as an alternative means to produce squeezed vacuum states. As is already well known in cavity QED experiments, the modes of the vacuum field and its ZPF amplitude are modified near the mirror with respect to those in free space due to the boundary condition imposed upon the quantum vacuum by the mirror. These modified vacuum modes have a sinusoidal spatial dependence on the distance from the mirror. The spatially amplitude-modulated vacuum fields can produce squeezed vacuum states of light in the same balanced homodyne detection configuration as for a typical quantum optical squeezed vacuum apparatus. Youn et al. showed that by using a single-mode coherent light (laser) as the local oscillator and combining it with the amplitude-modulated vacuum field thus produced as the signal, the apparatus will produce squeezed light whose vacuum fluctuations are reduced below that of the intrinsic vacuum fluctuations of the laser by as much as 30% at distances of $z = n\lambda$ ($n =$ positive integer) between the photodetectors and the mirror. At those positions, the apparatus reduces the vacuum fluctuations of the difference of two beams that come from the unused beam splitter port such that highly non-Poissonian fluctuations of the difference beam may come from totally correlated beams such as photon pairs (2-particle Fock states) that are generated by parametric amplification. Fig. 31 illustrates this simple apparatus.

Furthermore, the conducting mirror's squeezing effect will continue to show up over a macroscopic distance from the mirror, which means that the scanning and measuring of the negative vacuum energy density pulses in the squeezed light beam can take place at spatial locations far from the mirror. This is unlike the situation in typical laboratory quantum optical squeezed light devices, which rely on the standard method of quantum tomographic scanning and measuring the squeezed vacuum pulses in very tight quarters within a short distance from the signal beam on a quantum optics bench. Peter Milonni

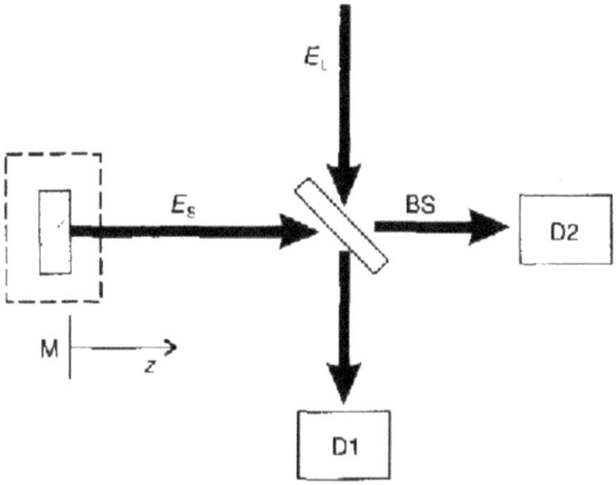

FIG. 31

BHD photodiodes D1, D2, and beam splitter BS (the diagonal rectangle in the center) [118]. E_S and E_L represent the signal and local oscillator laser, respectively. M is the conducting mirror and z is the direction of the tangential vacuum field-mode with wavevector normal to M. The electric field propagation and polarization are along the x- and z-axes (the x- and y-axes are not indicated in the figure).

(Los Alamos National Lab) advised G. J. Maclay and this author that this experiment can be done, but he warned that there is a subtlety involved with the type of photodiodes that one chooses for the BHD to properly measure the negative vacuum energy density pulses. It is not known whether anyone has constructed Youn et al.'s proposed apparatus and tested it on a quantum optics bench. An experimental program using their proposed device should be done to produce and measure squeezed light along with the additional optical components to separate out and concentrate the negative vacuum energy density pulses at a distant location. This apparatus should be less complicated to implement than state-of-the-art quantum optical squeezing devices that are constructed on very cramped quantum optics benches.

The ultimate concerns with any type of quantum optical squeezing scheme are: (1) whether the magnitude of negative vacuum energy density pulses is scalable; (2) whether it will be possible to separate out the negative vacuum energy density pulses from the positive vacuum energy density pulses and concentrate the former at a distant location; and (3) whether the quantum optical beam-combining of a linear array of numerous integer-multiple photon beam emitters is technically feasible and whether this apparatus and the magnitude of its negative vacuum energy density beam are also scalable. The prominent scalability drivers that must be investigated in future studies are how large one can engineer the numerical factor in parentheses in Eq. (16), what values r_{sq} and δ must have to make χ arbitrarily large, and whether there are engineering limits to scaling and optimizing these quantities. This is of major concern because state-of-the-art and theoretical quantum optical squeezing technologies constrain the magnitude of negative vacuum energy density pulses in Eq. (16) such that these are barely scalable and not presently capable of producing the rather large curved spacetime effects required to engineer a warp drive or traversable wormhole. Last, the concept illustrated in Fig. 30 requires theoretical development followed by an engineering feasibility study.

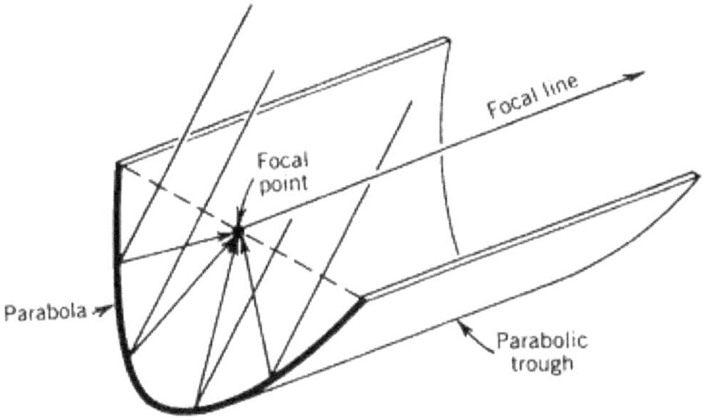

FIG. 32

Schematic diagram of a typical parabolic-cylindrical mirror.

William B. Stine, http://www.powerfromthesun.net/Book/chapter08/chapter08.html.

5.5.2 Parabolic-cylindrical mirror

Ford and Svaiter [85,86] used geometric optics and quantum field theory to argue that a perfectly reflecting parabolic-cylindrical mirror (shown in Fig. 32) can focus the vacuum ZPF modes of the quantum electromagnetic field, which can create large physical effects (enhanced Casimir forces) near the mirror's focal line, which are predicted to be experimentally observable. The large physical effects were found to come from only the short-wavelength vacuum ZPF modes according to their geometric optics analysis. Not only can the mirror focus a beam of light, but it can also focus something even in the absence of incoming light, which they dubbed "focusing a beam of dark" because the vacuum state of the quantum electromagnetic field is also a dark state. Another consequence of this focusing is a growth in the local total vacuum energy density and quantum vacuum expectation value of the mean-squared electric and magnetic field strengths, which grow as inverse powers of the distance from the focal line for small distances, as the focal line is approached. They found that for a perfectly reflecting mirror, these quantities diverge as the inverse-fourth power of the distance from the focal line, which is just like the distance dependence seen in Eqs. (11) and (12). Their analysis also accounted for mirror edge effects and finite mirror reflectivity.

Ford and Svaiter calculated analytical and numerical results for the quantum vacuum expectation value of the square of the electric field strength $\langle E^2 \rangle_{vac}$ (energy density of the electric field in the vacuum state) near the focal line of the parabolic-cylindrical mirror with an angular size restricted to $2\theta < 4\pi/3$. They also found that, for smaller mirrors of angular size $\theta < \pi/2$, $\langle E^2 \rangle_{vac} < 0$ everywhere, while for large mirrors of angular size $\pi/2 < \theta < 2\pi/3$, $\langle E^2 \rangle_{vac}$ can be positive or negative depending upon the direction from the focal line. In directions nearly perpendicular to the mirror's symmetry axis, they found that $\langle E^2 \rangle_{vac} > 0$. Because $\langle E^2 \rangle_{vac}$ for the electric field and $\langle B^2 \rangle_{vac}$ for the magnetic field are equal in Ford and Svaiter's geometric optics approximation, the local total energy density of the quantum vacuum electromagnetic field is just $\langle E^2 \rangle_{vac}$. They showed that, when $\langle E^2 \rangle_{vac} < 0$, the local total energy density of the quantum vacuum electromagnetic field is negative and columns of negative vacuum energy density will run parallel to the focal line of the parabolic-cylindrical mirror. Further theoretical work is required to develop the engineering scaling law(s) for the magnitude of the negative vacuum energy density along the focal line vs angular mirror size and any other relevant mirror parameters, and then make numerical predictions using such scaling law(s) to support engineering feasibility studies.

We speculate on a futuristic scenario in which several practicable ultrahigh-reflectivity "Ford-Svaiter mirrors" are arranged in an array that is designed to intersect their focal lines to produce a local region of intensely focused large-magnitude negative vacuum energy density, which could be shaped and controlled by reorienting one or more elements of the array to adjust the focal line intersection geometry. It is theoretically conceivable that future near-perfectly or perfectly reflecting metamaterial hypermirrors, or some sort of hyperlenses arranged in an array of some sort, could accomplish this same effect far more effectively with far less complexity and engineering difficulty than an array of Ford-Svaiter mirrors. This possibility will need to be theoretically explored to determine whether metamaterial hypermirrors and hyperlenses can respectively reflect and focus quantum vacuum ZPF. One could imagine a warp drive spacecraft using an array of externally mounted Ford-Svaiter mirrors (or putative metamaterial hypermirrors or hyperlenses) to produce, shape, and orient a bubble of negative vacuum energy (the warp bubble) that produces the warp drive spacetime geometry surrounding the spacecraft. For traversable wormholes, perhaps an array of Ford-Svaiter mirrors (or putative metamaterial hypermirrors or hyperlenses) is deployed to open and stabilize a traversable wormhole throat along with a small fleet of Ford-Svaiter mirror (or putative metamaterial hypermirrors or hyperlenses) spacecraft that enter the throat to extend its exit to the desired interstellar destination. The technical details for these scenarios will have to be worked out in future studies.

5.5.3 Moving mirror

Negative energy can be produced by a moving single reflecting boundary such as a plane mirror. A mirror moving with increasing acceleration emanates a flux of negative vacuum energy from its surface in the forward direction of its motion (see Fig. 33) [119,120]. This is essentially the simple case of an infinite plane conductor undergoing acceleration perpendicular to its surface. If the acceleration varies with time, the conductor will generally emit photons—in other words, create quanta (particles)—by exchanging energy with the quantum vacuum, even though it is neutral. This is an example of the well-known quantum phenomenon of parametric excitation, which in this case is caused by the distortion of the quantum vacuum electromagnetic ZPF modes by the moving mirror. The parameters of the quantum vacuum electromagnetic field oscillators (e.g., the frequency distribution function) change with time owing to the acceleration of the mirror [121]. Ironically, the moving mirror behaves just like a time-dependent background curved spacetime geometry (i.e., dynamical gravitational field) that creates quanta via the parametric excitation caused by the distortion of the quantum vacuum ZPF modes by the dynamical gravitational field [119]. However, a numerical analysis by this author has shown that the flux of negative vacuum energy produced by a moving mirror is exceedingly feeble for technologically achievable time-varying accelerations, so this is not the most effective way to produce negative vacuum energy to construct FTL spacetimes.

5.5.4 The Casimir effect

The quantum electrodynamic Casimir effect was previously discussed in Sections 1.3 and 5.2, and it is the most well-known way to produce a minute semiconfined region of extremely feeble negative vacuum energy in the lab. It turns out that there's a long list of different analog Casimir effects within quantum field theory, as shown in Fig. 25 [11–13,122,123]. For example, if one introduces a single

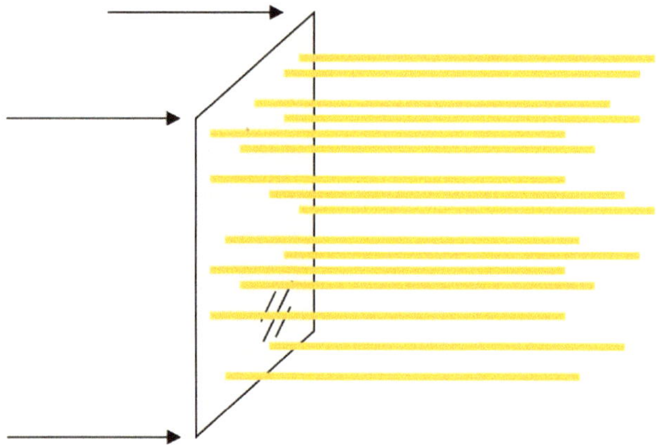

FIG. 33

Conceptual illustration of negative vacuum energy flux *(yellow)* emanating in the direction of motion of a moving mirror.

infinite plane conductor into the flat Minkowski spacetime vacuum by bringing it adiabatically from infinity so that whatever quantum fields are present suffer no excitation but remain in their ground state, then the quantum vacuum (electromagnetic field) stress induced by the presence of the infinite plane conductor produces a Casimir effect. In this case, as well as in the case of the cavity/waveguide Casimir effect (see Fig. 34), the spacetime manifold is made incomplete by the introduction of the plane conductor boundary condition(s). The vacuum region put under stress by the presence of the plane conductor(s) is called a Casimir vacuum.

Analogs of the Casimir effect exist for the other quantum fields. When considering the vacuum state of other fields, one must consider boundary conditions that are analogous to the perfect-conductor boundary conditions for the electromagnetic field at the surfaces of the plates. Other fields are not electromagnetic in nature (i.e., they are non-Maxwellian), so the perfect-conductor boundary conditions do not apply to them. It turns out that complete manifolds exhibit what is called the topological Casimir effect for any non-Maxwellian fields. In order to define boundary conditions for other fields, we replace the conductor boundary conditions (plates) and flat Minkowski spacetime by a manifold of the form $\mathfrak{R} \times \sum$ (i.e., a product space), where $\mathfrak{R}$ is the real line defining the time dimension for this particular product space and $\sum$ is a flat 3D manifold (3-manifold) having any one of the following topologies: $\mathfrak{R}^2 \times \mathrm{S}^1$, $\mathfrak{R} \times \mathrm{T}^2$, T^3, $\mathfrak{R} \times \mathrm{K}^2$, etc., with $\mathfrak{R}$ identified as the real line that defines any linear space dimension (e.g., $\mathfrak{R} =$ line, $\mathfrak{R}^2 =$ 2D plane, etc.), T^n being the n-torus, K^2 the 2D Klein bottle, S^1 the circle, etc.

The case $\sum = \mathfrak{R}^2 \times \mathrm{S}^1$ has the closest resemblance to the quantum electrodynamic Casimir effect, the difference being that instead of imposing physical conductor boundary conditions with plate separation distance d along the $x^3 = z$ axis perpendicular to the plates, one imposes periodic boundary conditions with period d along z. When imposing this topological constraint on the field theoretic calculation of the Casimir effect for the case of linear massless fields (LMF), one finds that the expression for the corresponding Casimir vacuum energy density is [13]:

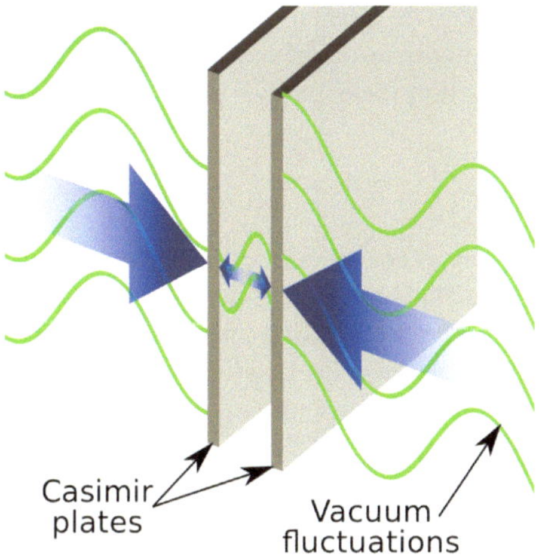

FIG. 34

An illustration of a typical laboratory quantum electrodynamic Casimir cavity/waveguide.

C. C. Emok at Commons, https://commons.wikimedia.org/wiki/File:Casimir_plates.png.

$$\hat{\rho}_{\mathrm{LMF}} = -A_{\mathrm{LMF}} \frac{\hbar c}{d^4}, \tag{17}$$

where $A_{\mathrm{LMF}} = \pm d_f(\pi^2/90)$, d_f is the number of degrees of freedom (e.g., helicity states) per spatial point, and the plus sign holds for boson fields (giving negative energy density) and the minus sign for fermion fields (giving positive energy density).

If one were to admit spin structure in the manifolds described above and the quantum field is spinorial, then there would be another important subtlety that would need to be taken into account when evaluating $\langle T_{\mu\nu} \rangle_{\mathrm{vac}}$. However, this introduces an additional complexity involving the relationship between the spin structure and the global structure (i.e., the configuration space or fiber bundle) of the quantum field in question, whereby the topology not only of the base manifold but of the fiber bundle itself has an effect on $\langle T_{\mu\nu} \rangle_{\mathrm{vac}}$. In addition to this, there are (compactified) extra-space dimensional quantum field (i.e., D-brane or "brane world") analogs of the Casimir effect. A detailed consideration of the various quantum field and brane world analogs of the Casimir effect for producing FTL spacetimes is beyond the scope of this chapter and will be left for future investigation. One proposal that exploits a brane world analog of the Casimir effect to construct a warp drive was described in Section 3.2 [42,43].

We also point out that the methods used to obtain $\hat{\rho}_{\mathrm{CV}}$ between the parallel plane conductors of a Casimir cavity/waveguide can also be used when the conductors are not parallel but are joined together along a line of intersection. If the conductors have curved surfaces instead, then one obtains results that are similar to the intersecting conductor case. These geometries have also been evaluated for the case of dielectric media. These cases will not be considered further since there are technical subtleties involved that complicate the calculations and application of the different approaches. This topic will also be left for future investigation.

Morris et al. [24] proposed using the quantum electrodynamic Casimir effect to construct a spherically symmetric traversable wormhole throat of radius r_{throat} by designing a cavity made of perfectly conducting spherically concentric thin plates with a plate separation d defined by:

$$d = \left(\frac{\pi^3}{30}\right)^{1/4}\left(r_{\text{throat}}\sqrt{\frac{\hbar G}{c^3}}\right)^{1/2} = \left(4.05 \times 10^{-18}\right)\sqrt{r_{\text{throat}}}. \tag{18}$$

Eq. (18) has units of meter because $\sqrt{\hbar G/c^3}$ defines the Planck length. To counteract the collapse of the cavity due to the attractive Casimir force acting between the concentric plates, the plates will have equal electric charge placed upon them to establish adequate Coulomb repulsion. In a detailed analysis, the electrostatic energy required to support the Coulomb repulsion between the plates would be considered separately. Eq. (18) shows that a wormhole with $r_{\text{throat}} = 1000\,\text{m}$ gives $d = 1.28 \times 10^{-16}$ m (smaller than a proton diameter). Using Eq. (12), we get $\hat{\rho}_{CV} = -1.62 \times 10^{36}$ J/m^3 for this concentric cavity configuration. By comparison, a wormhole with $r_{\text{throat}} = 1.5 \times 10^{11}\,\text{m}$ (or 1 AU) gives $d = 1.57 \times 10^{-12}$ m (or 35% smaller than the electron's Compton wavelength), which gives $\hat{\rho}_{CV} = -7.17 \times 10^{19}$ J/m^3. There is no technology known today or that can be imagined in the future that can construct a concentric spherical cavity having such minuscule plate separations. In addition, these example plate separations are unrealistic because the Casimir effect switches over to the nonretarded field behavior ($\sim d^{-3}$) of van der Waals forces when plate separations go below the wavelength ($\approx 15\,\text{nm}$) where they are no longer perfectly conducting [124]. We will not consider this concept further. However, future investigation will be necessary to elucidate whether the quantum field analogs of the Casimir effect can predict a more reasonable technical solution for constructing a traversable wormhole.

5.6 Detecting negative vacuum energy in the lab

How would negative vacuum energy be detected so that one could control it to construct an FTL spacetime? A nonoptical scheme for detecting negative vacuum energy was proposed by Davies and Ottewill [77], who studied the response of switched particle detectors to static negative energy densities and negative energy fluxes. Their model is based on a free massless scalar field in (3+1)-dimensional flat Minkowski spacetime and utilized a simple generalization of the standard monopole detector, which is switched on and off to concentrate the measurements on periods of isolated negative energy density or negative energy flux. Their model included an explicit switching factor, whereby five different switching functions (based on data windowing theory) are defined and evaluated. Their results shed light on the response of detectors to pulses of negative energy of finite duration, and they showed that negative energy should have the effect of enhancing deexcitation (i.e., induce cooling) of the detector. Furthermore, they concluded that the enhanced cooling effect cannot be used to draw a thermodynamic conclusion because their modeling was restricted to first-order in perturbation theory. It is not possible at first-order to determine whether the enhanced cooling effect is due to the small violation of energy conservation expected in any process in which a general quantum state collapses to an energy eigenstate, or whether this predicts a systematic reduction in the energy of the detector, which has serious thermodynamic implications. Their results are model-dependent, and there is not always a simple relationship between the strength of the negative energy density/flux and the behavior of the monopole detector.

Slusher and collaborators [104–106] and Robinson [107,108] were the first to experimentally observe the alternating subvacuum pulses of negative vacuum energy density in squeezed light. Numerous other experiments followed that employed variations on the original experimental devices and techniques used to generate squeezed light and measure its subvacuum fluctuation noise pulses. Those early experimental devices later gave way to the development and use of BHDs. For example, Schneider et al. [125] describe their compact efficient source of amplitude-squeezed light. Their experiment used a semimonolithic degenerate $MgO:LiNbO_3$ optical parametric amplifier pumped by a frequency-doubled Nd:YAG laser at 532 nm. They employed injection-seeding of the amplifier by a 1064 nm wave to provide active stabilization of the cavity length and stable operation. At a pump power of 380 mW, their device detected a maximum noise reduction of 6.5 dB in the vacuum amplitude fluctuations of the 0.2 mW 1064 nm wave, while the average detected vacuum noise reduction in continuous operation over 14 min was 6.2 dB. They reported a squeezing of 7.2 dB in the emitted wave.

Hansen et al. [126] describe their experimental time-domain BHD device, which they developed as a pulsed BHD for precise measurement of the electric field quadratures of pulsed optical quantum states. A high level of common mode suppression (>85 dB) and low electronic noise (730 electrons per pulse) in their device provides a signal-to-noise ratio of 14 dB for measurement of the vacuum noise of individual pulses. Their device achieved a signal-to-noise ratio of 14 dB at a pulse repetition rate of up to 1 MHz, enabling high-accuracy quantum measurements to be carried out in a short time. They performed a quantum tomography of the coherent state as a test for their device, and the Wigner function and density matrix were reconstructed with 99.5% fidelity, while their detector exhibited 91% quantum efficiency. Their detection system can also be used for ultrasensitive balanced detection in continuous wave mode. Note, however, that no BHD or other quantum optics technology has been invented for the purpose of directly measuring the energy density of the alternating negative and positive vacuum energy pulses in squeezed light.

What has not been experimentally measured yet are the subvacuum fluctuations and corresponding negative energy density inside of a Casimir cavity. The negative Casimir vacuum energy density is static, so this is not a time- or frequency-domain problem. Marecki [127,128] proposed a modified BHD that can quantify the vacuum fluctuations of the quantum electric field (via the quantum vacuum expectation values of products of the electric field operators, the one- and two-point correlation functions of arbitrary quantum states of the electric field for fields restricted to the frequency of the coherent local oscillator) and the associated spectral density for the ground state of the quantum electric field in Casimir geometries. He predicted a position- and frequency-dependent pattern of BHD responses if a submicron-scale device of this type is placed inside of a Casimir cavity. The proposed BHD allows for the direct detection of subvacuum fluctuations and provides a spatial-frequency mapping of the negative vacuum energy density inside the cavity. This offers a potential new characterization of ground states in Casimir geometries, which would provide an understanding of the negative vacuum energy densities present in some regions in these geometries, which addresses the Davies-Ottewill negative energy detector hypothesis.

The importance of measuring the Casimir vacuum's spectral density is that it reveals much finer details about the quantum ground state than typically measured Casimir forces do. By exploring the freedom of choosing the locations of two photodiodes inside the cavity, as well as the polarizations, phases, and frequencies of the coherent local oscillator (LO), one can obtain a detailed characterization of the one- and two-point correlation functions of any state S of the quantum electric field. Therefore, an application of this type of BHD measurement amounts to a tomography of the ground state of the Casimir cavity vacuum.

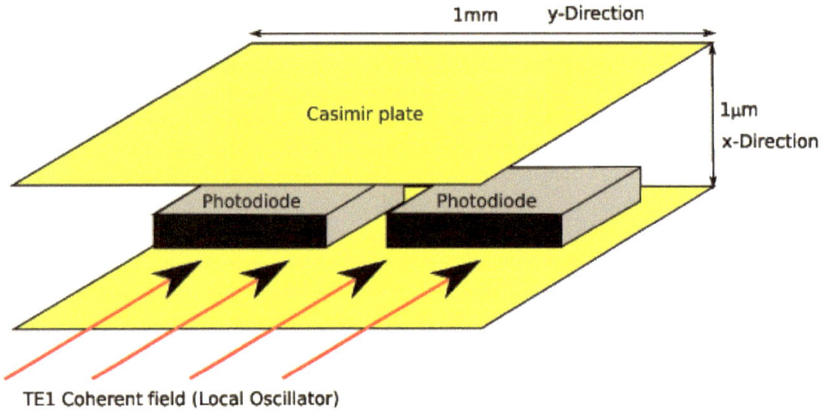

FIG. 35

Schematic diagram of Marecki's conceptual Casimir cavity quantum optical tomography experiment with submicron-scale BHD photodiodes inside the static Casimir vacuum region. The cavity/waveguide plate dimensions are 1 mm × 1 mm (or larger) and 1 μm plate separation distance. The BHD LO field was chosen to be the cavity's coherent transverse electric 01 field mode (indicated as TE1 by the *parallel red lines with black arrowheads*). The TE1 mode was selected because a standard scanning laser LO will not work in this configuration.

Courtesy P. Marecki.

For the experimental detection of the Casimir vacuum spectral density, the Casimir cavity plates are separated by 1 μm while the photodiodes inside the cavity must be smaller than the wavelength of the vacuum electromagnetic radiation that they are to be sensitive enough to measure (see Fig. 35). Marecki recommended that photodiodes with 0.1 μm × 100 μm active surface dimensions and good quantum efficiency should be used. However, there's the issue that photodiodes of this size will modify the distribution of the cavity/waveguide's transverse electric 01 (TE01) field modes (see Fig. 36). The longer the TE mode wavelength, the greater the modification (TE waves easily reflected). What happens behind the photodiodes is irrelevant in this configuration and the reflected TE mode waves are either not there or the situation is similar to what happens in BHD measurements of squeezed states of light. The presence of photodiodes inside the cavity breaks its translation-invariance; thus, the computation of a modified spectrum of vacuum fluctuations is more difficult to do (there might be other ideas for how to probe Casimir cavities, e.g., an experiment with nondestructive measurements on single photons and atoms, but these might be more difficult to perform). Given these considerations, it becomes apparent that the proposed photodiode dimensions are too big. A solution to this problem would be to use superconducting nanowires that can be designed to operate as a single photon photodetector and insert two of them into a Casimir cavity. Superconducting nanowire photodetectors have already been fabricated and experimentally demonstrated with close to unity quantum efficiencies and single photon absorption efficiencies >75% across the visual and IR band (see Ref. [128] and the references cited therein for more technical information). However, one drawback of these devices is that they are fabricated on substrates and photonic circuit chips, which is not useful for Marecki's BHD apparatus. The quantum optics industry continues working to develop and fabricate structurally robust free

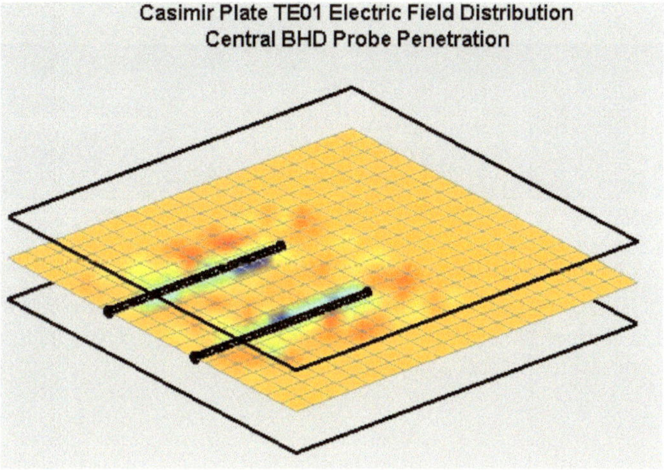

FIG. 36

Computational finite element analysis simulation showing the modified TE01 field mode distribution *(gridded gold-colored plane)* inside a Casimir cavity/waveguide (both cavity plates are depicted as transparent with *black square borders* for visual clarity) due to the presence of two BHD photodiodes with dimensions of 0.1 μm × 100 μm, which are centrally placed inside the cavity/waveguide. The modified TE01 field mode distribution is indicated by the *mottled coloring* surrounding the photodiodes.

Courtesy P. Marecki.

superconducting nanowire devices that can be connected to photonic circuit chips and substrates. Marecki (private communication, 2011) performed a computational finite element analysis simulation for a pair of freestanding superconducting nanowire photodetectors centrally placed inside the Casimir cavity and found that they did not modify the cavity's TE01 field mode distribution.

As shown in Fig. 35, a coherent state in the TE1 mode of the Casimir cavity with a very small wavenumber in the y-direction provides an appropriate LO. A BHD with such a LO and the photodiodes, located as shown in Fig. 37, would be sensitive only to the y-component of the quantum electric field. Fig. 38 shows a schematic of the BHD apparatus with a LO. In the figure, the linearly polarized signal field S (if present) is optically mixed with a coherent state (LO), which is polarized orthogonally to S, on the polarizing beam splitter (PBS1). The half-wave plate (HWP) reflects the planes of polarization with respect to its optical axis, thereby inducing a $\pi/4$ shift of the plane of polarization of the signal field S. The subsequent PBS2 separates the two orthogonally polarized signals, which are detected at the photodiodes PDx and PDy. The electronic charge (from the photocurrents produced by the PDs) collected at a point between the photodiodes in the electronic circuit provides a measure of the quantum expectation value of the observable corresponding to this electronic charge (i.e., the BHD current output, which is the difference of excitation probabilities of the two photodiodes) and its higher moments. Contrary to standard laboratory BHD technology, the correlations (quantum vacuum expectation value of the quantum electric field's two-point correlation function) between the two photodiodes are not small. Note that the setup is arranged in such a way that if S happened to be a monochromatic coherent state, then it would be phase-matched to the LO at the point x but shifted in phase by π at the point y.

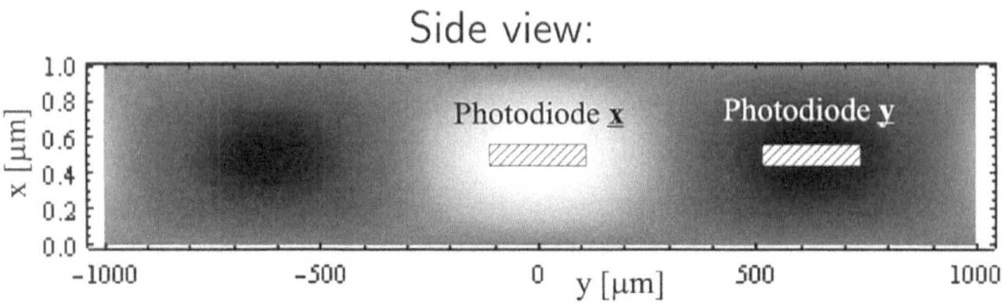

FIG. 37

Experimental setup of BHD photodiodes and LO field. This setup is drawn on the plot of the *y*-component of the electric field of the TE1 mode of the Casimir cavity/waveguide. The mode, serving as the LO, propagates in the *z*-direction perpendicular to the plot.

Courtesy P. Marecki.

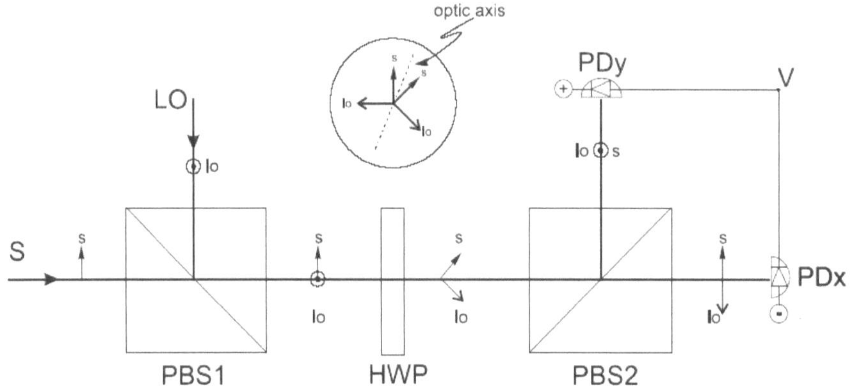

FIG. 38

Detailed breadboard schematic of Marecki's proposed Casimir cavity (not shown) quantum optical tomography BHD apparatus. S is the state of the quantum electric field going into the system for measurement, PBS is the polarizing beam splitter, HWP is the half-wave plate, LO is the coherent local oscillator field, and PD is the photodiode. The orientation of the S, LO (indicated as lo), and optically mixed S-lo fields are shown by the *small arrows and dot-in-circles* (indicating fields pointing out of the page).

Courtesy P. Marecki.

Fig. 39 displays Marecki's computational model plot of the predicted Casimir spectral density as a function of the distance from the cavity/waveguide plates x and the frequency ω. For a comparison with quantum optics literature, he plotted the normalized difference between the vacuum and ground state spectral density in the figure (see Refs. [126,127] for more detail). Note in the figure that, for $\omega < \pi c/a$, the Casimir spectral density vanishes while discontinuities in it appear at $\omega = n\pi c/a$. The predicted spectral density pattern is static, in other words, it is independent of the LO phase and in some regions corresponds to the suppression of vacuum fluctuations by at least 3 dB. Fig. 40 displays the corresponding computational model plot by Marecki of the predicted suppressed vacuum fluctuations in the

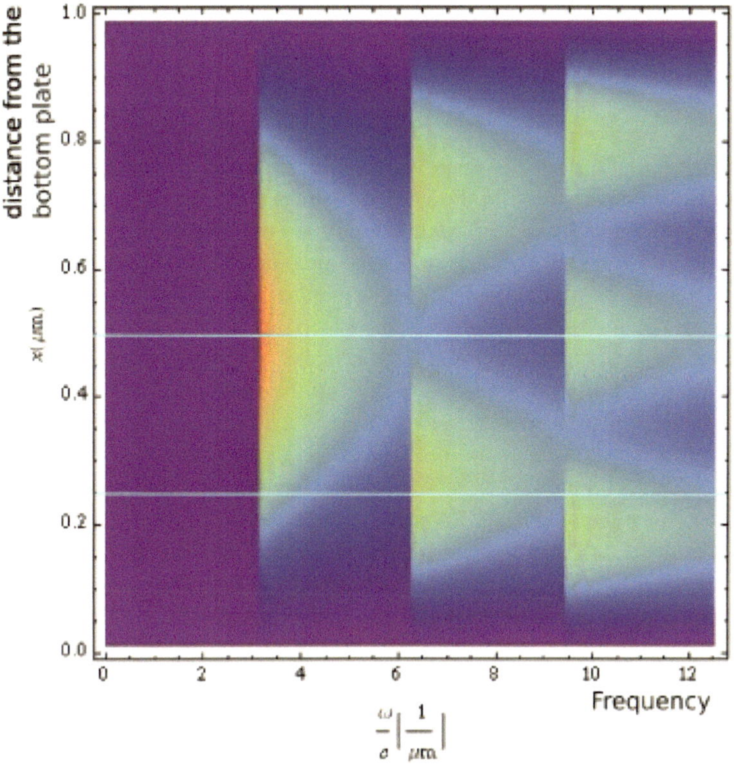

FIG. 39

Predicted Casimir spectral density. The spectral density is related to the expected output of a BHD with the LO polarized along the *y*-direction (parallel to the plates) for the ground state in the Casimir cavity. This is plotted as a function of the position $x \in [0, a]$ between the plates (separation $a = 1\,\mu\text{m}$ is assumed) and the frequency $\omega \in [0, 4\pi c/a]$. Negative values (subvacuum negative energy density) are shown in *deep purple*.

Courtesy P. Marecki.

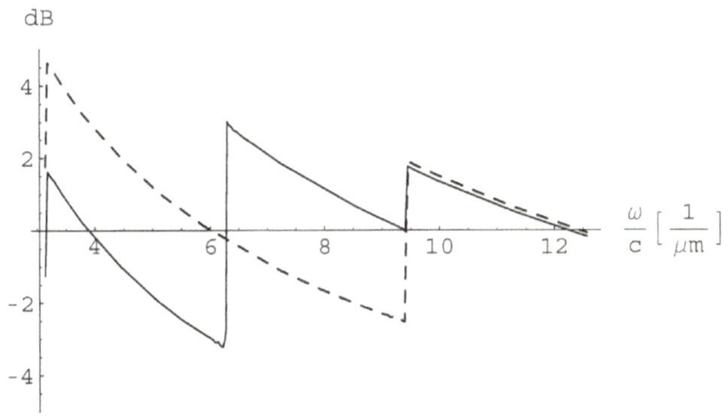

FIG. 40

Predicted suppression of vacuum fluctuations in dB. Vacuum fluctuations in the ground state (for field operators restricted to the frequency ω) relative to vacuum fluctuations (in the absence of the Casimir cavity) for a BHD at $x = 0.25\,\mu\text{m}$ *(solid line)* and $x = 0.5\,\mu\text{m}$ *(dashed line)* within the cavity. The frequency range is $\omega \in [0, 4\pi c/a]$.

Courtesy P. Marecki.

ground state of the Casimir cavity relative to undisturbed vacuum fluctuations (in absence of the Casimir cavity) in dB.

Calibration of the BHD is done by exposing it to an undisturbed vacuum using a LO of the same power as for the Casimir cavity configuration and then performing this same measurement inside the cavity. The pattern of vacuum fluctuation-reduction inside the cavity should be insensitive to the phase of the LO. The basic falsification criterion is that the measured vacuum state inside the cavity might be different from the cavity's ground (vacuum) state (e.g., it might be a thermal state).

Developing freestanding superconducting nanowire photodetectors is the key technical challenge to instrumenting Marecki's Casimir cavity BHD experiment. A related challenge is that the entire cavity will have to be immersed in a cryogenic fluid to keep the superconducting nanowire photodetectors cold enough to function at high photon absorption and quantum efficiencies. In addition to the cryogenic fluid management system, ancillary heat removal equipment might also be required. It is not clear at this time how a Casimir cavity will behave when immersed in a cryogenic fluid, so a study to address this issue will need to be done. One possibility of avoiding these technical complications is to go forward with Marecki's original experimental BHD design that uses contemporary microscale photodiode technology but apply it instead to the measurement of the negative vacuum energy density pulses produced by Youn et al.'s [118] simple free-space squeezed light apparatus (see Fig. 31 in Section 5.5.1).

6 The energetics of FTL propulsion

6.1 Traversable wormhole energetics

After Eq. (7) is cranked through Eq. (6), the nonzero components of $T_{\mu\nu}$ are solved, which serve as the energetics equations that provide the magnitude and spatial distribution of the exotic matter (negative vacuum energy) required to construct a spherically symmetric traversable wormhole. The components of $T_{\mu\nu}$ are complicated algebraic expressions, and the taxonomy of traversable wormholes is rather large, as shown in Fig. 10. Also, there is no generic one-size-fits-all expression for the $T_{\mu\nu}$ components that could be applied to any type of traversable wormhole that one wishes to study. For the case of the spherically symmetric Morris-Thorne traversable wormhole, the energy density $\rho_{\text{Ewh}}=T_{00}$ is given by [23,25,129]:

$$\rho_{\text{Ewh}}=\frac{b'c^4}{8\pi Gr^2}, \tag{19}$$

where $b'=\partial b(r)/\partial r$, which is a dimensionless quantity. The algebraic expressions for the other components $\tau_{\text{wh}}=-T_{\text{rr}}$ (radial tension minus radial pressure) and $p_{\text{wh}}=T_{\theta\theta}=T_{\varphi\varphi}$ (transverse pressure) are much worse. All other components of $T_{\mu\nu}$ vanish by the spherical symmetry property of Eq. (7). It is not obvious that ρ_{Ewh}, τ_{wh}, and p_{wh} possess enough information to tell one how to construct a Morris-Thorne traversable wormhole. A detailed analysis of these components shows that $(\rho_{\text{Ewh}}-\tau_{\text{wh}})<0$ near the throat and $(\rho_{\text{Ewh}}-\tau_{\text{wh}})|_{\text{r0}}\leq 0$ at the throat (where $r=r_0$), both of which violate all of the energy conditions [23,25,129]. Interestingly, the expression for τ_{wh} at the throat is very simple [23,25]:

$$\tau_{\text{wh}}\big|_{r_0}=\frac{c^4}{8\pi Gr_0^2}=\left(4.83\times 10^{42}\right)\times\left(\frac{1\,\text{m}}{r_0}\right)^2=\left(4.83\times 10^{10}\right)\times\left(\frac{1\,\text{LY}}{r_0}\right)^2. \tag{20}$$

Note that the MKS units for all components of $T_{\mu\nu}$ are J/m^3 (or N/m^2). If we consider a sample of ordinary bulk matter (OBM) and recall from solid state physics that typical interatomic binding energies are in the order of 1 eV (1.602×10^{-19} J) and typical interatomic spacings are in the order of 1 Å (10^{-10} m), then the maximum conceivable tension supportable by OBM is in the order of:

$$\tau_{\text{OBM}} \approx \frac{1\,\text{eV}}{(1\,\text{Å})^3} \approx 1.60 \times 10^{11}\,\frac{\text{J}}{\text{m}^3}. \tag{21}$$

$\tau_{\text{wh}}|_{r_0}$ will always be larger than τ_{OBM} unless the size of the traversable wormhole's throat is in the order of 1 LY. Eqs. (20) and (21) challenge our notions of what future space engineering technologies might become capable of. One should consult Refs. [23,25] to examine the $T_{\mu\nu}$ components for the other types of traversable wormholes.

Simple scaling laws can be derived for the negative energy density and total negative energy required to construct a spherically symmetric traversable wormhole throat as a function of three adjustable physical parameters. Here we cite the results from our own theoretical study of this issue. The scaling law for negative energy density is:

$$\rho_{\text{Ewh}} \sim -\frac{\kappa_g}{2\pi r_{\text{throat}}}, \tag{22}$$

where the scalar κ_g is the surface gravity of the throat. κ_g is calculated by taking the product of a timelike Killing vector (a symmetry vector field on a Riemannian or pseudo-Riemannian manifold that preserves the spacetime metric; each Killing vector corresponds to a physically observable quantity) with the covariant derivate (the derivative operator in curved spacetime geometry) of the timelike Killing vector and then equating the result of this product to the product of κ_g with the timelike Killing vector. Calculating this physical parameter is not needed for this simple scaling analysis. Eq. (22) shows that as $r_{\text{throat}} \rightarrow$ *large* for fixed κ_g and $\kappa_g \rightarrow$ *small* for fixed r_{throat}, then $\rho_{\text{Ewh}} \rightarrow$ *small* in magnitude and vice versa. The scaling law for the total negative energy contained within the wormhole throat is:

$$E_{\text{wh}} \sim -2\ell\kappa_g r_{\text{throat}}, \tag{23}$$

where ℓ is the thickness of the thin shell of negative energy density.

If any two of the parameters in Eq. (23) are kept fixed while the third one grows small or large, then E_{wh} will correspondingly grow small or large in magnitude. If any two parameters grow small or large simultaneously while the third parameter is kept fixed, then E_{wh} will correspondingly grow small or large in magnitude. If any two parameters grow small or large simultaneously while the third parameter varies in the opposite sense from the other two, then one expects that E_{wh} will correspondingly grow small or large in magnitude slowly. If all three parameters grow small or large simultaneously, then E_{wh} will correspondingly grow small or large in magnitude rapidly.

We saw in Sections 5.1 and 5.3 that traversable wormhole throats violate the energy conditions. How big a violation is required is determined by calculating the total amount of negative energy that will be needed to construct and stabilize a wormhole throat. This implies that a spatial distribution of negative energy needs to be concentrated at the throat. There are only two cases to consider here per Eq. (14): a traversable wormhole constructed from an arbitrarily small amount of negative energy density, which is prone to becoming nontraversable per the discussion below Eq. (14), and a traversable wormhole with a short throat that minimizes the travel time and distance for space travelers going through the throat to reach a distant exoplanet. In the latter case, a simple generic approximation for a short-throat traversable wormhole can be found by using the general relativistic thin-shell formalism, which gives the equivalent mass required to construct it [25,26]:

$$M_{\mathrm{wh}} = -\frac{r_{\mathrm{throat}}c^2}{G} = -(0.71M_{\mathrm{J}})\frac{r_{\mathrm{throat}}}{1\,\mathrm{m}}, \tag{24}$$

where M_{wh} is the (equivalent) mass in units of M_{J}, M_{J} is the mass of the planet Jupiter (1.90×10^{27} kg), and r_{throat} is a suitable measure of the linear dimension (radius, diameter, or width in meters) of the throat. One can also obtain the required energy E_{wh} by multiplying both sides of Eq. (24) by c^2.

Eq. (24) shows that a mass of $-0.71M_{\mathrm{J}}$ will be required to construct a short-throat wormhole of 1 m in size. As the wormhole throat's size increases, the mass required grows negative-large. Table 3 shows a tabulation of the required negative (equivalent) mass as a function of short-throat wormhole sizes. One should note that M_{wh} is not the total mass of the wormhole system as seen by remote static observers residing in asymptotically flat spacetime. The nonlinearity of the general relativistic field equations dictate that the total mass is zero (actually, the total net mass is positive, negative, or zero in the Newtonian approximation, depending on the details of the negative energy configuration constituting the wormhole system).

We know that we need exotic matter or negative energy density to construct and stabilize a traversable wormhole throat. We then ask what kind of wormhole can be constructed with less effort. To answer this question, Ida and Hayward [130] used the Gauss-Bonnet Theorem to derive the following inequality (in geometrized units) for a generic nondegenerate traversable wormhole spacetime geometry that imposes no symmetry requirement:

$$\oint\!\!\oint \left(G(\boldsymbol{u}, \boldsymbol{u}) + \frac{1}{2}\|K\|^2 \right) dA \leq 2\pi\chi_{\mathrm{E}}, \tag{25}$$

where the surface integral is taken over a traversable wormhole's closed compact oriented minimal surface (without boundary), dA is the integration measure (element of spatial surface area), $G(\boldsymbol{u}, \boldsymbol{u})$ is the Einstein tensor (written in the notation of Ref. [17]) of the traversable wormhole's spacetime as a function of $\boldsymbol{u}$ (the unit vector tangent to a timelike Killing vector), $\|K\|$ is the trace of the second fundamental form (the wormhole's mean curvature, a.k.a. extrinsic curvature, of its throat), and χ_{E} is the global topological invariant Euler characteristic of the wormhole spacetime. χ_{E} is defined in terms of the genus g, representing the number of handles or throats that a traversable wormhole could have:

Table 3 Equivalent negative mass M_{wh} required for a short-throat traversable wormhole of size r_{throat}.

r_{throat} (m)	M_{wh} [a]
1000	$-709.9\ M_{\mathrm{J}}$
100	$-71\ M_{\mathrm{J}}$
10	$-7.1\ M_{\mathrm{J}}$
1	$-0.71\ M_{\mathrm{J}}$
0.1	$-22.6\ M_{\oplus}$
0.01	$-2.3\ M_{\oplus}$

[a] *$M_J = 1.90 \times 10^{27}$ kg, $M_{\oplus}$ (mass of Earth) $= 5.98 \times 10^{24}$ kg.*

$\chi_{\rm E} = 2(1 - g)$. (Degenerate wormholes are wormholes with multiple throats that merge together, which Ida and Hayward did not consider in their energetics analysis.)

Ida and Hayward [130] mathematically reduced Eq. (25) to the following two simple formulas for the case of static traversable wormholes:

$$U \leq \frac{\chi_{\rm E}}{4} \text{ or } U \leq \frac{(1-g)}{2}, \tag{26}$$

where U quantifies the energy density (in geometrized units) that threads through and stabilizes the wormhole throat. For the case of dynamic traversable wormholes, the term $-r\kappa_g$ is added to the right-hand side of both expressions in Eq. (26), where $r = \sqrt{A/4\pi}$ is the proper radial distance defined by the wormhole throat's surface area A [130].

Eq. (26) will help us to decide if we want to construct a traversable wormhole having one throat, two throats, or more than two throats, and determine what energy cost will be incurred for each case. The following three cases are the results from Eq. (26) for nondegenerate static traversable wormholes having:

(a) 1-handle/throat (i.e., flat torus or spherical wormhole topology) with $g = 1$, thus $\chi_{\rm E} = 0$ and $U \leq 0$.
(b) 2-handles/throats with $g = 2$, thus $\chi_{\rm E} = -2$ and $U \leq -1/2$.
(c) 3-handles/throats with $g = 3$, thus $\chi_{\rm E} = -4$ and $U \leq -1$; and so on.

By inspection of cases (a)–(c), it is clear that as the number of wormhole handles/throats increases, the amount of negative energy density required to construct the wormhole will grow larger in magnitude. This is an undesirable demand on our putative negative energy generator. Case (a) defines the most desirable engineering solution we can hope for: a 1-handle/throat traversable wormhole that at a minimum requires zero or (arbitrarily) little negative energy density to construct. However, we already discussed earlier that minimal negative energy density wormholes are likely unstable enough to collapse, so the actual negative energy density required to construct a 1-handle/throat wormhole will be larger in magnitude in accordance with Eq. (19) or (24).

With regards to interstellar "stargates" that are often depicted as traversable wormhole-like FTL transportation modes in various science fiction films and television series, it turns out that there is a traversable wormhole solution from Eq. (6) that qualifies as a stargate, which is otherwise called the flat-face traversable wormhole. Its solution is derived from the thin-shell (a.k.a. junction condition or surface layer) formalism applied to Eq. (6) [17,131]. The procedure to derive what we call the "stargate solution" is to take two copies of flat Minkowski spacetime and remove from each copy identical regions of the form $\Omega \times \Re$, where Ω is a compact 3D spacelike hypersurface, $\Re$ is a timelike straight line (time axis), and these two incomplete spacetimes along the timelike boundaries $\partial\Omega \times \Re$ are identified. The resulting spacetime is geodesically complete and possesses two asymptotically flat regions connected by a traversable wormhole throat. The throat of the wormhole is just the junction $\partial\Omega$ (a compact 2D spacelike hypersurface) at which the two original Minkowski spacetimes are identified (see Figs. 41 and 42).

The resulting spacetime is Riemann-flat everywhere except possibly at the throat/stargate. Also, the $T_{\mu\nu}$ in this spacetime is concentrated at the throat/stargate with a δ-function singularity there. This is a consequence of the fact that the spacetime metric at the throat/stargate is continuous, but not differentiable, while the connection is discontinuous, thus causing the Riemann curvature to possess a δ-function singularity (causing undesirable gravitational tidal forces) there. The magnitude of this

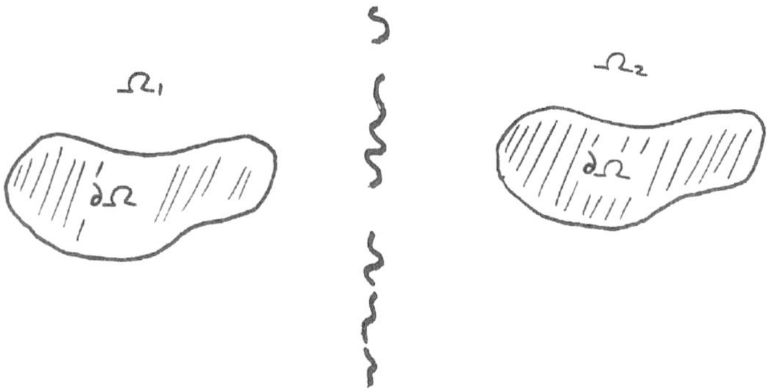

FIG. 41

Illustration of a simultaneous view of two remote compact regions, Ω_1 and Ω_2, of Minkowski spacetime used to create the traversable wormhole throat $\partial\Omega$. The time axis is suppressed in this illustration.

Reproduced with permission from M. Visser, Traversable wormholes: some simple examples, Phys. Rev. D 39 (10) (1989) 3182–3184.

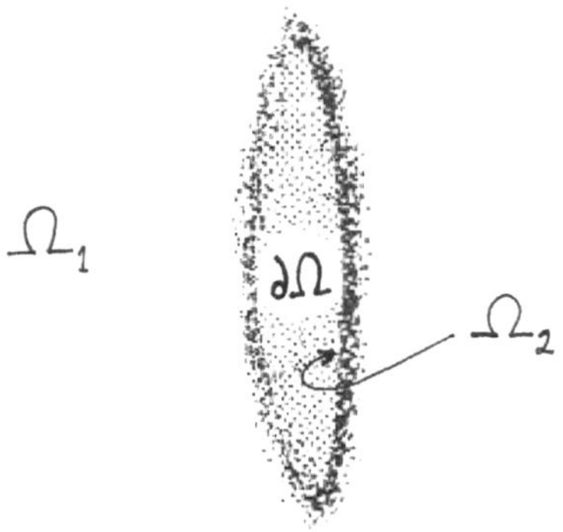

FIG. 42

This is the same illustration as Fig. 41 but as viewed by an observer sitting in region Ω_1 who looks through the wormhole throat $\partial\Omega$ and sees remote region Ω_2 *(dotted area inside the circle)* on the other side. $\partial\Omega$ conceptualizes the flat-face traversable wormhole throat, or stargate.

δ-function singularity can be calculated in terms of the second fundamental form on both sides of the throat/stargate, which we presume to be generated by a localized thin shell of negative energy. The second fundamental form represents the extrinsic curvature of the $\partial\Omega$ hypersurface (i.e., the throat/stargate), telling how it is curved with respect to the enveloping 4D spacetime.

It is a standard result of the thin-shell/junction condition formalism that Eq. (6) can be recast in terms of the surface energy-momentum-stress tensor S^i_j of a thin shell of negative energy localized in $\partial\Omega$ (note: we are exploiting the symmetry of the wormhole with respect to interchange of the two flat spacetime regions Ω_1 and Ω_2):

$$S^i_j = -\frac{c^4}{4\pi G}\left(K^i_j - \delta^i_j K^k_k\right), \tag{27}$$

where $i, j = 0, 1, 2$ are the tensor indices corresponding to the time dimension and the two spatial dimensions of the thin shell's 2D spatial hypersurface embedded in the 3D (time-space-space) space, and $\delta^i_j = diag(1, 1, 1)$ is the 3D unit matrix. K^i_j is the second fundamental form and K^k_k is its trace (shown as $\|K\|$ in Eq. 25). The thin shell's S^i_j may be interpreted in terms of the surface energy density σ and principal surface tensions ϑ_1 and ϑ_2 [16,17,131,132]:

$$S^i_j = \begin{pmatrix} -\sigma & 0 & 0 \\ 0 & -\vartheta_1 & 0 \\ 0 & 0 & -\vartheta_2 \end{pmatrix}. \tag{28}$$

K^i_j is defined as [16,17,131,132]:

$$K^i_j = \begin{pmatrix} \kappa_0 & 0 & 0 \\ 0 & \kappa_1 & 0 \\ 0 & 0 & \kappa_2 \end{pmatrix} = \begin{pmatrix} 0 & 0 & 0 \\ 0 & {}^1/_{\rho_1}\, 000\, {}^1/_{\rho_2} \end{pmatrix}, \tag{29}$$

where the quantities κ_0, κ_1, and κ_2 measure the extrinsic curvature of the thin shell of negative energy.

Typically, K^i_j has a superscripted $\pm$ index associated with an overall algebraic $\pm$ sign on the two matrices in Eq. (29), which comes from the fact that a unit normal vector points outward from one side of the thin shell and points inward on the other side. Here, we dropped the $\pm$ sign because of symmetry of the unit normal vectors on both sides of the thin shell. Because the throat/stargate is a spacelike hypersurface, we can exclude timelike hypersurfaces and their components in the calculations. Therefore, we set $\kappa_0 = 0$ in Eq. (29) because it is the timelike extrinsic curvature for the timelike hypersurface of the thin shell. As seen in Eq. (29), κ_1 and κ_2 are related to the two principal radii of curvature ρ_1 and ρ_2 (the eigenvalues of K^i_j) of the 2D spacelike hypersurface $\partial\Omega$, the throat/stargate. Fig. 43 shows an illustration of a generic thin shell of matter (mass-energy) with ρ_1 and ρ_2 indicated. It should be noted

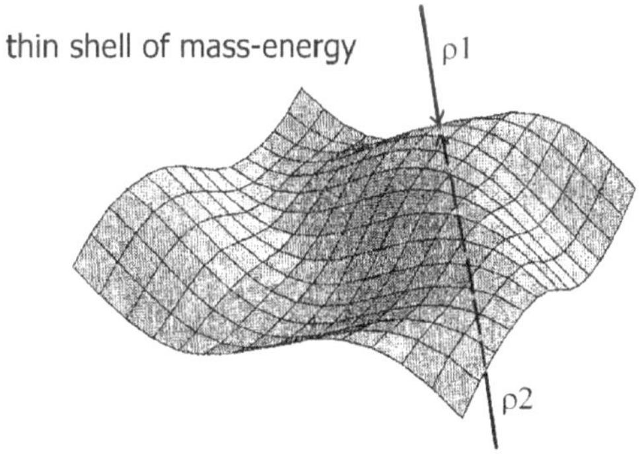

FIG. 43

Illustration of a generic thin shell of localized matter (mass-energy), which constructs the 2D spacelike hypersurface $\partial\Omega$ (a.k.a. the stargate). The two principal radii of curvature ρ_1 and ρ_2 for this hypersurface are indicated.

that a convex surface has positive radii of curvature while a concave surface has negative radii of curvature. The trace of Eq. (29) is:

$$K_k^k = \frac{1}{\rho_1} + \frac{1}{\rho_2}. \tag{30}$$

By inserting Eqs. (28)–(30) into Eq. (27), performing the matrix algebra, and then multiplying both sides of the result by -1, we arrive at the following expression:

$$\begin{pmatrix} \sigma & 0 & 0 \\ 0 & \vartheta_1 & 0 \\ 0 & 0 & \vartheta_2 \end{pmatrix} = -\frac{c^4}{4\pi G}\begin{pmatrix} \frac{1}{\rho_1} + \frac{1}{\rho_2} & 0 & 0 \\ 0 & 1/\rho_2 & 0 \\ 0 & 0 & 1/\rho_2 \end{pmatrix}, \tag{31}$$

which by inspection can be broken down into the following three separate equations:

$$\sigma = -\frac{c^4}{4\pi G}\left(\frac{1}{\rho_1} + \frac{1}{\rho_2}\right), \tag{32a}$$

$$\vartheta_1 = -\frac{c^4}{4\pi G}\frac{1}{\rho_2}, \tag{32b}$$

$$\vartheta_2 = -\frac{c^4}{4\pi G}\frac{1}{\rho_1}. \tag{32c}$$

Eqs. (32a)–(32c) are the components of the surface energy-momentum-stress tensor S_j^i for the arbitrary thin shell of matter localized in $\partial\Omega$, as illustrated in Fig. 43. The negative signs in Eqs. (32a)–(32c) show that (for $\partial\Omega$ convex) we are dealing with negative surface energy density and negative principal surface tensions. These quantities are needed to keep the stargate open and stable against collapse. It should be noted that, for the analysis in this section, we assumed an ultrastatic stargate (i.e., $g_{00} \equiv 1$ implies that the redshift function $\phi(r) = 0$) and dispensed with the assumption of spherical symmetry.

We can now build a stargate such that space travelers stepping through it encounter no negative surface energy density and no negative principal surface tensions. This remarkable feature requires our stargate to be perfectly flat. To make the stargate flat requires that we choose the throat $\partial\Omega$ to have at least one flat face. For a flat hypersurface, the principal radii of curvature must be $\rho_1 = \rho_2 = \infty$ according to standard geometry so that Fig. 43 looks like Fig. 42. Substituting these values into Eqs. (32a)–(32c) gives the flat stargate solution:

$$\sigma = \vartheta_1 = \vartheta_2 = 0, \tag{33}$$

which is a remarkably counterintuitive result for the energetics of this type of traversable wormhole.

A further consequence of Eq. (33) is that $K_j^i = 0$, making the Riemann curvature and energy-momentum-stress tensors ($R_\beta^\alpha \sim K_\beta^\alpha$ and $T_\beta^\alpha \sim K_\beta^\alpha$) zero at the throat such that the associated δ-function singularities disappear there. This means that space travelers going through the flat stargate will feel no tidal gravitational forces and see no exotic matter (negative surface energy density and tensions) while going through it. A traveler stepping through it will be immediately shunted into the other remote spacetime region or into another universe (note that Eq. 6 does not fix the topology of spacetime, so it is possible that traversable wormholes are interuniverse as well as intra-universe hyperspace tunnels). One could imagine constructing a flat stargate by generating a thin shell or surface layer of negative energy/tensions much like stretching a thin film of soap across a loop of wire. How to engineer the construction of this stargate remains unknown.

6.2 Warp drive energetics

The nonzero components of $T_{\mu\nu}$ are solved by substituting the metric tensor components (found by inspection) from Eq. (8) into the left-hand side of Eq. (6). The nonzero $T_{\mu\nu}$ components serve as the energetics equations that provide the magnitude and spatial distribution of the exotic matter (negative vacuum energy) required to construct Alcubierre's warp drive bubble. Once again, these tensor components are complicated algebraic expressions, and the taxonomy of warp drives is large (and still growing), as shown in Fig. 10. Just like the situation for traversable wormholes, there are no generic one-size-fits-all expressions for the $T_{\mu\nu}$ components that could be applied to any type of warp drive that one wishes to study. For Alcubierre's warp drive bubble, the energy density $\rho_{\mathrm{Ewarp}} = T_{00}$ in an orthonormal basis (a set of coordinate system basis vectors is orthonormal if all its vectors have unit norm/length and are pairwise orthogonal) is given by [29,129]:

$$\rho_{\mathrm{Ewarp}} = -\frac{c^4}{32\pi G}\frac{v_{sh}^2\left(y^2+z^2\right)}{\left[\left(x-x_{sh}\right)^2+y^2+z^2\right]}\left[\frac{df\left(r_{sh}(t)\right)}{dr}\right]^2. \tag{34}$$

The other nonzero diagonal and off-diagonal components of $T_{\mu\nu}$ are algebraically worse. It is not obvious that Eq. (34) and the other nonzero components of $T_{\mu\nu}$ possess enough information to tell one how to construct a warp drive bubble. Warp drive spacetimes violate the WEC and NEC (also the DEC and SEC). The amount of negative energy density required to create a warp bubble gives a measure of the WEC/NEC violation. Because the nonzero components of $T_{\mu\nu}$ for the Alcubierre and Natário warp drives are too complex to use for comparative analysis, we instead use a surprisingly simple generic formula that expresses the total negative energy E_{warp} required to construct a warp bubble around an interstellar warp spacecraft (with nonzero rest-mass) [21]:

$$E_{\mathrm{warp}} = -\frac{v_{sh}^2 c^4 R^2 \sigma}{G} = -\left(1.21\times 10^{44}\right)v_{sh}^2 R^2 \sigma, \tag{35}$$

where v_{sh} has values $(0, \infty)$.

Eq. (35) characterizes how much negative energy one needs to localize within the warp bubble wall in the Alcubierre and Natário warp drives. One can also obtain the equivalent total negative mass M_{warp} by dividing both sides of Eq. (35) by c^2. Table 4 presents a tabulation of the total negative energy required to construct an Alcubierre or Natário warp bubble as a function of selected values of v_{sh}. The consequence of Eq. (35) is that if one wants to travel at FTL speeds, then the warp bubble's total energy requirement for construction will be an enormous negative number. This remains true even if one engineers an arbitrarily low sublightspeed warp bubble.

One further complication arises from the serious flaw in the Alcubierre and other warp drive concepts that was discovered by Lobo and Visser [21]. They pointed out that in all versions of the warp drive, the spacecraft at the center of the warp bubble is massless and moves on a geodesic. The spacecraft is always treated as a test particle in all warp drive spacetime models. They corrected this flaw by constructing a more realistic model of the warp drive spacetime geometry by applying linearized gravity to the weak field warp drive case and testing the energy conditions to first and second order of v_{sh}.

Table 4 Total negative energy required to construct the Alcubierre or Natário warp bubble vs v_{sh} for $R = 50\,\text{m}$ and $\sigma = 10^3\,\text{m}^{-1}$.

v_{sh} (dimensionless)	E_{warp} (J)
10^{-5} (=3 km/s)	-3.03×10^{40}
10^{-4} (=30 km/s)	-3.03×10^{42}
0.01 (=3000 km/s)	-3.03×10^{46}
0.5 (=150,000 km/s)	-7.59×10^{49}
1 (=c)	-3.03×10^{50}
2 ($2c$ = 600,000 km/s)	-1.21×10^{51}
10 ($10c = 3.0 \times 10^6$ km/s)	-3.03×10^{52}
100 ($100c = 3.0 \times 10^7$ km/s)	-3.03×10^{54}

Note for comparison that the Sun's rest-energy is 1.79×10^{47} J.

The fundamental basis of their analysis is that it specifically includes a nonzero spacecraft mass that gravitationally interacts with the warp bubble, which consequentially established a severely low upper bound on the warp speed, as shown by Eq. (15). Results of their analysis verified that all warp drive spacetimes violate all the energy conditions and will continue to do so for arbitrarily low ($\ll c$) warp bubble speed. They also found that the energy condition violations in this class of spacetimes is generic to the form of the geometry under consideration and is not a side effect of its superluminal property.

Furthermore, Schuster et al. [133] and Santiago et al. [134] provided a full proof of the NEC violations in all generic Natário warp drive spacetimes using implicit arguments in the form of a proof by contradiction in which their analyses firmly established that published claims of nonviolation are "irretrievably erroneous." Their detailed analyses showed that superluminal warp drive solutions do not have a positive energy source for all observers, and their results also apply to the warp drive solutions of Lentz [54] and Fell and Heisenberg [56]. There are erroneous claims that assert avoidance or "softening" of the energy condition violations found in other recently published warp drive solutions [53,135–139]. Visser [26,75], Barceló and Visser [63], Visser and Barceló [67], and Lobo [129] emphasize that these violations were first established over 25 years ago to varying degrees of generality.

Based on the preceding discussion, plus Eqs. (15) and (35) and Table 4, it appears that for all conceivable contemporary and near-future laboratory experiments in which negative vacuum energy density is created in very feeble amounts, the warp bubble speed will be absurdly low. The detailed analyses done by Loup [35,36,40] and Perniciano [39] don't appear to improve this situation very much. It therefore appears unlikely that warp drive concepts will prove to be technologically practicable unless new ways to generate astronomical amounts of negative vacuum energy density are discovered and Eq. (15) could be circumvented somehow. In what follows, we briefly discuss the results of the four main energy-condition-violating alternative warp drive concepts that were discussed in Section 3.2 to determine whether any of them are capable of drastically reducing their energetics requirement to a degree that might be achievable by future negative energy generator technology.

6.2.1 Krasnikov warp drive (KWD)

Everett and Roman [32] calculated $T_{\mu\nu}$ and used the QI to derive the estimated total negative equivalent mass (multiply both sides by c^2 to get total energy) required to construct Krasnikov's warp tube:

$$M_{\mathrm{KWD}} \approx T_{00} V \approx -\frac{\alpha R_{\max} D}{\epsilon} \approx -\alpha 10^{18} M_{galaxy}, \tag{36}$$

where geometrized units are assumed in the middle two expressions, V is the volume of the thin warp tube wall that contains negative energy density, $\alpha \ll 1$, $R_{\max}$ is the maximum radius of the warp tube, D is the length of the tube (measured from departure point to destination point), $\epsilon \approx \ell_P/\Theta^2 = 100\ell_P$ is the proper radius of curvature of the warp tube wall, ℓ_P is the Planck length (see Eq. 18), $\Theta \ll 1$, and $M_{galaxy} \approx 10^{12}$ solar masses.

Eq. (36) is not a reliable estimate because we showed in Section 5.2 that the QI formulation is erroneous. However, it is worth looking at the two examples that Everett and Roman worked out to illustrate the extremely large magnitude of equivalent negative mass required to construct the KWD tube for interstellar travel. For their first example, they chose $\alpha = 0.01$, $R_{\max} = \mathrm{D} = 1\,\mathrm{m}$, and estimated that $M_{\mathrm{KWD}} \approx -10^{16}\ M_{galaxy}$ just to travel a distance of 1 m across spacetime. For a travel distance of 4.37 LY ($D \approx 4 \times 10^{16}\,\mathrm{m}$) from Earth to Alpha Centauri, $M_{\mathrm{KWD}} \approx -10^{32}\ M_{galaxy}$. These numbers are horrible but very similar to what one would get if Alcubierre's warp drive was used for these examples. The negative energy required to construct the KWD cannot be drastically reduced, thus it is not practicable to engineer.

6.2.2 Van Den Broeck warp drive

The topological geometrical design of Van Den Broeck's warp drive is very complicated as described in Section 3.2. In his only example warp drive design, the outermost surface of the warp bubble (Region IV in Fig. 14) will have an area corresponding to $R = 3 \times 10^{-15}\,\mathrm{m}$ while the inner diameter of its "pocket" is 200 m [33]. The other parameter he defines is the thickness of Region IV, $\Delta \leq 100 v_{sh} \ell_P$. Pfenning and Ford [38] used the QI to estimate the total equivalent negative mass (multiply both sides by c^2 to get total energy) contained within Region IV [33]:

$$M_{\mathrm{IV}} = -\frac{1}{12} v_{sh}^2 \left[\frac{(R + \Delta/2)^2}{\Delta} + \frac{\Delta}{12} \right] \simeq -\left(6.3 \times 10^{29}\right) v_{sh}, \tag{37}$$

where geometrized units are assumed in the middle expression and v_{sh} is the usual dimensionless warp speed. Van Den Broeck used the QI to calculate the total equivalent negative mass (multiply both sides by c^2 to get total energy) contained within Region II (see Fig. 14), and here we cite only the numerical result because there is no simple parametric expression for this quantity [33,38]:

$$M_{\mathrm{II}-} = -1.4 \times 10^{30}\,\mathrm{kg}, \tag{38}$$

where the derivation involved the definition: $\tilde{\Delta} = 10^{-15}$ m. Because $B(r_{sh})$ varies across Region II, there will be a corresponding positive equivalent mass $M_{\mathrm{II}+} = 4.9 \times 10^{30}\,\mathrm{kg}$ [33]. Note for comparison that the Sun's rest-mass is $1.99 \times 10^{30}\,\mathrm{kg}$. All three of these numerical estimates are not reliable because we showed in Section 5.2 that the QI formulation is erroneous. It remains to be seen whether the negative energy required to construct Van Den Broeck's warp drive can be drastically reduced. Also, Loup's [40] detailed analysis of this warp drive doesn't seem to improve upon this situation very much. However, see the discussion in the Hart et al. warp drive section that follows.

6.2.3 Natário warp drive

To no one's surprise, the total negative energy E_{NWD} required to construct Natário's warp drive is given by Eq. (35) [21]. This is simply because Natário's warp drive is a generalization of Alcubierre's warp drive. See Table 4 for numerical examples. There's no ability to drastically reduce the negative energy required to construct this warp drive. Loup's [35,36,40] extensive analyses were unable to improve this situation very much.

6.2.4 Hart et al. warp drive

Unfortunately, Hart et al. [37] did not do an energetics analysis for their modified Van Den Broeck warp drive. Instead, they cited Krasnikov's [93] reanalysis of Van Den Broeck's warp drive energetics, which completely avoided using the erroneous QI. Krasnikov demonstrated that Van Den Broeck's complicated warp drive with a macroscopic-sized pocket can be constructed by using an astonishingly minute -10^{-3} g (or -10^{-6} kg, multiply by c^2 to get the total energy) of equivalent mass if classical scalar fields are used as the source of exotic matter instead of quantum fields [67,99,140]. Notice the two types of classical scalar fields (massive or massless nonminimally coupled and massless conformally coupled) shown in Fig. 25 that violate all the energy conditions (see also, the Surprise 2 bullet in Section 5.1). This appears to represent a tremendous breakthrough for the total negative energy required to construct Van Den Broeck's warp drive.

It is not known how to make the two known types of classical scalar fields in the lab. However, the Higgs boson field is a cosmological-scale quantum vacuum scalar field that has been repeatedly produced in particle collisions at the Large Hadron Collider. It is endowed with a nonvanishing quantum vacuum expectation value of its quantum field operator, which theoretically defines its (experimentally verified) rest-mass-energy of 125 GeV throughout the universe. Thus it is a cosmological vacuum field condensate with superconducting-like properties. Spontaneous symmetry breaking of the electroweak force (the unified QED and weak nuclear forces) by this vacuum field condensate gives mass to three of the electroweak vector bosons (W^+, W^-, Z^0). The fourth electroweak vector boson is the massless photon that serves as the mathematical quantum basis state that defines the quantum field for each of the three massive vector bosons. This vacuum field condensate also couples to the three leptons and six quarks, which has the effect of giving each of them their unique mass. Theoretical and experimental studies should be performed to determine whether the Higgs field violates the energy conditions and, if that turns out to be the case, then figure out if it's technologically feasible to make it without the need for large particle colliders.

7 Construction of traversable wormholes and warp drives

GTR does not include engineering assembly instructions for traversable wormholes and warp drives because it only provides a recipe to design the essential spacetime geometry, the appropriate boundary conditions, and the material ingredients required to open and stabilize them against collapse. There's a whole taxonomy of traversable wormholes to choose from. What's missing from GTR is the engineering method of construction, which can only be speculated on in absence of any concrete hypotheses. How does one deploy, shape, and control a specialized spatial distribution of negative energy density/flux and/or negative pressure/stress to construct these FTL propulsion concepts? Does one need to pull a Planck scale wormhole from out of the putative quantum spacetime foam (a multiply connected spacetime comprised of fluctuating wormholes at the Planck scale) by using negative energy to inflate it up to macroscopic scale, or will wormhole engineers have to produce extremely large spacetime curvatures to "punch a big hole in spacetime"? What's more, does one stabilize either of these approaches using spatially distributed negative energy density/flux and/or negative pressure/stress? To construct a macroscopic traversable wormhole throat requires the topology of spacetime to change. Is this even possible?

It turns out that the answer is yes. Carlip [141] reported that a topological approach to modeling 2D quantum gravity theories, which predicted black holes and dark energy, showed that at least one formulation of the theory permits the topology of spacetime to change on quantum and macroscopic scales, thus traversable wormholes are allowed. It is expected that this result will be extensible to any (3+1)-dimensional quantum gravity theory. We already know that quantum superstring theory and its various incarnations allow for the topology of spacetime to change at the Planck scale as well as in the trans-Planckian regime, but it is not yet known whether these theories allow this to occur at macroscopic scale within our physical (3+1)-dimensional spacetime. Additionally, how does one "stretch" the traversable wormhole throat to form the hyperspace shortcut between the departure and destination locations? One idea is to construct one side of the wormhole throat and keep it permanently stationary somewhere near the Earth while an interstellar spacecraft transports (at $< c$) the other side of the throat to the destination star's exoplanetary system, where it will be kept stationary. Per our discussion in Section 1.2, this method of construction will necessarily entail relativistic spaceflight involving many light-years of travel, undesirable relativistic time dilation, and large amounts of provisions and propellant that the interstellar wormhole construction spacecraft will have to carry with it to get to the exoplanetary system at the destination star. How does one "stretch out" the wormhole throat in this way? GTR is silent on this question too.

For warp drives, GTR provides only a recipe to design the essential spacetime geometry, the appropriate boundary conditions, and the material ingredients required to construct a warp bubble. Alcubierre designed an idealized warp drive that's impractical to implement because of its design flaws. However, his theory work motivated a minor cottage industry within the GTR community to take a serious look at this topic and find solutions to the flaws in his warp drive model. This line of academic research led to the publication of numerous modified-Alcubierre and alternative warp drive models that are designed to mitigate or avoid the flaws in Alcubierre's warp drive, which has spawned a growing taxonomy of warp drive types. All of them require the spatial distribution of negative energy density/flux and/or negative pressure/stress. There are also the new, more recently proposed warp drive models that claim to mitigate or avoid using exotic matter. As we discussed in the previous section, these were found to be erroneous. How does one deploy, shape, and control a specialized spatial distribution of negative energy density/flux and/or negative pressure/stress to construct any of the types of energy-condition-violating warp drives? Just like the case for the taxonomy of traversable wormholes, GTR is silent on this question for the taxonomy of warp drives.

Unfortunately, although we can discuss the various spacetime metric designs for the taxonomies of traversable wormholes and warp drives and the exotic matter required to construct them, we simply have no idea how to technologically generate the redshift and shape functions for traversable wormholes and the shape functions and topological coefficients for warp drives. So, unfortunately, all the published discussions and theoretical solutions for traversable wormholes and warp drives remain within the realm of hypotheses mixed with some conjecture.

Fortunately, we have a much firmer grasp on what exotic matter is, what its physical properties are or what they must be, what the taxonomy of exotic matter looks like, which types of exotic matter are technologically feasible to produce in the lab, and how much of the feasible exotic matter is required to construct traversable wormholes and warp drives. The last item remains under study and debate since many GTR physicists are still tackling putative spacetime geometrical and topological mechanisms that might drastically reduce the amount of negative energy required to construct any of the types of traversable wormholes and warp bubbles. Associated with this concern is the alternative of investigating whether the feasible types of exotic matter (negative quantum vacuum energy) can be technologically scaled-up to astronomically or subastronomically large amounts to implement a wormhole or warp drive. The answer to these concerns might lie somewhere in between the two options: designing FTL spacetimes having a much lower negative energy requirement to construct and developing a negative energy generator technology that is scalable. Sections 5.5 and 5.6 showed that contemporary laboratory technology for producing squeezed vacuum states of light is maturing to some degree, while the harvesting and detection of its negative energy density pulses and the scaling of their magnitude is still at the lowest technology readiness level. The technical issue of capturing and storing negative vacuum energy is not considered in this chapter because free-space sources or projectors of negative vacuum energy appear to be a more desirable option for producing FTL spacetimes than stored/confined sources of negative vacuum energy, and because there is very little technical literature that addresses how to capture and store/confine negative vacuum energy (e.g., see the discussion in Ref. [79]). This technical issue requires much theoretical investigation in the future.

The author is convinced that the answers to all of these questions and concerns can only be found by performing many decades' worth of dedicated theoretical and empirical studies to narrow down the most feasible general relativistic FTL solutions, determine whether additional supplementary theories will be required to answer the question of how to physically construct FTL spacetimes, and develop a scalable negative vacuum energy generator technology for laboratory exploration. It would be a worthwhile exercise to throw the GTR field equations, all of the known energy-condition-violating traversable wormhole and warp drive solutions, QED and quantum optics theories, and all of the known negative vacuum energy generator concepts into an artificial intelligence/machine learning program to see what it spits out as its optimal solution to answer our questions. It is unknown whether alternative theories of gravity (e.g., Jordan-Brans-Dicke scalar-tensor, scalar-vector-tensor, higher curvature Gauss-Bonnet, Dilatonic Einstein-Gauss-Bonnet, Einstein-Cartan, etc.) allow warp drive solutions. While there is one class of quantum gravity theories that supports a warp drive solution [42,43], alternative gravity and quantum gravity theories are known to allow traversable wormhole solutions. All of this remains within the realm of theoretical speculation until one of these theories becomes the universally adopted replacement for GTR and quantum field theory. For this reason, alternative gravity and quantum gravity theories will not be considered further.

8 The problems of navigation, guidance, control, and stability in FTL propulsion concepts

8.1 Navigation, guidance, and control Laws

GTR does not provide any mathematical recipe to develop navigation, guidance, and control for FTL interstellar flight using a traversable wormhole or a warp drive. FTL motion through interstellar space is different for traversable wormholes and warp drives. Whereas interstellar space travelers that use a warp drive are literally sliding across spacetime at FTL speed while they reside at relative rest within their warp bubble, those that use a traversable wormhole are simply passing through a hyperspace shortcut through spacetime using conventional or relativistic rocket propulsion technology to traverse its throat at sublight speed while remote static observers view the travelers' interstellar flight as occurring at FTL speed.

The technical description of taking a trip through the Morris-Thorne traversable wormhole is given by the proper time and proper distance of travel through its throat as measured by the space travelers who always move on a timelike curve through the throat, while the (radial) travel velocity through the throat is $v = v(r) < c$. The proper time $\Delta\tau$ of travel through the wormhole throat as measured by the space travelers is [23,25]:

$$\Delta\tau = \int (\gamma v)^{-1} d\ell, \tag{39}$$

where γ is defined in Section 1.2, $d\ell$ is the element of proper distance, and the integration is taken from the wormhole entrance to its exit. Correspondingly, the travelers' proper distance $\Delta\ell$ of travel through the wormhole throat is $\Delta\ell = v\Delta\tau$. With respect to the clocks of remote static observers, the space travelers' proper time Δt of travel through the wormhole throat is [23,25]:

$$\Delta t = \int \left(ve^{\phi(r)}\right)^{-1} d\ell, \tag{40}$$

and their proper distance of travel through the throat is $\Delta\ell = v\Delta t$. $\phi(r)$ is defined beneath Eq. (7), and the integration is taken over the same limits as for Eq. (39).

The Appendix in Ref. [23] provides three case studies for traversable wormholes with different exotic matter and throat geometry specifications, and the corresponding proper travel times and distances are calculated. Whatever time-distance values one desires for their interstellar journey will depend largely on the type of exotic matter that one chooses to assemble the wormhole with and the particular type of wormhole throat geometry one chooses, which serves as further input to throat stability analysis, determining the tidal gravitational forces exerted on the spacecraft along the way, etc. Note, however, that there will be no proper time and proper distance of travel through a wormhole-stargate as measured by space travelers and remote static observers because the former will simply be shunted into the remote region of spacetime at their destination as soon as they step through the stargate.

For FTL interstellar travel by warp drive, the interstellar spacecraft moves on a timelike trajectory by design [29]. This trajectory establishes a very simple relationship between the element of coordinate time t and the element of proper time τ, $d\tau = dt$, which means that the coordinate time is equal to the proper time in this type of FTL spacetime geometry. Since the coordinate time is also equal to the proper time as measured by remote static observers, the warp spacecraft suffers no relativistic time dilation as it travels. The proper acceleration along the spacecraft's trajectory will always be zero, while the coordinate acceleration can be an arbitrary function of time. For the case of a one-way journey from Earth to some stellar destination a distance D away, the total elapsed proper time as measured onboard the spacecraft, which is equal to the total elapsed coordinate time T, is [29]:

$$\tau \simeq T \simeq 2\sqrt{\frac{D}{a}}, \tag{41}$$

where a is the spacecraft's coordinate acceleration. T can be made as small as one desires simply by increasing the value of a, which is how the spacecraft's FTL motion is seen by remote static observers.

The control laws, cockpit/bridge visual display, positional reference/guide stars, and star maps for navigation and guidance through interstellar space have not been developed for either type of FTL propulsion concept. Navigation, guidance, and control systems will have to be developed separately for warp drives and traversable wormholes. Hoag and Wrigley [142] studied navigation and guidance for relativistic interstellar missions and Vulpetti [143,144] studied relativistic interstellar navigation using the 3D rocket equation. These studies could provide a starting point to begin an equivalent study for FTL interstellar missions.

Of further interest is how the forward and aft starfields appear to the FTL spacecraft bridge crew that visually monitor their flight progress using windows and electronic visual displays. Ford and Roman [79] and Clark et al. [145] showed that, for Alcubierre's warp drive, the angular deflection and redshift of photons propagating through the distortion of the warp bubble is such that the starfield in the forward and aft fields of view will appear to cluster much closer to the forward and aft directions of motion than they would to an observer at rest. The starfield viewed in the forward direction will be strongly blueshifted while the starfield viewed in the aft direction will be strongly redshifted. This optical aberration is similar to that in STR for the case of relativistic interstellar rocket travel [146–148]. This suggests that visual guide/reference stars and typical star maps will be useless for warp drive starship navigation. Real-time electronic visual displays will be required to display accurate virtual starfields and maps, and they must have computer algorithms that perform real-time adjustments to account for the effects of FTL aberration to display visually accurate views and maps. The other warp drive concepts have not been analyzed to discover what type of FTL aberration they produce.

The view through a traversable wormhole is even worse. The negative energy density threading a wormhole throat generates repulsive gravity, which deflects any light rays going through and around it. The entrance to the Morris-Thorne wormhole would look like a sphere that contained the incredibly shrunken and distorted mirror image of a whole other universe or a remote region within our universe. This would look just like a glass Christmas tree ornament, which is just a spherical mirror that reflects, in principle, the entire universe around it. The spherical wormhole entrance/exit (a.k.a. throat) is called a hypersphere because it is the 3D hyperspace surface of our (3+1)-dimensional spacetime. A wormhole-stargate would not distort the image of the remote space region or other universe that's viewed through it because the negative energy density and negative tension at the throat are zero as seen and felt by any light and matter passing through it.

8.2 Controllability, communications, and stability

The discussion in this section focuses only on warp drive propulsion concepts because the technical issues to be raised are not applicable to traversable wormholes.

Hiscock [149], Finazzi et al. [150], Barceló et al. [151], and Clark et al. [145] evaluated a modified (massless point-particle spacecraft) Alcubierre warp drive in a toy (1+1)-dimensional spacetime and concluded that this is unstable against the quantum back-reaction effects (nonlocal particle production and local vacuum polarization energy density) produced by the warp bubble's spacetime curvature, which has the following consequences:

(i) intense thermal Hawking radiation floods the warp bubble, thus endangering the spacecraft passengers and destabilizing the warp bubble;
(ii) local vacuum polarization energy density counteracts the destabilizing thermal Hawking radiation, but the system doesn't get ahead at superluminal speed;
(iii) conical-shaped forward and aft event horizons form in front and aft of the spacecraft, thus causally disconnecting the interior of the warp bubble from its forward edge such that the bridge crew cannot steer the warp bubble or turn it on/off, and communications cannot be sent from the spacecraft to the spacetime region exterior to the warp bubble;
(iv) each of the previous problems goes away at sublight speed, and the warp drive becomes stable and controllable.

The forward and aft event horizons in item (iii) are analogous to the Mach cones associated with supersonic air flow. Lee and Cleaver [152] pointed out a different set of problems associated with controlling, navigating, and maintaining stability of Alcubierre's warp drive. Their analysis found in part that the Alcubierre warp bubble suffers from extreme energy loss and eventual destabilization at superluminal and subluminal speeds due to the extreme blueshifting of interstellar photons entering the forward side of the warp bubble. However, in a comprehensive analysis of all these technical issues, Barceló et al. [153] found a solution that cures all of them, which we will discuss in detail later in this section.

As we discussed in Section 3.2, the warp drive metric studies by Hart et al. [37], Krasnikov [30], Loup [35,36,40], and Perniciano [39] cured the pathologies in items (i)–(iii) of the previous paragraph by use of novel geometric and topological design features, which were incorporated into Alcubierre's, Van Den Broeck's, and Natário's warp drive spacetime metrics. However, no analysis has been done to reproduce items (i)–(iii) in the physical (3+1)-dimensional spacetime. This is because Hiscock [149], Finazzi et al. [150], Barceló et al. [151], and Clark et al. [145] admitted that their toy (1+1)-dimensional warp drive metric analysis is not extensible to the physical (3+1)-dimensional spacetime, since there are two spatial degrees of freedom missing from their metric (to greatly simplify their calculations), the source of negative energy in their toy model is the conformal anomaly for an uncharged massless scalar field, and there are unresolved problems associated with calculating the exotic matter source's $\langle T_{\mu\nu}\rangle_{\text{vac}}$ in the physical (3+1)-dimensional spacetime. Another problem with their analyses is that gravity doesn't exist in (1+1)-dimensions.

The situation as it stands now is that all theoretical results and interpretations for any toy (1+1)-dimensional warp drive spacetime models are simply not relevant to the physical (3+1)-dimensional spacetime because most of the intuition from (1+1)-dimensional spacetime does not carry over to the physical (3+1)-dimensional spacetime. For instance, in (1+1)-dimensional spacetime the exotic matter source's $\langle T_{\mu\nu}\rangle_{\text{vac}}$ has local components that can be computed in the sense that a change of spacetime geometry at any given spacetime location will induce a corresponding change in $\langle T_{\mu\nu}\rangle_{\text{vac}}$ at that same spacetime location. This is called quantum back-reaction in semiclassical quantum gravity theory (in this theory, the right-hand side of Eq. 6 is quantized while the left-hand side is not quantized). This same scenario is not true in the physical (3+1)-dimensional spacetime, which is the essence of the well-known nonlocality effects in quantum field theory. In (3+1)-dimensional spacetime, one must do the quantum vacuum ZPF mode sums to calculate the components of $\langle T_{\mu\nu}\rangle_{\text{vac}}$ associated with nonlocal particle production and this is almost always impossible to do. These conclusions are also true for the case of traversable wormholes, so a separate discussion on the effects of quantum back-reaction on traversable wormhole spacetimes is not required.

An estimate for the magnitude of the components of $\langle T_{\mu\nu}\rangle_{\text{vac}}$ in the physical (3+1)-dimensional spacetime is to just calculate the vacuum polarization energy density component, which is a back-reaction effect that is local and depends on the local curvature that characterizes the spacetime at a given location of interest. The vacuum polarization will be collocated with the spacetime curvature associated with the warp bubble or the traversable wormhole throat. An estimate for the vacuum polarization energy density ρ_{vacpol} is [20,31]:

$$\rho_{\text{vacpol}} \approx -\frac{2\pi^2 \hbar c}{\ell_c^4} \approx -\frac{2\pi^2 \hbar G^2 M_{\text{warp}}^2}{c^3 r^6} \approx -\left(3.43 \times 10^{-79}\right)\frac{M_{\text{warp}}^2}{r^6}, \tag{42}$$

where $\ell_c \approx (r^3 c^2/GM_{\text{warp}})^{1/2}$ is the local radius of curvature (defined by the inverse square-root of the typical Riemann curvature tensor component in a local orthonormal frame) of M_{warp}, M_{warp} is the absolute value of the equivalent mass of the total negative energy contained within the thin shell that forms the wormhole throat or warp bubble wall, and r is the radial distance from M_{warp}.

It is of interest to point out that ρ_{vacpol} is algebraically negative, and the first expression on the left-hand side of Eq. (42) is characteristic of expressions for the Casimir vacuum energy density (e.g., see the discussions in Sections 1.3, 5.2, and 5.5.4). Of further interest is that Hochberg and Kephart [154] proposed that the spacetime curvature-induced vacuum ZPF field distortion (a.k.a. gravitational squeezing of the vacuum, akin to quantum optically squeezed vacuum state of light) in the region of space proximally surrounding collapsed stars could create and stabilize a traversable wormhole. This author showed that the negative vacuum polarization energy density arising from a gravitationally squeezed vacuum is also quantified by Eq. (42) [20].

In Eq. (42), the r^{-6} dependence of ρ_{vacpol} indicates that the bulk of the negative vacuum polarization energy will be concentrated within the immediate vicinity of the thin shell that forms the wormhole throat or warp bubble wall, and the numerical coefficient indicates that the magnitude of this energy density will be extremely minute even for values of M_{warp} that are typical for traversable wormholes and warp drives. Eq. (42) implies that vacuum polarization will have a negligible impact on the wormhole throat or warp bubble unless M_{warp} is of cosmological-scale and/or r is of quantum-scale. Also, Eq. (42) serves as a rough estimate for the magnitude of the nonlocal components (i.e., particle production) of $\langle T_{\mu\nu}\rangle_{\text{vac}}$ in the physical (3+1)-dimensional spacetime, which directly impacts the controllability of the warp bubble. Based on the foregoing discussion, the magnitude of this impact is quite negligible.

And finally, Barceló et al. [153] analyzed the instability present in the Alcubierre warp drive through its relation to the behavior of null geodesics (paths followed by photons in spacetime). This instability is comprised of the accumulation of interstellar photons and vacuum ZPF modes of the background quantum field on the forward side of the warp bubble, which leads to an exponential growth in the background quantum field's vacuum energy density that's analogous to that found at the inner event horizon of Schwarzschild black holes. They argued that the strong instability found in the toy

(1+1)-dimensional Alcubierre warp drive can be suppressed in higher spacetime dimensions by choosing the appropriate shape and trajectory of the warp bubble. Their analysis determined that the warp bubble should have an aerodynamic shape to deflect incoming photons away from its unstable point in the shortest time possible. The warp drive's trajectory can also be designed to further promote this deflection by introducing a slight zig-zag component to its direction of motion that includes alternating intervals of superluminal and subluminal warp bubble speeds.

It is of particular interest to note that Barceló et al. [153] found in their toy (1+1)-dimensional Alcubierre warp drive model that the forward wall of the warp bubble behaves like an inner black hole event horizon. In fact, they found that in higher spacetime dimensions the warp bubble does not have a closed inner event horizon or any closed trapped surfaces. They found instead that the warp bubble can be interpreted as an interpolation between an inner event horizon point (the forward side of the bubble) and an outer event horizon point (the aft side of the bubble). In between the two is a causal structure that resembles an ergoregion (the region between the infinite redshift surface and the event horizon of Schwarzschild black holes) from which communication signals can escape.

Barceló et al.'s [153] study showed that the accumulation of incoming interstellar photons and corresponding blueshift instability is limited to only single points on the warp bubble wall when the shape of the bubble is smooth. Their study presents strong evidence that the instabilities do not preclude aerodynamic-like configurations of the Alcubierre warp bubble from being able to sustain a net superluminal speed. This means that the warp spacecraft can communicate with, and control, the warp bubble, as well as communicate with the spacetime exterior to it. Surprisingly, they found that a convex warp bubble shape with a very sharp peak offers the least resistance for superluminal and subluminal travel in the presence of a background quantum field, even in its vacuum state. This configuration resembles the aerodynamic planforms used for supersonic aircraft, which minimize the aerodynamic drag and frontal air compression (pressure) they experience at Mach speeds. They also found that a warp bubble having a flat or concave peak configuration would experience much more resistance and greater blueshift instability.

There remains the severe upper bound on warp speed imposed by Eq. (15) that nobody in the general relativistic FTL community is presently addressing. This is the remaining theoretical "800-pound gorilla in the room" that's largely ignored by many practitioners in this community.

9 Conclusion

In this chapter, we briefly surveyed the types of advanced relativistic propulsion concepts followed by an extensive technical review and analysis of the general relativistic FTL propulsion concepts that are applicable to interstellar missions. We delineated a substantial number of crucial technical details along with their accompanying make or break issues, having either tentative or firm results. The tentative results suggest a path for future research to put certain crucial technical issues on a firmer footing. Nuclear fusion and matter-antimatter annihilation propulsion, or some hybrid combination of these, appear to be the best options for making relativistic interstellar flight achievable; although these technologies are afflicted by the relativistic rocket equation, which requires huge amounts of propellant to propel the interstellar spacecraft and its propellant. A potential way around this problem is to somehow acquire or otherwise produce the nuclear fusion and matter-antimatter fuels in situ along the journey. Beamed (or directed) energy propulsion might be a third possibility. Yuri Milner's Breakthrough

Starshot project is developing low mass nano-spacecraft on a wafer (a.k.a. computer chip) that will be propelled to the Proxima Centauri star system at 20% of light speed by using a GW-class laser phased array, comprised of an array of kW-class lasers, to push a swarm of laser sails [155,156].

However, the main thrust of this chapter is to explore the physics and technologies of general relativistic FTL propulsion for interstellar flight. We identified the two primary types of FTL spacetimes found in GTR that can be constructed in principle: traversable wormholes and warp drives. There is a large taxonomy of traversable wormholes and a growing taxonomy of warp drives. The construction of these specialized spacetimes requires exotic matter that violates all of the long-held classical and quantum energy conditions now known to be erroneous due to flawed assumptions and formulation, which were found prior to have been theoretically and experimentally violated by a large number of classical and quantum matter fields that qualify as forms of exotic matter.

The specific type of exotic matter that meets the GTR requirements for the construction of FTL spacetimes must have the properties of negative energy density/flux, or negative pressure/stress, or the energy density must be less than its pressure/stress. There's also a taxonomy of classical and quantum matter fields that have these properties. We found that QED theory provides the template for engineering and producing one source of negative energy: squeezed vacuum states of light, the negative vacuum energy associated with the modified quantized electromagnetic field in its vacuum state. Quantum optical squeezed light devices and Casimir cavities are the two laboratory technologies that produce negative vacuum energy, both of which produce feeble amounts of it. We surveyed other speculative and theoretically well-developed QED theory-predicted quantum optics concepts and found that the Ford-Svaiter parabolic-cylindrical mirror might be capable of producing scalable negative vacuum energy. The requirement for negative energy portends the need for a quantum optics technology that produces negative vacuum energy that can be scaled up to astronomical amounts or less. We also explored proposals for directly detecting/imaging negative vacuum energy in the lab, which is a necessary sensor technology to have if one wishes to control negative vacuum energy for the construction and control of any type of FTL spacetime.

To produce the type of extraordinarily large spacetime curvature effects that create any type of traversable wormhole or warp drive forces one to reckon with the fact that the ultra-minute coupling constant $8\pi G/c^4 \sim 10^{-43}$ multiplying $T_{\mu\nu}$ in Eq. (6) guarantees that the magnitude of the components of $T_{\mu\nu}$ must be astronomically huge. This seems to be a deal killer for constructing most types of warp drive and traversable wormhole, but in general, traversable wormholes require at a minimum a near-zero magnitude of negative energy via the energy condition constraint that applies to their genre of FTL spacetime geometries. However, the latter become unstable to collapse, while for the former, there is only one known type of warp drive in which the magnitude of negative energy required to construct it is low enough to be lab-scale even though the classical scalar field that has this property is so far not known to exist in nature. The energetics analysis of the warp drive and traversable wormhole concepts that we reviewed in this chapter confirms that their negative energy requirements are severe for the most part. On the other hand, the stargate solution is an ultra-static, ultra-short throat traversable wormhole with a simple geometrical property that gives it zero surface energy density and zero surface tensions as seen by space travelers moving through it.

For any warp drive concepts derived from GTR, Eq. (15) shows that there is a severe upper bound to the warp speed given any selected values of the warp bubble's radius and wall thickness, and spacecraft mass. If Lobo and Visser's [21] analysis is correct, then this absurdly low upper limit to warp speed appears to be a deal killer for all GTR-based warp drive concepts. It is alarming that the theoretical warp drive community doesn't appear to be actively addressing this issue.

Eq. (6) does not provide any instructions for how to physically construct any type of traversable wormhole or warp drive. There are freely specifiable redshift and shape functions for the former and freely specifiable shape and topological functions (plus the constant positive nonzero warp bubble radius and wall thickness parameters) for the latter. The spatial distribution of negative vacuum energy/flux and/or negative vacuum pressure/stress is mathematically tied to these freely specifiable parameters via Eq. (6). One can invert the wormhole or warp drive components of $T_{\mu\nu}$ to find algebraic expressions for the spatial distributions of their freely specifiable parameters as functions of the negative energy density and pressure/stress. However, to physically construct these parameters to "punch a hole" through spacetime for traversable wormholes or construct a complicated warp bubble structure remains unknown. Eq. (6) also does not provide instructions for how to implement guidance, navigation, and control for FTL spacecraft traversing interstellar space using a traversable wormhole or warp drive. This remains an area of speculation until substantial theoretical work is done to determine from Eq. (6) how guidance, navigation, and control can be implemented. We also saw that there is no relativistic time dilation effect that negatively impacts space travelers using a traversable wormhole or a warp drive for interstellar travel. However, time dilation with an associated time travel capability, which was not discussed in this chapter, can be incorporated into the FTL spacetime geometries if one so desires.

We found that, unlike traversable wormhole spacetimes, warp drive spacetimes suffer from a suite of technical problems that impairs communication of the spaceship with its warp bubble to control it, as well as communicate with the spacetime exterior to the warp bubble. Another serious problem is the destabilization of the warp bubble caused by highly blueshifted quantum matter and interstellar photons impacting the forward side of the warp bubble. The good news is that all these technical problems were found to be wrong because they are based on terribly flawed mathematical modeling in an unphysical (1+1)-dimensional spacetime. A complete cure for these problems was provided by the novel design of an aerodynamic-shaped warp bubble in the physical (3+1)-dimensional spacetime.

It is readily apparent that many more years or even decades of further research will be necessary to address these theoretical and technological issues and achieve full resolution to them. Thus, general relativistic FTL propulsion will remain at Technology Readiness Level 1 for many decades to come unless disruptive theoretical and technological innovations emerge to provide a viable path forward to technologically implementing one or more types of FTL propulsion. It's also very possible that GTR physicists have not yet discovered the full parameter space of FTL spacetime solutions, where novel new solutions having less onerous energy requirements remain to be discovered. Meanwhile, the genre of general relativistic FTL propulsion concepts continues to grow and mature as time progresses, while the associated theoretical debates continue as well. We suggest the following critical theory and technology paths, which should be followed by interested investigators to find a successful and technologically viable FTL propulsion concept:

1. Negative vacuum energy generator: explore new ways to produce it and scale-up the magnitude of negative vacuum energy density/flux:
   a. Conduct a full theoretical, parametric, and technological exploration of the Ford-Svaiter parabolic-cylindrical mirror.
   b. Investigate whether there are theoretical and technologically achievable superradiance effects or mechanisms acting on the QED vacuum ZPF modes that could drastically enhance the production and magnitude of negative vacuum energy density/flux.
   c. Explore other options to theoretically and experimentally develop via quantum field theory and quantum optics.

- **d.** Use AI/machine learning to probe the parameter space of QED vacuum state configurations to find optimally feasible negative energy density/flux solutions with achievable engineering implementations and scalability.

2. Detecting/measuring/imaging negative vacuum energy density/flux: develop and test Marecki's proposed quantum optical tomography-BHD apparatus for directly measuring negative vacuum energy density pulses in squeezed light:

- **a.** Develop and test Marecki's apparatus for a Ford-Svaiter mirror and a Casimir cavity (very hard to do at present).
- **b.** Expand Davies and Ottewill's [77] study to explore negative energy detection by using elementary particles, Bose-Einstein condensates, Ryderg atoms, atomic optical molasses, etc., in a cavity QED, a superconducting cavity, or other apparatus.

3. Gain new theoretical insights into the construction, navigation, guidance, and control of FTL spacetimes by developing and performing lab experiments that use the transformational optics of metamaterials to produce analog electromagnetic models of a warp drive and a traversable wormhole:

- **a.** Perform the usual course of theoretical computational engineering-physics studies.
- **b.** Use AI/machine learning to probe the parameter space of traversable wormhole and warp drive solutions to find optimally feasible solutions with technologically achievable energetics requirements and engineering implementations.

We hope that this chapter serves as a useful guide for the interested researcher who wishes to pursue this topic on their own.

Acknowledgments

The author would like to thank the Institute for Advanced Studies at Austin and H. E. Puthoff for having supported this work and for encouraging our exploration into general relativistic FTL propulsion during 2004–19. The author wishes to honor the memory of his mentor, R. L. Forward, and the contributions he made to breakthrough propulsion physics over 44 years. We also wish to extend thanks and appreciation to G. J. Maclay, G. B. Cleaver, J. S. Lee, B. Mattingly, W. Julius, C. Watson, M. Gorban, M. G. Millis, D. Bushnell, R. M. Wald, M. Visser, F. Loup, T. W. Kephart, P. C. W. Davies, L. H. Ford, and C. Maccone for many useful discussions held during the previous 25 years.

References

[1] Cosmic Milestone: NASA Confirms 5,000 Exoplanets, Jet Propulsion Laboratory, CalTech, March 21, 2022. https://www.jpl.nasa.gov/news/cosmic-milestone-nasa-confirms-5000-exoplanets. (Accessed 1 April 2022).

[2] C. Maccone, SETI and SEH (statistical equation for habitables), Acta Astronaut. 68 (1–2) (2011) 63–75.

[3] R.H. Frisbee, Limits of interstellar flight technology, in: M.G. Millis, E.W. Davis (Eds.), Frontiers of Propulsion Science, Progress in Astronautics and Aeronautics, vol. 227, AIAA Press, Reston, VA, 2009, pp. 31–126 (2nd printing with corrections in 2012), (Chapter 2).

[4] E.F. Mallove, G.L. Matloff, The Starflight Handbook: A Pioneer's Guide to Interstellar Travel, John Wiley & Sons, Inc., New York, 1989, p. 54.
[5] R.T. Weidner, R.L. Sells, Elementary Modern Physics, third ed., Allyn and Bacon, Inc., Boston, MA, 1980. pp. 34–35, 64–65.
[6] R.L. Forward, A transparent derivation of the relativistic rocket equation, in: Paper ASEE 95-3060, 31st AIAA/ASME/SAE/ASEE Joint Propulsion Conference and Exhibit, San Diego, CA, July 10-12, 1995.
[7] K. Scharnhorst, On propagation of light in the vacuum between plates, Phys. Lett. B 236 (3) (1990) 354–359.
[8] J.I. Latorre, P. Pascual, R. Tarrach, Speed of light in non-trivial vacua, Nucl. Phys. B 437 (1) (1995) 60–82.
[9] P.A.M. Dirac, Is there an æther? Nature 168 (1951) 906–907.
[10] P.A.M. Dirac, A new classical theory of electrons, Proc. R. Soc. Lond. Ser. A 209 (1951) 291–296.
[11] B.S. DeWitt, Quantum field theory in curved spacetime, Phys. Rep. Sect. C Phys. Lett. 19 (6) (1975) 295–357.
[12] B.S. DeWitt, Quantum gravity: the new synthesis, in: S.W. Hawking, W. Israel (Eds.), General Relativity: An Einstein Centenary Survey, Cambridge University Press, Cambridge, 1979, pp. 680–745.
[13] B.S. DeWitt, The Casimir effect in field theory, in: A. Sarlemijn, J. Sparnaay (Eds.), Physics in the Making, Essays on Developments in 20th Century Physics, in Honour of H. B. G. Casimir, North-Holland Elsevier Science Publishing, New York, 1989, pp. 247–272.
[14] M. Urban, et al., The quantum vacuum as the origin of the speed of light, Eur. Phys. J. D 67 (2013) 58. 1–6.
[15] G. Leuchs, L.L. Sánchez-Soto, A sum rule for charged elementary particles, Eur. Phys. J. D 67 (2013) 57. 1–5.
[16] E.W. Davis, Teleportation Physics Study, Final Report AFRL-PR-ED-TR-2003-0034, Air Force Research Laboratory, Air Force Materiel Command, Edwards AFB, CA, 2004, pp. 3–11.
[17] C.W. Misner, K.S. Thorne, J.A. Wheeler, Gravitation, W. H. Freeman and Co., San Francisco, CA, 1973.
[18] R.M. Wald, General Relativity, The University of Chicago Press, Chicago, IL, 1984, pp. 11–28.
[19] C. Nash, S. Sen, Topology and Geometry for Physicists, Academic Press, New York, 1989. pp. 25–50, 164.
[20] E.W. Davis, Faster-than-light approaches in general relativity, in: M.G. Millis, E.W. Davis (Eds.), Frontiers of Propulsion Science, Progress in Astronautics and Aeronautics, vol. 227, AIAA Press, Reston, VA, 2009, pp. 471–507 (2nd printing with corrections in 2012), (Chapter 15).
[21] F.S.N. Lobo, M. Visser, Fundamental limitations on 'warp drive' spacetimes, Class. Quantum Gravity 21 (2004) 5871–5892.
[22] W. Rindler, Gödel, Einstein, Mach, Gamow, and Lanczos: Gödel's remarkable excursion into cosmology, Am. J. Phys. 77 (6) (2009) 498–510.
[23] M.S. Morris, K.S. Thorne, Wormholes in spacetime and their use for interstellar travel: a tool for teaching general relativity, Am. J. Phys. 56 (5) (1988) 395–412.
[24] M.S. Morris, K.S. Thorne, U. Yurtsever, Wormholes, time machines, and the weak energy conditions, Phys. Rev. Lett. 61 (13) (1988) 1446–1449.
[25] M. Visser, Lorentzian Wormholes: From Einstein to Hawking, AIP Press, New York, 1995.
[26] M. Visser, Section 3: Warp drives and worm holes: what's fact and what's fantasy? in: R. Koczor (Ed.), Physics for the Third Millennium, Presentation Materials From a Series of Lectures Given February 9-12, 1998, NASA George C. Marshall Space Flight Center, Huntsville, AL, 1998.
[27] D. Hochberg, M. Visser, Geometric structure of the generic static traversable wormhole throat, Phys. Rev. D 56 (8) (1997) 4745–4755.
[28] R. Rucker, The Fourth Dimension: A Guided Tour of the Higher Universes, Houghton Mifflin Co., Boston, MA, 1984. pp. 113, 117–121.
[29] M. Alcubierre, The warp drive: hyper-fast travel within general relativity, Class. Quantum Gravity 11 (1994) L73–L77.
[30] S.V. Krasnikov, Hyperfast travel in general relativity, Phys. Rev. D 57 (8) (1998) 4760–4766.

[31] E.W. Davis, Faster-than-light space warps, status and next steps, J. Br. Interplanet. Soc. 66 (3/4) (2013) 68–84.
[32] A.E. Everett, T.A. Roman, Superluminal subway: the Krasnikov tube, Phys. Rev. D 56 (4) (1997) 2100–2108.
[33] C.V.D. Broeck, A 'warp drive' with more reasonable total energy requirements, Class. Quantum Gravity 16 (1999) 3973–3979.
[34] J. Natário, Warp drive with zero expansion, Class. Quantum Gravity 19 (2002) 1157–1165.
[35] F. Loup, An Extended Version of the Natário Warp Drive Equation Based in the Original 3 + 1 ADM Formalism Which Encompasses Accelerations and Variable Velocities (Ph.D. thesis), Residencia de Estudantes Universitas, Lisboa, Portugal, 2017. HAL Open Science Archive, hal-01655423, https://hal.archives-ouvertes.fr/hal-01655423. (Accessed 5 July 2022).
[36] F. Loup, Six Different Natário Warp Drive Spacetime Metric Equations, Research Report, Residencia de Estudantes Universitas, Lisboa, Portugal, 2018. HAL Open Science Archive, hal-01862911, https://hal.archives-ouvertes.fr/hal-01862911. (Accessed 5 July 2022).
[37] C.B. Hart, et al., On the Problems of Hazardous Matter and Radiation at Faster Than Light Speeds in the Warp Drive Space-Time, 2002, Cornell Univ. Library arXiv.org e-Print Archive, arXiv:gr-qc/0207109, https://arxiv.org/abs/gr-qc/0207109. (Accessed 5 July 2022).
[38] M.J. Pfenning, L.H. Ford, The unphysical nature of 'warp drive', Class. Quantum Gravity 14 (1997) 1743–1751.
[39] G. Perniciano, Modified Alcubierre Warp Drive II, Department of Physics of the University of Cagliari, Italy, 2016. HAL Open Science Archive, hal-01240285v3, https://hal.archives-ouvertes.fr/hal-01240285v3. (Accessed 5 July 2022).
[40] F. Loup, The Analysis of Chris Van Den Broeck Applied to the Natário Warp Drive Spacetime Using the Original Alcubierre Shape Function to Generate the Broeck Spacetime Distortion: The Natario-Broeck Warp Drive, Research Report, Residencia de Estudantes Universitas, Lisboa, Portugal, 2017. HAL Open Science Archive, hal-01456561, https://hal.archives-ouvertes.fr/hal-01456561. (Accessed 11 July 2022).
[41] H.G. White, E.W. Davis, The Alcubierre warp drive in higher dimensional spacetime, in: M.S. El-Genk (Ed.), Proceedings of the STAIF-2006: 3rd Symposium on New Frontiers and Future Concepts, AIP Conference Proceedings, vol. 813, AIP Press, New York, 2006, pp. 1382–1389.
[42] R. Obousy, Supersymmetry Breaking Casimir Warp Drive, 2005, Cornell Univ. Library arXiv.org e-Print Archive, arXiv:gr-qc/0512152, http://arxiv.org/gr-qc/0512152.pdf. (Accessed 27 December 2005).
[43] R. Obousy, G. Cleaver, Supersymmetry breaking Casimir warp drive, in: M.S. El-Genk (Ed.), Proceedings of the STAIF-2007: 4th Symposium on New Frontiers and Future Concepts, AIP Conference Proceedings, vol. 880, AIP Press, New York, 2007, pp. 1163–1169.
[44] A.D. Sakharov, Vacuum quantum fluctuations in curved space and the theory of gravitation, Sov. Phys. Dokl. 12 (11) (1968) 1040–1041.
[45] M. Visser, Sakharov's induced gravity: a modern perspective, Mod. Phys. Lett. A 17 (15) (2002) 977–991.
[46] H.E. Puthoff, SETI, the velocity-of-light limitation, and the Alcubierre warp drive: an integrating overview, Phys. Essays 9 (1) (1996) 156–158.
[47] H.E. Puthoff, Polarizable-vacuum (PV) approach to general relativity, Found. Phys. 32 (6) (2002) 927–943.
[48] H.E. Puthoff, S.R. Little, M. Ibison, Engineering the zero-point field and polarizable vacuum for interstellar flight, J. Br. Interplanet. Soc. 55 (2002) 137–144.
[49] H.E. Puthoff, E.W. Davis, C. Maccone, Levi-Civita effect in the polarizable vacuum (PV) representation of general relativity, Gen. Relativ. Gravit. 37 (3) (2005) 483–489.
[50] H.E. Puthoff, Advanced space propulsion based on vacuum (spacetime metric) engineering, J. Br. Interplanet. Soc. 63 (2010) 82–89.
[51] A.M. Volkov, A.A. Izmest'ev, G.V. Skrotskiĭ, The propagation of electromagnetic waves in a Riemannian space, Sov. Phys. JETP 32 (4) (1971) 686–689.

[52] F. de Felice, On the gravitational field acting as an optical medium, Gen. Relativ. Gravit. 2 (12) (1971) 347–357.
[53] A. Bobrick, G. Martire, Introducing physical warp drives, Class. Quantum Gravity 38 (10) (2021) 105009. 1–22.
[54] E.W. Lentz, Breaking the warp barrier: hyper-fast solitons in Einstein-Maxwell-plasma theory, Class. Quantum Gravity 38 (7) (2021) 075015. 1–14.
[55] R. Arnowitt, S. Deser, C.W. Misner, The dynamics of general relativity, in: L. Witten (Ed.), Gravitation: An Introduction to Current Research, Wiley Publishing Co., New York, 1962–2004, pp. 227–264. Reprinted on Cornell Univ. Library arXiv.org e-Print Archive, arXiv:gr-qc/0405109, https://arxiv.org/pdf/gr-qc/0405109.pdf. (Accessed 24 July 2022), (Chapter 7).
[56] S.D.B. Fell, L. Heisenberg, Positive energy warp drive from hidden geometric structures, Class. Quantum Gravity 38 (15) (2021) 155020. 1–16.
[57] E.B. Christoffel, Ueber die Transformation der homogenen Differentialausdrücke zweiten Grades, J. Reine Angew. Math. 70 (1869) 46–70. Reprinted on The European Digital Mathematics Library (EuDML), https://eudml.org/doc/148073. (Accessed 17 April 2017).
[58] B. Mattingly, et al., Curvature invariants for Lorentzian traversable wormholes, Universe 6 (1) (2020) 11. 1–9.
[59] J. Overduin, et al., Curvature invariants for charged and rotating black holes, Universe 6 (22) (2020) 6020022. 1–9.
[60] J. Carminati, R.G. McLenaghan, Algebraic invariants of the Riemann tensor in a four-dimensional Lorentzian space, J. Math. Phys. 32 (11) (1991) 3135–3140.
[61] B. Mattingly, et al., Curvature invariants for the Alcubierre and Natário warp drives, Universe 7 (2) (2021) 21–53.
[62] B. Mattingly, et al., Curvature invariants for the accelerating Natário warp drive, Particles 3 (3) (2020) 642–659.
[63] C. Barceló, M. Visser, Twilight for the energy conditions? Int. J. Mod. Phys. D 11 (2002) 1553–1560.
[64] C.K. Watson, et al., An invariant characterization of the Levi-Civita spacetimes, Symmetry 13 (8) (2021) 1469–1489.
[65] S.W. Hawking, G.F.R. Ellis, The Large-Scale Structure of Space-Time, Cambridge University Press, Cambridge, 1973. pp. 88–91, 95–96.
[66] E. Curiel, A Primer on Energy Conditions, Cornell Univ. Library arXiv.org e-Print Archive, arXiv:physics.hist-ph/1405.0403, 2014, https://arxiv.org/pdf/1405.0403.pdf. (Accessed on 12 October 2017).
[67] M. Visser, C. Barceló, Energy Conditions and their Cosmological Implications, 2000, Plenary talk presented at Cosmo99, Trieste, Italy, Sept./Oct. 1999, Cornell Univ. Library arXiv.org e-Print Archive, arXiv:gr-qc/0001099 https://arxiv.org/pdf/gr-qc/0001099.pdf. (Accessed 22 April 2000).
[68] M. Visser, Wormholes, baby universes, and causality, Phys. Rev. D 41 (4) (1990) 1116–1124.
[69] F.J. Tipler, Energy conditions and spacetime singularities, Phys. Rev. D 17 (10) (1978) 2521–2528.
[70] H. Epstein, V. Glaser, A. Jaffe, Nonpositivity of the energy density in quantized field theories, Nuovo Cimento 36 (1965) 1016–1022.
[71] M. Visser, Gravitational vacuum polarization. I. Energy conditions in the Hartle-Hawking vacuum, Phys. Rev. D 54 (1996) 5103–5115.
[72] M. Visser, Gravitational vacuum polarization. II. Energy conditions in the Boulware vacuum, Phys. Rev. D 54 (1996) 5116–5122.
[73] M. Visser, Gravitational vacuum polarization. III. Energy conditions in the (1+1)-dimensional Schwarzschild spacetime, Phys. Rev. D 54 (1996) 5123–5128.
[74] M. Visser, Gravitational vacuum polarization. IV. Energy conditions in the Unruh vacuum, Phys. Rev. D 56 (1997) 936–952.
[75] M. Visser, Energy conditions in the epoch of galaxy formation, Science 276 (1997) 88–90.

[76] L. Ford, T.A. Roman, Restrictions on negative energy density in flat spacetime, Phys. Rev. D 55 (4) (1997) 2082–2089.
[77] P.C.W. Davies, A. Ottewill, Detection of negative energy: 4-dimensional examples, Phys. Rev. D 65 (2002) 104014. 1–10.
[78] L.H. Ford, Quantum coherence effects and the second law of thermodynamics, Proc. R. Soc. Lond. A Math. Phys. Sci. 364 (1978) 227–236.
[79] L.H. Ford, T.A. Roman, Negative energy, wormholes and warp drive, Sci. Am. 13 (2003) 84–91.
[80] L.H. Ford, Negative energy densities in quantum field theory, Int. J. Mod. Phys. A 25 (11) (2010) 2355–2363.
[81] L.H. Ford, Constraints on negative-energy fluxes, Phys. Rev. D 43 (12) (1991) 3972–3978.
[82] G.J. Maclay, E.W. Davis, Testing a quantum inequality with a meta-analysis of data for squeezed light, Found. Phys. 49 (2019) 797–815.
[83] A. Borde, L.H. Ford, T.A. Roman, Constraints on spatial distributions of negative energy, Phys. Rev. D 65 (2002) 084002. 1–15.
[84] K.D. Olum, N. Graham, Static negative energies near a domain wall, Phys. Lett. B 554 (2003) 175–179.
[85] L.H. Ford, N.F. Svaiter, Focusing vacuum fluctuations, Phys. Rev. A 62 (2000) 062105. 1–10.
[86] L.H. Ford, N.F. Svaiter, Focusing vacuum fluctuations. II, Phys. Rev. A 66 (2002) 062106. 1–13.
[87] A. Borde, A. Vilenkin, Violation of the weak energy condition in inflating spacetimes, Phys. Rev. D 56 (1997) 717–723.
[88] S. Winitzki, Null Energy Condition Violations in Eternal Inflation, 2001, talk given at COSMO-2001, Rovaniemi, Finland, Cornell Univ. Library arXiv.org e-Print Archive, arXiv:gr-qc/0111109v1, http://arxiv.org/gr-qc/0111109v1.pdf. (Accessed 29 December 2001).
[89] T. Vachaspati, Eternal Inflation and Energy Conditions in de Sitter Spacetime, 2003, talk presented at Davis Inflation Meeting 2003, Cornell Univ. Library arXiv.org e-Print Archive, arXiv:astro-ph/0305439v1, http://arxiv.org/astro-ph/0305439v1.pdf. (Accessed 1 February 2004).
[90] L.H. Ford, T.A. Roman, Quantum field theory constrains traversable wormhole geometries, Phys. Rev. D 53 (10) (1996) 5496–5507.
[91] S.V. Krasnikov, Counter example to a quantum inequality, Gravit. Cosmol. 46 (2006) 195–198.
[92] S.V. Krasnikov, The Wormhole Hazard, 2003, Cornell Univ. Library arXiv.org e-Print Archive, arXiv:gr-qc/0302107 https://arxiv.org/pdf/gr-qc/0302107.pdf. (Accessed 20 March 2008).
[93] S.V. Krasnikov, Quantum inequalities do not forbid spacetime shortcuts, Phys. Rev. D 67 (10) (2003) 104013. 1–9.
[94] C. Barceló, M. Visser, Traversable wormholes from massless conformally coupled scalar fields, Phys. Lett. B 466 (2–4) (1999) 127–134.
[95] C. Barceló, M. Visser, Scalar fields, energy conditions and traversable wormholes, Class. Quantum Gravity 17 (18) (2000) 3843–3864.
[96] P.K.F. Kuhfittig, Static and dynamic traversable wormhole geometries satisfying the Ford-Roman constraints, Phys. Rev. D 66 (2002) 024015. 1–8.
[97] P.K.F. Kuhfittig, Can a wormhole supported by only small amounts of exotic matter really be traversable? Phys. Rev. D 68 (2003) 067502. 1–4.
[98] M. Visser, S. Kar, N. Dadhich, Traversable wormholes with arbitrarily small energy condition violations, Phys. Rev. Lett. 90 (20) (2003) 201102. 1–4.
[99] F. Lobo, P. Crawford, Weak Energy Condition Violation and Superluminal Travel, Lecture Notes in Physics, vol. 617, Springer, Berlin, 2003, pp. 277–291.
[100] S. Kar, N. Dadhich, M. Visser, Quantifying energy condition violations in traversable wormholes, Pramana 63 (2004) 859–864.
[101] U. Leonhardt, T.G. Philbin, General relativity in electrical engineering, New J. Phys. 8 (2006) 247–264.
[102] U. Leonhardt, Essential Quantum Optics: From Quantum Measurements to Black Holes, Cambridge University Press, Cambridge, 2010, pp. 189–230.

[103] E.W. Davis, Review of gravity control within Newtonian and general relativistic physics, in: M.G. Millis, E. W. Davis (Eds.), Frontiers of Propulsion Science, Progress in Astronautics and Aeronautics, vol. 227, AIAA Press, Reston, VA, 2009, pp. 175–228 (2nd printing with corrections in 2012), (Chapter 4).
[104] R.E. Slusher, et al., Observation of squeezed states generated by four-wave mixing in an optical cavity, Phys. Rev. Lett. 55 (22) (1985) 2409–2412.
[105] R.E. Slusher, et al., Erratum: Observation of squeezed states generated by four-wave mixing in an optical cavity, Phys. Rev. Lett. 56 (7) (1986) 788.
[106] R.E. Slusher, B. Yurke, Squeezed light, Sci. Am. 254 (1986) 50–56.
[107] A.L. Robinson, Bell labs generates squeezed light, Science 230 (4728) (1985) 927–929.
[108] A.L. Robinson, Now four laboratories have squeezed light, Science 233 (4761) (1986) 280–281.
[109] C.M. Caves, Quantum-mechanical noise in an interferometer, Phys. Rev. D 23 (8) (1981) 1693–1708.
[110] M.J. Pfenning, Quantum Inequality Restrictions on Negative Energy Densities in Curved Spacetimes (Ph.D. dissertation), Tufts University, Medford, MA, 1998. Cornell Univ. Library arXiv.org e-Print Archive, arXiv:gr-qc/9805037, https://arxiv.org/pdf/gr-qc/9805037.pdf. (Accessed 11 June 1998).
[111] H. Vahlbruch, et al., Detection of 15 dB squeezed states of light and their application for the absolute calibration of photoelectric quantum efficiency, Phys. Rev. Lett. 117 (2016) 110801. 1–5.
[112] P.C.W. Davies, How to Build a Time Machine, Penguin Books, New York, 2001, pp. 82–85.
[113] J.R. Rabeau, et al., Fabrication of single nickel-nitrogen defects in diamond by chemical-vapor deposition, Appl. Phys. Lett. 86 (2005) 131926. 1–3.
[114] J.R. Rabeau, et al., Diamond chemical-vapor deposition on optical fibers for fluorescence waveguiding, Appl. Phys. Lett. 86 (2005) 134104. 1–3.
[115] P. Holewa, et al., Bright quantum dot single-photon emitters at telecom bands heterogeneously integrated on Si, ACS Photonics 9 (7) (2022) 2273–2279.
[116] P.J. Mosley, et al., Heralded generation of ultrafast single photons in pure quantum states, Phys. Rev. Lett. 100 (2008) 133601. 1–4.
[117] K. Bartkiewicz, et al., Experimental Construction of a Versatile Four-Photon Source, 2018, Cornell Univ. Library arXiv.org e-Print Archive, arXiv:quant-ph/1808.07073, https://arxiv.org/pdf/1808.07073.pdf. (Accessed 2 November 2018).
[118] S.-H. Youn, J.-H. Lee, J.-S. Chang, Modulation of the vacuum field in the beam splitter and mirror system, Opt. Quant. Electron. 27 (1995) 355–362.
[119] N.D. Birrell, P.C.W. Davies, Quantum Fields in Curved Space, Cambridge University Press, Cambridge, 1989. pp. 89, 102–109.
[120] W.R. Walker, Negative energy fluxes and moving mirrors in curved space, Class. Quantum Gravity 2 (1985) L37–L40.
[121] G.T. Moore, Quantum theory of the electromagnetic field in a variable-length one-dimensional cavity, J. Math. Phys. 11 (9) (1970) 2679–2691.
[122] K.A. Milton, The Casimir Effect: Physical Manifestations of Zero-Point Energy, World Scientific, New Jersey, 2001.
[123] L.S. Brown, G.J. Maclay, Vacuum stress between conducting plates: an image solution, Phys. Rev. 184 (1969) 1272–1279.
[124] R.L. Forward, Alternate Propulsion Energy Sources, Air Force Rocket Propulsion Laboratory, Final Report AFRPL TR-83-067, Air Force Space Tech. Ctr. Space Div., Air Force Systems Command, Edwards AFB, CA, 1983, pp. A1–A14.
[125] K. Schneider, et al., Generation of strongly squeezed continuous-wave light at 1064 nm, Opt. Express 2 (3) (1998) 59–64.
[126] H. Hansen, et al., Ultrasensitive pulsed, balanced homodyne detector: application to time-domain quantum measurements, Opt. Lett. 26 (21) (2001) 1714–1716.

[127] P. Marecki, Balanced homodyne detectors and Casimir energy densities, J. Phys. A Math. Theor. 41 (16) (2008) 164037. 1–8.
[128] P. Marecki, Balanced homodyne detectors in quantum field theory, Phys. Rev. A 77 (1) (2008) 012101. 1–7.
[129] F.S.N. Lobo (Ed.), Wormholes, Warp Drives and Energy Conditions, Fundamental Theories of Physics, vol. 189, Springer International Publisher, Cham, 2017.
[130] D. Ida, S.A. Hayward, How much negative energy does a wormhole need? Phys. Lett. A 260 (3–4) (1999) 175–181.
[131] M. Visser, Traversable wormholes: some simple examples, Phys. Rev. D 39 (10) (1989) 3182–3184.
[132] A.J. McConnell, Applications of Tensor Analysis, Dover Publications, New York, 1957, pp. 163–217.
[133] S. Schuster, J. Santiago, M. Visser, ADM mass in warp drive spacetimes, Gen. Relativ. Gravit. 55 (2023) 14. 1–19.
[134] J. Santiago, S. Schuster, M. Visser, Generic warp drives violate the null energy condition, Phys. Rev. D 105 (6) (2022) 064038. 1–20.
[135] O.L. Santos-Pereira, E.M.C. Abreu, M.B. Ribeiro, Dust content solutions for the Alcubierre warp drive spacetime, Eur. Phys. J. C 80 (8) (2020) 786. 1–8.
[136] O.L. Santos-Pereira, E.M.C. Abreu, M.B. Ribeiro, Fluid dynamics in the warp drive spacetime geometry, Eur. Phys. J. C 81 (2) (2021) 133. 1–22.
[137] O.L. Santos-Pereira, E.M.C. Abreu, M.B. Ribeiro, Charged dust solutions for the warp drive spacetime, Gen. Relativ. Gravit. 53 (2) (2021) 23. 1–21.
[138] O.L. Santos-Pereira, E.M.C. Abreu, M.B. Ribeiro, Perfect fluid warp drive solutions with the cosmological constant, Eur. Phys. J. Plus 136 (9) (2021) 902. 1–13.
[139] O.L. Santos-Pereira, E.M.C. Abreu, M.B. Ribeiro, Warp Drive Dynamic Solutions Considering Different Fluid Sources, 2021, Cornell Univ. Library arXiv.org e-Print Archive, arXiv:gr-qc/2111.01298, https://arxiv.org/pdf/2111.01298.pdf. (Accessed 11 May 2022).
[140] L.H. Ford, T.A. Roman, Classical scalar fields and the generalized second law, Phys. Rev. D 64 (2) (2001) 024023. 1–12.
[141] S. Carlip, Quantum gravity in flatland, Sci. Am. 306 (4) (2012) 40–47.
[142] D.G. Hoag, W. Wrigley, Navigation and guidance in interstellar space, Acta Astronaut. 2 (5–6) (1975) 513–533.
[143] G. Vulpetti, A problem in relativistic navigation: the three-dimensional rocket equation, J. Br. Interplanet. Soc. 31 (1978) 344–350.
[144] G. Vulpetti, Relativistic astrodynamics: non-rectilinear trajectories for star exploration flights, J. Br. Interplanet. Soc. 34 (1981) 477–485.
[145] C. Clark, W.A. Hiscock, S.L. Larson, Null geodesics in the Alcubierre warp drive spacetime: the view from the bridge, Class. Quantum Gravity 16 (1999) 3965–3972.
[146] J.M. McKinley, P. Doherty, In search of the 'starbow': the appearance of the starfield from a relativistic spaceship, Am. J. Phys. 47 (1979) 309–316.
[147] R.W. Stimets, E. Sheldon, The celestial view from a relativistic starship, J. Br. Interplanet. Soc. 34 (1981) 83–99.
[148] E. Sheldon, R.H. Giles, Celestial views from nonrelativistic and relativistic interstellar spacecraft, J. Br. Interplanet. Soc. 36 (1983) 99–114.
[149] W.A. Hiscock, Quantum effects in the Alcubierre warp-drive spacetime, Class. Quantum Gravity 14 (1997) L183–L188.
[150] S. Finazzi, S. Liberati, C. Barceló, Semiclassical instability of dynamical warp drives, Phys. Rev. D 79 (12) (2009) 124017. 1–15.
[151] C. Barceló, S. Finazzi, S. Liberati, On the Impossibility of Superluminal Travel: The Warp Drive Lesson, 2010, Cornell Univ. Library arXiv.org e-Print Archive, arXiv:gr-qc/1001.4960, https://arxiv.org/pdf/1001.4960.pdf. (Accessed 30 March 2010).

[152] J.S. Lee, G.B. Cleaver, Effects of external radiation on an Alcubierre warp bubble, Phys. Essays 29 (2) (2016) 201–206.
[153] C. Barceló, et al., Warp drive aerodynamics, J. High Energy Phys. 2022 (8) (2022) 288. 1–15.
[154] D. Hochberg, T.W. Kephart, Lorentzian wormholes from the gravitationally squeezed vacuum, Phys. Lett. B 268 (3–4) (1991) 377–383.
[155] N. Kulkarni, P. Lubin, Q. Zhang, Relativistic spacecraft propelled by directed energy, Astron. J. 155 (4) (2018) 155. 1–8.
[156] P. Lubin, W. Hettel, The path to interstellar flight, Acta Futura 12 (2020) 9–44.

Further reading

[157] Y. Chong, A brief review on the recent progress of superconducting nanowire single photon detectors, Prog. Supercond. Cryog. 19 (4) (2017) 22–25.
[158] G.L. Bennett, R.L. Forward, R.H. Frisbee, Report on the NASA/JPL workshop on advanced quantum/relativity theory propulsion, AIAA-95-2599, in: 31st AIAA/ASME/ASE/ASEE Joint Propulsion Conference and Exhibit, San Diego, CA, 1995.
[159] E.-A. Kontou, K. Sanders, Energy conditions in general relativity and quantum field theory, Class. Quantum Gravity 37 (19) (2020) 193001. 1–63.

Index

Note: Page numbers followed by *f* indicate figures and *t* indicate tables.

D

E

F

G

H

Z